Label-Free Biosensor

Label-Free Biosensor

Editors

Pengfei Zhang
Rui Wang

MDPI • Basel • Beijing • Wuhan • Barcelona • Belgrade • Manchester • Tokyo • Cluj • Tianjin

Editors
Pengfei Zhang
Chinese Academy of Sciences (CAS)
Beijing
China

Rui Wang
Southeast University
Nanjing
China

Editorial Office
MDPI
St. Alban-Anlage 66
4052 Basel, Switzerland

This is a reprint of articles from the Special Issue published online in the open access journal *Biosensors* (ISSN 2079-6374) (available at: https://www.mdpi.com/journal/biosensors/special_issues/lab_fr_bio).

For citation purposes, cite each article independently as indicated on the article page online and as indicated below:

LastName, A.A.; LastName, B.B.; LastName, C.C. Article Title. *Journal Name* **Year**, *Volume Number*, Page Range.

ISBN 978-3-0365-7874-3 (Hbk)
ISBN 978-3-0365-7875-0 (PDF)

Contents

About the Editors

Pengfei Zhang

Pengfei Zhang, Ph.D., is a professor and PI in the CAS Key Laboratory of Analytical Chemistry for Living Biosystems, Institute of Chemistry, Chinese Academy of Sciences (since 2022). His research focuses on label-free biosensors using evanescent detection schemes, particularly with single-molecule detection capabilities, and their combination with fluorescence imaging approaches to image the cell activities for multiparameter analysis. He received a Ph.D. in physics from Tsinghua University (2017) and a Bachelor's degree in physics from the University of Science and Technology of China (2011). He worked as an electron microscopy engineer in KYKY Technology Co., LTD. (2017-2018) and finished his postdoc studies under the supervision of Prof. Nongjian Tao and Dr. Shaopeng Wang in Biodesign Institute, Arizona State University (2018-2022).

Rui Wang

Rui Wang, Ph.D., is a professor in the School of Biological Science and Medical Engineering at Southeast University. With a doctoral degree in Chemistry earned in 2017 from Fudan University in China, he pursued postdoctoral research at Texas A&M University, followed by Arizona State University in the United States. In 2021, he joined Southeast University's State Key Laboratory of Digital Medical Engineering as a professor. Dr. Wang's research focuses on fluorescence imaging, Raman imaging and sensing, plasmonic nanocavities, and SPR technologies.

 biosensors

Editorial

Label-Free Biosensor

Pengfei Zhang [1,*] and Rui Wang [2]

1 Beijing National Laboratory for Molecular Sciences, CAS Key Laboratory of Analytical Chemistry for Living Biosystems, Institute of Chemistry, Chinese Academy of Sciences, Beijing 100190, China
2 State Key Laboratory of Digital Medical Engineering, School of Biological Science & Medical Engineering, Southeast University, Nanjing 210096, China; ruiwang938@seu.edu.cn
* Correspondence: pfzhang@iccas.ac.cn

Label-free biosensors have become an indispensable tool for analyzing intrinsic molecular properties, such as mass, and quantifying molecular interactions without interference from labels, which is critical for the screening of drugs, detecting disease biomarkers, and understanding biological processes at the molecular level. In the last decade, novel label-free imaging biosensors, including surface plasmon resonance microscopy, interferometric scattering microscopy, plasmonic scattering microscopy, and evanescent scattering microscopy, have further advanced this field by pushing beyond ensemble averages to reveal the statistical distributions of molecular properties and binding processes at the single-molecule level. These imaging techniques pave the way to understanding molecular interaction processes in great detail. Nonetheless, the demand for developing novel single-molecule, label-free sensing schemes that are cost-effective, easy to use, and especially applicable in commercial microscopy or other commercial label-free biosensors is ever increasing. Moreover, the application range of label-free biosensors is highly likely to expand, thus providing more powerful economical tools for clinical diagnosis, environmental monitoring, and industrial quality control. Therefore, this Special Issue, entitled “Label-free Biosensors”, focuses on recent advances in producing sensitive and easy-to-use label-free biosensors and their applications in diverse fields.

There are 11 peer-reviewed papers collected in this Special Issue. Six articles are included to show the development and applications of novel label-free biosensors. Xu et al. [1] developed a sequence-specific visualization method based on loop-mediated isothermal amplification for the detection of Salmonella—one of four key global causes of diarrhea—in milk. This method does not involve any sophisticated instrument; thus, it may be a useful tool in resource-limited areas. Mayer et al. [2] developed a sandwich-type electrochemical immunosensor for the quantitative detection of the carcinoembryonic antigen, an important tumor marker in clinical tests, using the redox-tagged, single-walled carbon nanohorns/thionine/AuNPs. The results showed that carbon nanohorns possess great potential for clinical diagnostics of CEA and other biomarkers. Cui et al. [3] developed a strong optical second-harmonic generation of a monolayer 2D semiconductor to observe interfacial molecular adsorption and desorption dynamics in a label-free manner. The proposed detection scheme principally undertakes a nanometer-scale spatial resolution across interfaces, which provides a powerful tool for understanding the spatiotemporal transports of matter and energy across interfaces. Nayak et al. [4] developed a label-free immunosensor by synthesizing carbon dots (CDs) with a one-step hydrothermal method and then covalently functionalizing the dots with antibodies for the sensing of progesterone hormone. This label-free immunosensor has emerged as a potential platform for simplified progesterone analysis due to the high selectivity performance and good recovery in different samples of spiked water. Zhao et al. [5] developed an aptamer-based fluorescence anisotropy sensor for rapid and sensitive detection of heavy metal cadmium ions, which is of great significance to food safety and environmental monitoring. This aptamer-based sensor works in a direct format for detection without the need for labeling in real water

Citation: Zhang, P.; Wang, R. Label-Free Biosensor. *Biosensors* **2023**, *13*, 556. https://doi.org/10.3390/bios13050556

Received: 15 May 2023
Accepted: 16 May 2023
Published: 18 May 2023

samples, showing broad application potential. He et al. [6] developed a smartphone-based biosensor for detecting human total hemoglobin concentration in vivo with high accuracy. The smartphone-based sensor utilizes the camera, memory, and computing power of the phone, thus largely reducing the cost. The authors demonstrated that this sensor could meet the accuracy requirements for non-invasive detection of hemoglobin concentration. Five reviews are also included to introduce recent advances in typical label-free biosensor families. Lei et al. [7] reviewed the development and applications of field-effect transistor (FET)-based biosensors, which have shown great technical potential in the biomarker detection platform. This mini review gives an overview of the design strategies of biosensors and then focuses on several representative aspects. Finally, the authors summarize the long-term prospects for the commercialization of FET sensing systems. Kuralay et al. [8] reviewed the label-free cancer diagnosis platforms using electrochemical methods for cancer diagnosis. The classification of the sensing platforms is generally presented according to their recognition element, and the most recent achievements of these attractive sensing substrates are described in detail. Cheung et al. [9] reviewed the biofunctionalization needs and strategies—which are inextricably linked to sensor performance—for silicon photonic (SiP) biosensors, a promising platform for robust and low-cost decentralized diagnostics. The authors evaluated the adsorption, bioaffinity, and covalent chemistries for immobilizing bioreceptors. Then, different biopatterning techniques were compared for spatially controlling and multiplexing the biofunctionalization of SiP sensors. Fang et al. [10] reviewed optical methods such as surface-enhanced Raman scattering spectroscopy, surface plasmon resonance, and dark-field microscopic imaging techniques for the rapid detection of pathogenic bacteria in a label-free manner. The advantages and disadvantages of these label-free technologies for bacterial detection are summarized in order to promote their application for rapid bacterial detection in source-limited environments and for drug resistance assessments. Feng et al. [11] reviewed the application of MXenes in electrochemical, optical, and other bioanalytical methods in recent years. The authors summarize and discuss problems in the field of biosensing and possible future directions of MXenes.

Label-free biosensors have been experiencing rapid development in recent decades and are becoming important tools for clinical diagnosis and environmental monitoring. A comprehensive review is thus necessary, and this is the main purpose of this Special Issues. Besides the review, this collection also presents several novel label-free detection techniques to demonstrate design strategies and operation protocols. Therefore, this Special Issue might be of particular value to beginners and graduate students who have just entered this field. We hope that they, through reading this collection of articles and reviews, might fully understand the principles and operation skills of label-free biosensors, and will go on to explore further the chemical and physical properties of the biosensors and their applications.

This Special Issue is the distillation of the authors' intelligence and efforts. We expect that this collection will help and inspire researchers who are working in the fields of chemistry and biological science, and we also hope that it will provide a reference source or serve as a textbook for undergraduate and graduate students who major in chemistry, chemical engineering, physics, materials, and biology, as well as those readers who are otherwise interested in label-free biosensors. As this collection covers a breadth of highly diverse content connected to multiple scientific issues of some complexity, errors and omissions may not be entirely avoided; thus, we sincerely appreciate criticism and comments from readers.

Acknowledgments: The authors are grateful for the opportunity to serve as guest editors of the Special Issue, "Label-Free Biosensors", as well as for the contributions of each of the authors who contributed to this Special Issue. The dedicated work of the Special Issue Editor of *Biosensors* and the efforts of the editorial and publishing staff are greatly appreciated.

Conflicts of Interest: The authors declare no conflict of interest.

References

1. Zeng, H.; Huang, S.; Chen, Y.; Chen, M.; He, K.; Fu, C.; Wang, Q.; Zhang, F.; Wang, L.; Xu, X. Label-Free Sequence-Specific Visualization of LAMP Amplified Salmonella via DNA Machine Produces G-Quadruplex DNAzyme. *Biosensors* **2023**, *13*, 503. [CrossRef]
2. Domínguez-Aragón, A.; Zaragoza-Contreras, E.A.; Figueroa-Miranda, G.; Offenhäusser, A.; Mayer, D. Electrochemical Immunosensor Using Electroactive Carbon Nanohorns for Signal Amplification for the Rapid Detection of Carcinoembryonic Antigen. *Biosensors* **2023**, *13*, 63. [CrossRef] [PubMed]
3. Xia, C.; Sun, J.; Wang, Q.; Chen, J.; Wang, T.; Xu, W.; Zhang, H.; Li, Y.; Chang, J.; Shi, Z.; et al. Label-Free Sensing of Biomolecular Adsorption and Desorption Dynamics by Interfacial Second Harmonic Generation. *Biosensors* **2022**, *12*, 1048. [CrossRef] [PubMed]
4. Disha; Kumari, P.; Patel, M.K.; Kumar, P.; Nayak, M.K. Carbon Dots Conjugated Antibody as an Effective FRET-Based Biosensor for Progesterone Hormone Screening. *Biosensors* **2022**, *12*, 993. [CrossRef] [PubMed]
5. Yu, H.; Zhao, Q. A Sensitive Aptamer Fluorescence Anisotropy Sensor for Cd2+ Using Affinity-Enhanced Aptamers with Phosphorothioate Modification. *Biosensors* **2022**, *12*, 887. [CrossRef] [PubMed]
6. Fan, Z.; Zhou, Y.; Zhai, H.; Wang, Q.; He, H. A Smartphone-Based Biosensor for Non-Invasive Monitoring of Total Hemoglobin Concentration in Humans with High Accuracy. *Biosensors* **2022**, *12*, 781. [CrossRef]
7. Hao, R.; Liu, L.; Yuan, J.; Wu, L.; Lei, S. Recent Advances in Field Effect Transistor Biosensors: Designing Strategies and Applications for Sensitive Assay. *Biosensors* **2023**, *13*, 426. [CrossRef]
8. Sanko, V.; Kuralay, F. Label-Free Electrochemical Biosensor Platforms for Cancer Diagnosis: Recent Achievements and Challenges. *Biosensors* **2023**, *13*, 333. [CrossRef] [PubMed]
9. Puumala, L.S.; Grist, S.M.; Morales, J.M.; Bickford, J.R.; Chrostowski, L.; Shekhar, S.; Cheung, K.C. Biofunctionalization of Multiplexed Silicon Photonic Biosensors. *Biosensors* **2023**, *13*, 53. [CrossRef] [PubMed]
10. Wang, P.; Sun, H.; Yang, W.; Fang, Y. Optical Methods for Label-Free Detection of Bacteria. *Biosensors* **2022**, *12*, 1171. [CrossRef] [PubMed]
11. Lu, D.; Zhao, H.; Zhang, X.; Chen, Y.; Feng, L. New Horizons for MXenes in Biosensing Applications. *Biosensors* **2022**, *12*, 820. [CrossRef] [PubMed]

MDPI

Article

Label-Free Sequence-Specific Visualization of LAMP Amplified *Salmonella* via DNA Machine Produces G-Quadruplex DNAzyme

Huan Zeng [1], Shuqin Huang [1], Yunong Chen [1,2], Minshi Chen [3], Kaiyu He [2], Caili Fu [1], Qiang Wang [2], Fang Zhang [1,*], Liu Wang [2,4,*] and Xiahong Xu [2]

1 College of Biological Science and Engineering, Fuzhou University, Fuzhou 350108, China
2 Institute of Agro-Product Safety and Nutrition, Zhejiang Academy of Agricultural Sciences, Hangzhou 310021, China
3 Technology Center of Fuzhou Customs District, Fuzhou 350015, China
4 Key Laboratory of Traceability for Agricultural Genetically Modified Organisms, Ministry of Agriculture and Rural Affairs, Hangzhou 310021, China
* Correspondence: fangzh921@fzu.edu.cn or zhangfang921@gmail.com (F.Z.); wangliually@126.com (L.W.)

Abstract: *Salmonella* is one of four key global causes of diarrhea, and in humans, it is generally contracted through the consumption of contaminated food. It is necessary to develop an accurate, simple, and rapid method to monitor *Salmonella* in the early phase. Herein, we developed a sequence-specific visualization method based on loop-mediated isothermal amplification (LAMP) for the detection of *Salmonella* in milk. With restriction endonuclease and nicking endonuclease, amplicons were produced into single-stranded triggers, which further promoted the generation of a G-quadruplex by a DNA machine. The G-quadruplex DNAzyme possesses peroxidase-like activity and catalyzes the color development of 2,2′-azino-di-(3-ethylbenzthiazoline sulfonic acid) (ABTS) as the readouts. The feasibility for real samples analysis was also confirmed with *Salmonella* spiked milk, and the sensitivity was 800 CFU/mL when observed with the naked eye. Using this method, the detection of *Salmonella* in milk can be completed within 1.5 h. Without the involvement of any sophisticated instrument, this specific colorimetric method can be a useful tool in resource-limited areas.

Keywords: loop-mediated isothermal amplification; colorimetric detection; sequence-specific; G-quadruplex; DNA machine

Citation: Zeng, H.; Huang, S.; Chen, Y.; Chen, M.; He, K.; Fu, C.; Wang, Q.; Zhang, F.; Wang, L.; Xu, X. Label-Free Sequence-Specific Visualization of LAMP Amplified *Salmonella* via DNA Machine Produces G-Quadruplex DNAzyme. *Biosensors* **2023**, *13*, 503. https://doi.org/10.3390/bios13050503

Received: 9 March 2023
Revised: 20 April 2023
Accepted: 25 April 2023
Published: 26 April 2023

1. Introduction

Foodborne disease is a growing public health problem worldwide, usually contaminating food through any stage of food production, delivery, and consumption chain [1]. There are nearly 1 in 10 people around the world falls ill after eating contaminated food, which leads to over 420,000 deaths every year [2]. Salmonellosis is a common intestinal infection caused by *Salmonella* spp. *Salmonella* has a high detection rate in raw milk, cheese, raw meat, raw eggs, fruits, and vegetables [3], leading to diarrhea, vomiting, abdominal pain, chills, fever, headache, etc. People will get a foodborne illness when they eat undercooked meat or eat other foods or beverages that are contaminated by raw meat or its juices. However, as the "gold standard method", the conventional culture method needs to be pre-accumulated, with selective separation and biochemical identification of the samples. It is accurate but time-consuming and very labor-intensive. It follows that it is necessary to control pathogen-based food poisoning outbreaks with an earlier, more rapid, and more sensitive method.

The nucleic acid test is a powerful technique for molecular diagnosis by analyzing the genetic sequence in organisms. Among them, isothermal amplification technology has drawn much attention because of the capability of on-site utilization. Compared to methods that require thermal cycling, isothermal amplification is performed at a single reaction temperature, so it is more rapid and more energy efficient. Loop-mediated isothermal amplification (LAMP) is one of the most promising and comprehensively applied isothermal amplification techniques. Developed by Tsugunori Notomi in 2000, LAMP is realized with four specially designed primers recognizing six distinct sequences on the target, which ensures its high specificity [4]. The cauliflower-like structured products possess abundant stem–loops that initiate the next cycle by hybridizing with the inner primer [4]. Therefore, the amplification can accumulate 10^9 copies of the target in less than an hour. By introducing loop primers, its amplification efficiency can be further improved, and the amplification process can be accomplished in half-hour [5]. The amplification process can also be real-time monitored by collecting the fluorescent signal with an exclusive instrument [6], which permits its wide application for the detection of foodborne pathogenic bacteria [7,8], infectious diseases [9,10], and genetically modified organisms investigated by the artificial mouth simulator [11,12]. In order to make full use of LAMP in point-of-care diagnostic platforms, it is preferable to analyze the amplicons visually. With the visual method, the nucleic acid test can be achieved at point-of-care testing with high convenience in resource-poor settings because it does not rely on big and heavy instruments.

Recently, some ingenious colorimetric methods have been developed by detecting the generated amplicons or by monitoring the variation of reaction compositions [13]. For example, some intercalating dyes can bind double-stranded amplicons and indicate positive amplification by changing color [14]. The generation of pyrophosphate ions [15,16], the pH variation [17], and the consumption of deoxyribonucleotides (dNTPs) [18] in positive amplifications also enable colorimetric detection by the naked eye. However, how to visually detect the sequence-specific products of LAMP is still a great challenge.

G-quadruplex structures are composed of two or more stacked guanine (G)-tetrad planes and a monovalent cation such as K^+ or Na^+, which are formed at specific G-rich regions in the genome, mRNA, and non-coding RNA, and G-quadruplex DNAzymes are stacked G-tetrads structure with peroxidase-like activity when binding hemin (iron (III)-protoporphyrin IX) [19]. They can catalyze the color change of substrate, such as 2,2′-azino-di-(3-ethylbenzthiazoline sulfonic acid) (ABTS) and 3,3′,5,5′-Tetramethylbenzidine. Taking advantage of the catalytic activity of the G-quadruplex DNAzyme, colorimetric sensors can be developed for the analysis of DNA amplification [20–22].

DNA machine is constructed from DNA self-assembly depending on the sequence-specific interactions between complementary sequences [23]. The base sequence of nucleic acids encodes substantial structural and functional information into biopolymers [24], such as the base pairing, the pH-induced self-assembly of the C-rich sequence into i-motif configurations [25], and the ion-induced self-organization of the G-rich sequence into the G-quadruplex [26]. With rational design, DNA machines can perform machine-like functions by autonomously generating expected sequences in the presence of the appropriate trigger [27,28]. The powerful amplification ability exhibits great potential in constructing biosensors [29–31].

Here, by combining G-quadruplex DNAzymes and DNA machine, we present a sequence-specific method for colorimetric detection of LAMP amplicons. Typically, the LAMP amplicons are digested into short fragments by a restriction endonuclease. A nicking endonuclease recognition site is introduced into the inner primer to facilitate the generation of sequence-specific single-stranded amplicons by repeatedly nicking and extending. This restriction endonuclease and nicking endonuclease-mediated amplification are termed as LAMP-Res-Nick. The ssDNA products from the LAMP-Res-Nick reaction can trigger cascade amplification via DNA machine to generate G-quadruplex. The peroxidase-like activity of G-quadruplex DNAzymes, when binding hemin, makes the colorimetric readouts possible.

2. Materials and Methods

2.1. Reagents and Oligonucleotides

Bst DNA polymerase (Large Fragment) was purchased from Vazyme Biotech Co., Ltd. (Nanjing, China). Nt.BstNBI, ScrFI, DraI, and agarose were obtained from New England Biolabs (Ipswich, MA, USA). Syto 9 was achieved from Thermo Fisher (Waltham, MA, USA). 4-(2-Hydroxyethyl) piperazine-1-ethanesulfonic acid (HEPES), hemin, dNTP mixture, and ABTS were all supplied by Sangon Biotech (Shanghai, China) Co., Ltd. LB agar and LB broth were offered by Beijing Land Bridge Technology Co., Ltd. (Beijing, China) 30% H_2O_2 was bought from Sigma-Aldrich, Inc. (Burlington, MA, USA).

The gene of invasion protein A (invA) was selected as the reference gene for amplification (GenBank accession no. NC_003197). Conventional LAMP primers were synthesized according to the previous report [32]. Inner primers incorporated with the recognition site (GAGTC) of nicking endonuclease, Nt.BstNBI, are termed as Nick-BIP and Nick-FIP. All the sequences were evaluated with IDT Oligo Analyzer 3.1 (Integrated DNA Technologies, Coralville, IA, USA). The oligonucleotides were synthesized by Sangon Biotech (Shanghai, China) Co., Ltd. Sequences used in this work are listed in Table S1, and the corresponding template sequence is displayed in Figure S1.

2.2. DNA Extraction and Purification

The bacteria were separated by streak method on the LB agar plate, and a single colony was selected for further culturing in LB broth overnight.

DNA was extracted with the Bacteria Genome DNA Isolation kit (Spin Column) (Bioteke Corporation, Beijing, China) and stored at −20 °C. The concentration and purity of the extracted DNA were determined by NanoDrop One (Thermo Fisher, Waltham, MA, USA) for counting the copy number.

2.3. Real-Time LAMP Assay

For real-time LAMP, 10 μL amplification buffer contained 1 μL *Salmonella* DNA, 1.6 μM FIP and BIP, 0.2 μM F3 and B3, 0.4 μM LF and LB, 1.4 mM dNTPs, 3.2 U Bst DNA Polymerase (Large Fragment), 1× ThermoPol Reaction Buffer, 6 mM $MgSO_4$, 1 mM SYTO 9. The amplification was performed on CFX 96 (Bio-Rad, Hercules, CA, USA) at 65 °C with fluorescence collected every 30 s. The products were analyzed with 3% agarose gel, and the gel results were recorded via gel image system (UVP, Upland, CA, USA).

2.4. Cascade Amplification of LAMP-Res-Nick and DNA Machine

For LAMP process, the inner primer FIP was replaced by Nick-FIP, and no Syto 9 was involved. The reaction was performed at 65 °C on an MSC-100 ThermoMixer (AllSheng, Hangzhou, China) for 20 min as the protocol described in Section 2.4. A total of 20 μL solution contained 10 μL of LAMP products, 50 mM Tris-HCl (pH 7.9), 100 mM NaCl, 6 mM $MgSO_4$, 100 μg/mL BSA, 10 U ScrFI. The reaction was incubated at 37 °C for 1 h followed at 95 °C for 10 min. Then, 1.5 μL 10× Isothermal Amplification Buffer II Pack, 4.8 U Bst 3.0 DNA Polymerase, 10 U Nt.BstNBI, 0.32 mM dNTPs were added to make the solution 25 μL. The reaction was performed at 58.8 °C for 10 min. Thereafter, 5 μL of 1 μM M-G was added, and the mixture was incubated at 58.8 °C for another 10 min. The final products were incubated at 95 °C for 5 min followed on ice for 10 min.

2.5. Colorimetric Detection by G-Quadruplex DNAzyme

For colorimetric detection, the cascade amplification products were mixed with 1× HEPES buffer (25 mM HEPES, 20 mM KCl, 200 nM NaCl, and 0.05% Triton X-100, pH 5.3), 200 nM hemin, and 1 μL of 1M HCl. Then 2 mM $ABTS^{2-}$ and 2 mM H_2O_2 were added in the final 100 μL system for naked-eye detection in 5 min. The RGB values were extracted by Image J.

2.6. Sensitivity and Specificity

In order to test the sensitivity, 1 µL of a series of 10-fold diluted DNA was employed as template. To test the specificity of the method, DNA extracted from *Salmonella* typhimurium CMCC(B)50115, Vibrio parahemolyticus KP9, Vibrio parahemolyticus ATCC 17802 and Escherichia fergusonii 19ZEF91003 were employed as the templates. DNA was amplified and colorimetrically detected by the protocol described.

2.7. Detection of Salmonella Spiked Milk

Select *Salmonella* typhimurium colony and transfer it into LB broth for culturing at 37 °C for 6 h. The bacteria solution was then diluted 10-fold with saline and plate cultured for counting the colony number. The diluted solution was spiked into sterile milk with 10 times dilution. *Salmonella* DNA was extracted by the Bacteria Genome DNA Isolation kit. A total of 1 µL of extracted DNA was amplified and detected by the method described. All experiments were repeated 3 times. Results were shown as mean ± standard deviation. Differences were assessed by ANOVA.

3. Results and Discussion

3.1. Proof of Principle

In this study, the colorimetric and sequence-specific method is realized by producing a G-quadruplex DNAzyme by the cascade amplification of restriction endonuclease- and nicking endonuclease-mediated LAMP (LAMP-Res-Nick) and DNA machine. As shown in Scheme 1, the recognition site of nicking endonuclease is incorporated in the inner primer (Nick-FIP) and acts as the spacer between F1c and F2. The LAMP process produces a great variety of stem–loop DNAs (I). These DNAs, which have different stem lengths and possess multiple loops, provide a great number of restriction endonuclease recognition sites. Thanks to the restriction endonuclease, these products of different structures are cleaved into short double-helix fragments with the nicking endonuclease recognition sites embedded in them. Thereafter, nicking endonuclease, Nt.BstNBI, will recognize these sites and produce a nick on the double-stranded products. Meanwhile, Bst polymerase will add new free nucleotides at the 3′ end of the nicking site and displace the original strands. The synergistic effect of Bst polymerase and Nt.BstNBI promotes repeatedly nicking and extension that innumerable single-stranded products are generated (III). In order to transform this DNA sequence information into color development, a DNA machine (M-G) is added to realize cascade amplification. The M-G compromises three parts, i.e., a complementary sequence of the generated ssDNA products at the 3′ terminus, a nicking endonuclease recognition site in the middle, and a C-rich sequence at the 5′ terminus. Once the ssDNAs hybridize with the DNA machine, a great deal of G-rich sequences that can form G-quadruplex structures with the presence of potassium ions are produced. The complex, formed by hemin binding to the G quadruplex, possesses the peroxidase-mimicking activity, which can catalyze the oxidation of the substrate $ATBS^{2-}$ to $ATBS^{-}$ by H_2O_2, thereby turning the solution green. On the contrary, in the absence of targets, the LAMP process cannot be initiated, and the G-rich sequence cannot be produced. Consequently, no color change is observed.

3.2. Effect of Nicking Endonuclease Recognition Site

The recognition site (GAGTC) of nicking endonuclease, Nt.BstNBI, was inserted in the inner primers (termed Nick-FIP and Nick-BIP). We used only one or two modified inner primers for real-time LAMP and compared them with conventional inner primers. As shown in Figure 1A, compared with conventional inner primers, the incorporation of a nicking endonuclease recognition site in the inner primer reduced the amplification efficiency. However, the taking off time only delayed 2 min. The amplification was further verified by gel electrophoresis (Figure 1B). All these positive amplifications, including the inner primers incorporated with a nicking endonuclease recognition site, generated ladder-like bands. In contrast, no obvious non-specific products were produced in all of the negative samples. These results indicate that the presence of a nicking endonuclease recognition site in the inner primer has little effect on the LAMP process. To prove the nicking endonuclease recognition site was successfully incorporated in the LAMP products, the products were incubated with the nicking endonuclease, Nt.BstNBI, for another 10 min. As shown in the electrophoresis image (Figure 1B), the ladder-like bands became smeared when the modified inner primers were used for amplification. In contrast, the products produced by the conventional primers were still ladder-like. This indicates the successful incorporation of the nicking endonuclease recognition site into the LAMP amplicons. Since only one modified inner primer was enough for introducing the nicking endonuclease recognition site into the LAMP products, we used modified FIP for the subsequent experiments.

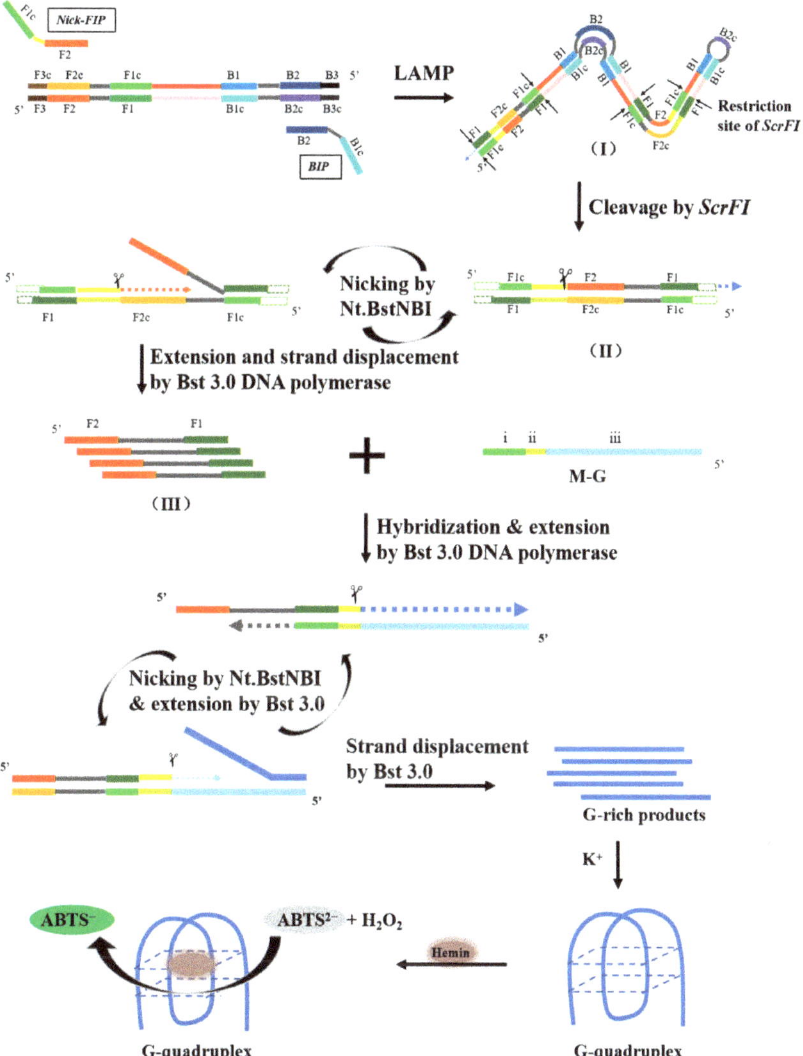

Scheme 1. Schematic illustration of the colorimetric and sequence-specific method for detection of LAMP products via DNA machine and G-quadruplex DNAzyme.

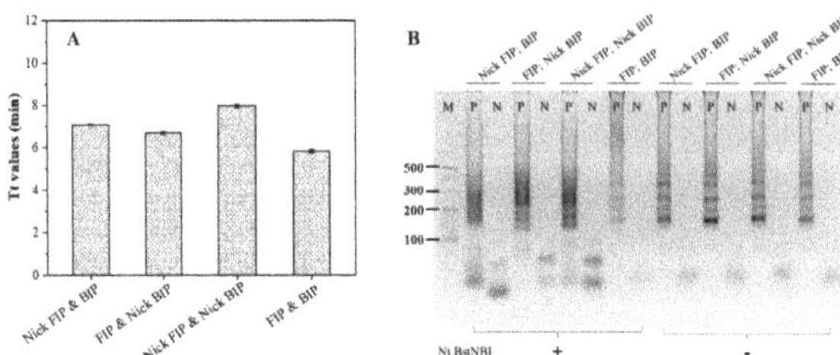

Figure 1. Investigate the effect of inner primers incorporated with a nicking endonuclease recognition site (termed as Nick-FIP and Nick-BIP) on LAMP process by (**A**) Tt values of real-time LAMP and (**B**) gel electrophoresis. Nt.BstNBI was added to incubate with the LAMP products for 10 min. N stands for no template control, P stands for positive control with 10^5 copies of *Salmonella* DNA, and M stands for 20 bp DNA ladder.

3.3. Effect of Restriction Endonuclease

According to the previous study, the nicking and extension process prefers producing short single-stranded DNA sequences. Since the LAMP products are cauliflower-like structures with multiple loops, they can provide multiple nicking sites. Therefore, the single-stranded DNA products are of different lengths because of the strong processivity of Bst polymerase and random nicking by Nt.BstNBI. We employed restriction endonuclease, ScrFI, and DraI, respectively, to cleave the LAMP products into small fragments which contained the same 3′ terminus. The restriction sites of ScrFI and DraI are illustrated in Figure S2A, and the expected cleaved products are indicated in Figure S2B. As shown in Figure 2, after digestion by ScrFI, the ladder-like bands disappeared. Instead, the products were enriched at around 100 bp and 200 bp, indicating the products were cleaved by ScrFI. By contrast, when the products were treated with DraI, the products were still ladder-like bands on the gel; this might be attributed to the poor compatibility of buffer for DraI and LAMP.

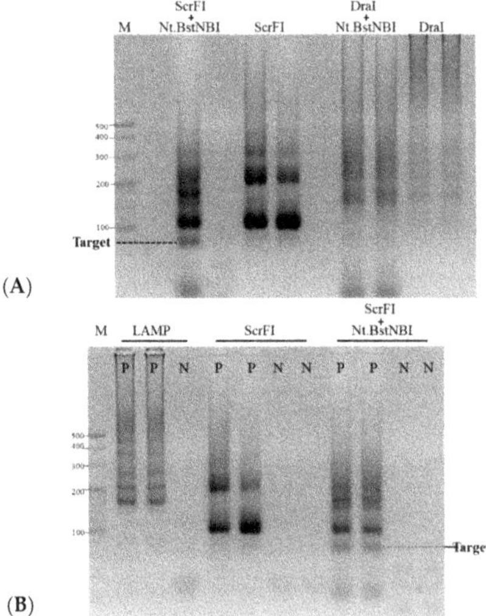

Figure 2. (**A**) In the presence of 10^5 copies of Salmonella DNA, LAMP products were cleaved by restriction endonucleases ScrFI and DraI, respectively, followed by treatment with nicking endonucleases Nt.BstNBI, then electrophoresis in 3% agarose gel. (**B**) 3% agarose gel electrophoresis showing variability in products generated after isothermal amplification and these products cleaved by ScrFI and Nt.BstNBI, N means no template control, N stands for no template control, P stands for positive control with 10^5 copies of Salmonella DNA, and M stands for 20 bp DNA ladder.

To further verify the cleavage of LAMP products by ScrFI, we added Nt.BstNBI into the system and observed the generation of 73 bp products (Figure 2). This was attributed to the synergistic effect of Bst polymerase and Nt.BstNBI, which promoted repeated nicking and extension from the nicking site on the LAMP products and displaced the sequence downstream of the nicking site. In contrast, when Nt.BstNBI was added to the samples treated by DraI, and the ladder-like bands became smeared. This was probably owing to the LAMP products not being digested before nicking, and the dissociated single-stranded sequence being of different lengths.

Based on these results, ScrFI was used for cleaving the LAMP products into short fragments in the following experiment.

3.4. Sensitivity

It is reported that *Salmonella* is one of the four key reasons that cause diarrhea diseases [33]. Around 3.4 million cases of diseases are caused by invasive nontyphoidal *Salmonella* annually [31]. To test the feasibility of the developed method, we chose *Salmonella* typhimurium as an example. DNA extracted from *Salmonella* was serially diluted as 4×10^1, 4×10^2, 4×10^3, and 4×10^4 copies/μL for sensitivity evaluation, and the template was amplified by LAMP-Res-Nick and DNA machine. The products were evaluated by G-quadruplex DNAzyme catalyzing the color development of ABTS. Since the e RGB pattern can effectively eliminate the error in human observation, the RGB value was also extracted and analyzed, and the RGB values were extracted by Image J. The RGB (red, green, and blue) is an important index for color expression. Each channel of red, green, and blue has 256 levels of brightness, of which level 0 means the darkest and 255, the brightest. The results are shown in Figure 3A. There was no color change in the negative control and the sample with 4×10^1 copies of DNA. In contrast, an obvious green was developed for samples containing 4×10^2, 4×10^3, and 4×10^4 copies of the template. The green and blue channels were a little more sensitive, which dropped from color density 184 (without target) to 148 (4×10^3 copies/μL target), and the best sensitivity was achieved for the red channel as color density dropped from 184 to 94. Therefore, the red channel was chosen for further assays as the optimal one. The results were compared with real-time LAMP (Figure 3B), and consistent results were obtained.

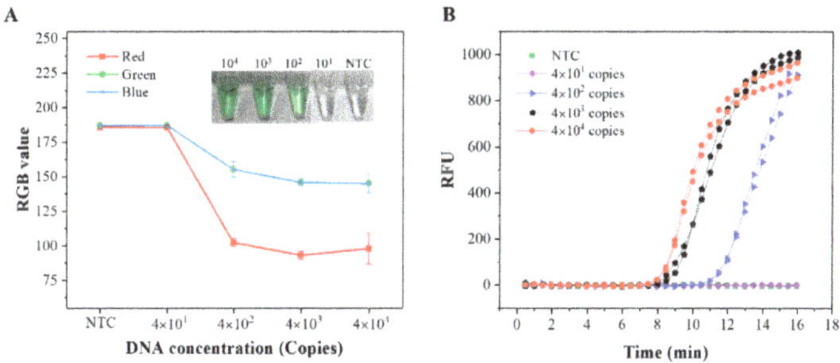

Figure 3. (**A**) RGB values of colorimetric assay of different concentrations of DNA (inset is the colorimetric photo). (**B**) Real-time LAMP assay of different concentrations of DNA. NTC indicates no template control. Error bars indicate standard deviations for $n = 3$.

3.5. Specificity

To test the specificity of the method, DNA extracted from *Salmonella* typhimurium CMCC(B)50115, Vibrio parahemolyticus KP9, Vibrio parahemolyticus ATCC 17802, and Escherichia fergusonii 19ZEF91003 was detected. As shown in Figure 4A, only DNA extracted from Salmonella typhimurium developed green. In contrast, other samples gave colorless results. The results were consistent with real-time LAMP (Figure 4B).

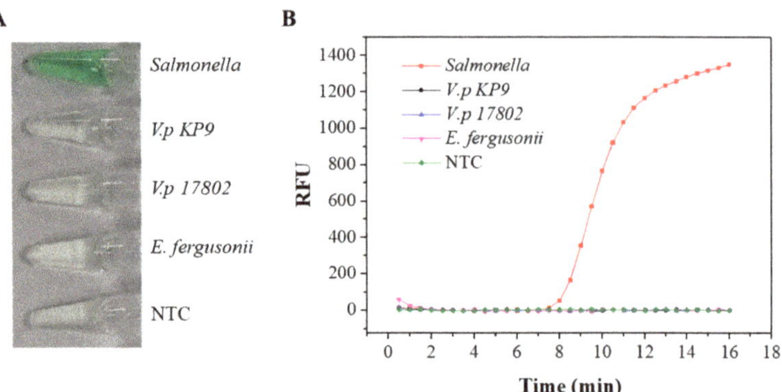

Figure 4. Specificity of (**A**) colorimetric assay and (**B**) real-time LAMP. The amount of *Salmonella* and nontarget strains (*V.p KP9*, *V.p 17802*, and *E. fergusonii*) were 10^3 copies, and no DNA template was used as NTC.

3.6. Detection of Salmonella in Milk

Salmonella is one of the main pathogens in raw milk, and healthy people of any age can be gravelly sick after drinking raw milk contaminated with *Salmonella*, not only these people with weakened immune systems. From 2013 to 2018, there are 75 outbreaks reported to CDC were linked to raw milk, which included 675 illnesses and 98 hospitalizations [34]. Therefore, it is meaningful to test the practicability of the method for the detection of *Salmonella* contamination in milk. Here, milk samples spiked with a series of concentrations of *Salmonella* typhimurium, ranging from 8×10^6 CFU/mL to 8×10^1 CFU/mL, were detected. As shown in Figure 5A, milk contaminated by 800 CFU/mL or more of *Salmonella* turned green. As shown in Figure 5B, when the *Salmonella* concentration was lower than 800 CFU/mL, the red channel was close to 180. When increasing the *Salmonella* concentration made, the value of the red channel decreased significantly, indicating the sensitivity for *Salmonella* detection in spiked milk was 800 CFU/mL. The results confirm that our sensor can be used for practical sample analysis. Though this method does not improve the sensitivity (Table 1), the method does not require an exclusive instrument, and the colorimetric results can ensure specificity.

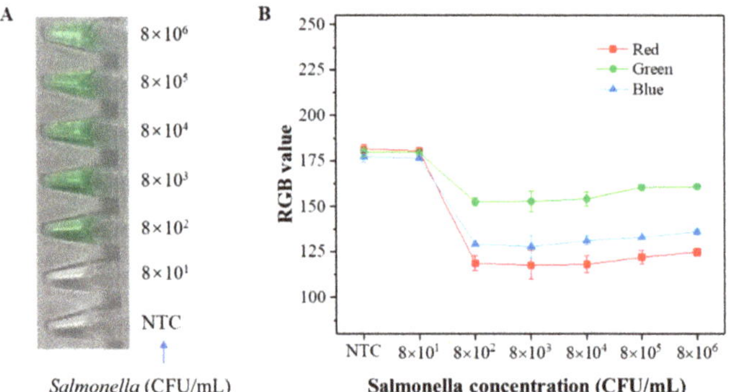

Figure 5. (**A**) Colorimetric detection of milk samples spiked with a series of concentrations of *Salmonella* typhimurium. (**B**) RGB values of the colorimetric assay. NTC indicates no template control. Error bars indicate standard deviations for $n = 3$.

Table 1. Comparison of the performance of our colorimetric method with other reported LAMP-based methods.

Method	Instrument or Device	Sensitivity	Results Determination	Method to Verify the Specificity	References
qLAMP	qPCR thermocycler	4 CFU/25 g (chicken)	Real-time fluorescence	Melting curve	[35]
LAMP-Turbidity	Real-time turbidimeter	6.1×10^3–6.1×10^4 CFU/g	Real-time turbidity	Agarose gel electrophoresis	[36]
LAMP-ELISA	A thermal cycler/water bath, a plater reader	10^3 CFU/mL (spiked meat sample)	Absorbance	Capture the amplicons with specific probes and detect by ELISA	[37]
In situ LAMP	A water bath, a fluorescence microscope	1 CFU/cm^2 (eggshells)	Microscopy	fluorescence microscope	[38]
Triplex LAMP	Genie III LAMP detector	64 CFU/g (chicken meat)	Real-time fluorescence	Melting curve	[39]
LAMP on a microfluidic compact disc	a digital RPM meter, a spinning motor, an IR thermometer	3.4×10^4 CFU/mL (spiked tomato)	Visual observation	N[a]	[40]
Real-time LAMP	Genie III LAMP detector	1.2–12 CFU/reaction	Real-time fluorescence	N[a]	[41]
LAMP on a chip	an eight-channel pump, a heater, and a small ESE log detector	50 cells/test (pork meat)	Real-time fluorescence	N[a]	[42]
Microfluidic LAMP	A rotary system consists of three heating blocks, a servo motor	50 CFU/mL (tap water or milk)	lateral flow strip	Capture the amplicons with antibody	[43]
Visual LAMP	A metal heater	800 CFU/mL for milk sample	Visual observation	DNA machine for transferring the target sequence to DNAzyme	This manuscript

N[a]: Not mentioned in the article.

4. Conclusions

In summary, the key issue to realizing colorimetric and sequence-specific detection of LAMP amplicons is to efficiently transduce the amplification signal into color development of ABTS. In this work, we employed a cascade amplification of restriction endonuclease- and nicking endonuclease-mediated LAMP (LAMP-Res-Nick) and DNA machine to generate G-quadruplex and employed the peroxidase-mimicking activity of G-quadruplex DNAzyme to catalyze color development. The sensitivity was comparable to real-time LAMP. The specificity was confirmed by testing four kinds of common foodborne bacteria. The method was verified to be feasible for the detection of 800 CFU/mL *Salmonella* spiked milk by the naked eye. The biggest advantage of this method is that sequence-specific colorimetric readouts can be obtained simply, with no sophisticated instrument required during the whole process, which is promising for point-of-care utilization.

Supplementary Materials: The following supporting information can be downloaded at: https://www.mdpi.com/article/10.3390/bios13050503/s1, Table S1: Sequences employed in this work.; Figure S1: Template sequence for LAMP-Res-Nick amplification of the invA gene of *Salmonella*; Figure S2: The effect of restriction endonuclease, ScrFI, and DraI, on cleaving LAMP products of invA gene of *Salmonella*. (A) Scheme illustrating the restriction sites of ScrFI and DraI on the long stem–loop structure of LAMP products. Red arrows indicate the restriction site of ScrFI, and the blue arrows indicate the restriction site of DraI. (B) The expected LAMP products extended from Nick-FIP. The numbers in the box denote the expected length of the amplification products.

Author Contributions: Writing-review and editing, H.Z.; writing-original draft preparation, S.H.; investigation, data curation, Y.C., M.C. and K.H.; funding acquisition, C.F., Q.W. and X.X.; writing-original draft preparation, supervision, and funding acquisition, L.W.; writing-review and editing, funding acquisition and conceptualization, F.Z. All authors have read and agreed to the published version of the manuscript.

Funding: This work was financially supported by the National Natural Science Foundation of China (32172285, 32172307); Project funded by China Postdoctoral Science Foundation (2022M710709); Natural Science Foundation of Fujian Province, China (2022J01100); Basic Public Welfare Research Program of Zhejiang Province, China (LGN21C200007); Fuzhou Agricultural Science and Technology Project (2021-N-133) and Open Project Program of the Key Laboratory of Marine Fishery Resources Exploitment & Utilization of Zhejiang Province (SL2021009).

Institutional Review Board Statement: Not applicable.

Informed Consent Statement: Not applicable.

Data Availability Statement: Data are contained within the article or supplementary material.

Acknowledgments: We thank Biao Tang, Yingping Xiao, Xiaojing Li, and Rui Wang for providing the bacteria for the experiment.

Conflicts of Interest: The authors declare no conflict of interest.

References

1. Shang, Y.; Sun, J.; Ye, Y.; Zhang, J.; Zhang, Y.; Sun, X. Loop-mediated isothermal amplification-based microfluidic chip for pathogen detection. *Crit. Rev. Food Sci. Nutr.* **2020**, *60*, 201–224. [CrossRef]
2. Foodborne Diseases. Available online: https://www.who.int/health-topics/foodborne-diseases (accessed on 23 December 2022).
3. Hugas, M.; Beloeil, P.A. Controlling Salmonella along the food chain in the European Union—Progress over the last ten years. *Eurosurveillance* **2014**, *19*, 208204. [CrossRef]
4. Notomi, T.; Okayama, H.; Masubuchi, H.; Yonekawa, T.; Watanabe, K.; Amino, N.; Hase, T. Loop-mediated isothermal amplification of DNA. *Nucleic Acids Res.* **2000**, *28*, E63. [CrossRef] [PubMed]
5. Nagamine, K.; Hase, T.; Notomi, T. Accelerated reaction by loop-mediated isothermal amplification using loop primers. *Mol. Cell. Probes* **2002**, *16*, 223–229. [CrossRef]
6. Oscorbin, I.P.; Belousova, E.A.; Zakabunin, A.I.; Boyarskikh, U.A.; Filipenko, M.L. Comparison of fluorescent intercalating dyes for quantitative loop-mediated isothermal amplification (qLAMP). *BioTechniques* **2016**, *61*, 20–25. [CrossRef]
7. Niessen, L.; Luo, J.; Denschlag, C.; Vogel, R.F. The application of loop-mediated isothermal amplification (LAMP) in food testing for bacterial pathogens and fungal contaminants. *Food Microbiol.* **2013**, *36*, 191–206. [CrossRef] [PubMed]
8. Anupama, K.P.; Chakraborty, A.; Karunasagar, I.; Karunasagar, I.; Maiti, B. Loop-mediated isothermal amplification assay as a point-of-care diagnostic tool for Vibrio parahaemolyticus: Recent developments and improvements. *Expert Rev. Mol. Diagn.* **2019**, *19*, 229–239. [CrossRef]
9. Parida, M.; Sannarangaiah, S.; Dash, P.K.; Rao, P.V.L.; Morita, K. Loop mediated isothermal amplification (LAMP): A new generation of innovative gene amplification technique; perspectives in clinical diagnosis of infectious diseases. *Rev. Med. Virol.* **2008**, *18*, 407–421. [CrossRef] [PubMed]
10. Mori, Y.; Notomi, T. Loop-mediated isothermal amplification (LAMP): A rapid, accurate, and cost-effective diagnostic method for infectious diseases. *J. Infect. Chemother.* **2009**, *15*, 62–69. [CrossRef] [PubMed]
11. Singh, M.; Pal, D.; Soo, P.; Randhawa, G. Loop-mediated isothermal amplification assays: Rapid and efficient diagnostics for genetically modified crops. *Food Control* **2019**, *106*, 106759. [CrossRef]
12. Pu, D.D.; Duan, W.; Huang, Y.; Zhang, L.L.; Zhang, Y.Y.; Sun, B.G.; Ren, F.Z.; Zhang, H.Y.; Tang, Y.Z. Characterization of the dynamic texture perception and the impact factors on the bolus texture changes during oral processing. *Food Chem.* **2021**, *339*, 128078. [CrossRef]
13. Zhang, M.; Ye, J.; He, J.-S.; Zhang, F.; Ping, J.; Qian, C.; Wu, J. Visual detection for nucleic acid-based techniques as potential on-site detection methods. A review. *Anal. Chim. Acta* **2020**, *1099*, 1–15. [CrossRef] [PubMed]
14. Zhang, Y.; Shan, X.; Shi, L.; Lu, X.; Tang, S.; Wang, Y.; Li, Y.; Alam, M.J.; Yan, H. Development of a fimY-based loop-mediated isothermal amplification assay for detection of Salmonella in food. *Food Res. Int.* **2012**, *45*, 1011–1015. [CrossRef]
15. Goto, M.; Honda, E.; Ogura, A.; Nomoto, A.; Hanaki, K.-I. Colorimetric detection of loop-mediated isothermal amplification reaction by using hydroxy naphthol blue. *BioTechniques* **2009**, *46*, 167–172. [CrossRef]

16. Zhang, F.; Wang, R.; Wang, L.; Wu, J.; Ying, Y. Tracing phosphate ions generated during DNA amplification and its simple use for visual detection of isothermal amplified products. *Chem. Commun.* **2014**, *50*, 14382–14385. [CrossRef]
17. Tanner, N.A.; Zhang, Y.; Evans, T.C. Visual detection of isothermal nucleic acid amplification using pH-sensitive dyes. *BioTechniques* **2015**, *58*, 59–68. [CrossRef] [PubMed]
18. Zhao, W.; Chiuman, W.; Lam, J.C.F.; Brook, M.A.; Li, Y. Simple and rapid colorimetric enzyme sensing assays using non-crosslinking gold nanoparticle aggregation. *Chem. Commun.* **2007**, *36*, 3729–3731. [CrossRef] [PubMed]
19. Liu, Z.; He, K.; Li, W.; Liu, X.; Xu, X.; Nie, Z.; Yao, S. DNA G-Quadruplex-Based Assay of Enzyme Activity. *Methods Mol. Biol.* **2017**, *1500*, 133–151.
20. Lee, J.-E.; Mun, H.; Kim, S.-R.; Kim, M.-G.; Chang, J.-Y.; Shim, W.-B. A colorimetric Loop-mediated isothermal amplification (LAMP) assay based on HRP-mimicking molecular beacon for the rapid detection of Vibrio parahaemolyticus. *Biosens. Bioelectron.* **2020**, *151*, 111968. [CrossRef] [PubMed]
21. Song, S.; Wang, X.; Xu, K.; Xia, G.; Yang, X. Visualized Detection of Vibrio parahaemolyticus in Food Samples Using Dual-Functional Aptamers and Cut-Assisted Rolling Circle Amplification. *J. Agric. Food Chem.* **2019**, *67*, 1244–1253. [CrossRef] [PubMed]
22. Zhu, L.; Xu, Y.; Cheng, N.; Xie, P.; Shao, X.; Huang, K.; Luo, Y.; Xu, W. A facile cascade signal amplification strategy using DNAzyme loop-mediated isothermal amplification for the ultrasensitive colorimetric detection of Salmonella. *Sens. Actuators B-Chem.* **2017**, *242*, 880–888. [CrossRef]
23. Bath, J.; Turberfield, A.J. DNA nanomachines. *Nat. Nanotechnol.* **2007**, *2*, 275–284. [CrossRef] [PubMed]
24. Modi, S.; Bhatia, D.; Simmel, F.C.; Krishnan, Y. Structural DNA nanotechnology: From bases to bricks, from structure to function. *J. Phys. Chem. Lett.* **2010**, *1*, 1994–2005. [CrossRef]
25. Gehring, K.; Leroy, J.-L.; Guéron, M. A tetrameric DNA structure with protonated cytosine-cytosine base pairs. *Nature* **1993**, *363*, 561–565. [CrossRef] [PubMed]
26. Davis, J.T.; Spada, G.P. Supramolecular architectures generated by self-assembly of guanosine derivatives. *Chem. Soc. Rev.* **2007**, *36*, 296–313. [CrossRef] [PubMed]
27. Zhang, H.; Fang, C.; Zhang, S. An Autonomous Bio-barcode DNA Machine for Exponential DNA Amplification and Its Application to the Electrochemical Determination of Adenosine Triphosphate. *Chem.-A Eur. J.* **2010**, *16*, 12434–12439. [CrossRef]
28. Beissenhirtz, M.K.; Elnathan, R.; Weizmann, Y.; Willner, I. The aggregation of Au nanoparticles by an autonomous DNA machine detects viruses. *Small* **2007**, *3*, 375–379. [CrossRef]
29. Zhao, Y.; Chen, F.; Zhang, Q.; Zhao, Y.; Zuo, X.; Fan, C. Polymerase/nicking enzyme synergetic isothermal quadratic DNA machine and its application for one-step amplified biosensing of lead (II) ions at femtomole level and DNA methyltransferase. *Npg Asia Mater.* **2014**, *6*, e131. [CrossRef]
30. Xu, J.; Qian, J.; Li, H.; Wu, Z.-S.; Shen, W.; Jia, L. Intelligent DNA machine for the ultrasensitive colorimetric detection of nucleic acids. *Biosens. Bioelectron.* **2016**, *75*, 41–47. [CrossRef]
31. Wu, Y.; Wang, L.; Zhu, J.; Jiang, W. A DNA machine-based fluorescence amplification strategy for sensitive detection of uracil-DNA glycosylase activity. *Biosens. Bioelectron.* **2015**, *68*, 654–659. [CrossRef]
32. Shao, Y.; Zhu, S.M.; Jin, C.C.; Chen, F.S. Development of multiplex loop-mediated isothermal amplification-RFLP (mLAMP-RFLP) to detect Salmonella spp. and Shigella spp. in milk. *Int. J. Food Microbiol.* **2011**, *148*, 75–79. [CrossRef]
33. Salmonella (Non-Typhoidal). Available online: https://www.who.int/news-room/fact-sheets/detail/salmonella-(non-typhoidal) (accessed on 20 February 2018).
34. Raw Milk Questions and Answers. Available online: https://www.cdc.gov/foodsafety/rawmilk/raw-milk-questions-and-answers.html (accessed on 4 January 2023).
35. Garrido-Maestu, A.; Fucinos, P.; Azinheiro, S.; Carvalho, J.; Prado, M. Systematic loop-mediated isothermal amplification assays for rapid detection and characterization of Salmonella spp., Enteritidis and Typhimurium in food samples. *Food Control* **2017**, *80*, 297–306. [CrossRef]
36. Chen, S.; Wang, J.; Stein, B.C.; Ge, R.E. Rapid Detection of Viable Salmonellae in Produce by Coupling Propidium Monoazide with Loop-Mediated Isothermal Amplification. *Appl. Environ. Microbiol.* **2011**, *77*, 4008–4016. [CrossRef] [PubMed]
37. Ravan, H.; Yazdanparast, R. Development and evaluation of a loop-mediated isothermal amplification method in conjunction with an enzyme-linked immunosorbent assay for specific detection of Salmonella serogroup D. *Anal. Chim. Acta* **2012**, *733*, 64–70. [CrossRef] [PubMed]
38. Ye, Y.; Wang, B.; Huang, F.; Song, Y.; Yan, H.; Alam, M.J.; Yamasaki, S.; Shi, L. Application of in situ loop-mediated isothermal amplification method for detection of Salmonella in foods. *Food Control* **2011**, *22*, 438–444. [CrossRef]
39. Kim, M.-J.; Kim, H.-J.; Kim, H.-Y. Direct triplex loop-mediated isothermal amplification assay for the point-of-care molecular detection of Salmonella genus, subspecies I, and serovar Typhimurium. *Food Control* **2021**, *120*, 107504. [CrossRef]
40. Sayad, A.A.; Ibrahim, F.; Uddin, S.M.; Pei, K.X.; Mohktar, M.S.; Madou, M.; Thong, K.L. A microfluidic lab-on-a-disc integrated loop mediated isothermal amplification for foodborne pathogen detection. *Sens. Actuators B Chem.* **2016**, *227*, 600–609. [CrossRef]
41. Hu, L.J.; Ma, L.M.; Zheng, S.M.; He, X.H.; Hammack, T.S.; Brown, E.W.; Zhang, G. Development of a novel loop-mediated isothermal amplification (LAMP) assay for the detection of Salmonella ser. Enteritidis from egg products. *Food Control* **2018**, *88*, 190–197. [CrossRef]

42. Sun, Y.; Quyen, T.L.; Hung, T.Q.; Chin, W.H.; Wolff, A.; Bang, D.D. A lab-on-a-chip system with integrated sample preparation and loop-mediated isothermal amplification for rapid and quantitative detection of Salmonella spp. in food samples. *Lab Chip* **2015**, *15*, 1898–1904. [CrossRef]
43. Park, B.H.; Oh, S.J.; Jung, J.H.; Choi, G.; Seo, J.H.; Kim, D.H.; Lee, E.Y.; Seo, T.S. An integrated rotary microfluidic system with DNA extraction, loop-mediated isothermal amplification, and lateral flow strip based detection for point-of-care pathogen diagnostics. *Biosens. Bioelectron.* **2017**, *91*, 334–340. [CrossRef]

Article

Electrochemical Immunosensor Using Electroactive Carbon Nanohorns for Signal Amplification for the Rapid Detection of Carcinoembryonic Antigen

Angélica Domínguez-Aragón [1,2], Erasto Armando Zaragoza-Contreras [2,*], Gabriela Figueroa-Miranda [1], Andreas Offenhäusser [1] and Dirk Mayer [1,*]

[1] Institute of Biological Information Processing, Bioelectronics (IBI-3), Forschungszentrum Jülich GmbH, 52428 Jülich, Germany

[2] Centro de Investigación en Materiales Avanzados, S.C. Miguel de Cervantes 120, Complejo Industrial Chihuahua, Chihuahua 31136, Mexico

* Correspondence: armando.zaragoza@cimav.edu.mx (E.A.Z.-C.); dirk.mayer@fz-juelich.de (D.M.)

Abstract: In this work, a novel sandwich-type electrochemical immunosensor was developed for the quantitative detection of the carcinoembryonic antigen, an important tumor marker in clinical tests. The capture antibodies were immobilized on the surface of a gold disk electrode, while detection antibodies were attached to redox-tagged single-walled carbon nanohorns/thionine/AuNPs. Both types of antibody immobilization were carried out through Au-S bonds using the novel photochemical immobilization technique that ensures control over the orientation of the antibodies. The electroactive SWCNH/Thi/AuNPs nanocomposite worked as a signal tag to carry out both the detection of carcinoembryonic antigen and the amplification of the detection signal. The current response was monitored by differential pulse voltammetry. A clear dependence of the thionine redox peak was observed as a function of the carcinoembryonic antigen concentration. A linear detection range from 0.001–200 ng/mL and a low detection limit of 0.1385 pg/mL were obtained for this immunoassay. The results showed that carbon nanohorns represent a promising matrix for signal amplification in sandwich-type electrochemical immune assays working as a conductive and binding matrix with easy and versatile modification routes to antibody and redox tag immobilization, which possesses great potential for clinical diagnostics of CEA and other biomarkers.

Keywords: electrochemical immunosensor; carcinoembryonic antigen; carbon nanohorns; redox-tag

Citation: Domínguez-Aragón, A.; Zaragoza-Contreras, E.A.; Figueroa-Miranda, G.; Offenhäusser, A.; Mayer, D. Electrochemical Immunosensor Using Electroactive Carbon Nanohorns for Signal Amplification for the Rapid Detection of Carcinoembryonic Antigen. *Biosensors* **2023**, *13*, 63. https://doi.org/10.3390/bios13010063

Received: 16 November 2022
Revised: 15 December 2022
Accepted: 22 December 2022
Published: 30 December 2022

1. Introduction

Cancer is a life-threatening disease with worldwide significance for the healthcare systems and a huge economic impact. Tumor biomarkers are important tools for the detection of cancer diseases, which either originate from tumor cells or emerge from the organism as a response to it. Alterations of their concentration in the body fluids may correlate qualitatively or quantitatively with the presence of cancer cells and therefore possess important clinical value for the early detection and diagnosis of the cancer diseases and thus the prognosis of the patient [1]. In fact, some biomarkers have been routinely used in clinical diagnosis including carcinoembryonic antigen (CEA), alpha-fetoprotein, prostate-specific antigen, carbohydrate antigen 125, carbohydrate antigen 153, carbohydrate antigen 199, and so on [2]. Among them, CEA, which is a set of glycoproteins of great relevance for cell adhesion during fetal development, has been considered a common cancer biomarker in clinical diagnosis since its expression declines after birth. CEA overexpression in blood serum in adult humans is usually related to the presence or progression of different types of cancer such as colorectal, liver, breast, ovarian or lung. In addition, CEA levels can also be monitored during chemotherapy to assess the progress and result of the treatment [3]. In healthy individuals, the concentration of CEA in blood serum should be less than

3 ng/mL [4]. Therefore, the development of simple and accurate methods for ultrasensitive monitoring of CEA is of great importance to help detect the presence of cancerous tumors without the need to use invasive or costly methods.

Immunoassays are important analytical techniques based on specific antigen-antibody interactions, which are widely used in clinical diagnosis. Numerous conventional immunoassays for CEA determination, including enzyme-linked immunosorbent assay (ELISA) [5], fluorescence immunoassay [1], and electrochemical immunoassay [3], have been reported. For instance, electrochemical immunosensors are considered being promising tools due to their simple instrumentation, portability, high sensitivity, low cost, and fast response. In conventional sandwich-type immunosensors, the detection antibodies are usually labeled with a peroxidase enzyme to generate the amperometric detection signal [6,7]. Although this method is sensitive and very useful, it has been shown that the biological tag can be replaced with nanomaterials for signal conversion and amplification [3].

Single-walled carbon nanohorns (SWCNH) are an emerging class of semiconducting nanocarbons, similar to carbon nanotubes, composed of single-layer of graphene wrapped to nanosized sheaths. SWCNH form spherical aggregates with a diameter of around 80–100 nm with different morphologies, such as dahlia, bud, and seed structures, rather than dispersing separately [8]. These unique morphologies provide special properties, such as a large surface area, small size, numerous internal nanospaces, high conductivity, and mechanical strength, making them ideal nanomaterials for application in electrochemical sensors. In addition, SWCNH can be functionalized by chemical oxidation to obtain a highly hydrophilic material and to multiply the number of binding sites on the surface for the coupling with biomolecules, metal particles, etc [9]. SWCNH can be directly used in electrochemical sensing for electrode preparation due to their superior conductivity or they can be adopted as a signal tag after their decoration with redox groups [10]. Due to the high surface area, a large number of signal molecules can be attached to the SWCNHs to facilitate strong signal amplification and consequently low detection limits.

In this work, a sandwich-type electrochemical immunosensor was developed. The capture antibodies (AntiCEA$_1$) were immobilized on the surface of a gold disk electrode, while detection antibodies (AntiCEA$_2$) were tethered to a nanocomposite based on SWCNHs functionalized with thionine (Thi) and gold nanoparticles (AuNPs) (SWCNH/Thi/AuNPs). Both types of antibody immobilization were carried out using the photochemical immobilization technique (PIT). In this technique, antibodies are exposed to UV irradiation, which leads to selective photoreduction of the disulfide bonds in specific cysteine-cysteine/tryptophan triads (Cys-Cys/Trp) [11]. The breaking of these Cys-Cys bonds produced free thiol groups, which can interact with the proximal gold surface, resulting in a covalent bonding of the antibody. Besides, PIT ensures control over the orientation of immobilized Abs, with their binding sites exposed to the solution phase and accessible for antigen coupling [12]. The main advantage of this immunosensing system is the reduction of the fabrication time that the PIT method provided; the immobilization of the antibodies required only 15 min. Meanwhile, the PIT technique used in the present work does not require any additional surface modification steps at the electrode, which decreased the total fabrication time to only 2.25 h, notably less than that of other reported assays.

Thionine was used as a redox tag for the amperometric detection scheme and a multitude of these redox molecules was attached to the large surface area of the SWCNH. The detection was carried out through differential pulse voltammetry (DPV), where the change in the current intensity of the redox peak of thionine was related to the concentration of the biomarker CEA. The SWCNH were thus employed as conductive, high surface area but still small binding matrix for the attachment of AuNP-antibody entities and redox tags.

2. Materials and Methods

2.1. Characterization

Cyclic voltammetry (CV) and differential pulse voltammetry (DPV) were measured on a multichannel potentiostat (CHI1030B, CH Instruments, Inc. Austin, USA.) with a three-electrode configuration. While electrochemical impedance spectroscopy was measured on a BioLogic potentiostat (SP-300, BioLogic Systems, Grenoble, France. The gold electrode (Au-disk, 2 mm diameter) was used as the working electrode, and a Pt wire and saturated Ag/AgCl electrode were used as the counter electrode and reference electrode, respectively. All potentials in this work are quoted with respect to the potential of the Ag/AgCl reference electrode.

The morphology and energy-dispersive X-ray spectroscopy (EDS) mapping was analyzed with a high-resolution Hitachi 7700 transmission electron microscope (TEM, Hitachi High-Technologies Corporation, Ibaraki, Jpan) and with a scanning electron microscope (SEM, Magellan 400, FEI, Hillsboro, OR, USA, and 1550VP, Carl. Zeiss SMT AG, Oberkochen, Germany).

2.2. Materials and Reagents

Potassium ferrocyanide $K_4[Fe(CN)_6]$, potassium ferricyanide $K_3[Fe(CN)_6]$, thionine acetate salt, gold nanoparticles (5 nm diameter), and oxidized carbon nanohorns were obtained from Sigma-Aldrich (Merck KGaA, Darmstadt, Germany). Phosphate buffer solution (PBS) (0.01 M) was prepared from sodium chloride (NaCl), potassium chloride (KCl), disodium phosphate Na_2HPO_4 and dipotassium phosphate (KH_2PO_4), AntiCEA$_1$, AntiCEA$_2$ and CEA were purchased from mybiosource.com.

2.3. Fabrication of SWCNH/Thionine/AuNPs Nanocomposite (SWCNH/Thi/AuNPs)

In this method, 2 mL of a SWCNH (I) dispersion (1 mg/mL) was mixed with 2 mL of thionine (4 mg/mL), stirring vigorously for 24 h at room temperature. The product was purified with Milli-Q water by centrifugation (12,000 rpm) to remove unbound thionine molecules. The SWCNH/Thi (II) was dispersed in 2 mL of Milli-Q water and then, 8 mL of AuNPs dispersion was added to the dispersion. The mixture was allowed to react for 48 h under magnetic stirring. Subsequently, the mixture was washed several times by centrifugation (12,000 rpm); the recovered solid was redispersed in 2 mL of 0.01 M PBS and stored at 4 °C.

2.4. Preparation of Detection Antibody Labeled SWCNH/Thi/AuNPs/AntiCEA$_2$

AntiCEA$_2$ was immobilized on SWCNH/Thi/AuNPs by covalent interaction between AntiCEA$_2$ and AuNPs. Briefly, 350 µL of AntiCEA$_2$ (121.42 µg/mL) was irradiated with a UV lamp (Trylight®, Promete Srl. Naples, Italy) for 30 s; afterward, it was mixed with 500 µL of SWCNH/Thi/AuNPs (III) by gently stirring for 15 min. The UV source consisted of two U-shaped low-pressure mercury lamps (6 W at 254 nm) in which a standard quartz cell could be easily housed. Considering the envelope geometry of the lamps and the cell proximity, the irradiation intensity used to produce the thiol group was approximately 0.3 W/cm^2 [12].

The obtained SWCNH/Thi/AuNPs/AntiCEA$_2$ (IV) nanocomposite was washed by centrifugation (12,000 rpm) with PBS to remove unbound material. Then, the product was redispersed in 500 µL of PBS 0.01 M. To avoid non-specific adsorption on the surface of the AuNPs, 350 µL of aqueous bovine serum albumin (BSA, Sigma Aldrich) solution (50 µg/mL) was added to the SWCNH/Thi/AuNPs/AntiCEA$_2$, shaking gently for 1 h. Finally, the SWCNH/Thi/AuNPs/AntiCEA$_2$/BSA (V) system was washed again, and the recovered material was redispersed in 500 µL of 0.01 M PBS and stored at 4 °C. For practical purposes, the term SWCNH/Thi/AuNPs/AntiCEA$_2$/BSA will be referred to as the nanoprobe (NaPro).

2.5. Assembly Process of the Immunosensor

First, the Au-disk was polished on a micro cloth using 0.3 µm and later 0.05 µm alumina. Then, it was electrochemically annealed by 100 cyclic voltammetry scans using H_2SO_4 0.5 M at a potential sweep of 0.35 to 1.5 V at 1 V s^{-1} (VI). The CV with H_2SO_4 did not only clean the Au surface, but also worked as a pretreatment to improve the electroactive area of the Au electrode, helping with the sensitivity of the immunosensor.

The immobilization of the $AntiCEA_1$ antibody, on the surface of the gold electrode, was carried out using the photochemical immobilization technique (PIT). Briefly, 350 μL of $AntiCEA_1$ (15 μg/mL) was irradiated with a UV lamp (Trylight®, Promete Srl) using a quartz cell for 30 s. The irradiated $AntiCEA_1$ was transferred to an Eppendorf tube, and the gold electrode was immediately dipped in the solution for 15 min. Subsequently, the electrode was rinsed with PBS, obtaining Au-disk/$AntiCEA_1$ (VII). Afterward, 25 μL of BSA (50 μg/mL) was deposited on the Au-disk/$AntiCEA_1$ and incubated for 30 min at room temperature to avoid non-specific absorption. Subsequently, the electrode was rinsed with PBS, obtaining Au-disk/$AntiCEA_1$/BSA (VIII). Then, 25 μL of CEA antigen was deposited at different concentrations and incubated for 45 min at room temperature. Afterward, the electrode was rinsed with PBS, obtaining Au-disk/$AntiCEA_1$/BSA/CEA. Finally, 30 μL of NaPro was deposited and incubated for 45 min at room temperature, and the electrode was rinsed with PBS, obtaining Au-disk/$AntiCEA_1$/BSA/CEA/NaPro. Figure 1 shows the assembly steps of the immunosensor.

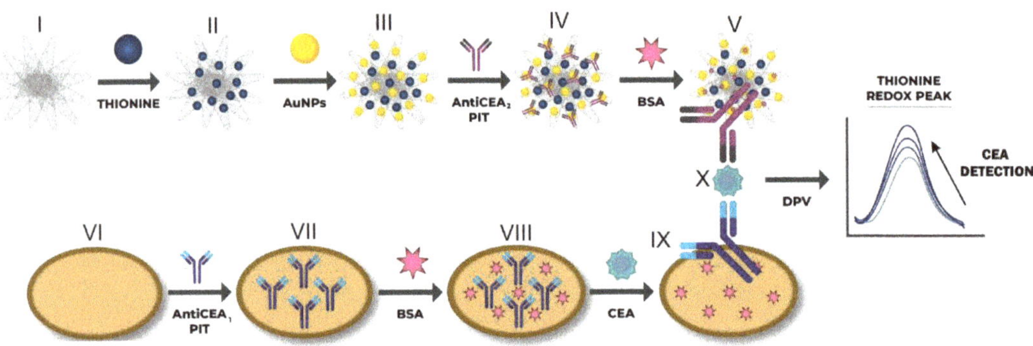

Figure 1. (I–V) Preparation of the nanoprobe consisting of SWCNH/Thi/AuNPs/$AntiCEA_2$. (VI–X) Immobilization of the $AntiCEA_2$ on the Au-disk electrode by PIT and assembly of the electrochemical immunosensor. The electrochemical detection is achieved by a dependence on the thionine redox peak as a function of the CEA concentration.

2.6. CEA Biomarker Detection

The modification step-wise process of the working electrode was characterized by cyclic voltammetry (CV) and electrochemical impedance spectroscopy (EIS) in $[Fe(CN)_6]^{3-/4-}$ 10 mM in PBS 0.01 M. CEA detection was carried out by differential pulse voltammetry (DPV) in PBS 0.01 M at pH 7.4, monitoring the redox peak of thionine around −0.25 V.

3. Results

3.1. SWCNH/Thi/AuNPs Characterization

The morphology of SWCNH and SWNH/Thi/AuNPs were characterized by HRTEM and STEM. Figure 2A,B shows that a single carbon nanohorn is around 2–5 nm in diameter and 40–50 nm in length. The individual nanohorns tend to aggregate forming the typical dahlia-like nanostructure with an approximate diameter of 80–100 nm [8,13]. The STEM images of SWCNH also confirmed the dahlia-like assemblies.

Figure 2C–F show the SWCNH/Thi/AuNPs, which demonstrated the AuNPs were homogeneously distributed and anchored on the pristine SWCNH surface, providing uniform binding sites for the attachment of antibodies to the nanohorns. The average size of the AuNPs was 5–10 nm. Notably, the structure of the SWCNH was not altered during the incorporation of the AuNPs.

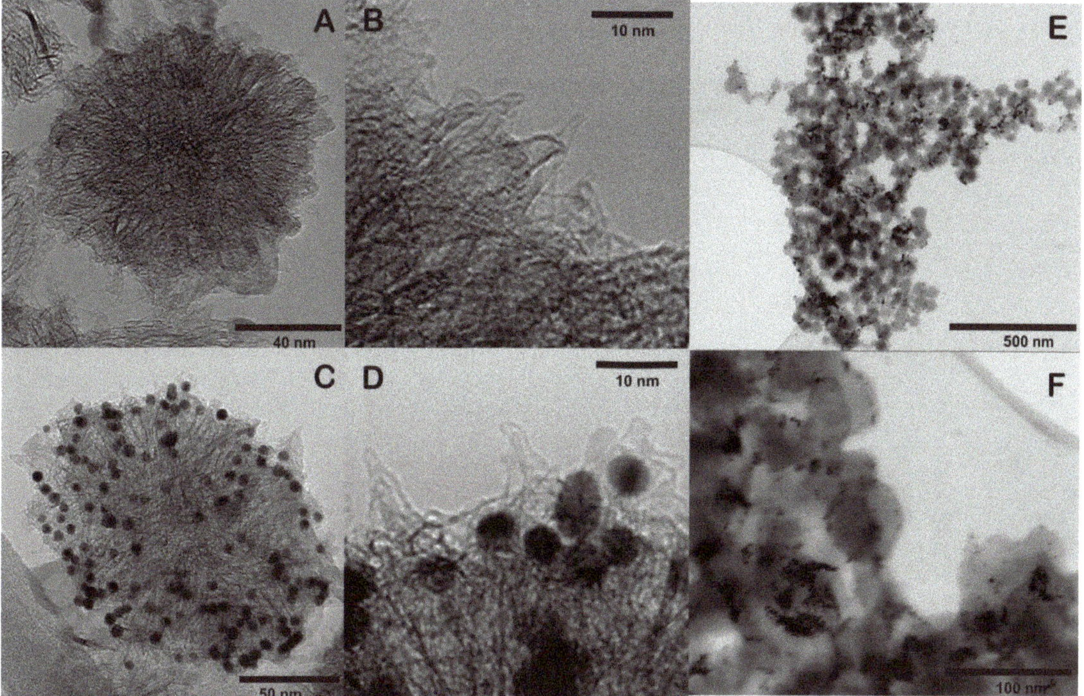

Figure 2. (**A**,**B**) HRTEM images of pristine SWCNH, (**C**,**D**) HRTEM images of SWCNH/Thi/AuNPs, (**E**,**F**) STEM images of SWCNH/Thi/AuNPs.

To verify the presence of thionine and AuNPs in SWCNH an EDS mapping was performed, and the elemental distribution within the SWNHs was verified using STEM-EDS (Figure 3). The SWCNH/Thi/AuNPs is mainly composed of carbon (Figure 3B), the presence of oxygen was also observed as a consequence of the oxidation treatment of SWCNH (Figure 3C), which likely contains a variety oxygen associated functional units such as hydroxyl and carboxyl groups [14]. Besides, the element sulfur was also observed (Figure 3D), which is attributed to the presence of thionine in the surface of the SWCNH since thionine contains a thiazinium group [15]. The presence of gold nanoparticles can be clearly seen on the SWCNH surface, which indicates that the nanoparticles observed by TEM and STEM are certainly gold nanoparticles, Figure 3E. These results prove that thionine and AuNPs were firmly tethered to the SWCNH with uniform distribution and cannot be washed off be rinsing.

Possible explanations for strong interaction between these components are on the one hand that thionine has a planar aromatic structure that facilitates strong π-π stacking interactions to the likewise aromatic SWCNH surface [16]. In addition to π- stacking interactions, also coupling via (electro)activated functional groups (-C=O- and COOH) of SWCNH can be involved in thionine linking via its amino-groups [17]. Additionally, the AuNPs were attached mainly to the SWCNH surface predominantly via unspecific adsorption, which could involve physisorption, π-π stacking, hydrophobic and electrostatistic interactions [18].

A glassy carbon electrode was modified with SWCNH/Thi/AuNPs and characterized by cyclic voltammetry (CV). Figure 3F shows an anodic peak around 0.21 V and a cathodic peak at 0 V, which are characteristic potentials of the reversible two-electron transfer process of thionine at acidic pH [19]. The difference between the anodic and cathodic peak is 210 mV, and the current ratio I_{pa}/I_{pc} is 2.41 mA. Therefore, the thionine redox reaction can be considered as quasi-reversible process [20].

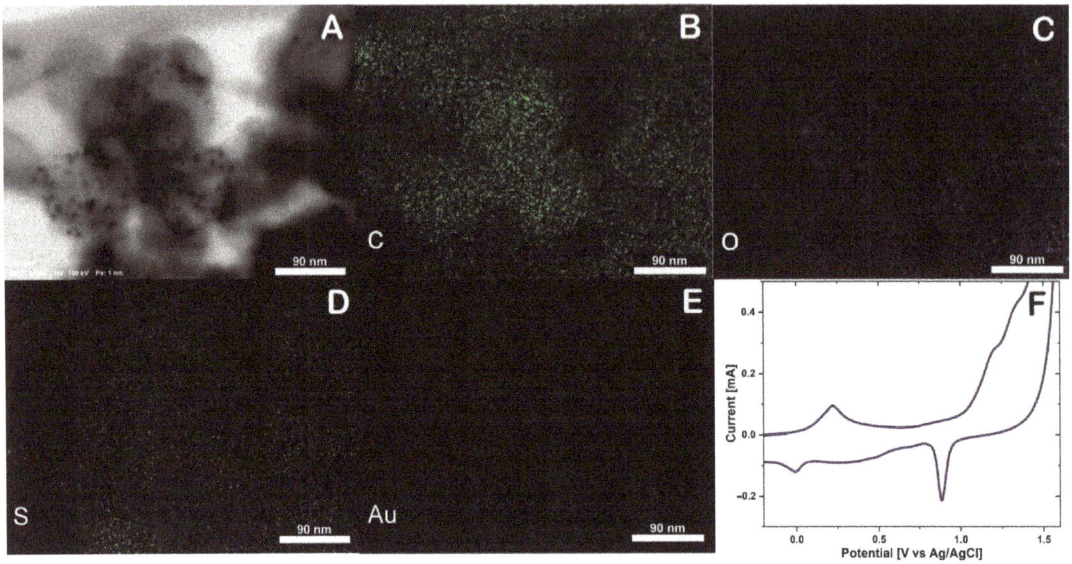

Figure 3. (**A**) Energy-dispersive X−ray spectroscopy (EDS) mapping of SWCNH/Thi/AuNPs, including SEM image (**B**) Carbon (C) element, (**C**) Oxigen (O) element, (**D**) Sulfur (S) element, (**E**) Gold (Au) element and (**F**) Cyclic voltammetry of glassy carbon electrode (GCE) modified SWCNH/Thi/AuNPs in H_2SO_4 0.5 M at 50 mVs^{-1}.

Noteworthy, the redox properties of thionine were not affected by the incorporation into the SWCNH. Moreover, a prominent cathodic peak was observed at approximately 0.9 V. This peak is characteristic for the reduction of gold in the reverse potential sweep after a considerable electro-oxidation in the forward sweep to potentials higher than 1.2 V, confirming the presence of the AuNPs [21].

The SWCNH/Thi/AuNPs nanocomposite showed distinct redox activity, supporting the incorporation of Thi on the SWCNH surface. Furthermore, it should be pointed out that the high conductivity of the SWCNH facilitated the electron transfer across this carbon material and thus the redox reaction of Thi molecules at the distal side of SWCNH, which enhanced the electrochemical current [22,23]. Likewise, the multitude of SWCNH associated AuNPs offers multiple sites for antibody immobilization, promising all in all high potential as signal tag for the fabrication of the sandwich-type immunosensor.

3.2. Optimization Test

To verify that the immobilization of $AntiCEA_1$ antibodies on the surface of the Au-disk by PIT was successful, the Au electrode was characterized by CV before and after immobilization of the antibodies, using a redox probe $Fe(CN)_6^{3-}$ /$Fe(CN)_6^{4-}$ (Figure 4).

The PIT method includes an exposure of the antibodies to UV irradiation, which leads to selective photoreduction of the typical disulfide bond of the antibodies in specific cysteine-cysteine/tryptophan triads (Cys-Cys/Trp). The breaking of these Cys-Cys bonds produces free thiol groups, which can interact with the proximal gold surface, resulting in a covalent Au-S bond between the antibody and the Au surface [12].

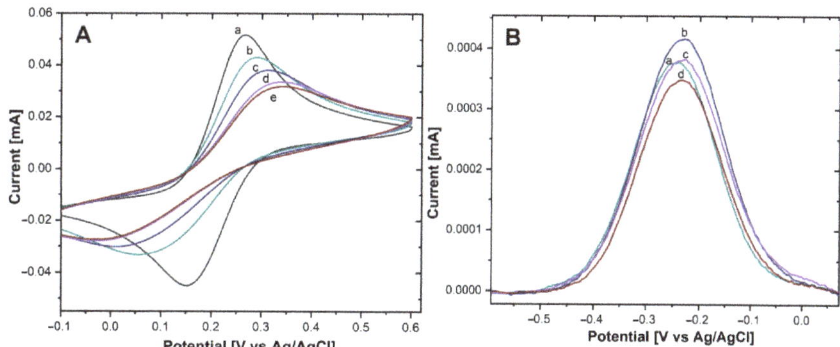

Figure 4. (**A**) CV of Au−disk with (a) 0, (b) 5 μg/mL, (c) 15 μg/mL, (d) 30 μg/mL, (e) 50 μg/mL of AntiCEA$_1$ in $Fe(CN)_6^{3-}/Fe(CN)_6^{4-}$ in PBS 0.01 M at pH 7.4. (**B**) DPV of Au−disk/AntiCEA$_1$/CEA/NaPro with (a) 5 μg/mL, (b) 15 μg/mL, (c) 30 μg/mL, (d) 50 μg/mL of AntiCEA$_1$ in PBS 0.01 M at pH 7.4.

The Au-disk electrode showed well-defined anodic and cathodic peaks, due to the reversible oxidation and reduction of the solution phase $Fe(CN)_6^{3-}/Fe(CN)_6^{4-}$ redox molecule, with a peak-to-peak difference (ΔE_P) of 112 mV (±2.16) and an anodic current intensity (I_P) of 54.1 μA ± (1.87). After antibody immobilization and as the AntiCEA$_1$ concentration increased, ΔE_P increased and I_P decreased, confirming that the antibodies were immobilized on the Au-disk since their covalent immobilization on the surface is acting as an insulating layer, causing slower electron transfer [24]. Besides, at the concentration of 30 μg/mL of AntiCEA$_1$, the surface of the Au-disk was practically saturated, since the change of I_P from 30 to 50 μg/mL was minimal. It should be noted that the higher the concentration of immobilized capture antibodies, the higher the impact on the available electroactive surface area of the Au-disk. In other words, there could be a tradeoff between receptor density for binding the target molecule and the efficiency of the charge transfer between the electroactive surface and thionine redox probes. Hence, it is important to find an optimal immobilization concentration that leads to a receptor surface coverage at which the biosensor generates the highest analytical signal. Consequently, the electrodes were modified with different concentrations of AntiCEA$_1$ and exposed to the complete detection system (CEA antigen, and the nanoprobe NaPro) at a constant concentration.

In Figure 4B, a DPV voltammogram is shown. A redox peak around −0.22 V can be observed, which is characteristic for the Thi_{ox}/Th_{red} redox couples [20]. This demonstrates that Thi was present on the NaPro and it underwent electron transfer reactions [25]. Since the amount of attached NaPro is related with the amount of CEA present in the electrode surface, Thi works as a redox tag for the electrochemical detection of CEA. The intensity of the redox peak of Thi is related to the concentration of the biomarker CEA. The highest current intensity was obtained with an AntiCEA$_1$ concentration of 15 μg/mL, therefore, this concentration was chosen for the further implementation of the immunosensor.

Moreover, the immobilization of the AntiCEA$_2$ on the SWCNH/Thi/AuNPs was also carried out by the PIT. To verify the immobilization effectivity, a glassy carbon electrode (GCE) was modified with the SWCNH/Thi/AuNPs/AntiCEA$_2$, testing AntiCEA$_2$ concentrations of 50 and 100 μg/mL. Figure 5B shows the DPV plots of the SWCNH/Thi/AuNPs. At around −0.22 V a peak was observed that can be attributed to the redox reactions of the present thionine. The redox peak possessed a current intensity of 14.79 μA for the antibody-free SWCNH/Thi/AuNPs. This current intensity decreased after immobilization of AntiCEA$_2$ to 5.44 μA (50 μg/mL) and 0.37 μA (100 μg/mL), confirming that AntiCEA$_2$ was successfully immobilized. Since the concentration of 100 μg/mL significantly decreased the redox peak of thionine by blocking the charge transfer, 50 μg/mL of AntiCEA$_2$ was chosen as the optimal concentration to maintain high analytical sensitivity.

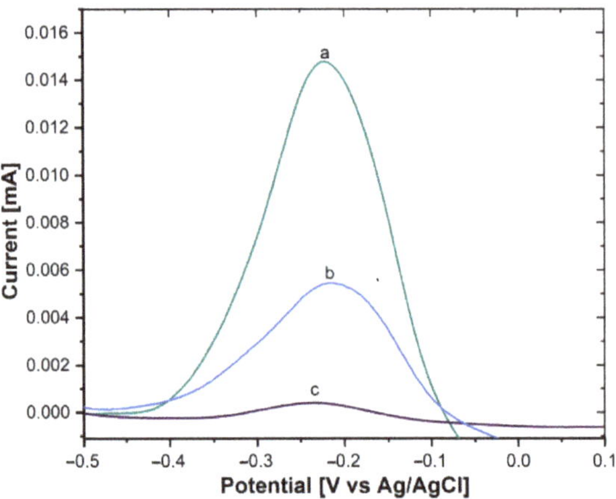

Figure 5. DPV of (a) GCE/SWCNH/Thi/AuNPs/0 μg/mL of AntiCEA$_2$, (b) GCE/SWCNH/Thi/AuNPs/50 μg/mL of AntiCEA$_2$, (c) GCE/SWCNH/Thi/AuNPs/100 μg/mL of AntiCEA$_2$ in PBS 0.01 M at pH 7.4.

3.3. Electrochemical Characterization by Fabrication Steps

CV and EIS were used to corroborate the immunosensor assembly process at each modification stage and to verify the binding of the biomarker CEA and the NaPro. Both characterizations provide information on the electron transfer process and specifically, the changes in charge transfer resistance caused by anchoring the insulating biomolecules on the gold electrode.

CV tests were carried out in a $Fe(CN)_6^{3-}/Fe(CN)_6^{4-}$ 10 mM solution. The ΔE_P and I_{Pa} values were determined for Au-disk (ΔE_P = 112.4 mV ± 2.16), I_{Pa} = 54.11 μA ± 1.87), Au-disk/AntiCEA$_1$ (ΔE_P = 335.9 mV ± 1.22, I_{Pa} = 32.7 μA ± 0.81), Au-disk/AntiCEA$_1$/BSA (ΔE_P = 403 mV ± 2.04, I_{Pa} = 21.97 μA ± 0.49), Au-disk/AntiCEA$_1$/BSA/AgCEA (ΔE_P = 498.11 mV ± 0.4, I_{Pa} = 16.57 μA ± 0.81), Au-disk/AntiCEA$_1$/AgCEA/NaPro (ΔE_P = 376.77 mV ± 6.94, I_{Pa} = 29.28 μA ± 0.61).

The height of the redox peaks consecutively decreases after the addition of AntiCEA$_1$, BSA, and CEA antigen, Figure 6A. This behavior is attributed to the fact that these biomolecules do not possess conductive properties, which on the one hand do not contribute to the transport of electrons and on the other hand block the diffusion of solution-phase redox probes to the surface of the electrode [26]. In the last step, where the NaPro is incorporated, the I_{Pa} increased and ΔE_P decreased, indicating that the addition of the NaPro improves the electroactivity, due to the good conductive properties of the SWCNH and AuNPs, similar effect was found in previous reports [27,28]. The corresponding changes observed at each stage confirm that each component was successfully implemented in the system.

Electrochemical impedance spectroscopy (EIS) is an effective tool to characterize the electrode-electrolyte interface properties. The charge transfer resistance (R_{ct}) can be calculated from the semicircular section of the Nyquist plot with the axis for the real part of the impedance in EIS at low frequencies [29].

$Fe(CN)_6^{3-}$ /$Fe(CN)_6^{4-}$ was used as a redox couple for the EIS experiments, Figure 6B. According to the Nyquist plot, the R_{ct} values were Au-disk (204.65 Ω ± 5.8), Au-disk/AntiCEA$_1$ (1679.19 Ω ± 15), Au-disk/AntiCEA$_1$/BSA (8361.51 Ω ± 84), Au-disk/AntiCEA1/BSA/CEA (22861.61 Ω ± 116), and Au-disk/AntiCEA$_1$/BSA/CEA/NaPro (2360.66 Ω ± 44). The addition of AntiCEA$_1$, BSA, and CEA antigen increased the diameter of the semicircle consecutively, indicating that these biomolecules enhanced the blocking of the charge transfer at the electrode interface. Interestingly, R_{ct} decreased with the incorporation of the NaPro due to the highly conductive nature of the carbon nanohorns. The result of

EIS coincided with the characteristics observed for CV; which demonstrates the successful implementation of a sandwich electrochemical immunosensor for the carcinoembryonic antigen detection.

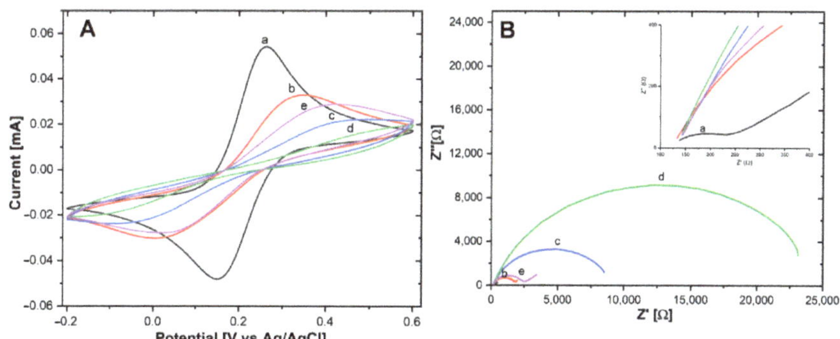

Figure 6. (**A**) CV and (**B**) EIS of (a) Au−disk, (b) Au−disk/AntiCEA$_1$, (c) Au−disk/AntiCEA$_1$/BSA, (d) Au−disk/AntiCEA$_1$/BSA/CEA, (e) Au−disk/AntiCEA$_1$/BSA/CEA/NaPro in 10 mM of $Fe(CN)_6^{3-}/Fe(CN)_6^{4-}$ in PBS 0.01 M at pH 7.4.

3.4. Analytical Performance of the Immunosensor

The performance of the immunosensor for the CEA biomarker detection was investigated using DPV. The CEA antigen detection was carried out in PBS 0.01 M at pH 7.4. Figure 7A shows the immunosensor response at different concentrations of CEA. The DPV signals increased as the CEA concentration rose.

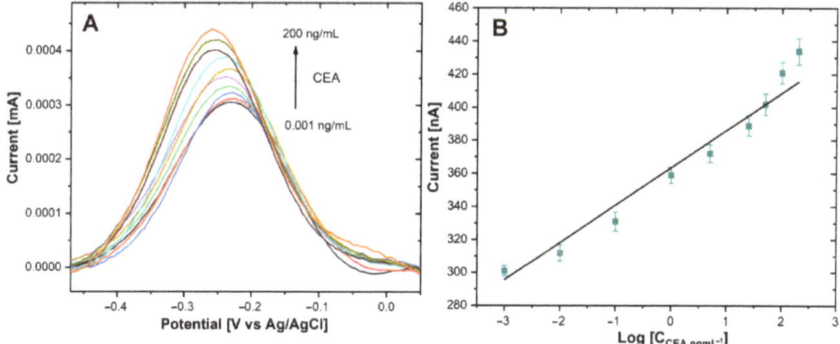

Figure 7. (**A**) DPV of Au−disk/AntiCEA$_1$/BSA/CEA/NaPro with different concentrations of CEA in PBS 0.01 M at pH 7.4. (**B**) The linear relationship between the current peak and the log concentration of CEA.

The sensing mechanism is attributed to the Thi used as a redox tag, since a multitude of these redox molecules was attached to the large surface area of the SWCNH. The amount of attached NaPro is related with the amount of CEA present in the electrode surface due to the formation of immunocomplex between CEA and AntiCEA$_2$. Therefore, the change in the current intensity of the redox peak of Thi is related to the concentration of the biomarker CEA.

The calibration curve (Figure 7B) showed a linear relationship between the current intensity of the thionine redox peak and the logarithm of the CEA concentration. The linear detection range extended from 0.001 to 200 ng/mL for CEA. The calibration curve equation was I_P (nA) = 24.726 log C_{CEA} (ng/mL) + 363.24 (R^2 = 0.964) and the limit of detection was calculated to be 0.1385 pg/mL defined as the mean of the blank signal and 3 times the relative standard deviation. It should be noted that the concentration of CEA in blood serum is typically 3 ng/mL [4] in healthy individuals; therefore, the proposed immunosensor covers the medical relevant concentration range of the CEA biomarker and potentially facilitates practical application to monitor this biomarker. The promising performance of this sensor could be attributed to the high signal amplification capabilities of the highly conductive SWCNH/Thi/AuNPs and their decoration with a high number of redox active thionine.

Compared with other previously reported methods in the literature (Table 1), our immunosensor advanced current detection technology in the combination of exhibiting a wider detection range and lower detection limits. It should also be noted that the preparation time of the previously reported systems is typically quite long because the incubation times for the immobilization of the antibodies can take several hours while here it required only 15 min thanks to the PIT activation and enhanced the immobilization via the thiol groups of the cysteines proteins. In addition, before immobilization, other methods require a modification of the electrode surface. Meanwhile, the PIT technique used in the present work does not require any additional surface modification steps, which decreases the total fabrication time to only 2.25 h, notably less than that of other reported techniques.

Besides, the immobilization of the antibodies by the PIT is very effective since it ensured control over the orientation of the immobilized Ab, with their binding sites exposed for the formation of the antigen-antibody immune complex [12,24,30]. Indeed, Funari et al. [11] investigated the immobilization and orientation of antibodies (Abs) photoactivated by PIT. In their experiments, the photoactivated antibodies were immobilized on ultrasmooth template stripped gold films and investigated by atomic force microscopy (AFM) at the level of individual molecules. They found smaller contact area and larger heights measured in the surfaces with the antibodies immobilized by PIT than the ones immobilized by physisorption. Therefore, the activated antibodies tend to be more upright compared with nonirradiated ones, thereby providing better exposure to the binding sites. The immobilization and orientation of antibodies photoactivated by PIT enhance the binding capabilities of antibody receptors, which is a critical aspect of immunosensor development because both the number and the orientation of the immobilized biomolecules are closely related to the detection efficiency of the device [31].

Table 1. Comparison of the proposed immunosensor with previous similar works.

Signal Tag	Fabrication Time (h)	Linear Range (ng/mL)	Detection Limit (pg/mL)	Reference
Ti_3C_2@CuAu-LDH	15.5	0.0001–80	0.033	[26]
PdNPs–V_2O_5/MWCNTs	2.5	0.0005–25	0.17	[32]
AuNP-HRP	5.32	0.01–80	2.36	[33]
NiPtAu-rGO	4	0.001–100	0.27	[34]
Au@SiO_2/Cu_2O	15.4	0.00001–80	0.0038	[35]
CPS@PANI@Au	8.8	0.006–12	1.56	[36]
SWCNH/Thi/AuNPs	2.25	0.001–200	0.138	This work

3.5. Selectivity

The high and evolutionary evolved specificity of antibodies is one advantage of immunoassays over competing biosensor concepts. To evaluate the specificity of the electrochemical immunosensor, a selectivity analysis was performed, spiking possible interfering agents such as bovine serum albumin (BSA), human serum albumin (HSA), or CA15-3 antigen to the blank sample solution (containing the nanoprobe without the presence of CEA). The tests were performed separately by incubating the Au-disk/AntiCEA$_1$/BSA electrode surface in 50 ng/mL CEA, 50 ng/mL BSA, 50 ng/mL HSA, 50 U/mL CA15-3, and blank solution (0 ng/mL CEA). Although the interfering substances were applied under the same conditions as the real analyte, the response currents were much lower compared to the response toward CEA (Figure 8). This result indicates that these substances do not interfere with the target detection and the high selectivity of the antibodies was conserved during the implementation of the immunosensor, resulting in an immunosensor with excellent selectivity for CEA.

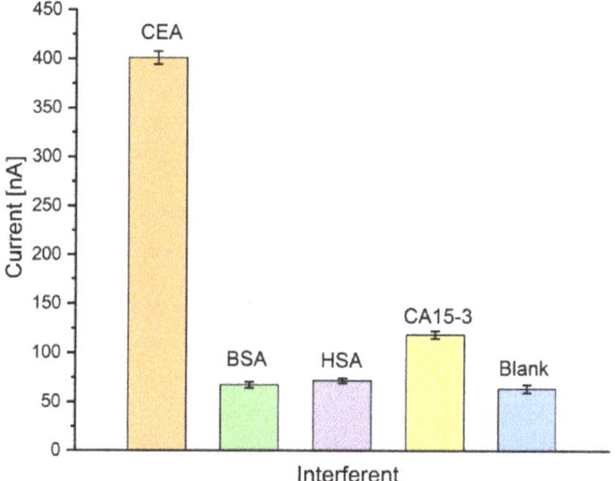

Figure 8. Current responses of the immunosensor to CEA (50 ng/mL) and interfering substances BSA (50 ng/mL), HSA (50 ng/mL), CA15-3 (50 U/mL) and blank in PBS 0.01 M at pH 7.4 (n = 3).

3.6. Real Sample Testing

To investigate the performance of the immunosensor for detection in real clinical samples, human serum samples with known CEA concentrations were analyzed, Table 2. The standard addition method was used to corroborate electrochemical detection. CEA concentrations were calculated from the calibration curve and the tests were repeated three times for each sample. Table 2 shows the recovery (%) of the serum samples found in the range of 95 to 113%. The successful results demonstrate high accuracy and the feasibility of using the immunosensor for the electrochemical detection of CEA in real clinical samples. Therefore, the results confirm the potential of the proposed method to be implemented in the clinical field for the detection and monitoring of the carcinogenic biomarker CEA in patients.

Table 2. Results of the recovery of the immunosensor in serum samples in 0.01 M PBS.

Added CEA (ng/mL)	Found CEA (ng/mL)	Recovery (%)	% RSD (n = 3)
1	1.13	113.16	14.5
5	4.57	91.52	11.84
10	9.51	95.13	13.46

4. Conclusions

In this work, a sandwich-type electrochemical immunosensor was developed for the quantitative determination of the CEA biomarker using a signal amplification strategy based on carbon nanohorns. The fast photochemical immobilization technique (PIT) was employed for both capture and detection antibodies to tether them onto the gold electrode and the SWCNH/Thi/AuNPs, respectively, which facilitated short assay assembly times of less than three hours. The immunosensor showed a low detection limit of 0.1385 pg/mL, a linear detection range from 0.001–200 ng/mL, and high selectivity. The remarkable performance was attributed on the one hand to the antibodies being covalently bound to the gold surfaces by PIT, controlling the orientation of their active sites. On the other hand, the large surface area, high conductivity, and manifold thionine redox activity of the SWCNH/Thi/AuNP nanocomposite enhanced the amperometric sensor signal, which resulted in a high sensitivity of the device. Therefore, the proposed strategy of PIT antibody immobilization and SWCNH/Thi/AuNP nanocomposite-based signal amplification can be used as a versatile strategy for the clinical detection of the CEA biomarker and could potentially be extended for the clinical detection of other relevant biomarkers.

Author Contributions: A.D.-A.: investigation, methodology, formal analysis, writing—original draft. E.A.Z.-C.: conceptualization, validation, writing—review and editing. G.F.-M.: validation, methodology. A.O.: validation, review, and editing. D.M.: conceptualization, writing—review and editing, supervision, funding acquisition. All authors have read and agreed to the published version of the manuscript.

Funding: This research was funded by the German Academic Exchange Service (DAAD) by an scholarship granted to Angélica Domínguez Aragón (grant number 91791585) to carry out the research stay at the Forschungszentrum Jülich (Research Center of Jülich). Furthermore, she received a Ph.D. scholarship from the National Council for Science and Technology of Mexico (CONACYT) (grant number 701397).

Institutional Review Board Statement: The study was conducted in accordance with the Declaration of Helsinki, and approved by the Kommission für Ethik in der Forschung (KEF) of the Research Center Jülich (protocol code Humanserum and date of approval: 06.Jan.2022).

Informed Consent Statement: Not applicable.

Data Availability Statement: The data presented in this study are available on request from the corresponding author.

Acknowledgments: We thank Mateo Martinez, Elke Brauweiler, Raul Ochoa and Paola Anchondo for their valuable collaboration during this research.

Conflicts of Interest: The authors declare no conflict of interest.

References

1. Jiang, J.; Xia, J.; Zang, Y.; Diao, G. Electrochemistry/Photoelectrochemistry-Based Immunosensing and Aptasensing of Carcinoembryonic Antigen. *Sensors* **2021**, *21*, 7742. [CrossRef]
2. Ullah, M.F.; Aatif, M. The footprints of cancer development: Cancer biomarkers. *Cancer Treat. Rev.* **2009**, *35*, 193–200. [CrossRef] [PubMed]
3. Hasanzadeh, M.; Shadjou, N.; Lin, Y.; de la Guardia, M. Nanomaterials for use in immunosensing of carcinoembryonic antigen (CEA): Recent advances. *TrAC—Trends Anal. Chem.* **2017**, *86*, 185–205. [CrossRef]
4. Costa, R.E.; Agustín, C.-G. Screen-printed Electrochemical Immunosensors for the Detection of Cancer and Cardiovascular Biomarkers. *Electroanalysis* **2016**, *28*, 1700–1715. [CrossRef]
5. Lequin, R.M. Enzyme Immunoassay (EIA)/Enzyme-Linked Immunosorbent Assay (ELISA). *Clin. Chem.* **2005**, *51*, 2415–2418. [CrossRef]
6. Ricci, F.; Adornetto, G.; Palleschi, G. A review of experimental aspects of electrochemical immunosensors. *Electrochim. Acta* **2012**, *84*, 74–83. [CrossRef]
7. Felix, F.S.; Angnes, L. Electrochemical immunosensors—A powerful tool for analytical applications. *Biosens. Bioelectron.* **2018**, *102*, 470–478. [CrossRef]
8. Liu, X.; Ying, Y.; Ping, J. Structure, synthesis, and sensing applications of single-walled carbon nanohorns. *Biosens. Bioelectron.* **2020**, *167*, 112495. [CrossRef]
9. Carli, S.; Lambertini, L.; Zucchini, E.; Ciarpella, F.; Scarpellini, A.; Prato, M.; Castagnola, E.; Fadiga, L.; Ricci, D. Single walled carbon nanohorns composite for neural sensing and stimulation. *Sens. Actuators B Chem.* **2018**, *271*, 280–288. [CrossRef]
10. Zhao, C.; Wu, J.; Ju, H.; Yan, F. Multiplexed electrochemical immunoassay using streptavidin/nanogold/carbon nanohorn as a signal tag to induce silver deposition. *Anal. Chim. Acta* **2014**, *847*, 37–43. [CrossRef]
11. Funari, R.; Della Ventura, B.; Altucci, C.; Offenhäusser, A.; Mayer, D.; Velotta, R. Single Molecule Characterization of UV-Activated Antibodies on Gold by Atomic Force Microscopy. *Langmuir* **2016**, *32*, 8084–8091. [CrossRef] [PubMed]

12. Minopoli, A.; Della Ventura, B.; Lenyk, B.; Gentile, F.; Tanner, J.A.; Offenhäusser, A.; Mayer, D.; Velotta, R. Ultrasensitive antibody-aptamer plasmonic biosensor for malaria biomarker detection in whole blood. *Nat. Commun.* **2020**, *11*, 1–10. [CrossRef] [PubMed]
13. Dutta, K.; De, S.; Das, B.; Bera, S.; Guria, B.; Ali, S.; Chattopadhyay, D. Development of an Efficient Immunosensing Platform by Exploring Single-Walled Carbon Nanohorns (SWCNHs) and Nitrogen Doped Graphene Quantum Dot (N-GQD) Nanocomposite for Early Detection of Cancer Biomarker. *Cite This ACS Biomater. Sci. Eng* **2021**, *7*, 5541–5554. [CrossRef]
14. Domínguez-Aragón, A.; Dominguez, R.B.; Peralta-Pérez, M.D.R.; Armando Zaragoza-Contreras, E. Catalase biosensor based on the PAni/cMWCNT support for peroxide sensing. *E-Polymers* **2021**, *21*, 476–490. [CrossRef]
15. Xu, H.; Wang, Y.; Wang, L.; Song, Y.; Luo, J.; Cai, X. A Label-Free Microelectrode Array Based on One-Step Synthesis of Chitosan-Multi-Walled Carbon Nanotube-Thionine for Ultrasensitive Detection of Carcinoembryonic Antigen. *Nanomaterials* **2016**, *6*, 132. [CrossRef] [PubMed]
16. Han, J.; Ma, J.; Ma, Z. One-step synthesis of graphene oxide-thionine-Au nanocomposites and its application for electrochemical immunosensing. *Biosens. Bioelectron.* **2013**, *47*, 243–247. [CrossRef] [PubMed]
17. Mahbubur Rahman, M.; Lee, J.J. Sensitivity control of dopamine detection by conducting poly(thionine). *Electrochem. Commun.* **2021**, *125*, 107005. [CrossRef]
18. Duc Chinh, V.; Speranza, G.; Migliaresi, C.; Van Chuc, N.; Minh Tan, V.; Phuong, N.T. Synthesis of gold nanoparticles decorated with multiwalled carbon nanotubes (Au-MWCNTs) via cysteaminium chloride functionalization. *Sci. Rep.* **2019**, *9*, 5667. [CrossRef]
19. Zhuo, Y.; Yuan, R.; Chai, Y.; Tang, D.; Zhang, Y.; Wang, N.; Li, X.; Zhu, Q. A reagentless amperometric immunosensor based on gold nanoparticles/thionine/Nafion-membrane-modified gold electrode for determination of α-1-fetoprotein. *Electrochem. Commun.* **2005**, *7*, 355–360. [CrossRef]
20. Hashemnia, S.; Eskanari, M. Preparation and Electrochemical Characterization of an Enzyme Electrode Based on Catalase Immobilized onto a Multiwall Carbon Nanotube-Thionine Film. *J. Chin. Chem. Soc.* **2014**, *61*, 903–909. [CrossRef]
21. Zhang, Y.; Figueroa-Miranda, G.; Wu, C.; Willbold, D.; Offenhäusser, A.; Mayer, D. Electrochemical dual-aptamer biosensors based on nanostructured multielectrode arrays for the detection of neuronal biomarkers. *Nanoscale* **2020**, *12*, 16501. [CrossRef] [PubMed]
22. Karousis, N.; Suarez-Martinez, I.; Ewels, C.P.; Tagmatarchis, N. Structure, Properties, Functionalization, and Applications of Carbon Nanohorns. *Chem. Rev.* **2016**, *116*, 4850–4883. [CrossRef] [PubMed]
23. Huang, W.; Xiang, G.; Jiang, D.; Liu, L.; Liu, C.; Liu, F.; Pu, X. Electrochemical Immunoassay for Cytomegalovirus Antigen Detection with Multiple Signal Amplification Using HRP and Pt-Pd Nanoparticles Functionalized Single-walled Carbon Nanohorns. *Electroanalysis* **2016**, *28*, 1126–1133. [CrossRef]
24. Cimafonte, M.; Fulgione, A.; Gaglione, R.; Papaianni, M.; Capparelli, R.; Arciello, A.; Censi, S.B.; Borriello, G.; Velotta, R.; Ventura, B. Della Screen printed based impedimetric immunosensor for rapid detection of Escherichia coli in drinking water. *Sensors* **2020**, *20*, 274. [CrossRef]
25. Shobha Jeykumari, D.R.; Ramaprabhu, S.; Sriman Narayanan, S. A thionine functionalized multiwalled carbon nanotube modified electrode for the determination of hydrogen peroxide. *Carbon N. Y.* **2007**, *45*, 1340–1353. [CrossRef]
26. Zhang, M.; Mei, L.; Zhang, L.; Wang, X.; Liao, X.; Qiao, X.; Hong, C. Ti3C2 MXene anchors CuAu-LDH multifunctional two-dimensional nanomaterials for dual-mode detection of CEA in electrochemical immunosensors. *Bioelectrochemistry* **2021**, *142*, 107943. [CrossRef]
27. Zhu, Q.; Chai, Y.; Yuan, R.; Zhuo, Y. Simultaneous detection of four biomarkers with one sensing surface based on redox probe tagging strategy. *Anal. Chim. Acta* **2013**, *800*, 22–28. [CrossRef]
28. Zhang, Z.; Yang, M.; Wu, X.; Dong, S.; Zhu, N.; Gyimah, E.; Wang, K.; Li, Y. A competitive immunosensor for ultrasensitive detection of sulphonamides from environmental waters using silver nanoparticles decorated single-walled carbon nanohorns as labels. *Chemosphere* **2019**, *225*, 282–287. [CrossRef]
29. Pänke, O.; Balkenhohl, T.; Kafka, J.; Schäfer, D.; Lisdat, F. *Impedance Spectroscopy and Biosensing BT—Biosensing for the 21st Century*; Renneberg, R., Lisdat, F., Eds.; Springer: Berlin/Heidelberg, Germany, 2008; pp. 195–237. ISBN 978-3-540-75201-1.
30. Minopoli, A.; Della Ventura, B.; Campanile, R.; Tanner, J.A.; Offenhäusser, A.; Mayer, D.; Velotta, R. Randomly positioned gold nanoparticles as fluorescence enhancers in apta-immunosensor for malaria test. *Microchim. Acta* **2021**, *188*, 1–9. [CrossRef]
31. Trilling, A.K.; Harmsen, M.M.; Ruigrok, V.J.B.; Zuilhof, H.; Beekwilder, J. The effect of uniform capture molecule orientation on biosensor sensitivity: Dependence on analyte properties. *Biosens. Bioelectron.* **2013**, *40*, 219–226. [CrossRef]
32. Han, J.; Jiang, L.; Li, F.; Wang, P.; Liu, Q.; Dong, Y.; Li, Y.; Wei, Q. Ultrasensitive non-enzymatic immunosensor for carcinoembryonic antigen based on palladium hybrid vanadium pentoxide/multiwalled carbon nanotubes. *Biosens. Bioelectron.* **2016**, *77*, 1104–1111. [CrossRef] [PubMed]
33. Miao, J.; Wang, X.; Lu, L.; Zhu, P.; Mao, C.; Zhao, H.; Song, Y.; Shen, J. Electrochemical immunosensor based on hyperbranched structure for carcinoembryonic antigen detection. *Biosens. Bioelectron.* **2014**, *58*, 9–16. [CrossRef] [PubMed]
34. Tian, L.; Liu, L.; Li, Y.; Wei, Q.; Cao, W. Ultrasensitive sandwich-type electrochemical immunosensor based on trimetallic nanocomposite signal amplification strategy for the ultrasensitive detection of CEA. *Sci. Rep.* **2016**, *6*, 30849. [CrossRef]

35. Liao, X.; Wang, X.; Ma, C.; Zhang, L.; Zhao, C.; Chen, S.; Li, K.; Zhang, M.; Mei, L.; Qi, Y.; et al. Enzyme-free sandwich-type electrochemical immunosensor for CEA detection based on the cooperation of an Ag/g-C3N4-modified electrode and Au@SiO2/Cu2O with core-shell structure. *Bioelectrochemistry* **2021**, *142*, 107931. [CrossRef] [PubMed]
36. Song, D.; Zheng, J.; Myung, N.V.; Xu, J.; Zhang, M. Sandwich-type electrochemical immunosensor for CEA detection using magnetic hollow Ni/C@SiO_2 nanomatrix and boronic acid functionalized CPS@PANI@Au probe. *Talanta* **2021**, *225*, 122006. [CrossRef]

Article

Label-Free Sensing of Biomolecular Adsorption and Desorption Dynamics by Interfacial Second Harmonic Generation

Chuansheng Xia [1], Jianli Sun [1], Qiong Wang [1], Jinping Chen [1], Tianjie Wang [2], Wenxiong Xu [1], He Zhang [1], Yuanyuan Li [2], Jianhua Chang [2], Zengliang Shi [1], Chunxiang Xu [1,*] and Qiannan Cui [1,*]

[1] State Key Laboratory of Bioelectronics, School of Biological Science and Medical Engineering, Southeast University, Nanjing 210096, China
[2] School of Electronic and Information Engineering, Nanjing University of Information Science and Technology, Nanjing 210044, China
* Correspondence: xcxseu@seu.edu.cn (C.X.); qiannan@seu.edu.cn (Q.C.)

Abstract: Observing interfacial molecular adsorption and desorption dynamics in a label-free manner is fundamentally important for understanding spatiotemporal transports of matter and energy across interfaces. Here, we report a label-free real-time sensing technique utilizing strong optical second harmonic generation of monolayer 2D semiconductors. BSA molecule adsorption and desorption dynamics on the surface of monolayer MoS_2 in liquid environments have been all-optically observed through time-resolved second harmonic generation (SHG) measurements. The proposed SHG detection scheme is not only interface specific but also expected to be widely applicable, which, in principle, undertakes a nanometer-scale spatial resolution across interfaces.

Keywords: second harmonic generation; bovine serum albumin; heterointerface; adsorption and desorption; biomolecule

Citation: Xia, C.; Sun, J.; Wang, Q.; Chen, J.; Wang, T.; Xu, W.; Zhang, H.; Li, Y.; Chang, J.; Shi, Z.; et al. Label-Free Sensing of Biomolecular Adsorption and Desorption Dynamics by Interfacial Second Harmonic Generation. *Biosensors* **2022**, *12*, 1048. https://doi.org/10.3390/bios12111048

Received: 25 October 2022
Accepted: 18 November 2022
Published: 20 November 2022

1. Introduction

Biomolecular activities at interfaces are fundamental phenomena of lives. Interpreting the interfacial dynamics of biomolecules is important for constructing accurate disease models [1–3], performing effective drug screening [4,5] and understanding spatiotemporal transport of matter/energy for life systems. As a spatial region with a thickness usually smaller than 10 nm, label-free probing of the interfacial dynamics of biomolecules is challenging. Thus far, only limited label-free probing techniques have been developed, such as surface plasmon resonance [6,7], optical fiber sensors [8], time resolved sum-frequency generation [9,10], surface-enhanced Raman spectroscopy and so on [11–15]. These methods have significantly improved the performance of interfacial biosensing in terms of high sensitivity, high resolution and real-time observation, and have greatly deepened the understanding of the interfacial dynamics at the molecular level.

Surface plasmon resonance microscopy, elegantly utilizing the localized interactions between interfacial electromagnetic fields and biomolecules, possesses molecular-level sensitivity and interfacial spatial resolutions beyond the optical diffraction limit. Inspired by the physical scheme of surface plasmon resonance microscopy, we intend to develop a complementary optical spectrum sensing technique which can realize interfacial biosensing specificity for microfluidic chips in a label-free manner. In our opinion, interfacial second harmonic generation (SHG), a second-order nonlinear optical effect induced by broken inversion symmetry of an interface, is promising to fulfill this goal. Unfortunately, second-order susceptibility of biomolecule interfaces is usually rather small, resulting in a weak SHG signal. In practice, a PMT of a single pixel is usually required to magnify the weak SHG signals. The consequence is that it is difficult to monitor the biomolecular dynamics of an interface in real-time. An alternative way to overcome such difficulty is to significantly increase the power of a fundamental femtosecond laser, which imposes a high risk of damage to the biomaterials and biostructures.

Two dimensional (2D) semiconducting monolayers, such as monolayer MoS_2, with broken inversion symmetry, possessing large second-order nonlinear optical susceptibility, can produce strong SHG signals under femtosecond laser pulse excitations, which have been comprehensively investigated in the recent decade [16–18]. In our opinion, the strong SHG of these 2D semiconducting monolayers can serve as excellent biosensors if biomolecules can interact with them and form heterointerfaces in a liquid environment. Intuitively, the formation of a heterointerface on the surface of 2D semiconducting monolayers can change the inversion symmetry and then lead to a change in SHG signals. Since strong SHG signals of 2D semiconducting monolayers can be readily detected by regular spectrometer or CMOS sensors, it will be possible to develop a real-time sensing technique to probe biomolecule dynamics at 2D interfaces.

To establish progress, we report a label-free interfacial biomolecular sensing technique by monitoring biomolecular adsorption and desorption processes on the surface of monolayer MoS_2 through time-resolved SHG. Chitosan nanoclusters and bovine serum albumin (BSA) molecules have been used to form effective Coulomb attractions with the negatively charged surface of monolayer MoS_2. By measuring the SHG intensity changes as a function of time, we realize the label-free real-time sensing of biomolecular adsorption and desorption processes all-optically in liquid environments. Our results open new avenues of label-free interfacial biosensing, taking advantage of the strong optical SHG of monolayer 2D semiconductors.

2. Results and Discussion

Monolayer MoS_2 is an ideal platform to construct biosensors for interfacial molecule adsorption and desorption processes, considering that the monolayer MoS_2 lattice undertakes a sub-nanometer thickness with broken inversion symmetry. Inspired by the pioneering work of strong SHG observations in monolayer MoS_2 in 2013 [16], monolayer MoS_2 can be considered as a sub-nanometer thick nonlinear optical source emitting SHG with extremely space-confined dipole moments, which can facilitate the interfacial sensing and imaging of ultrahigh spatial resolution across interfaces. Moreover, large specific surface areas and the abundant binding sites of monolayer MoS_2 can further enable effective interaction with biomolecules. To be specific, negatively charged surfaces of monolayer MoS_2 samples grown by chemical vapor deposition (CVD) [19–23] are expected to induce strong Coulomb attractions in liquid environments for positively charged biomolecules, which, in our opinion, are very promising for realizing label-free, real-time sensing of molecule adsorption and desorption processes.

Figure 1a shows the experimental setup detecting the SHG spectra of monolayer MoS_2 embedded in a microfluidic chip. The wavelength of a fundamental laser is centered at 780 nm, with a pulse width of about 60 fs and a repetition frequency of about 80 MHz. The fundamental laser is reflected by the dichroic mirror and then focused on the MoS_2 monolayer by the objective lens (NA = 0.55). SHG of the monolayer MoS_2 is collected by the same objective lens. A short-pass filter is deployed behind a dichroic mirror to eliminate the reflected fundamental residual. The SHG signals are simultaneously sent to a spectrometer and a CCD by a beam splitter. In our measurements, the power of the fundamental laser was low enough to avoid optical damage of monolayer MoS_2. The fine structure of the microfluidic chip is illustrated in Figure 1b. The main body of the microfluidic chip is fabricated by 3D printing with an optical glass window on top. A sapphire substrate containing monolayer MoS_2 is attached to the optical glass window. The microfluidic channel enables unidirectional flow of liquid solutions of biomolecules to form a laminar flow. As a result, a homogeneous 3D fluid–2D solid interaction is constructed to facilitate adsorption and desorption of biomolecules on the surface of monolayer MoS_2. Monolayer MoS_2 grown by CVD on double-sided polished sapphire substrates (Six Carbon Technology, Shenzen, China) were used as received. To confirm the monolayer nature of these samples, optical characterizations were carried out before microfluidic experiments. Figure 1c presents optical absorption and photoluminescence (PL) spectra of monolayer MoS_2. Lorentzian fitting of the PL spectrum points to a resonant peak centered at about 669 nm, which agrees well with the A-exciton resonance peak of the optical absorption spectrum [24,25]. Employing the experimental setup of Figure 1a, we measured SHG (centered at 390 nm) and fundamental (centered at 780 nm) spectra of monolayer MoS_2, as indicated by the purple squares and red dots in Figure 1d, respectively. Solid lines are Gaussian fittings. Full width at half maxima (FWHM) of fundamental and SHG were fitted to be about 12.8 nm and 5.5 nm. Meanwhile, no SHG signals were observed when the fundamental was focused on the substrate, as indicated by the spectrum of sapphire (black triangles) in Figure 1d. Fundamental power dependence of SHG spectra was obtained by varying the power of the 780-nm femtosecond laser, and the result is presented in Figure 1e. Fitting with a square function (solid purple line) matches well with experimental SHG results (purple dots), suggesting a quadratic dependence of SHG power with respect to fundamental power [16]. These observations validate that our MoS_2 samples are monolayers, and strong SHG can be readily recorded by our experimental setup equipped with a regular spectrometer.

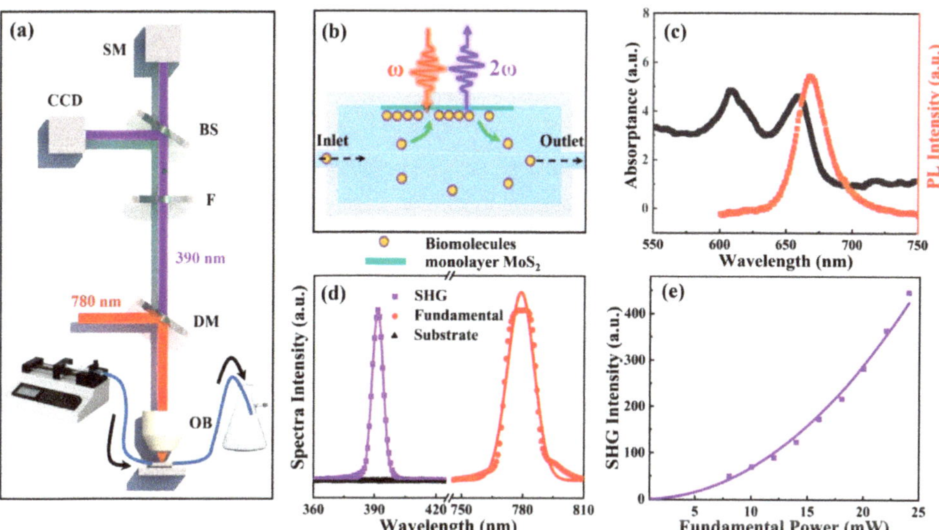

Figure 1. (**a**) Experimental setup of SHG detection with a microfluidic chip. SM: Spectrometer, BS: Beam splitter, F: Filter, DM: Dichroic mirror, OB: Objective lens. (**b**) Schematic diagram of the microfluidic chip. (**c**) Optical absorbance and photoluminescence (PL) spectra of monolayer MoS_2. (**d**) The fundamental laser (red) and SHG spectra of monolayer MoS_2 (purple) and sapphire substrate (black). (**e**) Fundamental power dependence of SHG in monolayer MoS_2.

To justify the electrostatic adsorption effect of monolayer MoS_2, we employed positively charged chitosan to interact with monolayer MoS_2. Chitosan was dissolved into a water solution of acetic acid, configuring a solution with a mass fraction of 0.8 mg/mL. Before adding the chitosan solution, a pixel-to-pixel SHG mapping of a monolayer MoS_2 sample in the air was taken by scanning a 2D translation stage (Physik Instrumente, P-51, Karlsruhe, Germany). The SHG image of this sample was presented in Figure 2a. The spatial scanning step was set at 300 nm, and the grey value of each pixel was obtained by integrating the SHG spectrum counts within a wavelength ranging from 370 nm to 410 nm. For monolayer MoS_2 region, strong SHG leads to a distribution of triangle reflecting the spatial profile of monolayer MoS_2 lattice. For the substrate region, the absence of SHG points to a black background. Subsequently, 1μL of chitosan solution (one drop) was added onto the same monolayer MoS_2 sample. When the solution completely evaporated at room temperature, pixel-to-pixel SHG mapping of the monolayer MoS_2 sample was repeated. The obtained SHG image was presented in Figure 2b. Similarly, by repeating the procedures of dropping, drying and SHG mapping, Figure 2c,d are SHG images of the same monolayer MoS_2 sample covered by two and three drops of chitosan solutions, respectively. We anticipate that the electrostatic adsorption process can be initiated by the Coulomb forces between the negatively charged monolayer and chitosan. Accumulated amounts of chitosan are expected to form chitosan nanoclusters on the surface of monolayer MoS_2, as illustrated by the schematic diagram in Figure 2e. By carefully comparing the SHG images before (Figure 2a) and after (Figure 2b–d) adding chitosan solutions, it is clear that adsorbed chitosan nanoclusters on monolayer MoS_2 can enhance SHG's intensity. To directly reveal these differences, we subtracted the SHG image in Figure 2a from that in Figure 2b–d. The corresponding differential SHG images are plotted in Figure 2f–h. These differential SHG images present randomly distributed SHG enhancements, suggesting adsorbed chitosan nanoclusters were randomly distributed, as well. We expect that this phenomenon originated from the process wherein a solution dropping action causes a rearrangement of molecules on the surface of monolayer MoS_2. Rather, the differential SHG intensity shows an increasing trend from Figure 2f to Figure 2h, as added amounts of chitosan were increased. Especially, for certain edge regions of monolayer MoS_2, SHG enhancement effects turn out to be stronger, suggesting that the edge region with more charged active sites tends to facilitate chitosan adsorptions [26].

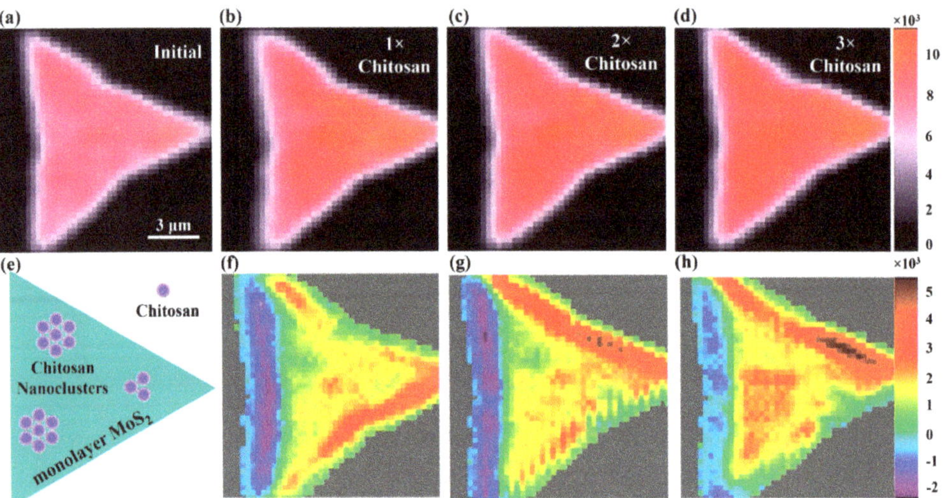

Figure 2. SHG mapping of monolayer MoS_2 adsorbing chitosan nanoclusters. (**a**) SHG mapping of monolayer MoS_2 (**a**) without chitosan nanoclusters and (**b–d**) with adsorbed chitosan nanoclusters of increasing concentrations. (**e**) Schematic diagram of chitosan nanocluster adsorption on surface of monolayer MoS_2. (**f–h**) Differential SHG images of (**b–d**) with respect to (**a**).

To visualize the adsorbed chitosan nanoclusters on the surface of monolayer MoS_2, we measured the AFM image after dropping and drying a chitosan solution for monolayer MoS_2 samples on the sapphire substrate. As shown in Figure 3a, it is clear that chitosan nanoclusters adsorbed on monolayer MoS_2 (triangle region) form many white dots. Furthermore, on the sapphire substrate, some chitosan nanoclusters can still be absorbed, but their density of distribution as well as size are smaller than the case of monolayer MoS_2. The physical reason is that positively charged chitosan molecules tend to be more efficiently adsorbed by the negatively charged monolayer MoS_2 through attractive Coulomb forces [19–23]. The height profile along the red arrow in Figure 3a is plotted in Figure 3b. Observed height values of CNs on the sapphire substrate (distance range: from 0 to 1.4 μm) suggest an averaged thickness of about 7 nm. It is obvious that height values of CNs on the monolayer MoS_2 (distance range: from 1.4 to 3.5 μm) could be as large as about 13 nm. The measured thickness of monolayer MoS_2 is less than 1 nm (about 0.9 nm) according to Figure 3b, which agrees well with previous measurement [27]. In addition, size distributions of adsorbed chitosan nanoclusters on monolayer MoS_2 and on the sapphire substrate were analyzed, as shown in Figure 3c, where two representative regions, marked in Figure 3a, were selected. Histograms of adsorbed chitosan nanoclusters in red and green of Figure 3c are statistics of region A and B of Figure 3a, respectively. After fitting these histograms with a Gaussian function, it turns out that the averaged diameter of adsorbed chitosan nanoclusters on substrate is about 41 nm. In comparison, on monolayer MoS_2, the averaged diameter of adsorbed chitosan nanoclusters is about 78 nm, validating that Coulomb attraction forces between monolayer MoS_2 and chitosan nanoclusters enhance the adsorption processes. The size FWHM of adsorbed chitosan nanoclusters on monolayer MoS_2 is about 42 nm, which is about twice that (about 22 nm) on the substrate. At this point, we can conclude that chitosan nanoclusters can be effectively adsorbed on monolayer MoS_2 by electrostatic attractions and mediate SHG intensity after drying. However, whether such an interfacial adsorption effect can induce enough SHG intensity change for biomolecules flowing in a liquid environment is still unknown.

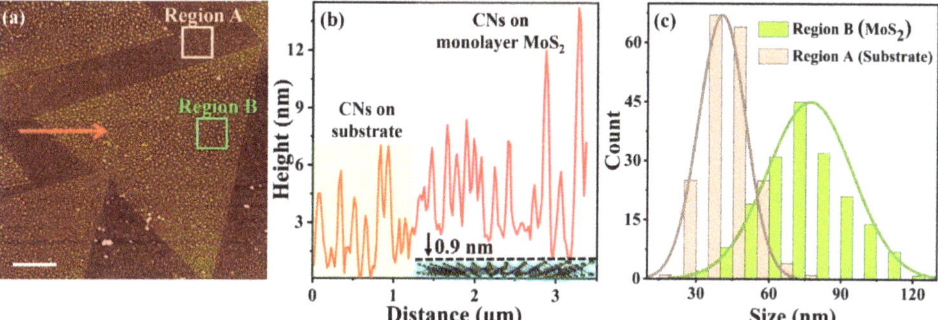

Figure 3. (**a**) Atomic force microscopy (AFM) image of monolayer MoS_2 with adsorbed chitosan nanoclusters (CNs). The scale bar is 2 μm. (**b**) Height profile along the red arrow in (**a**). (**c**) Size histogram of chitosan nanoclusters distributed on the surface of monolayer MoS_2 and sapphire substrate.

To demonstrate the feasibility of our SHG technique towards real-time sensing for flowing biomolecules in liquid environments, we selected bovine serum albumin (BSA) molecules and performed time-resolved SHG spectra measurements employing the experimental setup in Figure 1a,b. The proposed experimental schemes of BSA adsorption and desorption are presented in Figure 4a,b, respectively. Simply speaking, protonated BSA molecules are positively charged, so that the adsorption process is expected when a negatively charged monolayer MoS_2 tends to apply attractive Coulomb forces. Then, by controlling the pH of the liquid environment, protonated BSA molecules can gain electrons, and positive charges of adsorbed BSA molecules will be neutralized. Laminar flow in the microfluidic channel will take away these interfacial BSA molecules and trigger a desorption process. Before placing the monolayer MoS_2, the microfluidic chip and tubes were carefully cleaned. The BSA solution (1 μg/mL) was configured in PBS buffer solution (pH = 3.6) using BSA (5%, Yuanye Bio-Technology, Shanghai, China). The power of the fundamental laser was fixed at 8 mW. By focusing the fundamental laser tightly on the center of a monolayer MoS_2 sample by a 50× objective lens (NA = 0.55), we ensured that the size of the focal spot (1.7 μm) was much smaller than the lateral size of the monolayer

MoS_2 sample (about 15 μm). By finely tuning the axial position of monolayer MoS_2 sample, we maximized the intensity of SHG spectrum recorded by the spectrometer. A computer program was coded to record the SHG spectra every 500 ms, which integrated all non-zero SHG counts between 370 nm and 410 nm.

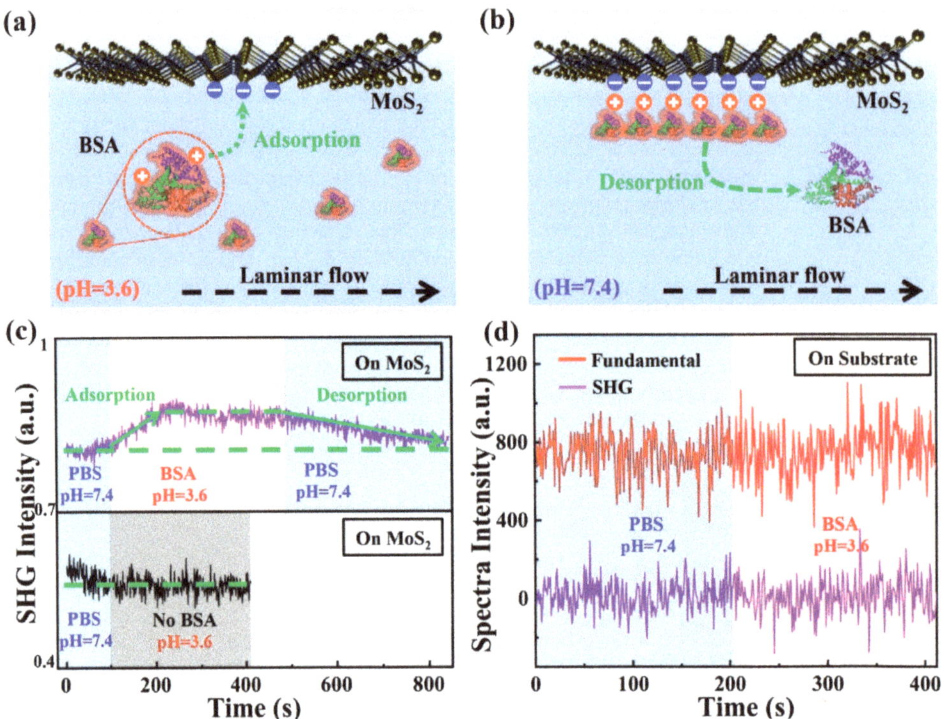

Figure 4. (**a**,**b**) Schematic diagram of BSA adsorption and desorption processes on surface of monolayer MoS_2. (**c**) Time-resolved SHG signals (purple) of BSA adsorption and desorption processes on monolayer MoS_2. The lower panel is time-resolved SHG signals (black) when there is no BSA. (**d**) Time-resolved fundamental (red) and SHG (purple) signals of substrate.

In our measurements, a fluid pump sent solutions into the microfluidic channel at a constant flowing rate of 22 μL/s. In the beginning, we flushed the microfluidic chip with a PBS solution (pH = 7.4) until the SHG signal of monolayer MoS_2 became stable in flowing conditions, as indicated by the time-resolved SHG signals (purple spectra) before 90 s in Figure 4c. At the 90 s mark, the BSA solution (pH = 3.6) was sent into the microfluidic channel. The intensity of SHG signals started to increase and, approximately, maintained a constant after the 200 s point. The increasing evolution of SHG signals between 90 s and 200 s is caused by the BSA molecule's adsorption on the surface of monolayer MoS_2. Then, at 480 s, the PBS solution (pH = 7.4) was sent to trigger a BSA molecule desorption process. Interestingly, the intensity of SHG signals started to decrease and, eventually, recovered to a constant magnitude at about 850 s, which is equal to the scenario when no BSA was added (before the 90 s mark). Furthermore, a control experiment was performed by shifting the fundamental laser focus onto a nearby monolayer MoS_2 sample and replacing the BSA solution with a PBS solution without BSA molecules, while other experimental conditions were kept the same. As indicated by the lower panel of Figure 4c, at 90s, the intensity of SHG signals (black spectra) when the fundamental laser was focusing on monolayer MoS_2 remained a constant. This result strongly validates that ions or other molecule components in the PBS solutions would not induce a detectable intensity change of monolayer MoS_2 SHG signals for our experimental setup. The baseline decrease is attributed to the lattice orientation difference in CVD MoS_2 from sample to sample. To evaluate the contributions of adsorbed BSA molecules to the refractive index change as well as intensity change of SHG signals, we focused the fundamental laser on the sapphire substrate. By replacing the short-pass optical filters in front of the spectrometer, a small portion of the fundamental laser (780 nm) was allowed to pass. Hence, we can measure fundamental and SHG signals at the same time. The BSA adsorption experiments were repeated by adding BSA solutions at 200 s. As shown in Figure 4d, the spectra intensity of fundamental (780 nm) remained a constant after adding BSA molecules, ruling out the possible effect of interfacial refractive index change. More importantly, the spectra intensity of SHG remained zero, indicating that SHG contributions of the sapphire substrate as well as adsorbed BSA generated can be neglected compared with the monolayer MoS_2. The low SHG conversion efficiency of sub-nanometer thick monolayer MoS_2 and high noise-level of our spectrometer lead to a relatively low signal to noise ratio. This issue can be further improved by increasing the integration time of each SHG spectrum and optimizing the design of microfluidic chips.

To address the physical mechanism of observed SHG signal changes accompanying the BSA adsorption process, we can consider the second-order polarization of the interface with the follow model [28,29]: $E_{2\omega} \propto P^{(2)} = \chi^{(2)}\ E_\omega E_\omega + \chi^{(3)}\ E_\omega E_\omega \varnothing_0$, where $\varnothing_0$ is the interfacial electric field and $\chi^{(2)}$ and $\chi^{(3)}$ are the second-order optical susceptibility of monolayer MoS_2 and third-order optical susceptibility of the interface, respectively. In our case, we believe that the spatial distribution of adsorbed BSA molecules on the surface monolayer MoS_2 is random. Specifically, since the initial charge distribution of monolayer MoS_2 is inhomogeneous in a 2D plane defined by the flat substrate, the orientations and 3D stacking orders of adsorbed BSA molecules are expected to be highly random. As a result, directions of interfacial electric fields between the monolayer MoS_2 and adsorbed BSA molecules in disorder will no longer be strictly perpendicular to the 2D plane. Therefore, the angle between the wave vector of the fundamental laser and the direction of interfacial electric fields will not be zero, so that the $\varnothing_0$ term can induce a non-zero second-order polarization field. When BSA molecules are dynamically adsorbed, the total magnitude of second-order polarization fields formed by superposition of polarization fields from monolayer MoS_2 and interfacial electric fields will depend on their initial phase difference.

3. Conclusions

We have comprehensively demonstrated that the interfacial SHG of monolayer MoS_2 can be utilized for label-free biomolecule sensing. Through static SHG mapping experiments, we show that the intensity of SHG in monolayer MoS_2/adsorbed chitosan nanocluster heterostructures can be mediated due to electrostatic attractions. With a time-resolved SHG measuring system equipped with a microfluidic chip, we further realize label-free sensing of BSA adsorption and desorption dynamics in real time through the SHG intensity change of monolayer MoS_2 in liquid environments, which has been tailored by Coulomb interactions between BSA molecules and monolayer MoS_2. Our work provides a complementary mean of label-free interfacial biomolecule sensing, which, in principle, undertakes molecular-level spatial resolution across the interfaces for applications including, but not limited to, in vitro medicine evaluation.

Author Contributions: Q.C. and C.X. (Chunxiang Xu) conceived the idea, designed the experiments and supervised the project. C.X. (Chuansheng Xia), Q.C., J.S., Q.W., J.C. (Jinping Chen), W.X. and H.Z. performed the experiments. C.X. (Chuansheng Xia), Q.C., T.W., Y.L., J.C. (Jianhua Chang), Z.S. and C.X. (Chunxiang Xu) analyzed the data. C.X. (Chuansheng Xia) and Q.C. drafted the paper with input from all authors. All authors have read and agreed to the published version of the manuscript.

Funding: This work was supported by National Key Research and Development Plan of China (Nos. 2017YFA0700500 and 2018YFA0209101), National Natural Science Foundation of China (Nos.

11734005, 62075041, 61904082, 61821002, 61875089, and 62175114) and Natural Science Foundation of Jiangsu Province (BK20190765).

Institutional Review Board Statement: Not applicable.

Informed Consent Statement: Not applicable.

Data Availability Statement: Not applicable.

Conflicts of Interest: The authors declare no conflict of interest.

References

1. Libby, P. The Changing Landscape of Atherosclerosis. *Nature* **2021**, *592*, 524–533. [CrossRef] [PubMed]
2. Soehnlein, O. Multiple Roles for Neutrophils in Atherosclerosis. *Circ. Res.* **2012**, *110*, 875–888. [CrossRef] [PubMed]
3. Ashrafian, H.; Zadeh, E.H.; Khan, R.H. Review on Alzheimer's Disease: Inhibition of Amyloid Beta and Tau Tangle Formation. *Int. J. Biol. Macromol.* **2021**, *167*, 382–394. [CrossRef] [PubMed]
4. Moffat, J.G.; Rudolph, J.; Bailey, D. Phenotypic Screening in Cancer Drug Discovery-Past, Present and Future. *Nat. Rev. Drug Discov.* **2014**, *13*, 588–602. [CrossRef]
5. Astashkina, A.; Mann, B.; Grainger, D.W. A Critical Evaluation of in Vitro Cell Culture Models for High-Throughput Drug Screening and Toxicity. *Pharmacol. Ther.* **2012**, *134*, 82–106. [CrossRef]
6. Zeng, S.; Baillargeat, D.; Ho, H.P.; Yong, K.T. Nanomaterials Enhanced Surface Plasmon Resonance for Biological and Chemical Sensing Applications. *Chem. Soc. Rev.* **2014**, *43*, 3426–3452. [CrossRef]
7. Zhang, P.; Ma, G.; Dong, W.; Wan, Z.; Wang, S.; Tao, N. Plasmonic Scattering Imaging of Single Proteins and Binding Kinetics. *Nat. Methods* **2020**, *17*, 1010–1017. [CrossRef]
8. Floris, I.; Adam, J.M.; Calderón, P.A.; Sales, S. Fiber Optic Shape Sensors: A Comprehensive Review. *Opt. Lasers Eng.* **2021**, *139*, 106508–106525. [CrossRef]
9. Xu, Z.; Zhang, Y.; Wu, Y.; Lu, X. Spectroscopically Detecting Molecular-Level Bonding Formation between an Epoxy Formula and Steel. *Langmuir* **2022**, *38*, 13261–13271. [CrossRef]
10. Xu, Z.; Zhang, Y.; Wu, Y.; Lu, X. Molecular-Level Correlation between Spectral Evidence and Interfacial Bonding Formation for Epoxy Adhesives on Solid Substrates. *Langmuir* **2022**, *38*, 5847–5856. [CrossRef]
11. Li, H.; Guo, Y.; Xiao, L.; Chen, B. A Fluorometric Biosensor Based on H2O2-Sensitive Nanoclusters for the Detection of Acetylcholine. *Biosens. Bioelectron.* **2014**, *59*, 289–292. [CrossRef] [PubMed]
12. Su, Q.; Wu, F.; Xu, P.; Dong, A.; Liu, C.; Wan, Y.; Qian, W. Interference Effect of Silica Colloidal Crystal Films and Their Applications to Biosensing. *Anal. Chem.* **2019**, *91*, 6080–6087. [CrossRef] [PubMed]
13. Xiao, L.; Wei, L.; He, Y.; Yeung, E.S. Single Molecule Biosensing Using Color Coded Plasmon Resonant Metal Nanoparticles. *Anal. Chem.* **2010**, *82*, 6308–6314. [CrossRef] [PubMed]
14. Peltomaa, R.; Glahn-Martínez, B.; Benito-Peña, E.; Moreno-Bondi, M.C. Optical Biosensors for Label-Free Detection of Small Molecules. *Sensors* **2018**, *18*, 4126. [CrossRef]
15. Xu, L.J.; Lei, Z.C.; Li, J.; Zong, C.; Yang, C.J.; Ren, B. Label-Free Surface-Enhanced Raman Spectroscopy Detection of DNA with Single-Base Sensitivity. *J. Am. Chem. Soc.* **2015**, *137*, 5149–5154. [CrossRef]
16. Kumar, N.; Najmaei, S.; Cui, Q.; Ceballos, F.; Ajayan, P.M.; Lou, J.; Zhao, H. Second Harmonic Microscopy of Monolayer MoS_2. *Phys. Rev. B* **2013**, *87*, 161403–161409. [CrossRef]
17. Xia, C.; Cui, Q.; Ding, H.; Chen, J.; Wang, R.; Zhang, L.; Yang, Y.; Wang, X.; Xu, W.; Shi, Z.; et al. Crescent-shaped Shadow of Second Harmonic Generation in Dielectric Microsphere/TMD Monolayer Heterostructures. *J. Phys. D Appl. Phys.* **2022**, *55*, 325301. [CrossRef]
18. Li, Y.; Rao, Y.; Mak, K.F.; You, Y.; Wang, S.; Dean, C.R.; Heinz, T.F. Probing Symmetry Properties of Few-Layer MoS_2 and h-BN by Optical Second-Harmonic Generation. *Nano Lett.* **2013**, *13*, 3329–3333. [CrossRef]
19. Hong, J.; Hu, Z.; Probert, M.; Li, K.; Lv, D.; Yang, X.; Gu, L.; Mao, N.; Feng, Q.; Xie, L.; et al. Exploring Atomic Defects in Molybdenum Disulphide Monolayers. *Nat. Commun.* **2015**, *6*, 6293–6300. [CrossRef]
20. Schmidt, H.; Wang, S.; Chu, L.; Toh, M.; Kumar, R.; Zhao, W.; Castro Neto, A.H.; Martin, J.; Adam, S.; Özyilmaz, B.; et al. Transport Properties of Monolayer MoS_2 Grown by Chemical Vapor Deposition. *Nano Lett.* **2014**, *14*, 1909–1913. [CrossRef]
21. Zhang, W.; Huang, J.K.; Chen, C.H.; Chang, Y.H.; Cheng, Y.J.; Li, L.J. High-Gain Phototransistors Based on a CVD MoS_2 Monolayer. *Adv. Mater.* **2013**, *25*, 3456–3461. [CrossRef] [PubMed]
22. Lee, Y.H.; Zhang, X.Q.; Zhang, W.; Chang, M.T.; Lin, C.; Chang, K.; Yu, Y.C.; Wang, J.T.W.; Chang, C.S.; Li, L.J.; et al. Synthesis of Large-Area MoS_2 Atomic Layers with Chemical Vapor Deposition. *Adv. Mater.* **2012**, *24*, 2320–2325. [CrossRef] [PubMed]
23. Chang, Y.H.; Zhang, W.; Zhu, Y.; Han, Y.; Pu, J.; Chang, J.K.; Hsu, W.T.; Huang, J.K.; Hsu, C.L.; Chiu, M.H.; et al. Monolayer $MoSe_2$ Grown by Chemical Vapor Deposition for Fast Photodetection. *ACS Nano* **2014**, *8*, 8582–8590. [CrossRef] [PubMed]
24. Mak, K.F.; Lee, C.; Hone, J.; Shan, J.; Heinz, T.F. Atomically Thin MoS_2: A New Direct-Gap Semiconductor. *Phys. Rev. Lett.* **2010**, *105*, 136805–136909. [CrossRef]
25. Splendiani, A.; Sun, L.; Zhang, Y.; Li, T.; Kim, J.; Chim, C.Y.; Galli, G.; Wang, F. Emerging Photoluminescence in Monolayer MoS_2. *Nano Lett.* **2010**, *10*, 1271–1275. [CrossRef]

26. Yin, X.; Ye, Z.; Chenet, D.A.; Ye, Y.; O'Brien, K.; Hone, J.C.; Zhang, X. Edge Nonlinear Optics on a MoS_2 Atomic Monolayer. *Science* **2014**, *344*, 488–490. [CrossRef]
27. Li, Y.; Qi, Z.; Liu, M.; Wang, Y.; Cheng, X.; Zhang, G.; Sheng, L. Photoluminescence of Monolayer MoS_2 on $LaAlO_3$ and $SrTiO_3$ Substrates. *Nanoscale* **2014**, *6*, 15248–15254. [CrossRef]
28. Achtyl, J.L.; Vlassiouk, I.v.; Surwade, S.P.; Fulvio, P.F.; Dai, S.; Geiger, F.M. Interaction of Magnesium Ions with Pristine Single-Layer and Defected Graphene/Water Interfaces Studied by Second Harmonic Generation. *J. Phys. Chem. B* **2014**, *118*, 7739–7749. [CrossRef]
29. Jena, K.C.; Covert, P.A.; Hore, D.K. The Effect of Salt on the Water Structure at a Charged Solid Surface: Differentiating Second- and Third-Order Nonlinear Contributions. *J. Phys. Chem. Lett.* **2011**, *2*, 1056–1061. [CrossRef]

MDPI

Article

Carbon Dots Conjugated Antibody as an Effective FRET-Based Biosensor for Progesterone Hormone Screening

Disha [1,2], Poonam Kumari [1,2], Manoj K. Patel [2,3], Parveen Kumar [4] and Manoj K. Nayak [1,2,*]

1 Materials Science and Sensor Applications, CSIR-Central Scientific Instruments Organisation (CSIR-CSIO), Sector 30-C, Chandigarh 160030, India
2 Academy of Scientific and Innovative Research (AcSIR), Ghaziabad 201002, India
3 Manufacturing Science and Instrumentation, CSIR-Central Scientific Instruments Organisation (CSIR-CSIO), Sector 30-C, Chandigarh 160030, India
4 Exigo Recycling Pvt. Ltd., Noida 201309, India
* Correspondence: mknayak@csio.res.in

Abstract: In this work, carbon dots (CDs) were synthesized by a one-step hydrothermal method using citric acid and ethylene diamine, and covalently functionalized with antibodies for the sensing of progesterone hormone. The structural and morphological analysis reveals that the synthesized CDs are of average size (diameter 8–10 nm) and the surface functionalities are confirmed by XPS, XRD and FT-IR. Further graphene oxide (GO) is used as a quencher due to the fluorescence resonance energy transfer (FRET) mechanism, whereas the presence of the analyte progesterone turns on the fluorescence because of displacement of GO from the surface of CDs effectively inhibiting FRET efficiency due to the increased distance between donor and acceptor moieties. The linear curve is obtained with different progesterone concentrations with 13.8 nM detection limits (R^2 = 0.974). The proposed optical method demonstrated high selectivity performance in the presence of structurally resembling interfering compounds. The PL intensity increased linearly with the increased progesterone concentration range (10–900 nM) under the optimal experimental parameters. The developed level-free immunosensor has emerged as a potential platform for simplified progesterone analysis due to the high selectivity performance and good recovery in different samples of spiked water.

Keywords: endocrine; hormonal imbalance; progesterone; biorecognition; immunosensor; fluorescence resonance energy transfer (FRET) bioassay

Citation: Disha; Kumari, P.; Patel, M.K.; Kumar, P.; Nayak, M.K. Carbon Dots Conjugated Antibody as an Effective FRET-Based Biosensor for Progesterone Hormone Screening. *Biosensors* **2022**, *12*, 993. https://doi.org/10.3390/bios12110993

Received: 7 October 2022
Accepted: 6 November 2022
Published: 9 November 2022

1. Introduction

A biologically active hormone called progesterone occurs as a natural estrogenic steroid compound, derived from a cholesterol biosynthesis [1]. This multifunctional hormone is associated with the reproductive aspects and sexual growth, the maintenance of the menstrual cycle and pregnancy. Additionally, progesterone is having crucial pharmaceutical applications where it is implemented as oral contraceptives, menopausal hormone therapy, infertility, etc. [2]. Progesterone has been reported to be a significant carcinogen, its abnormally high levels can be toxic to both men and women and as tumorgenic, affecting human health by interfering with the natural hormonal activity due to excessive exposure during gestation. Indeed, steroid estrogens have gained significant attention in recent years due to their rapid rise in levels posing serious threats to soil, plants, water resources and humans [3]. Over the years, the analysis of progesterone in environmental samples has been considerably dominant because of its endocrine disrupting activity [1,4,5]. Accordingly, the detection of progesterone hormone is becoming essential for environmental analysis.

In recent years, several methods have influenced the analysis of progesterone by using chromatographic techniques like gas chromatography (GC) [1] and liquid chromatography (LC) coupled with mass spectrometry (LC-MS) [6] and high performance liquid chromatography (HPLC) [7]. However, these techniques require repeated measurements, long and cumbersome derivatization processes for sample preparation, and trained operators, leading to their inappropriate field applications. Additionally, the enzyme linked immunosorbent assay (ELISA), radioimmune assays, etc., come with their lingering challenges of exorbitant instrumentation prices and limited stability [8]. At present, the electrochemical techniques have also been emerged as a potential approach where steroid hormones are being quantified even at low levels of concentration with high sensitivity [2,9,10]. Additionally, molecularly imprinted polymers (MIPs)-based approaches have produced many opportunities for the detection, quantification and removal of estrogenic steroid hormones [11–13]. However, the molecular imprinting technology remained challenging and needs to be simplified in their synthesis, maintaining shape and porosity before and after template removal and binding procedures [1,14]. In this context, Table 1 represents the literature where CDs being exploited in progesterone determination in these past years are yet to be explored in an interference analysis or specificity performance with their analogous hormones or biomolecules such as β-estradiol, testosterone, cortisol, bisphenol A, etc. for applications in complex matrices [15–18].

Table 1. The summarised reported literature on sensing progesterone hormone in biological as well as environmental samples.

Sr. No.	Nanomaterials	Detection Method	Limit of Detection (LOD)	Linear Range	Detection Time	Sample	Year	Reference
1.	Pyrrole	MIP-GC	0.62 ng/mL	0.62 to 1.87 ng/mL	30 min	Blood and hospital water	2015	[1]
2.	Protein label GO-thionine	Electrochemical	6.3×10^{-5} ng/mL	0.02 to 20 ng/mL	100 min	Milk	2017	[9]
3.	GQDs-NiO-Au composite	Electrochemical	0.57 pg/mL (1.86 pM)	0.01 to 1000 nM	60 min	Human serum	2020	[10]
4.	CDs	Fluorescence	10.25 nM	0 to 200 μM	-	In vitro	2020	[15]
5.	CDs-GO	Photo-electrochemical	0.17 nM	0.5 nM to 180 nM	-	Human serum	2020	[16]
6.	QDs embedded hydrogel	Fluorescence	55 nM	78 to 632 nM	-		2020	[17]
7.	CDs	Fluorescence	9.3 nM	0 to 130 μM	1 min	Human serum	2021	[18]
8.	PdNPs/P-3TAA	MIP-electrochemical	-	0.1 nM to 110 nM	-		2022	[14]

Abbreviations: GC: Gas chromatography, MIP: Molecularly imprinted polymer, GQDs: Graphene quantum dots, NiO: Nickel oxide, Au: Gold, PdNPs/P-3TAA: Palladium nanoparticles/Poly (thiophene-3-acetic acid).

Simultaneously, the fluorescence spectroscopic methods provided unique advantages for bioanalysis applications owing to the simplicity of their operation, stability and sensitivity [19,20]. A renowned technology called fluorescence resonance energy transfer (FRET) garnered a considerable interest for the sensing analysis of biological molecules. FRET is described as a phenomenon in which the energy transfer occurs non-radiatively from the excited fluorophore donor to the acceptor of the ground state by dipole-dipole coupling [21,22]. The efficiency and sensitivity of the FRET are significantly enhanced by the combination of intermolecular dipole–dipole interactions between donor planarized geometries and acceptors in the resulting FRET process. In particular, chemically stable aqueous carbon dots (CDs) have superior advantages over the fluorescent organic dyes and other semiconductor quantum dots. Likewise, Junkai Ren et al. have presented the structural design and optimization of CD-based nanocomposites to meet the custom application aimed at avoiding quenching effects and improving the range of optical solid-state responses. Having the precise control over CDs' structural and chemical characteristics, solid-state can develop high performance devices by competing semiconductors and quantum dots [23]. Since CDs apparently have tunable fluorescence, nontoxicity, remarkable photostability, biocompatibility and ease to be modified with biomolecules, are therefore used worldwide for fluorescent detection strategies [24,25].

In contrast with the conventional CDs and other semiconductors, the fascinating physicochemical properties of nitrogen-doped fluorescent CDs such as biocompatibility and water solubility have opened a new horizon in the bio sensing era based on FRET applications. Moreover, the presence of abundant amine and carboxylic acid groups at the surface of nanoparticles provides a number of binding regions for conjugating antibodies. The emerging applications of nitrogen-doped CDs in FRET-based biosensors impart enhanced quenching efficiency to the system which is advantageous for developing a potential fluorescent technique [26]. The use of nanoparticles' distinctive optical quenching characteristics such as gold nanoparticles (Au NPs), transition metal dichalcogenides (TMDCs), graphene oxide (GO), etc., as an effective fluorescence quencher in FRET-based biosensors has been illustrated in the literature [22,26,27]. However, GO exhibits an outstanding quenching ability as an acceptor, owing to its wide electronic bands of absorption as well as the sp^2 hybridized monolayer structure [28]. Optical immunoassays using fluorescence have reached great interest in nanotechnology for the development of biosensors because of their sensitivity and selectivity. The literature has also presented number of reports for monitoring steroid hormones using FRET as a transduction part for biosensing approach [29,30]. However, the detection of FRET-based progesterone hormone using a donor-acceptor combination of CD and GO is in the early stages of investigation. Recently, a significant and sensitive FRET aptasensor for progesterone based on carbon dots has been reported by H. Cui et al. with a reduced detection limit 3.3×10^{-11} M [31]. However, the detection range is limited to 120 nM in milk, which is not suitable for aqueous samples where progesterone is also present at higher levels, i.e., >120 nM [32].

Herein, we presented the use of an optical platform in order to develop a FRET-based, label-free, and selective progesterone immunoassay through the photoluminescence system. In this context, the CDs were prepared from the precursors, citric acid and ethylenediamine using a one-step hydrothermal process. Consequently, graphene oxide was synthesized from our preceding report [2]. We use a simple one-step process to get antibody-conjugated CDs for assessment of the progesterone hormone. Moreover, CDs, GO and antibody-conjugated CDs were successfully characterised with XRD, FT-IR techniques. The as-synthesized amine doped CDs were highly fluorescent and could be implemented directly to immobilize end-on oriented antibodies. Further, we reported an optical technique in which π-π interactions between CDs and GO effectively quenched the fluorescence of CDs and antibody-conjugated CDs (Ab-CDs) in the absence of the target analyte i.e., progesterone. However, in the presence of progesterone molecules, the latter bound specifically to the corresponding antibodies on the CDs. As a result, increasing the distance between the quencher and the donor pair of the FRET process, recovers the fluorescent CDs. The analysis of the developed progesterone fluorescent assay based on GO and CDs demonstrated a remarkable selectivity in the presence of several analogous biomolecules. The current FRET-based biosensor for progesterone detection has demonstrated a straightforward and cost-effective methodology with high selectivity in complex aquatic samples. Moreover, the approach is devoid of keeping long incubations and provides opportunities in various other bioanalytical applications.

2. Experimental Sections

2.1. Materials and Solutions

Citric acid, Ethylenediamine, 1-ethyl-3-(3dimethylamino-propyl) carbodiimide (EDC), *N*-hydroxysuccinimide (NHS), Bisphenol A, β-estradiol, Testosterone, Cortisol, Progesterone hormone and its monoclonal antibody and other reagents were obtained from Sigma-Aldrich, Bangalore, India. All experimental solutions including phosphate buffered saline (PBS, 10 mM, pH 7.4) were carefully made with double distilled water (DDI).

2.2. Instrumentation

A Nicolet iS10 (Thermo-Scientific, Portland, OR, USA), Fourier Transform Infrared (FT-IR) was used to obtain an infrared absorption spectrum of solid samples. A Jeol/JEM 2100 (Tokyo, Japan), high-resolution transmission electron microscope (HR-TEM) and X-ray diffractometer (XRD) of Bruker AXS, Karlsruhe, Germany, D8 Advance (Cu Kα λ = 1.5406 Å) were used to obtain the morphology of carbonaceous nanomaterials and XRD pattern respectively. X-ray photoelectron spectra (XPS) was observed using a Thermo-Scientific (Portland, OR, USA) K-Alpha XPS for the chemical structure and composition of CDs. The fluorescence and whole of the optical experiments of CDs were recorded using a Cary Eclipse fluorescence spectrophotometer (AGILENT, Santa Clara, CA, USA) equipped with a xenon lamp using right angle geometry. The excitation and emission slit widths were set at 4 nm each. The fluorescence measurements covered excitation wavelengths from 320 to 380 nm and emission wavelengths from 10 nm above each excitation wavelength to 600 nm in 1 nm increments. The absorption spectra of CDs and GO were obtained using (PerkinElmer Lamda 850, Shelton, CT, USA) UV-vis spectrophotometer.

2.3. Synthesis of Carbon Dots

Nitrogen-doped CDs were synthesized using a one-step hydrothermal method using citric acid as a carbon precursor and ethylenediamine as a nitrogen dopant [19]. Briefly, 2 g of citric acid and 2.55 mL of ethylenediamine were dispersed in 20 mL of DI water and then transferred to the Teflon bottle in order to heat the reaction mixture at 200 °C for 5 h. The obtained pale-yellow solution was naturally cooled and filtered through a 0.22 µm filter membrane and dialysed for 24 h with a 0.5 K molecular weight cut-off (MWCO) dialysis bag by changing water at intervals of 12 h to obtain a solution of carbon dots. The nitrogen-doped CD solution was diluted and stored at 4 °C before further experimental use. GO were prepared using the modified Hummer method as described earlier [2].

Synthesis of Antibody Conjugated Carbon Dots (Ab-CDs)

CDs–antibody conjugated nanomaterials were prepared according to a reported method with some modifications [33]. The EDC–NHS-induced bioconjugation reaction was used to covalently bind antibodies to the functional groups of CDs. The 0.125 mL antibody solution (50 µg/mL) was incubated with 0.2 mL 100 mM (1:1) molar ratio of EDC–NHS solution at pH 7.4 for $\frac{1}{2}$ hour to activate the COOH groups on the antibodies. The activated antibodies were then incubated with 0.4 mL of CDs (2 mg/mL) for 2 h at 37 °C to form CD–antibody complexes. In this process, CDs and progesterone antibodies were conjugated through strong amide bonds between the amine groups of CD nanoparticles and the carboxylic acid groups of antibodies. The aforementioned solution was dialysed and stored at 4 °C prior to use.

3. Results and Discussion

3.1. Characterization

As per our previous report, GO exhibiting excellent dispersibility was obtained from modified Hummer's method [2]. The experimental data on GO synthesis and characterization are discussed in the supplementary information (Figures S1 and S2 in supplementary materials). CDs, synthesized by hydrothermally treating citric acid and ethylene diamine also reveal good aqueous solubility. Hence, the nanomaterials were characterized using several analytical techniques including UV, PL, FTIR, XPS, XRD, HRTEM, etc., to confirm the chemical structure and characteristics of the surface functional groups and optical behaviour. The structural morphology of the CD samples was elucidated using HR–TEM and shown in the Figure 1a, revealed that CDs are spherically shaped and have average particle sizes, ~8 nm. Furthermore, CDs absorbed on layers of GO is clearly observed in Figure 1b. In addition, the histogram for the particle size distribution of CDs' dispersion is provided in the Figure S4. FTIR spectra of *N*-doped CDs and Ab–CDs were recorded to identify the surface functional groups (Figure 2). The FTIR spectra of CDs exhibited broad vibration bands at 3465 and 3227 cm^{-1} corresponding to the functional groups, O–H and N–H stretching, respectively. The peaks observed at 2818–2929, 2084–2364, 1627 and 1592 cm^{-1} are due to the stretching of the C–H bond, the nitrile C≡N bond, the amide C=O and the N–H bending, respectively. Besides, the peaks at 1350–1384, 1100–1266, and 617–669 cm^{-1} are assigned to C–O stretching or C–H bending, C–N stretching and C–C bending vibrations in the CDs [34–38].

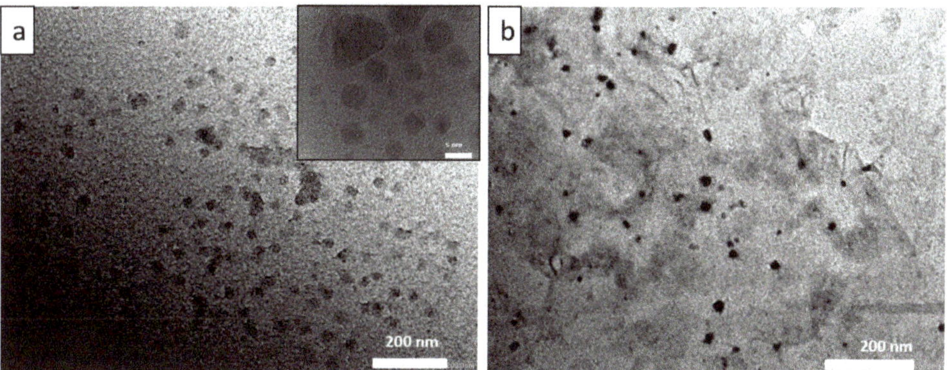

Figure 1. The representative HRTEM images of CDs at different scales (**a**) and CDs absorbed on GO surface (**b**).

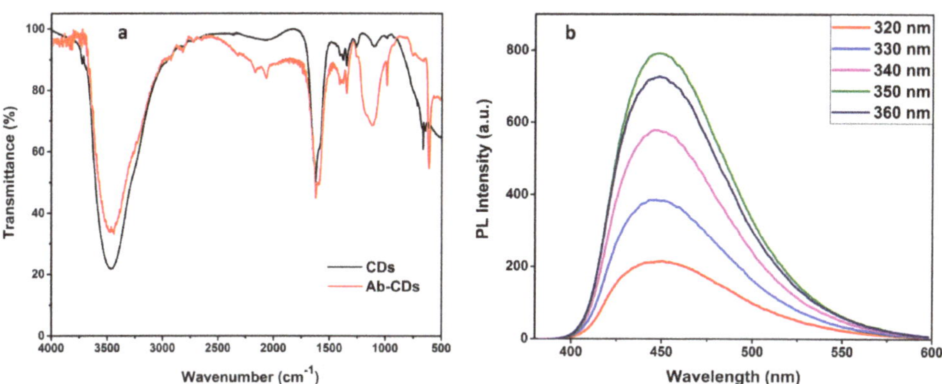

Figure 2. FTIR spectra of CDs and Ab−conjugated CDs (**a**) and PL graph of CDs at different excitation wavelengths (**b**).

In addition, the PL analysis of as-prepared CDs is displayed in Figure 2b at a range of 320–360 nm as excitation wavelength. Since there is no significant trend in red or blue shifting of emission peaks with the increase in excitation wavelength in PL, we can clearly state the optical property of CDs as excitation-independent [39]. Moreover, the strongest fluorescence emission, at 350 excitation wavelength was demonstrated at 450 nm. However, the decrease in PL intensity along with a slight red shift in emission peak can be observed at a higher wavelength range, i.e., 380–450 nm (Figure S3). This might be due to the different trap states of surface groups and size dispersion of CDs [40].

The XPS analysis was carried out for investigating the chemical as well as elemental composition of CDs which confirmed three typical peaks at 285.4 eV, 400 eV and 531.8 eV of C1s, N1s and O1s, respectively, as shown in Figure 3a. Additionally, the high resolved spectra of C1s displays a further three peaks of binding energy at 284.6, 285.9 and 287.6 eV, representing the sp^2 graphitic structure (C=C), existing C–N/C–O and C=O/C=N respectively (Figure 3b). Similarly, characteristic peaks of O1s and N1s are presented in Figure 3c,d where N1s revealed two peaks ascribing to the graphitic carbon nitride C–N and N–H groups at the CDs' surface. Similarly, characteristic peaks of O1s and N1s are presented in Figure 3c,d where N1s revealed two peaks ascribing to the graphitic carbon nitride C–N and N–H groups on the surface of CDs and coincides as well with FTIR results. The XPS confirms the relative % of C, O and N of CDs, i.e., 62.8%, 23.8% and 13.3%, respectively, which endows more binding sites to the nanomaterial [34,41].

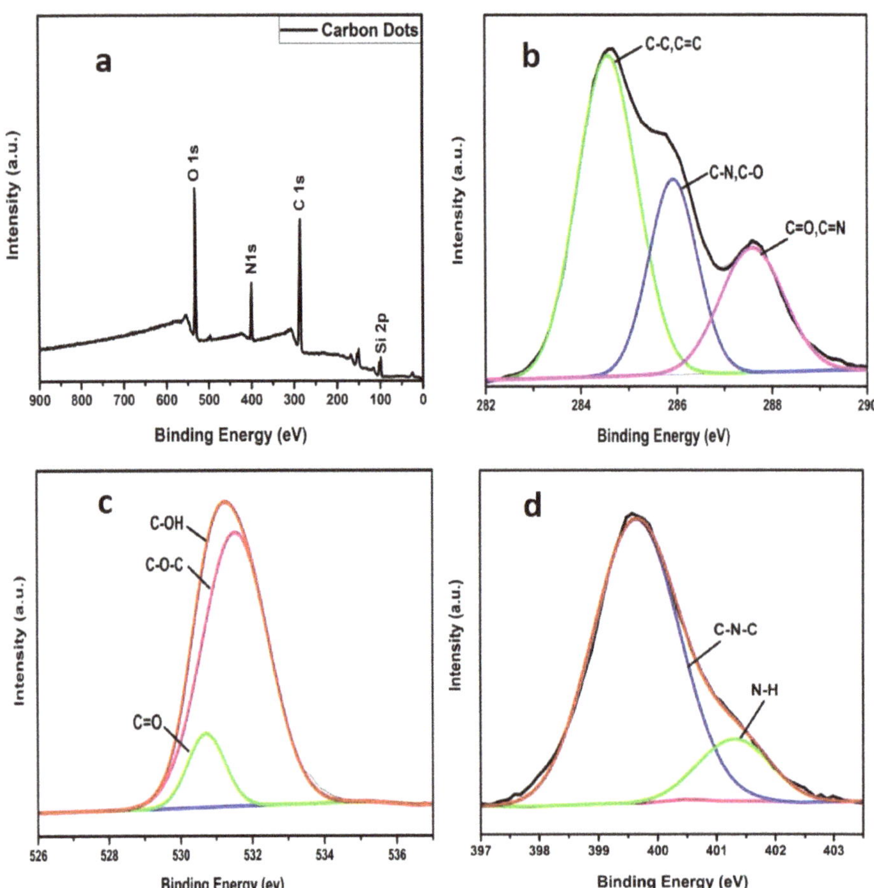

Figure 3. XPS graph of as prepared CDs (**a**), obtained spectra of C1s (**b**), O1s (**c**) and N1s (**d**) peaks with high resolution.

Figure 4a plots the absorbance as well as fluorescence spectra, where CDs exhibited a strong blue light under the ultraviolet light. The UV-Vis absorption spectrum of CDs' aqueous suspension exhibits two absorption peaks at ~240 nm owing to the C=C bond (π-π^* transition) and 340 nm of C=O bond (n-π^* transitions), respectively [42]. Additionally, the absorption band expanded in the visible region at 432–450 nm is ascribed to the nitrogen-doping CDs [40]. Furthermore, the as-obtained CDs solution was pale-yellow (under visible light) and strongly emitted a blue fluorescence when kept in UV light (Figure 4 inset). The fluorescence efficiency is compared with the standard procedure currently used to determine the relative quantum yield against the standard quinine sulphate is 0.55 in 0.1 M H_2SO_4. The quantum yield of CDs was estimated to be 8.5% using quinine sulphate as standard. The results from XRD analysis inferred clearly that CDs have an amorphous carbon phase as there are no sharp peaks existing in the spectra but a broad diffraction peak at 2θ = 18°, Figure 4b [34,41].

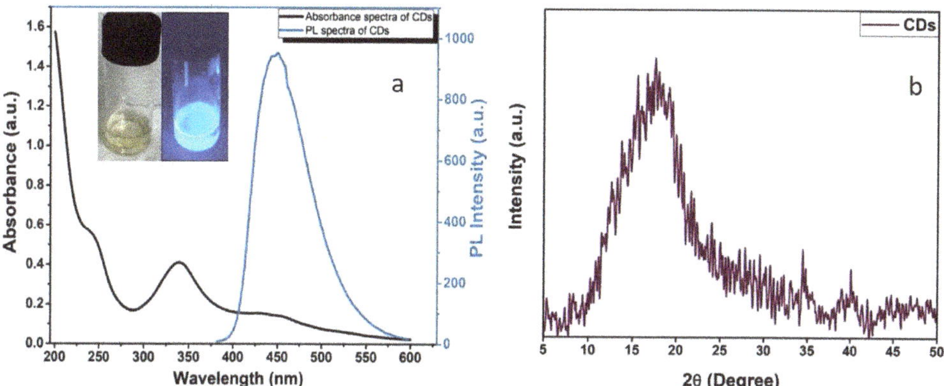

Figure 4. UV-PL graph of CDs and inset images of as-prepared pale-yellow coloured carbon dots solution in normal and UV light (**a**) and XRD pattern of CDs (**b**).

The confirmation of binding the antibodies to the functional groups bearing CDs is well obtained by the UV-Vis and PL spectral analysis. Figure 5a suggests the absorption band is around 260 nm, present in CDs–antibody complexes, which is ascribed by the amino acids of the antibodies. Concentration of conjugated antibody on CDs is also estimated from the UV graph [33] (Supplementary data). Subsequently, Figure 5b indicates the decreased PL intensity with a slight blue shift ~5 nm attributing to the existence of a proteinaceous layer of conjugated antibodies. In addition, the FTIR spectra of the antibody conjugated N-doped CDs (Ab–CDs) showed broad vibration bands at 3444–3477 and 3235 cm^{-1} corresponding to the functional groups, O–H and N–H stretching respectively. The peaks observed at 2726–2928, 2071–2328, 1627 and 1601 cm^{-1} are due to the stretching of the C–H bond, the nitrile bond C≡N, the amide bond C=O and the N–H bending, respectively. Similarly, the peaks at 1353–1385, 1117–1265 and 616–764 cm^{-1} are assigned to C–O stretching or C–H bending, C–N stretching and C–C bending vibrations in the Ab–CDs. The FTIR spectra also indicate that the increase in the intensity of amide >C=O, nitride C=N, amine C–N, and aliphatic C–H bonds of the antibody conjugated Ab–CDs in comparison to CDs. The UV–Vis absorption and FTIR spectra successfully confirmed the antibody conjugation on the amine functionalized CDs of *N*-doped CDs [42,43].

Parametric Performance Optimization and Quenching Efficiency

As illustrated in Figure 6a, GO has a broad absorption band, allowing easy pairing and overlapping with the fluorescence emission spectra of CDs, which is advantageous for the FRET process. Consequently, GO quenched the fluorescence of the adsorbed CDs on the GO surface. Following the optimization of the GO nanosheet as Ab–CQDs emission acceptors in the FRET biosensor, several GO concentrates were examined. A total of 100 μL Ab–CDs solution was mixed with 20 μL of each GO concentrations from 1–6 μg/mL, and then recorded the fluorescence spectra for optimizing the GO amount required in effectively quenching the fluorescence of CDs. A concentration of 20 μL GO (4 μg/mL) was enough to quench the 100 μL Ab–CDs solution, so that 20 μL was selected for subsequent experiments (Figure 6b,c).

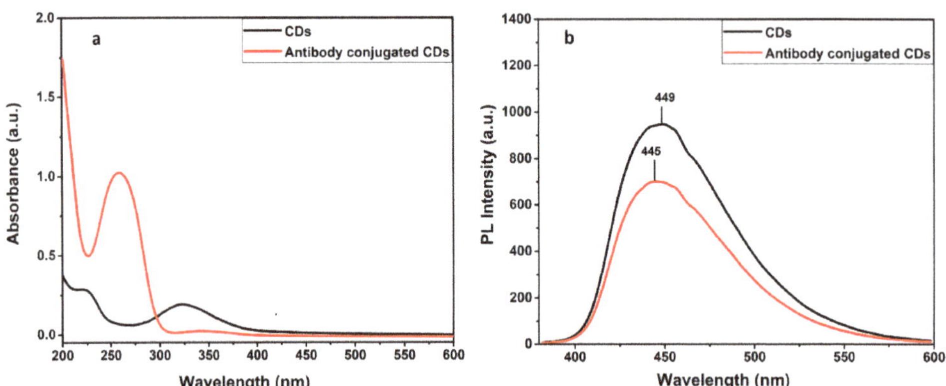

Figure 5. Absorbance (**a**) and PL emission spectra of CDs and antibody conjugated CDs (**b**).

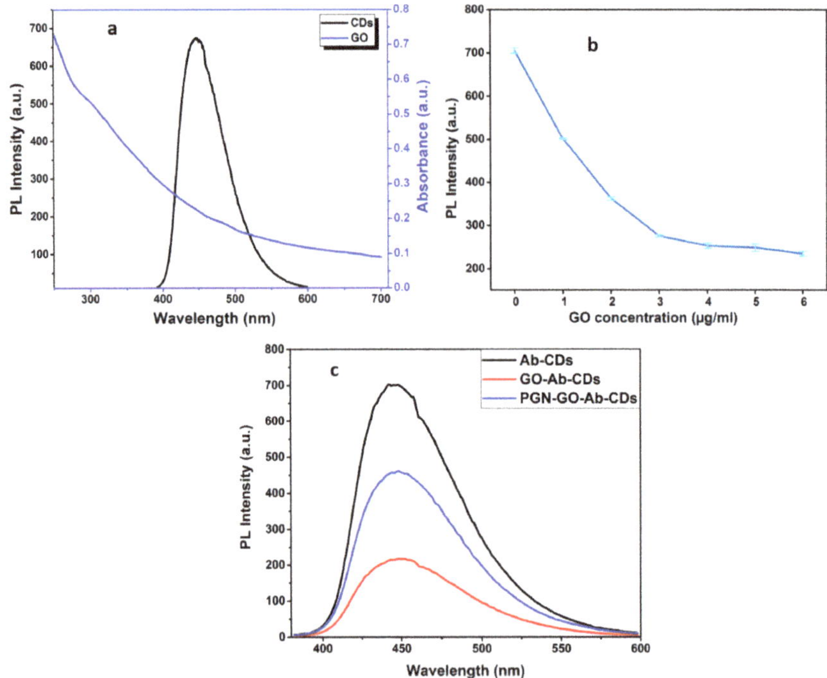

Figure 6. The overlapping absorbance of GO and emission of carbon dots (**a**), Investigation of the GO concentration for quenching efficiency (**b**) and Resultant PL spectra with and without the presence of quencher, GO (**c**).

3.2. Development of FRET-Based Progesterone Hormone Immunosensor: Detection and Performance Analysis

An amount of 100 µL of the CDs–antibody complex solution was incubated with various concentrations of progesterone (50 µL each) from 10 nM to 900 nM for 30 min, and then this assembling solution was advanced to record PL spectra after the addition of 20 µL (4 µg/mL) GO. The total volume of 0.2 mL was composed with the PBS buffer and incubated for 5 min before recording the fluorescence spectra for batch analysis. A prerequisite for fulfilling FRET is that the donor emission spectrum must overlap with the acceptor absorption spectrum. An additional FRET prerequisite is the distance between the donor and the acceptor which should be less than 10 nm. FRET is a non-radioactive energy transfer process in which the fluorescence intensity of the donor is much lower or sometimes quenched due to the transfer of energy to the acceptor. The results show that the dominant process of fluorescence quenching is attributed to FRET occurs between CDs as a donor and GO as an acceptor which turns off strong luminescent CDs. By adding progesterone, the distance between the Ab–CDs complexes and the GO surface was increased effectively, thereby reducing the FRET effect that restored the fluorescence of the CDs.

To demonstrate the PL response of the FRET-based immunosensor for the progesterone detection, the fluorescence spectra were recorded at an excitation wavelength of 350 nm for each variable progesterone concentration. As Figure 7 clearly shows, the PL intensity of the Ab–CD complex is somewhat attenuated by GO interactions (F_0). While the presence of progesterone has given rise to specific bindings with antibodies that have especially increased the required distance of the FRET process between the quencher pair GO and CDs. A change in PL intensity (F-F_0) occurred at the various concentrations of progesterone as a result of the formation of the CDs–antibody-progesterone complexes. The reason for fluorescence restoration may be that CQDs–antibody–progesterone complexes are not in immediate proximity to the GO quencher. In the absence of progesterone in solution, no characteristic change is seen in the emission peak. However, the change in PL intensity gradually increased from 10 nM to 900 nM in a linear trend (Figure 7). The detection limit of 13.8 nM was calculated using the standard deviation (S) of 3 blank readings in the formula, i.e., 3S/(SLOPE = 0.290). Consequently, the significant CDs-based FRET immunoassay was successfully employed for the selective detection of the target progesterone hormone.

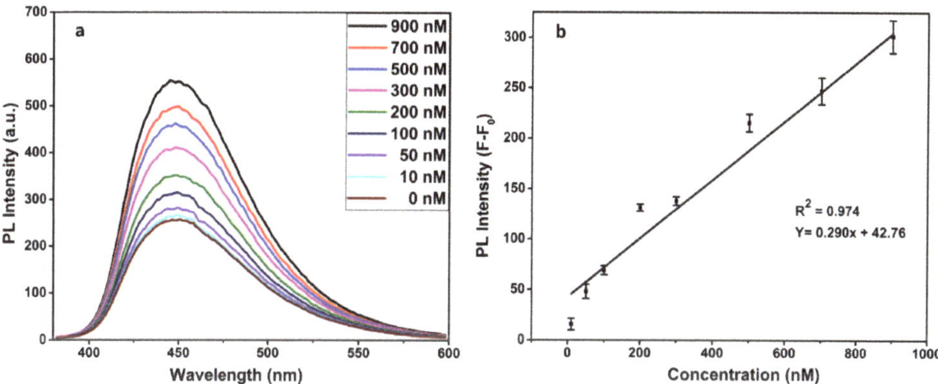

Figure 7. Change in PL intensity with different concentrations of progesterone hormone (**a**) and representative of a linear regression (**b**).

3.2.1. Analysing the pH Effect

For observing the effect of pH, which is a key factor affecting the antibody bindings, buffer solutions of different pHs ranging from 4.0 to 8.0 were used for measuring the PL response. As Figure 8a infers, there is no noticeable PL response under acidic and alkaline conditions and displays maximum response at pH 7.4. This is likely a charge reversal due to the impact of pH on antibody activity and antigen uptake. Consequently, the pH of 7.4 was used as an optimized condition in this study.

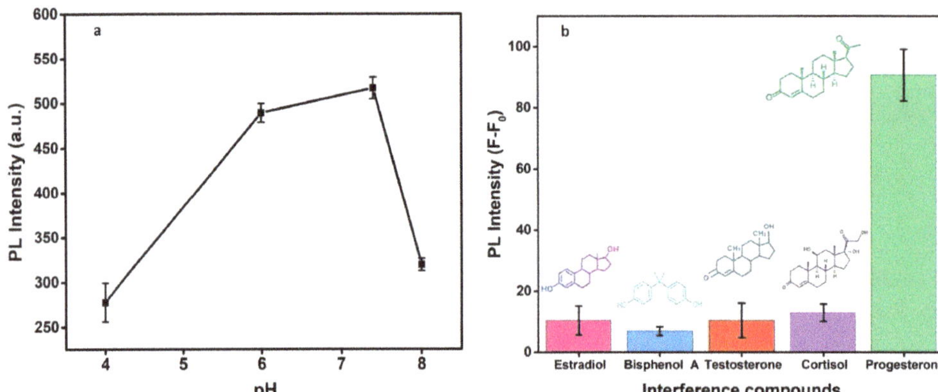

Figure 8. pH effect on the fluorescence of proposed PGN biosensor (**a**) and change in PL intensity with different interference compounds (**b**).

3.2.2. Specificity and Selectivity of the FRET Immunosensor

Several interfering or analogous biomolecules such as β-estradiol, bisphenol A, testosterone, and cortisol were tested for the specificity as an essential parameter of the present immunosensor. As shown in Figure 8b, there is no significant change in the PL response with the potentially interfering substances at 100 nM of concentration each. However, there is a significant change in PL intensity at the same concentration of the PGN hormone. The selectivity performance of the developed optical immunoassay is ascribed to the specific antigen–antibody reactions between progesterone and its corresponding antibodies covalently conjugated on CDs. Interfering samples displayed little influence on the detection system owing to the strong immunoreactions between progesterone and Ab–CQDs, suggesting that progesterone detection could be successful in complex matrices.

3.2.3. Evaluation of FRET-Based Immunosensor Performance in Real Samples

For verifying the reliability and practicability of our fluorescence biosensor, we used the spiked water samples. The supernatant of water samples was added with progesterone of two known concentrations (100 & 200 nM) and analysed further. The 100 nM- and 200 nM-spiked samples have shown considerable rates of recovery with 95.6 (RSD = 6.5) and 98.4% (RSD = 4.5) as recovery percentages, respectively.

4. Conclusions

The summary of the proposed optical methodology demonstrates the success of implementing nitrogen-doped CDs prepared from hydrothermal synthesis and GO benefits the FRET process for the detection of progesterone in aqueous samples. The prepared Ab–CDs possessed a good fluorescent quantum yield of 8.5% with quinine sulphate as a reference and an excellent photo as well as pH stabilities. The PL intensity of the CDs–antibodies conjugated system is quenched by the π-π stacking and electrostatic interactions of GO, resulting into a turn-off state. However, by adding progesterone, the strong antigenic bindings of antibodies increased the distance between the CDs–antibody complexes and GO which recovers and further modifies the PL intensity. The variation in PL intensity ranges linearly with progesterone concentrations from 10–900 nm. Based on the findings, the progesterone immunoassay indicated good analytical performance with a detection limit of 13.8 nM. Since the work is devoid of any cumbersome labelling substrate or an enzyme, it is well suited for the routine analysis of environmental water systems. Furthermore, we anticipate that this fluorescence-based rapid bio sensing approach offers various potential applications in environmental as well as complicated clinical areas.

Supplementary Materials: The following are available online at: https://www.mdpi.com/article/10.3390/bios12110993/s1. Figure S1: HR-TEM image of GO. Figure S2: FTIR and Raman spectra of GO (a & b). Figure S3: Fluorescence spectra of CDs at higher excitation wavelengths (from 350 nm to 450 nm) (a) and corresponding normalized spectra (b). Figure S4: Histogram displaying the size distribution of CDs.

Author Contributions: D.—Conceptualization, investigation, methodology, data retention, analysis of adsorption plots, writing the first draft; P.K. (Poonam Kumari)—Data curation, review and editing of data; M.K.P.—Visualization, reviewing and editing; P.K. (Parveen Kumar)—Co-supervision, review and editing and M.K.N.—Conceptualization, supervision, resources, data analysis, writing—critical content and publishing. All authors have read and agreed to the published version of the manuscript.

Funding: CSIR-Central Scientific Instruments Organization (CSIO), Chandigarh, India, supported this work (CSIR project grant number HCP0031 WP 1.1).

Institutional Review Board Statement: Not applicable.

Informed Consent Statement: Not applicable.

Acknowledgments: The authors express their gratitude to the UGC and the Director of CSIR-CSIO for providing research grants, facilities, and fellowships.

Conflicts of Interest: The authors declare no conflict of interest.

References

1. Nezhadali, A.; Es'Haghi, Z.; Khatibi, A. Selective extraction of progesterone hormones from environmental and biological samples using a polypyrrole molecularly imprinted polymer and determination by gas chromatography. *Anal. Methods* **2016**, *8*, 1813–1827. [CrossRef]
2. Disha; Kumari, P.; Nayak, M.K.; Kumar, P. An electrochemical biosensing platform for progesterone hormone detection using magnetic graphene oxide. *J. Mater. Chem. B* **2021**, *9*, 5264–5271. [CrossRef] [PubMed]
3. Adeel, M.; Song, X.; Wang, Y.; Francis, D.; Yang, Y. Environmental impact of estrogens on human, animal and plant life: A critical review. *Environ. Int.* **2017**, *99*, 107–119. [CrossRef] [PubMed]
4. Cáceres, C.; Bravo, C.; Rivas, B.; Moczko, E.; Sáez, P.; García, Y.; Pereira, E. Molecularly imprinted polymers for the selective extraction of bisphenol a and progesterone from aqueous media. *Polymers* **2018**, *10*, 679. [CrossRef]
5. Alnajrani, M.N.; Alsager, O.A. Lateral flow aptasensor for progesterone: Competitive target recognition and displacement of short complementary sequences. *Anal. Biochem.* **2019**, *587*, 113461. [CrossRef]
6. Shi, C.-X.; Chen, Z.-P.; Chen, Y.; Yu, R.-Q. Quantitative analysis of hormones in cosmetics by LC-MS/MS combined with an advanced calibration model. *Anal. Methods* **2015**, *7*, 6804–6809. [CrossRef]
7. Almeida, C.; Nogueira, J. Determination of steroid sex hormones in water and urine matrices by stir bar sorptive extraction and liquid chromatography with diode array detection. *J. Pharm. Biomed. Anal.* **2006**, *41*, 1303–1311. [CrossRef]
8. Alhadrami, H.A.; Chinnappan, R.; Eissa, S.; Rahamn, A.A.; Zourob, M. High affinity truncated DNA aptamers for the development of fluorescence based progesterone biosensors. *Anal. Biochem.* **2017**, *525*, 78–84. [CrossRef]
9. Dong, X.-X.; Yuan, L.-P.; Liu, Y.-X.; Wu, M.-F.; Liu, B.; Sun, Y.-M.; Shen, Y.-D.; Xu, Z.-L. Development of a progesterone immunosensor based on thionine-graphene oxide composites platforms: Improvement by biotin-streptavidin-amplified system. *Talanta* **2017**, *170*, 502–508. [CrossRef]
10. Samie, H.A.; Arvand, M. Label-free electrochemical aptasensor for progesterone detection in biological fluids. *Bioelectrochemistry* **2020**, *133*, 107489. [CrossRef]
11. Guć, M.; Schroeder, G. Molecularly imprinted polymers and magnetic molecularly imprinted polymers for selective determination of estrogens in water by ESI-MS/FAPA-MS. *Biomolecules* **2020**, *10*, 672. [CrossRef] [PubMed]
12. Ning, F.; Peng, H.; Li, J.; Chen, L.; Xiong, H. Molecularly imprinted polymer on magnetic graphene oxide for fast and selective extraction of 17β-estradiol. *J. Agric. Food Chem.* **2014**, *62*, 7436–7443. [CrossRef]
13. Lee, M.-H.; O'Hare, D.; Guo, H.-Z.; Yang, C.-H.; Lin, H.-Y. Electrochemical sensing of urinary progesterone with molecularly imprinted poly (aniline-co-metanilic acid) s. *J. Mater. Chem. B* **2016**, *4*, 3782–3787. [CrossRef] [PubMed]
14. Cherian, A.R.; Benny, L.; George, A.; Sirimahachai, U.; Varghese, A.; Hegde, G. Electro fabrication of molecularly imprinted sensor based on Pd nanoparticles decorated poly-(3 thiophene acetic acid) for progesterone detection. *Electrochim. Acta* **2022**, *408*, 139963. [CrossRef]
15. Cao, L.; Yu, L.; Yue, J.; Zhang, Y.; Ge, M.; Li, L.; Yang, R. Yellow-emissive carbon dots for "off-and-on" fluorescent detection of progesterone. *Mater. Lett.* **2020**, *271*, 127760. [CrossRef]
16. Zhu, Y.; Xu, Z.; Gao, J.; Ji, W.; Zhang, J. An antibody-aptamer sandwich cathodic photoelectrochemical biosensor for the detection of progesterone. *Biosens. Bioelectron.* **2020**, *160*, 112210. [CrossRef]

17. Chen, M.; Grazon, C.; Sensharma, P.; Nguyen, T.T.; Feng, Y.; Chern, M.; Baer, R.; Varongchayakul, N.; Cook, K.; Lecommandoux, S. Hydrogel-Embedded Quantum Dot–Transcription Factor Sensors for Quantitative Progesterone Detection. *ACS Appl. Mater. Interfaces* **2020**, *12*, 43513–43521. [CrossRef]
18. Zan, M.; An, S.; Cao, L.; Liu, Y.; Li, L.; Ge, M.; Liu, P.; Wu, Z.; Dong, W.-F.; Mei, Q. One-pot facile synthesis of yellow-green emission carbon dots for rapid and efficient determination of progesterone. *Appl. Surf. Sci.* **2021**, *566*, 150686. [CrossRef]
19. Hamd-Ghadareh, S.; Salimi, A.; Fathi, F.; Bahrami, S. An amplified comparative fluorescence resonance energy transfer immunosensing of CA125 tumor marker and ovarian cancer cells using green and economic carbon dots for bio-applications in labeling, imaging and sensing. *Biosens. Bioelectron.* **2017**, *96*, 308–316. [CrossRef]
20. Niu, X.; Zhong, Y.; Chen, R.; Wang, F.; Liu, Y.; Luo, D. A "turn-on" fluorescence sensor for Pb2+ detection based on graphene quantum dots and gold nanoparticles. *Sens. Actuators B Chem.* **2018**, *255*, 1577–1581. [CrossRef]
21. Tian, J.; Wei, W.; Wang, J.; Ji, S.; Chen, G.; Lu, J. Fluorescence resonance energy transfer aptasensor between nanoceria and graphene quantum dots for the determination of ochratoxin A. *Anal. Chim. Acta* **2018**, *1000*, 265–272. [CrossRef] [PubMed]
22. Mohammadi, S.; Salimi, A.; Hamd-Ghadareh, S.; Fathi, F.; Soleimani, F. A FRET immunosensor for sensitive detection of CA 15-3 tumor marker in human serum sample and breast cancer cells using antibody functionalized luminescent carbon-dots and AuNPs-dendrimer aptamer as donor-acceptor pair. *Anal. Biochem.* **2018**, *557*, 18–26. [CrossRef] [PubMed]
23. Ren, J.; Stagi, L.; Innocenzi, P. Fluorescent carbon dots in solid-state: From nanostructures to functional devices. *Prog. Solid State Chem.* **2021**, *62*, 100295. [CrossRef]
24. Zhou, L.; Lin, Y.; Huang, Z.; Ren, J.; Qu, X. Carbon nanodots as fluorescence probes for rapid, sensitive, and label-free detection of Hg 2+ and biothiols in complex matrices. *Chem. Commun.* **2012**, *48*, 1147–1149. [CrossRef] [PubMed]
25. Sharma, V.; Tiwari, P.; Mobin, S.M. Sustainable carbon-dots: Recent advances in green carbon dots for sensing and bioimaging. *J. Mater. Chem. B* **2017**, *5*, 8904–8924. [CrossRef]
26. Fu, X.; Sheng, L.; Yu, Y.; Ma, M.; Cai, Z.; Huang, X. Rapid and universal detection of ovalbumin based on N, O, P-co-doped carbon dots-fluorescence resonance energy transfer technology. *Sens. Actuators B Chem.* **2018**, *269*, 278–287. [CrossRef]
27. Wang, L.; Xu, D.; Jiang, L.; Gao, J.; Tang, Z.; Xu, Y.; Chen, X.; Zhang, H. Transition metal dichalcogenides for sensing and oncotherapy: Status, challenges, and perspective. *Adv. Funct. Mater.* **2021**, *31*, 2004408. [CrossRef]
28. Guo, J.; Chan, E.W.; Chen, S.; Zeng, Z. Development of a novel quantum dots and graphene oxide based FRET assay for rapid detection of invA gene of Salmonella. *Front. Microbiol.* **2017**, *8*, 8. [CrossRef]
29. Zhang, G.; Li, T.; Zhang, J.; Chen, A. A simple FRET-based turn-on fluorescent aptasensor for 17β-estradiol determination in environmental water, urine and milk samples. *Sens. Actuators B Chem.* **2018**, *273*, 1648–1653. [CrossRef]
30. Wei, Q.; Zhang, P.; Pu, H.; Sun, D.-W. A fluorescence aptasensor based on carbon quantum dots and magnetic Fe3O4 nanoparticles for highly sensitive detection of 17β-estradiol. *Food Chem.* **2022**, *373*, 131591. [CrossRef]
31. Cui, H.; Lu, H.; Yang, J.; Fu, Y.; Huang, Y.; Li, L.; Ding, Y. A Significant Fluorescent Aptamer Sensor Based on Carbon Dots and Graphene Oxide for Highly Selective Detection of Progesterone. *J. Fluoresc.* **2022**, *32*, 927–936. [CrossRef] [PubMed]
32. Jenkins, R.L.; Wilson, E.M.; Angus, R.A.; Howell, W.M.; Kirk, M. Androstenedione and progesterone in the sediment of a river receiving paper mill effluent. *Toxicol. Sci.* **2003**, *73*, 53–59. [CrossRef] [PubMed]
33. Bhatnagar, D.; Kumar, V.; Kumar, A.; Kaur, I. Graphene quantum dots FRET based sensor for early detection of heart attack in human. *Biosens. Bioelectron.* **2016**, *79*, 495–499. [CrossRef] [PubMed]
34. Ding, H.; Wei, J.-S.; Xiong, H.-M. Nitrogen and sulfur co-doped carbon dots with strong blue luminescence. *Nanoscale* **2014**, *6*, 13817–13823. [CrossRef]
35. Fedoseeva, Y.V.; Lobiak, E.V.; Shlyakhova, E.V.; Kovalenko, K.A.; Kuznetsova, V.R.; Vorfolomeeva, A.A.; Grebenkina, M.A.; Nishchakova, A.D.; Makarova, A.A.; Bulusheva, L.G. Hydrothermal activation of porous nitrogen-doped carbon materials for electrochemical capacitors and sodium-ion batteries. *Nanomaterials* **2020**, *10*, 2163. [CrossRef]
36. Abdullah Issa, M.; Abidin, Z.Z.; Sobri, S.; Rashid, S.; Adzir Mahdi, M.; Azowa Ibrahim, N.; Pudza, M.Y. Facile Synthesis of Nitrogen-Doped Carbon Dots from Lignocellulosic Waste. *Nanomaterials* **2019**, *9*, 1500. [CrossRef]
37. Ye, W.; Zhang, Y.; Hu, W.; Wang, L.; Zhang, Y.; Wang, P. A sensitive FRET biosensor based on carbon dots-modified nanoporous membrane for 8-hydroxy-2′-Deoxyguanosine (8-OHdG) detection with Au@ Zif-8 nanoparticles as signal quenchers. *Nanomaterials* **2020**, *10*, 2044. [CrossRef]
38. Raj, P.; Lee, S.-y.; Lee, T.Y. Carbon dot/naphthalimide based ratiometric fluorescence biosensor for hyaluronidase detection. *Materials* **2021**, *14*, 1313. [CrossRef]
39. Duan, L.; Du, X.; Zhao, H.; Sun, Y.; Liu, W. Sensitive and selective sensing system of metallothioneins based on carbon quantum dots and gold nanoparticles. *Anal. Chim. Acta* **2020**, *1125*, 177–186. [CrossRef]
40. Đorđević, L.; Arcudi, F.; Cacioppo, M.; Prato, M. A multifunctional chemical toolbox to engineer carbon dots for biomedical and energy applications. *Nat. Nanotechnol.* **2022**, *17*, 112–130. [CrossRef]
41. Zhang, P.; Li, W.; Zhai, X.; Liu, C.; Dai, L.; Liu, W. A facile and versatile approach to biocompatible "fluorescent polymers" from polymerizable carbon nanodots. *Chem. Commun.* **2012**, *48*, 10431–10433. [CrossRef]
42. Chattopadhyay, S.; Choudhary, M.; Singh, H. Carbon dots and graphene oxide based FRET immunosensor for sensitive detection of Helicobacter pylori. *Anal. Biochem.* **2022**, *654*, 114801. [CrossRef]
43. Singh, H.; Singh, S.; Bhardwaj, S.K.; Kaur, G.; Khatri, M.; Deep, A.; Bhardwaj, N. Development of carbon quantum dot-based lateral flow immunoassay for sensitive detection of aflatoxin M1 in milk. *Food Chem.* **2022**, *393*, 133374. [CrossRef]

 *biosensors* 

Article

A Sensitive Aptamer Fluorescence Anisotropy Sensor for Cd^{2+} Using Affinity-Enhanced Aptamers with Phosphorothioate Modification

Hao Yu [1,2] and Qiang Zhao [1,2,3,*]

1 State Key Laboratory of Environmental Chemistry and Ecotoxicology, Research Center for Eco-Environmental Sciences, Chinese Academy of Sciences, Beijing 100085, China
2 University of Chinese Academy of Sciences, Beijing 100049, China
3 School of Environment, Hangzhou Institute for Advanced Study, University of Chinese Academy of Sciences, Hangzhou 310000, China
* Correspondence: qiangzhao@rcees.ac.cn

Abstract: Rapid and sensitive detection of heavy metal cadmium ions (Cd^{2+}) is of great significance to food safety and environmental monitoring, as Cd^{2+} contamination and exposure cause serious health risk. In this study we demonstrated an aptamer-based fluorescence anisotropy (FA) sensor for Cd^{2+} with a single tetramethylrhodamine (TMR)-labeled 15-mer Cd^{2+} binding aptamer (CBA15), integrating the strengths of aptamers as affinity recognition elements for preparation, stability, and modification, and the advantages of FA for signaling in terms of sensitivity, simplicity, reproducibility, and high throughput. In this sensor, the Cd^{2+}-binding-induced aptamer structure change provoked significant alteration of FA responses. To acquire better sensing performance, we further introduced single phosphorothioate (PS) modification of CBA15 at a specific phosphate backbone position, to enhance aptamer affinity by possible strong interaction between sulfur and Cd^{2+}. The aptamer with PS modification at the third guanine (G) nucleotide (CBA15-G3S) had four times higher affinity than CBA15. Using as an aptamer probe CBA15-G3S with a TMR label at the 12th T, we achieved sensitive selective FA detection of Cd^{2+}, with a detection limit of 6.1 nM Cd^{2+}. This aptamer-based FA sensor works in a direct format for detection without need for labeling Cd^{2+}, overcoming the limitations of traditional competitive immuno-FA assay using antibodies and fluorescently labeled Cd^{2+}. This FA method enabled the detection of Cd^{2+} in real water samples, showing broad application potential.

Keywords: cadmium ions; aptamer; fluorescence; biosensors; fluorescence polarization

Citation: Yu, H.; Zhao, Q. A Sensitive Aptamer Fluorescence Anisotropy Sensor for Cd^{2+} Using Affinity-Enhanced Aptamers with Phosphorothioate Modification. *Biosensors* **2022**, *12*, 887. https://doi.org/10.3390/bios12100887

Received: 16 September 2022
Accepted: 14 October 2022
Published: 17 October 2022

1. Introduction

Heavy metal pollutant cadmium ion (Cd^{2+}) is released into the environment through natural and anthropogenic activities such as industrial emissions, agricultural fertilization, metallurgy, and mining [1,2]. Emissions of Cd^{2+} into the water, soil, and air can lead to serious pollution and deposition of fauna and flora, and Cd^{2+} can be eventually accumulated in the human body through the food chain [1–5]. Cd^{2+} is highly toxic and causes adverse effects on human health, such as itai-itai disease, hypertension, and cancers [1–4]. The U.S. Environmental Protection Agency (EPA) sets the maximum concentration of Cd^{2+} as 5.0 μg/L in drinking water. As Cd^{2+} contamination and Cd^{2+} exposure are widespread, Cd^{2+} detection is important and necessary for food safety, environmental monitoring, and health risk assessment [6]. Conventional methods for Cd^{2+} detection include atomic absorption spectrometry, inductively coupled plasma atomic emission spectrometry, inductively coupled plasma mass spectrometry, etc. [6–8]. Despite high sensitivity and accuracy, certain limitations include expensive instruments, complicated sample pretreatment, and time-consuming operations, and the process is not suitable for rapid onsite monitoring of Cd^{2+}. Therefore, there is an urgent need to develop simple, rapid and cost-effective methods to detect trace amounts of Cd^{2+}. Biosensors for Cd^{2+} can meet these demands, and have attracted considerable attention [6,7].

Aptamers are single-stranded oligonucleotides that bind to targets with high specificity and affinity [9,10]. As new recognition elements in biosensors, aptamers show advantages over antibodies in terms of low cost, good thermal stability, facile production, easy modification with functional groups, and binding-induced conformation change. Aptamer-based sensors have enabled detection of various targets such as metal ions, small molecules, proteins, and cells, due to the appealing features of aptamers [11–14]. Since aptamers for Cd^{2+} were identified [15,16], aptamer-based sensors and assays for Cd^{2+} have become possible and have provided new approaches for detection of Cd^{2+}, including electrochemistry, colorimetry, and fluorescent methods [15–21].

Fluorescence polarization (FP)/anisotropy (FA) is an attractive fluorescence technique due to its sensitivity, simplicity, and homogeneous and high-throughput analysis, and is often used in environmental monitoring, drug discovery, food analysis, and affinity-binding research [22–27]. As a ratiometric method, FA assay can eliminate the influence of fluorescence fluctuation and photobleaching, and shows high reproducibility [22–24]. Combining the advantages of aptamer and FP/FA analysis, the use of aptamers in FP/FA assays further improves applications of FA analysis, allowing development of versatile formats of FA assays for various targets [23,24,26,27]. Among the available aptamer-based FA assays, the FA assay using tetramethylrhodamine (TMR)-labeled aptamers is a simple non-competitive method, which relies on target-binding-induced change of TMR–nucleotide (e.g., TMR-G) interaction, without the need for fluorescent labeling of targets [27]. FP/FA methods for the detection of Cd^{2+} are limited and their development is challenging. The immunoantibody-based FP method is a competitive assay and requires the Cd^{2+} complex to be labeled with fluorophore and the antibody for Cd^{2+} chelate [28], with limitations encountered in the preparation of antibodies against Cd^{2+} and fluorescence tracers of Cd^{2+} [28,29].

In this paper, we report a simple aptamer FA sensor in a direct format for rapid sensitive detection of Cd^{2+} using a single TMR labeled 15-mer Cd^{2+} binding aptamer (CBA15). By screening FA responses of different labeling sites of TMR (3′ end, 5′ end, and internal T bases), we identified that the aptamer-labeled TMR on the 12th T base showed remarkable FA change upon Cd^{2+} binding. In addition, to improve assay sensitivity, a high-affinity aptamer is desirable. We further demonstrated a strategy of introduction a single phosphorothioate (PS) modification to the aptamer, by replacing one of the phosphate oxygen atoms with sulfur, to greatly enhance aptamer affinity by tightening the binding between Cd^{2+} and the aptamer, with strong interaction between Cd^{2+} and sulfur [30]. We found that when PS modification was introduced to the linker between the third nucleotide G and the fourth nucleotide G of CBA15 (CBA15-G3S), a stronger aptamer affinity to Cd^{2+} was obtained, with a K_d about 47 nM representing a more than four-fold improvement in affinity compared with CBA15 without PS modification. We employed CBA15-G3S with optimal TMR label at the 12th T (CBA15-G3S-T12-TMR) in the FA sensor, and achieved more sensitive FA detection of Cd^{2+} with a detection limit at the nM level. The aptamer FA sensor allowed detection of Cd^{2+} in tap water and lake water samples, showing applicability for analysis of Cd^{2+} in a complex sample matrix.

2. Materials and Methods

2.1. Chemical and Reagents

$CdCl_2$ was ordered from J & K Chemicals (Beijing, China). $Pb(Ac)_2$, $Mn(Ac)_2$, $NiCl_2$, $CuCl_2$, and $ZnSO_4$ were ordered from Sangon Biotech (Shanghai, China). $HgCl_2$ was purchased from National Pharmaceutical Group. $MgCl_2$ and NaCl were obtained from Sinopharm (Shanghai, China). All of the unlabeled DNA aptamers, DNA aptamers with PS modification, and DNA aptamers with single tetramethylrhodamine (TMR) labeled at different sites (3′ end, internal thymine bases (T), and 5′ end) were synthesized and purified by Sangon Biotech (Shanghai, China). The sequences are listed in Table 1. The binding buffer used in this experiment was prepared with ultrapure water (18.2 MΩ·cm) from Purelab Ultra Elga Labwater (Buckinghamshire, England). Other reagents used in this experiment were of analytical grade.

Table 1. List of the DNA oligonucleotides.

Name	Sequences
CBA15	5′-CGG GTT CAC AGT CCG-3′
CBA15-3′-TMR	5′-CGG GTT CAC AGT CCG-TMR-**3′** [a]
CBA15-T5-TMR	5′-CGG G**T**(TMR)T CAC AGT CCG-3′
CBA15-T6-TMR	5′-CGG GT**T**(TMR) CAC AGT CCG-3′
CBA15-T12-TMR	5′-CGG GTT CAC AG**T**(TMR) CCG-3′
CBA15-5′-TMR	**5′**-TMR-CGG GTT CAC AGT CCG-3′
CBA15-G3S-T12-TMR	5′-CGG$_{PS}$ GTT CAC AG**T**(TMR) CCG-3′ [b]

[a] The labeling sites of TMR are shown in bold. [b] The labeling site of PS modification in the backbone is shown with PS, indicating that the backbone between G3 and G4 had a PS modification.

2.2. Isothermal Titration Calorimetry Measurement

Isothermal titration calorimetry (ITC) analysis was carried out at 25 °C using a MicroCal PEAQ-ITC (Malvern, Malvern, UK) to determine aptamer affinity. The binding buffer for ITC analysis contained 20 mM Tris-HCl (pH 7.5) and 20 mM NaCl. During ITC measurements, the reference power was set to 10 µcal/s and the stirring speed of the syringe was 750 rpm. Cd^{2+} solution (200 µM) from the injection syringe was titrated into the aptamer solution (20 µM) in a sample cell. After 60 s initial delay, the experiment began with the first 0.4 µL of Cd^{2+} solution and 19 successive 2.0 µL of Cd^{2+} solution every 100 s. The binding curves were obtained by integrating the heat pulse areas of each titration. Dissociation constants (K_ds), enthalpy change (ΔH), and entropy change (TΔS) were obtained by fitting the one-site binding model with the packaged MicroCal PEAQ-ITC analysis software.

2.3. Fluorescence Anisotropy Measurement

Certain concentrations of Cd^{2+} were mixed with dye-labeled aptamers (20 nM) in the binding buffer of 20 mM Tris-HCl (pH 7.5) and 20 mM NaCl. After incubation of the sample solution for 10 min at 25 °C, unless otherwise stated, fluorescence anisotropy (FA) measurements were conducted on a Synergy™ H1 microplate reader (BioTek, Highland Park, IL, USA) with excitation at 530 nm and emission at 590 nm. Duplicate samples were tested for detection of Cd^{2+}, and after three measurements of the same sample solution, the average data were used.

2.4. Detection of Cd^{2+} in Complex Sample Matrix

Tap water and lake water (Beijing Olympic Forest Park, China) were filtered through a 0.22 µm membrane, and then the treated water samples were diluted 20-fold with the binding buffer. Different concentrations of Cd^{2+} spiked in the complex sample matrix were tested by the aptamer-based FA assay, using the procedure described above.

3. Results and Discussion

3.1. FA Sensor for Cd^{2+} Using TMR-Labeled Aptamers

Figure 1 shows the principle of the aptamer FA sensor for Cd^{2+} detection using TMR-labeled aptamers. It has been reported that TMR-G interaction may limit the local rotation of TMR and affect FA values [26,27,31]. When Cd^{2+} binds specifically to the aptamer, the conformation of the aptamer changes, which alters the TMR-G interaction and the FA signals of the TMR label. Therefore, the quantitative analysis of Cd^{2+} can be achieved by measuring changes in the FA signals of the TMR-labeled aptamer. In this method, an appropriate labeling site for TMR at the aptamer is required. To identify an appropriate position at the aptamer for TMR labeling, enabling the TMR-labeled aptamer to show sensitive FA response to Cd^{2+}, we tested a series of sites of the 15-mer Cd^{2+}-binding aptamer (CBA15), including 3′ end, internal thymine bases (5 T, 6 T, 12 T), and 5′ end (Table 1). CBA15 was truncated from a 21-mer aptamer sequence against Cd^{2+} [15,21]. We measured the FA responses of different TMR-labeled CBA15 to Cd^{2+} in the binding buffer containing 20 mM Tris-HCl (pH 7.5) and 20 mM NaCl. As shown in Figure 2A, most of these aptamer probes showed FA values higher than 0.200 when Cd^{2+} was absent, indicating possible TMR–nucleotide (e.g., G) interaction, causing high FA. In the presence of 500 nM Cd^{2+}, the FA values of the aptamer probes decreased, except the probe CBA15-3′TMR (Figure 2B), suggesting that most of the TMR-labeled aptamers were FA-responsive to Cd^{2+}. The FA decrease was due to Cd^{2+}-binding-induced aptamer conformation change weakening the TMR-G interaction.

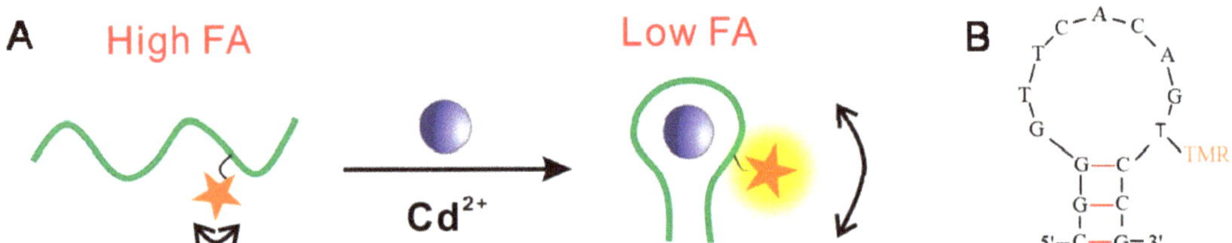

Figure 1. (**A**) Schematic diagram of fluorescence anisotropy sensor for Cd^{2+} using TMR-labeled aptamers. (**B**) The predicted secondary structure of the aptamer CBA15, with single TMR labeled on the 12th thymine base.

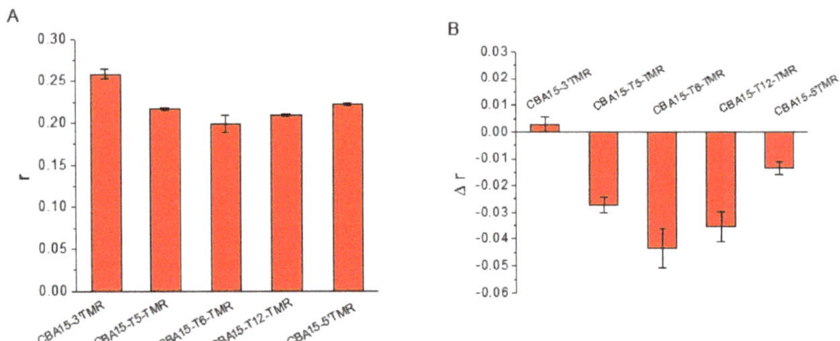

Figure 2. (**A**) FA responses of CBA15 with TMR labels at different sites (20 nM). (**B**) FA changes of TMR-labeled CBA15 caused by 500 nM Cd^{2+}. Δr was obtained by subtracting the FA values of the blank sample from the FA values of 500 nM Cd^{2+}.

We further tested the FA responses of different TMR-labeled aptamers upon addition of varying concentrations of Cd^{2+}. As shown in Figure 3, CBA15-T5-TMR, CBA15-T6-TMR, and CBA15-T12-TMR showed significant FA decrease upon Cd^{2+} binding. CBA15-5′TMR exhibited slight FA decrease, while the CBA15-3′TMR showed small FA increase upon addition of Cd^{2+}. When TMR was labeled on the internal 12th thymine (T) bases in the sequence, the corresponding aptamer probe CBA15-T12-TMR showed the largest FA signal change upon Cd^{2+} binding. Therefore, CBA15-T12-TMR was employed for FA detection of Cd^{2+}.

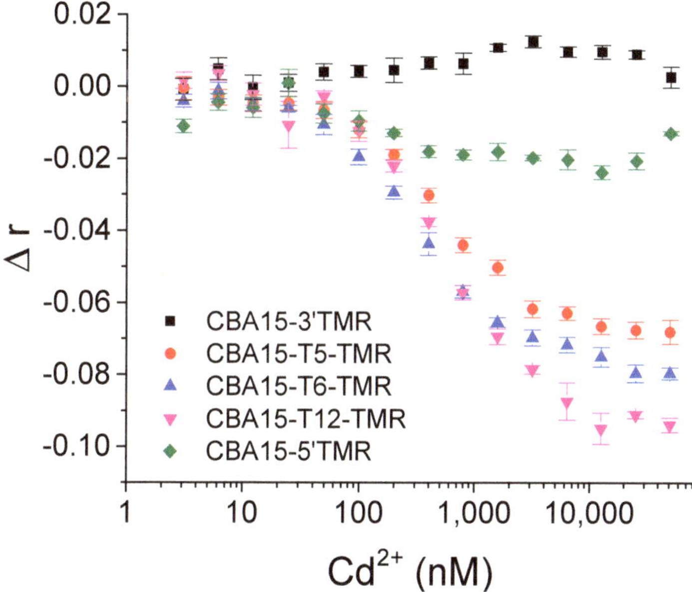

Figure 3. FA changes (Δr) of different TMR-labeled aptamers upon binding with various concentrations of Cd^{2+}. Δr was obtained by subtracting the FA values of blank samples from the FA values of various concentrations of Cd^{2+}.

3.2. Enhancing Aptamer Affinity with Phosphorothioate Modification

In order to improve the sensitivity of the aptamer FA sensor, aptamers with higher affinity are desirable. We attempted to introduce a single PS modification at a specific backbone site of the aptamer CBA15 (Table S1), to enhance affinity for the possible strong interaction between Cd^{2+} and the adjacent sulfur [30]. We determined the affinity of aptamers with PS modification at different labeling sites, by ITC analysis. After investigating a series of backbone sites, we found that CBA15-G3S with PS modification between the third G and fourth G of CBA15 showed higher binding affinity to Cd^{2+}, with K_ds of 46.6 ± 12.7 nM (Figure 4A), while the other PS-modified aptamers had K_ds ranging from 105 nM to 389 nM (Table S2). Compared with CBA15 without PS modification (Figure 4B, K_d = 216.0 ± 43.3 nM), CBA15-G3S had about four times higher affinity than CBA15 to Cd^{2+}. The results showed that PS modification at a favorable backbone site on the aptamer indeed greatly enhanced the aptamer affinity. An aptamer probe with higher affinity that can generate larger FA signal change upon Cd^{2+} binding is preferred for sensitive detection of Cd^{2+}. Thus, we introduced TMR at the 12th T of CBA15-G3S to obtain a CBA15-G3S-T12-TMR probe, and applied it for FA detection of Cd^{2+}.

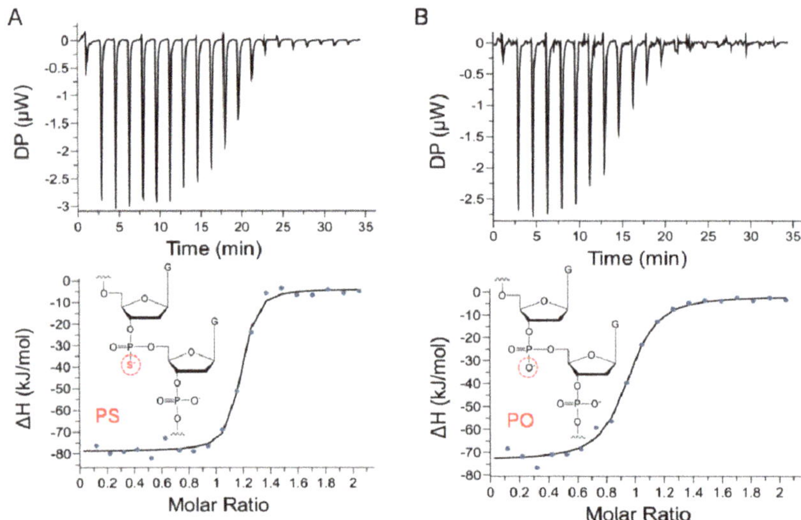

Figure 4. ITC analysis of aptamers (**A**) CBA15-G3S and (**B**) CBA15 with Cd^{2+}. The top graph shows raw data for ITC titration, and the bottom graph displays the binding curve obtained by integrating the heats of each spike. The difference between PS modification and the phosphate (PO) group in the backbone of the aptamer is shown.

3.3. Optimization of Aptamer FA Sensor for Cd^{2+}

To achieve better performance of the aptamer FA sensor for Cd^{2+}, we investigated the influence of NaCl concentration, $MgCl_2$ concentration, and pH of binding buffer on FA responses of CBA15-G3S-T12-TMR to Cd^{2+}. We first tested the effect of NaCl in binding buffer (Figure 5). In the absence of Cd^{2+}, the FA value of the blank sample slowly increased with the addition of NaCl. When Cd^{2+} was present, the FA value sharply increased from 0.109 to 0.184 with the increase of NaCl. The FA change (Δr) showed a high value at 20 mM NaCl. With the increase of NaCl from 20 mM to 200 mM, the FA change gradually became smaller. The results indicate that a high ionic strength of buffer is unfavorable for Cd^{2+} binding with aptamers. Therefore, we chose to use 20 mM NaCl in binding buffer for subsequent experiments.

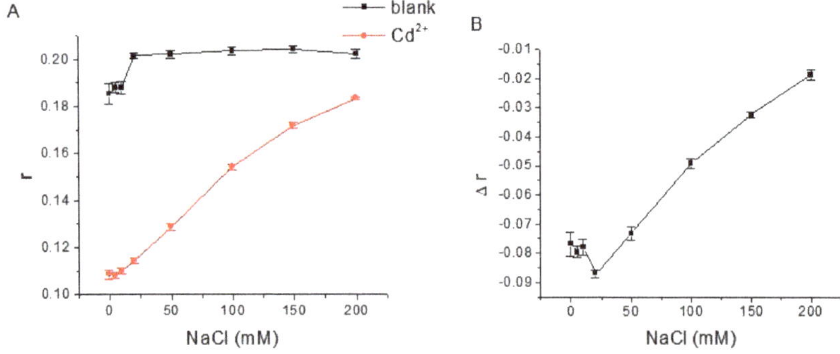

Figure 5. (**A**) Effects of NaCl concentration on FA responses of blank sample solution and the solution containing 1000 nM Cd^{2+} with 20 nM CBA15-G3S-T12-TMR. (**B**) The relationship between FA changes (Δr) caused by Cd^{2+} and NaCl concentration.

We further tested the influence of $MgCl_2$ concentration in the binding buffer. As displayed in Figure S1, with the addition of $MgCl_2$ from 0 to 5 mM, the FA value of the blank sample decreased from 0.197 to 0.174, and then changed slowly when the concentration of $MgCl_2$ was higher than 5 mM. In the presence of Cd^{2+}, the FA responses of CBA15-G3S-T12-TMR significantly increased with the addition of $MgCl_2$. Figure S1B shows that the addition of $MgCl_2$ caused decreased FA change (Δr), suggesting that $MgCl_2$ was not necessary for larger FA change. Thus, the binding buffer with 20 mM NaCl was preferred for the subsequent investigations.

We also assessed the influence of the pH of binding buffer solution on the FA responses of CBA15-G3S-T12-TMR (Figure S2). When the pH was lower than 7.5, a relatively small FA change caused by Cd^{2+} was observed. In order to obtain a larger FA change, binding buffer solution at pH 7.5 was chosen for our FA assay.

3.4. Detection of Cd^{2+} with Aptamer FA Sensor

Under the optimized experimental conditions, we successfully achieved detection of Cd^{2+} by FA analysis with the aptamer probe CBA15-G3S-T12-TMR. As shown in Figure 6A, the FA value of the probe gradually decreased with the addition of Cd^{2+}. The dynamic detection range was from 6.1 nM to 6250 nM, covering about three orders of magnitude. The nonlinear fitting equation of the response of FA to different concentrations of Cd^{2+} was obtained by GraphPad Prism software as $y = (-0.0918 \pm 0.0011)x/((171.8 \pm 9.362) + x) + (0.1913 \pm 0.0007)$ ($R^2 = 0.9987$). The detection limit of Cd^{2+} was 6.1 nM (S/N = 3). This method meets the requirements for assessing Cd^{2+} in drinking water as set by the U.S. Environmental Protection Agency (EPA) (5.0 μg/L, corresponding to 44 nM). For comparison, we also tested the FA response of the aptamer without PS modification (CBA15-T12-TMR) (Figure S3). From the FA response curves, we estimated that the K_d values of CBA15-G3S-T12-TMR and CBA15-T12-TMR were 147.4 ± 6.7 nM ($R^2 = 0.9973$) and 880.4 ± 25.5 nM ($R^2 = 0.9988$), respectively, by non-linear fitting with GraphPad Prism software [28]. The detection limit of CBA15-T12-TMR to Cd^{2+} was determined to be 24.4 nM (S/N = 3). The results indicated that introduction of PS modification at the specific site of the aptamer greatly enhanced the binding affinity of the aptamer, and improved its FA sensitivity to Cd^{2+}, while the aptamer CBA15-G3S-T12-TMR enabled a lower detection limit. Compared with other aptamer-based methods for Cd^{2+} detection [15,16,18,21,32–34], our method had better or comparable sensitivity (Table S3). Although the detection limit of our FA sensor was higher than that of the reported method [35], our strategy is simple, rapid, and requires only one TMR-labeled aptamer. In our method, quantification of Cd^{2+} can be achieved by mixing the TMR-labeled aptamer and Cd^{2+} solution, without need for separation and immobilization.

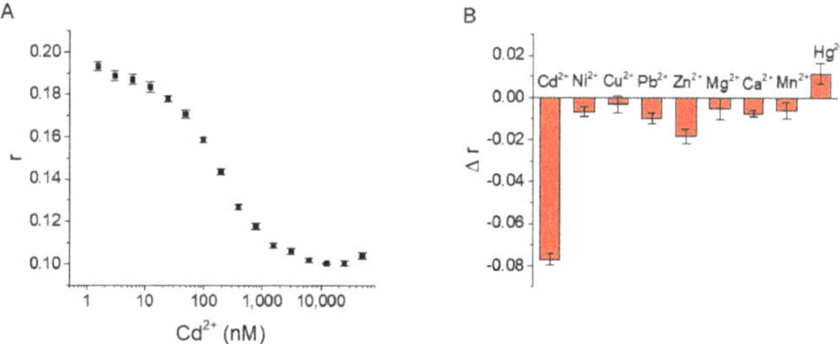

Figure 6. (**A**) FA detection of Cd^{2+} using CBA15-G3S-T12-TMR. (**B**) Selectivity test of the FA sensor using CBA15-G3S-T12-TMR for Cd^{2+} detection. A 20 nM aptamer probe was used, and the concentrations of tested metal ions were 1000 nM.

3.5. Selectivity Test and Practical Sample Analysis

We further tested the selectivity of the FA sensor for Cd^{2+} detection using CBA15-G3S-T12-TMR. As displayed in Figure 6B, other divalent metal ions including Cu^{2+}, Ni^{2+}, Pb^{2+}, Zn^{2+}, Hg^{2+}, Mn^{2+}, Mg^{2+}, and Ca^{2+} did not cause significant FA changes. The results showed that our assay was selective for Cd^{2+} detection.

To explore whether this FA sensor is applicable to real samples, we detected Cd^{2+} spiked in complex sample matrix with binding buffer. As displayed in Figure S4, the FA responses of CBA15-G3S-T12-TMR in 20-fold diluted tap water and 20-fold diluted lake water showed almost similar performances with binding buffer. The detection limits of Cd^{2+} in real water samples were 6.1 nM. Cd^{2+} spiked in 20-fold diluted lake water and 20-fold diluted tap water exhibited good recoveries of 89.1–121.3% and 81.7–127.1%, respectively (Table S4). The results confirm that our aptamer FA sensor can be used for practical sample analysis.

4. Conclusions

In summary, we have reported a simple aptamer fluorescence anisotropy sensor for detection of Cd^{2+} using a TMR-labeled high-affinity aptamer with specific PS modification at the backbone. After screening different labeling sites on the aptamer, we identified the aptamer probe with a TMR labeled on the 12th T base of a 15-mer DNA aptamer (CBA15) that showed large FA signal change in response to Cd^{2+} binding. We also demonstrated that introduction of PS modification at the specific backbone site (the third G) of CBA15 greatly enhanced the binding affinity of the aptamer, with approximately four-fold improvement. The TMR-labeled aptamer probe with PS modification allowed sensitive selective detection of Cd^{2+}, and the detection limit reached 6.1 nM Cd^{2+}. Our aptamer FA sensor provides a direct method for Cd^{2+} detection without need for competition, separation, and immobilization. This FA method offers the advantages of simplicity, sensitivity, robustness, rapidity, and high throughput, showing the potential for analysis of Cd^{2+} in various applications, especially for fast onsite detection of Cd^{2+}. This aptamer FA sensor circumvents the limitations of antibody-based FA assays for Cd^{2+} and certain methods using expensive instruments (e.g., ICP-MS). The introduction of PS modification provides an effective way to improve the affinity of aptamers for Cd^{2+}, and the aptamers with higher affinity will have wide applications in developing biosensors. This strategy will be helpful for the affinity enhancement of aptamers for use with other targets.

Supplementary Materials: The following supporting information can be downloaded at: https://www.mdpi.com/article/10.3390/bios12100887/s1, Table S1. List of the anti-Cd^{2+} aptamer sequences with PS modification at different labeling sites. Table S2. Binding affinity of PS modified aptamers and unlabeled aptamer characterized by ITC. Table S3. Comparison of some aptamer based methods for Cd^{2+} detection. Table S4 Detection of Cd^{2+} spiked in complex sample matrix. Figure S1. (A) Effect of $MgCl_2$ concentration on FA responses of CBA15-G3S-T12-TMR in the absence or in the presence of Cd^{2+}. (B) The relationship between FA changes (Δr) and $MgCl_2$ concentration. Figure S2. (A) Effect of pH of binding buffer on FA responses of CBA15-G3S-T12-TMR in the absence or in the presence of Cd^{2+}. (B) The FA changes (Δr) caused by Cd^{2+} at various pH of the binding buffer. Figure S3. Comparison of FA detection Cd^{2+} with CBA15-G3S-T12-TMR and CBA15-T12-TMR. Figure S4. Detection of Cd^{2+} in the binding buffer, 20-fold diluted tap water or 20-fold diluted lake water with the aptamer FA sensor by using CBA15-G3S-T12-TMR.

Author Contributions: Conceptualization, Q.Z.; Data curation, H.Y.; Funding acquisition, Q.Z.; Investigation, H.Y.; Methodology, H.Y.; Resources, Q.Z.; Supervision, Q.Z.; Visualization, H.Y.; Writing–original draft, H.Y.; Writing–review and editing, H.Y. and Q.Z. All authors have read and agreed to the published version of the manuscript.

Funding: This work was supported from the National Natural Science Foundation of China (grant number: 22074156 and 21874146).

Institutional Review Board Statement: Not applicable.

Informed Consent Statement: Not applicable.

Data Availability Statement: Not applicable.

Conflicts of Interest: The authors declare no conflict of interest.

References

1. Newbigging, A.M.; Yan, X.W.; Le, X.C. Cadmium in soybeans and the relevance to human exposure. *J. Environ. Sci.* **2015**, *37*, 157–162. [CrossRef]
2. Zhao, D.; Wang, P.; Zhao, F.J. Dietary cadmium exposure, risks to human health and mitigation strategies. *Crit. Rev. Environ. Sci. Technol.* **2022**, *7*, 1–25. [CrossRef]
3. Mahajan, M.; Gupta, P.K.; Singh, A.; Vaish, B.; Singh, P.; Kothari, R.; Singh, R.P. A comprehensive study on aquatic chemistry, health risk and remediation techniques of cadmium in groundwater. *Sci. Total Environ.* **2022**, *818*, 151784. [CrossRef]
4. Balali-Mood, M.; Naseri, K.; Tahergorabi, Z.; Khazdair, M.R.; Sadeghi, M. Toxic mechanisms of five heavy metals: Mercury, lead, chromium, cadmium, and arsenic. *Front. Pharmacol.* **2021**, *12*, 643972. [CrossRef] [PubMed]
5. Hu, J.H.; Chen, G.L.; Xu, K.; Wang, J. Cadmium in cereal crops: Uptake and transport mechanisms and minimizing strategies. *J. Agric. Food Chem.* **2022**, *70*, 5961–5974. [CrossRef] [PubMed]
6. Chauhan, S.; Dahiya, D.; Sharma, V.; Khan, N.; Chaurasia, D.; Nadda, A.K.; Varjani, S.; Pandey, A.; Bhargava, P.C. Advances from conventional to real time detection of heavy metal(loid)s for water monitoring: An overview of biosensing applications. *Chemosphere* **2022**, *307*, 136124. [CrossRef] [PubMed]
7. Maghsoudi, A.S.; Hassani, S.; Mirnia, K.; Abdollahi, M. Recent advances in nanotechnology-based biosensors development for detection of arsenic, lead, mercury, and cadmium. *Int. J. Nanomed.* **2021**, *16*, 803–832. [CrossRef]
8. Cao, Y.P.; Deng, B.Y.; Yan, L.Z.; Huang, H.L. An environmentally-friendly, highly efficient, gas pressure-assisted sample introduction system for ICP-MS and its application to detection of cadmium and lead in human plasma. *Talanta* **2017**, *167*, 520–525. [CrossRef]
9. Ellington, A.D.; Szostak, J.W. In vitro selection of RNA molecules that bind specific ligands. *Nature* **1990**, *346*, 818–822. [CrossRef] [PubMed]
10. Tuerk, C.; Gold, L. Systematic evolution of ligands by exponential enrichment-RNA ligands to bacteriophage-T4 DNA-polymerase. *Science* **1990**, *249*, 505–510. [CrossRef] [PubMed]
11. Li, F.; Zhang, H.Q.; Wang, Z.X.; Newbigging, A.M.; Reid, M.S.; Li, X.F.; Le, X.C. Aptamers facilitating amplified detection of biomolecules. *Anal. Chem.* **2015**, *87*, 274–292. [CrossRef]
12. Prante, M.; Segal, E.; Scheper, T.; Bahnemann, J.; Walter, J. Aptasensors for Point-of-Care Detection of Small Molecules. *Biosensors* **2020**, *10*, 108. [CrossRef]
13. Farzin, L.; Shamsipur, M.; Sheibani, S. A review: Aptamer-based analytical strategies using the nanomaterials for environmental and human monitoring of toxic heavy metals. *Talanta* **2017**, *174*, 619–627. [CrossRef]
14. Chang, C.C. Recent advancements in aptamer-based surface plasmon resonance biosensing strategies. *Biosensors* **2021**, *11*, 233. [CrossRef]
15. Wang, H.Y.; Cheng, H.; Wang, J.; Xu, L.J.; Chen, H.X.; Pei, R.J. Selection and characterization of DNA aptamers for the development of light-up biosensor to detect Cd(II). *Talanta* **2016**, *154*, 498–503. [CrossRef]
16. Wu, Y.G.; Zhan, S.S.; Wang, L.M.; Zhou, P. Selection of a DNA aptamer for cadmium detection based on cationic polymer mediated aggregation of gold nanoparticles. *Analyst* **2014**, *139*, 1550–1561. [CrossRef]
17. Fakude, C.T.; Arotiba, O.A.; Mabuba, N. Electrochemical aptasensing of cadmium (II) on a carbon black-gold nanoplatform. *J. Electroanal. Chem.* **2020**, *858*, 113796. [CrossRef]
18. Chen, Z.Y.; Liu, C.; Su, X.Y.; Zhang, W.; Zou, X.B. Signal on-off ratiometric electrochemical sensor based on semi-complementary aptamer couple for sensitive cadmium detection in mussel. *Sens. Actuator B Chem.* **2021**, *346*, 130506. [CrossRef]
19. Gan, Y.; Liang, T.; Hu, Q.W.; Zhong, L.J.; Wang, X.Y.; Wan, H.; Wang, P. In-situ detection of cadmium with aptamer functionalized gold nanoparticles based on smartphone-based colorimetric system. *Talanta* **2020**, *208*, 120231. [CrossRef]
20. Liu, Y.; Zhang, D.W.; Ding, J.N.; Hayat, K.; Yang, X.J.; Zhan, X.J.; Zhang, D.; Lu, Y.T.; Zhou, P. A facile aptasensor for instantaneous determination of cadmium ions based on fluorescence amplification effect of MOPS on FAM-labeled aptamer. *Biosensors* **2021**, *11*, 133. [CrossRef]
21. Yu, H.; Zhao, Q. Rapid sensitive fluorescence detection of cadmium (II) with pyrene excimer switching aptasensor. *J. Environ. Sci.* **2022**, *in press*. [CrossRef]
22. Hall, M.D.; Yasgar, A.; Peryea, T.; Braisted, J.C.; Jadhav, A.; Simeonov, A.; Coussens, N.P. Fluorescence polarization assays in high-throughput screening and drug discovery: A review. *Methods Appl. Fluores.* **2016**, *4*, 022001. [CrossRef] [PubMed]
23. Zhao, Q.; Tao, J.; Uppal, J.S.; Peng, H.Y.; Wang, H.L.; Le, X.C. Nucleic acid aptamers improving fluorescence anisotropy and fluorescence polarization assays for small molecules. *Trac-Trends Anal. Chem.* **2019**, *110*, 401–409. [CrossRef]
24. Zhang, H.; Yang, S.; De Ruyck, K.; Beloglazova, N.V.; Eremin, S.A.; De Saeger, S.; Zhang, S.; Shen, J.; Wang, Z. Fluorescence polarization assays for chemical contaminants in food and environmental analyses. *Trac-Trends Anal. Chem.* **2019**, *114*, 293–313. [CrossRef]

25. Li, C.L.; Mi, T.J.; Conti, G.O.; Yu, Q.; Wen, K.; Shen, J.Z.; Ferrante, M.; Wang, Z.H. Development of a screening fluorescence polarization immunoassay for the simultaneous detection of fumonisins B-1 and B-2 in maize. *J. Agric. Food Chem.* **2015**, *63*, 4940–4946. [CrossRef]
26. Li, Y.P.; Zhao, Q. Aptamer structure switch fluorescence anisotropy assay for aflatoxin B1 using tetramethylrhodamine-guanine interaction to enhance signal change. *Chin. Chem. Lett.* **2020**, *31*, 1982–1985. [CrossRef]
27. Zhao, Q.; Lv, Q.; Wang, H.L. Identification of allosteric nucleotide sites of tetramethylrhodamine-labeled aptamer for noncompetitive aptamer-based fluorescence anisotropy detection of a small molecule, ochratoxin A. *Anal. Chem.* **2014**, *86*, 1238–1245. [CrossRef]
28. Johnson, D.K. A fluorescence polarization immunoassay for cadmium(II). *Anal. Chim. Acta* **1999**, *399*, 161–172. [CrossRef]
29. Zhu, X.; Xu, L.; Lou, Y.; Yu, H.; Li, X.; Blake, D.A.; Liu, F. Preparation of specific monoclonal antibodies (MAbs) against heavy metals: MAbs that recognize chelated cadmium ions. *J. Agric. Food Chem.* **2007**, *55*, 7648–7653. [CrossRef]
30. Huang, P.J.J.; Liu, J.W. Rational evolution of Cd^{2+}-specific DNAzymes with phosphorothioate modified cleavage junction and Cd^{2+} sensing. *Nucleic Acids Res.* **2015**, *43*, 6125–6133. [CrossRef]
31. Zhang, D.P.; Shen, H.J.; Li, G.H.; Zhao, B.L.; Yu, A.C.; Zhao, Q.; Wang, H.L. Specific and sensitive fluorescence anisotropy sensing of guanine-quadruplex structures via a photoinduced electron transfer mechanism. *Anal. Chem.* **2012**, *84*, 8088–8094. [CrossRef]
32. Zhou, B.; Yang, X.Y.; Wang, Y.S.; Yi, J.C.; Zeng, Z.; Zhang, H.; Chen, Y.T.; Hu, X.J.; Suo, Q.L. Label-free fluorescent aptasensor of Cd^{2+} detection based on the conformational switching of aptamer probe and SYBR green I. *Microchem. J.* **2019**, *144*, 377–382. [CrossRef]
33. Zhu, Y.F.; Wang, Y.S.; Zhou, B.; Yu, J.H.; Peng, L.L.; Huang, Y.Q.; Li, X.J.; Chen, S.H.; Tang, X.; Wang, X.F. A multifunctional fluorescent aptamer probe for highly sensitive and selective detection of cadmium(II). *Anal. Bioanal. Chem.* **2017**, *409*, 4951–4958. [CrossRef]
34. Lai, B.; Wang, H.T.; Su, W.T.; Wang, Z.P.; Zhu, B.W.; Yu, C.X.; Tan, M.Q. A phosphorescence resonance energy transfer-based "off-on" long afterglow aptasensor for cadmium detection in food samples. *Talanta* **2021**, *232*, 122409. [CrossRef]
35. Zhou, D.H.; Wu, W.; Li, Q.; Pan, J.F.; Chen, J.H. A label-free and enzyme-free aptasensor for visual Cd^{2+} detection based on split DNAzyme fragments. *Anal. Methods* **2019**, *11*, 3546–3551. [CrossRef]

 *biosensors*

Article

A Smartphone-Based Biosensor for Non-Invasive Monitoring of Total Hemoglobin Concentration in Humans with High Accuracy

Zhipeng Fan [1], Yong Zhou [2], Haoyu Zhai [1], Qi Wang [1] and Honghui He [1,*]

[1] Guangdong Research Center of Polarization Imaging and Measurement Engineering Technology, Shenzhen Key Laboratory for Minimal Invasive Medical Technologies, Institute of Biopharmaceutical and Health Engineering, Tsinghua Shenzhen International Graduate School, Tsinghua University, Shenzhen 518055, China

[2] Shenzhen Maidu Technology Co., Ltd., Shenzhen 518000, China

* Correspondence: he.honghui@sz.tsinghua.edu.cn

Abstract: In this paper, we propose a smartphone-based biosensor for detecting human total hemoglobin concentration in vivo with high accuracy. Compared to the existing biosensors used to measure hemoglobin concentration, the smartphone-based sensor utilizes the camera, memory, and computing power of the phone. Thus, the cost is largely reduced. Compared to existing smartphone-based sensors, we developed a highly integrated multi-wavelength LED module and a specially designed phone fixture to reduce spatial errors and motion artifacts, respectively. In addition, we embedded a new algorithm into our smartphone-based sensor to improve the measurement accuracy; an L*a*b* color space transformation and the "a" parameter were used to perform the final quantification. We collected 24 blood samples from normal and anemic populations. The adjusted R^2 of the prediction results obtained from the multiple linear regression method reached 0.880, and the *RMSE* reached 9.04, which met the accuracy requirements of non-invasive detection of hemoglobin concentration.

Keywords: biosensor; smartphone; noninvasive; multi-wavelength; L*a*b* color space; total hemoglobin concentration

Citation: Fan, Z.; Zhou, Y.; Zhai, H.; Wang, Q.; He, H. A Smartphone-Based Biosensor for Non-Invasive Monitoring of Total Hemoglobin Concentration in Humans with High Accuracy. *Biosensors* **2022**, *12*, 781. https://doi.org/10.3390/bios12100781

Received: 19 August 2022
Accepted: 18 September 2022
Published: 21 September 2022

1. Introduction

Globally, more than one-third of pregnant women aged 15–49 suffer from anemia [1]. Anemia in pregnant women during pregnancy can cause complications, which can lead to death in severe cases. At the same time, the disease cycle of anemia is long, and it is not easy to treat or recover from; hence, it needs to be monitored and tested frequently. At present, the clinical solution in hospitals is mainly based on collecting blood sample from veins and then using the ferric cyanide method to measure the hemoglobin concentration [2]. Although this detection method has high measurement accuracy, it also has its limitations. The method requires the collection of human blood samples, during which both patients and medical staff are at risk of infection from exposure to blood and needles. In addition, professional medical personnel and equipment are required to perform the detection operation whose process is cumbersome and has low efficiency. Therefore, the method is inconvenient for the detection of hemoglobin concentration in daily life.

Currently, there are several pieces of non-invasive equipment on the market for daily hemoglobin concentration testing [3,4]. However, the measurement accuracy of these devices can hardly reach medical standards [5–7], and they are often not very portable. Compared with these medical testing devices, smartphones show irreplaceable portability and convenience. It is predicted that by the year of 2023, there will be more than four billion smartphone users globally, making smartphones play a potentially huge role in medical testing in non-invasive optical measurements. Recently, the application of smartphones in human physiological parameters measurement and monitoring has become increasingly common, such as smartphone-based detection platforms for molecular diagnostics [8], clinical diagnosis of albumin-related diseases [9,10], heart rate detection [11–13], blood oxygen saturation detection [14–16], respiratory rate detection [17,18], and anemia detection [19–21]. The advantages of using smartphones for detection are as follows: there is no need to carry other hardware devices, and the powerful CPU and memory of the smartphone provide a good hardware platform for the calculation and analysis of physiological parameters.

Here, we propose an accurate smartphone-based biosensor for detecting human total hemoglobin concentration. This sensor utilizes the camera, memory, and computing power of the smartphone. We also develop a highly integrated multi-wavelength LED module and a specially designed phone fixture to reduce spatial errors and motion artifacts, respectively. In addition, we embed a new algorithm into the smartphone-based sensor to improve the measurement accuracy; an L*a*b* color space transformation is used, and the "a" parameter is proposed to perform the final quantification. A total of 24 blood samples are collected from normal and anemic populations. The results show that the hardware and algorithm proposed in this study can meet the accuracy requirements of non-invasive detection of hemoglobin concentration in vivo.

2. Materials and Methods

2.1. Modified Beer–Lambert Law and Multiwavelength Selection

The Beer–Lambert law [22] indicates that when a beam of light passes through a medium, the intensity of the detected light is proportional to the concentration of the medium C and the propagation distance of the beam L. The radius of human blood vessels increases and decreases periodically with the beating of the heart, so the photoelectric volume diagram of the human body can be detected. However, this model is based on the hypothesis that only light is absorbed. This hypothesis is not applicable in a human body, where the attenuation of near-infrared light caused by scattering dominates relative to absorption (roughly 80% scattering vs. 20% absorption). In this case, the primary problem of using a photoelectric detector to receive the outgoing light is that the outgoing photons are not completely collected, as shown in Figure 1a. In addition, some of the light cannot reach the detector, so the actual optical attenuation is not accurately defined.

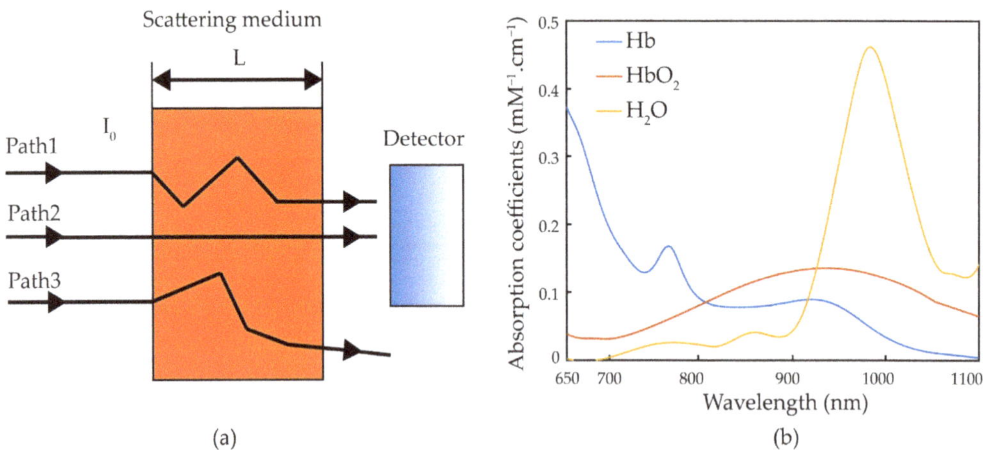

Figure 1. (**a**) Schematic of scattering, absorption of light in tissue, and detection. (**b**) Absorption coefficients of oxyhemoglobin, deoxy-hemoglobin, and water at 650–1100 nm.

In addition, the multiple scattering of light in the tissue leads to an increase in the path L of light propagation, which also has a great influence on the measurement results. The modified Beer–Lambert law [23] takes into account the effect of scattering in the tissue, as shown in Equation (1),

$$A_\lambda = (\varepsilon_{1_\lambda} \cdot c_1 + \varepsilon_{2_\lambda} \cdot c_2 + ...\varepsilon_{n_\lambda} \cdot c_n) \cdot d \cdot DPF + G. \quad (1)$$

where A_λ is light attenuation (in natural logarithm), also known as optical density [24] directly related to the concentration of substances and the distance that light passes through; ε_{n_λ} is the molar extinction coefficient of different substances; c_n is the concentration of different substances; *DPF* is the differential path length factor [25,26], representing the proportion of the increase in the optical path length due to scattering; $d{\cdot}DPF$ represents the effective path length of light; and the factor *G* is the scattering factor, representing the effect of the nature and geometry of the tissue.

Blood absorbs different wavelengths of light differently. Figure 1b shows the absorption curves of oxyhemoglobin, anaerobic hemoglobin, and water [27–29]. In the near-infrared optical window of 600–1000 nm, oxyhemoglobin and anaerobic hemoglobin have an iso-absorption point at 810 nm [30]. Hence, researchers generally choose 810 nm as the base signal for non-invasive measurement of hemoglobin concentration. The absorption of water near 970 nm and 1050 nm dominates, and the influence of water can be removed by using the signal at these wavelengths.

2.2. Smartphone Measurement Device

Here, we used a light source containing 5 LEDs of different wavelengths to obtain the photoplethysmography (PPG) signals from a fingertip. The measurement wavelengths were 660 nm, 810 nm, 900 nm, 970 nm, and 1050 nm. We chose the iso-absorption point of oxyhemoglobin and deoxyhemoglobin at 810 nm. Based on the therapeutic window, 660 nm and 900 nm were added to differentiate oxyhemoglobin and deoxyhemoglobin [31,32]. Moreover, near 970 nm and 1050 nm, the absorption of water was obvious; thus, we added these two wavelengths to exclude the influence of water absorption.

When a smartphone is held manually, the motion artifacts due to breathing and movement are prominent. Therefore, a 3D printed fix device is necessary, as shown in Figure 2a,b, which show the working smartphone system on the fixture. We also designed the arrangement of the lamps as shown in Figure 2c. As the absorption of the smartphone sensor at the wavelengths of 970 nm and 1050 nm is weaker than other wavelengths, we placed these two LEDs in the outermost positions of the fingertip, so the optical path was relatively short, and the signals were stronger. We designed the PCBA by ourselves and then handed it over to the factory to solder the patch to reduce the position error caused by human movement. The smartphone model (Huawei mate20pro, Shenzhen, China) used in this study has the acquisition pixel of 1920 × 1080. Its acquisition frame rate is 60 Hz, which is much larger than the PPG signal frequency of the human body (10 Hz). Hence, the sampling rate satisfies Shannon's sampling law.

2.3. Data Collection

During the signal detection process, the volunteer's finger remained motionless, and each LED was lit up to 20 s. As shown in Figure 3, in one cycle, all five LEDs were lit up separately. Thus, a video file with a total length of 100 s was collected.

Here we collected the PPG data on 12 normal adults (8 male volunteers, 4 female volunteers) and 12 hospitalized patients with anemia (6 male volunteers, 6 female volunteers). All the data were obtained from the volunteers' right index fingers. The volunteers' arms were placed on the table to ensure comfort and stability during the data acquisition, and a MATLAB-based program was used to check whether the frame rate was correct after the acquisition process. The volunteers' heart rate, blood pressure, and age were also recorded.

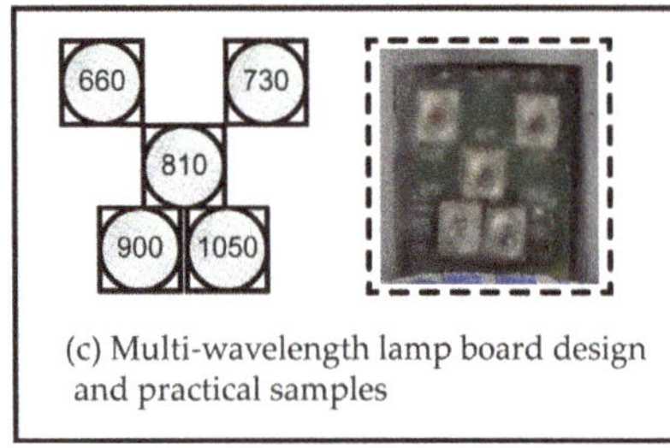

Figure 2. Hardware device for data collection: (**a**) Smartphone fixture with 660 nm LED on. (**b**) Smartphone on the fixture. (**c**) Arrangement of five-wavelength LEDs.

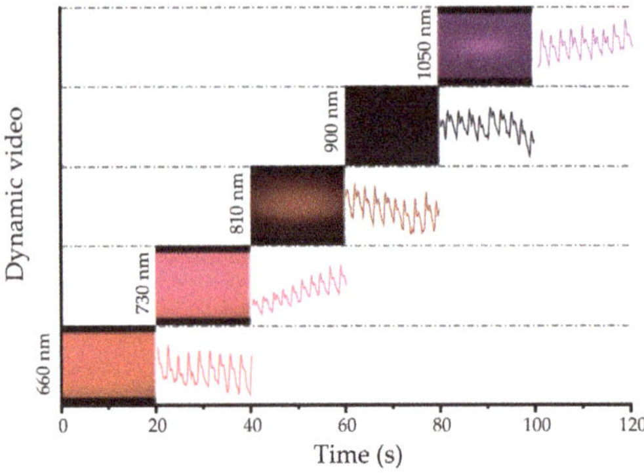

Figure 3. Schematic of the PPG data measurement and the waveforms display in sequence.

2.4. Color Space Transformation

The basic form of a time-continuous PPG video file is shown in Figure 4a. It is composed of color picture sequences of RGB channels. Here, the MATLAB program was used to extract images of each frame, and then an area of 1000 × 1000 pixels in the middle of the image was selected as the region of interest (ROI).

The RGB three-channel data of the ROI were extracted and averaged. The final obtained PPG signal is shown in Figure 4b. It can be seen that the signal noises of channels B and G are large, and the signals are unstable. However, the main information of the PPG signal image is concentrated in channel R, which is also more stable.

Previous studies have shown that the L*a*b* parameter provides a measurement of skin color perception [33] and, thus, mimics how skin is perceived by a dermatologist or the general population [34,35]. Chardon et al. proposed that in a three-dimensional L*a*b* space, all skin tones of light-skinned subjects are within a "banana"-shaped volume called skin tone volume. Skin redness (erythema reaction) can be represented as a displacement on the L*-a* plane. Since erythema is mainly caused by dilatation and congestion of local dermal capillaries of the skin, parameter "a" also reflects blood-related changes. CIE L*a*b* color space is shown in Figure 4c, where L* indicates light intensity related to the "luminous reflectance" (quantity of reflected light weighted with the spectral response of the human eye) and takes values from 0 (black) to 100 (white), a* indicates the color of the object on a scale that goes from green (−128) to red (128), and b* indicates the color of the object on a scale that goes from blue (−128) to yellow (128). The signals transformed from RGB to CIE L*a*b* color space by the RGB2Lab program are shown in Figure 4d, and the channel a signal was separately extracted as shown in Figure 4e. For all the volunteers' data, the channel a signal was used as the initial signal of PPG calculation.

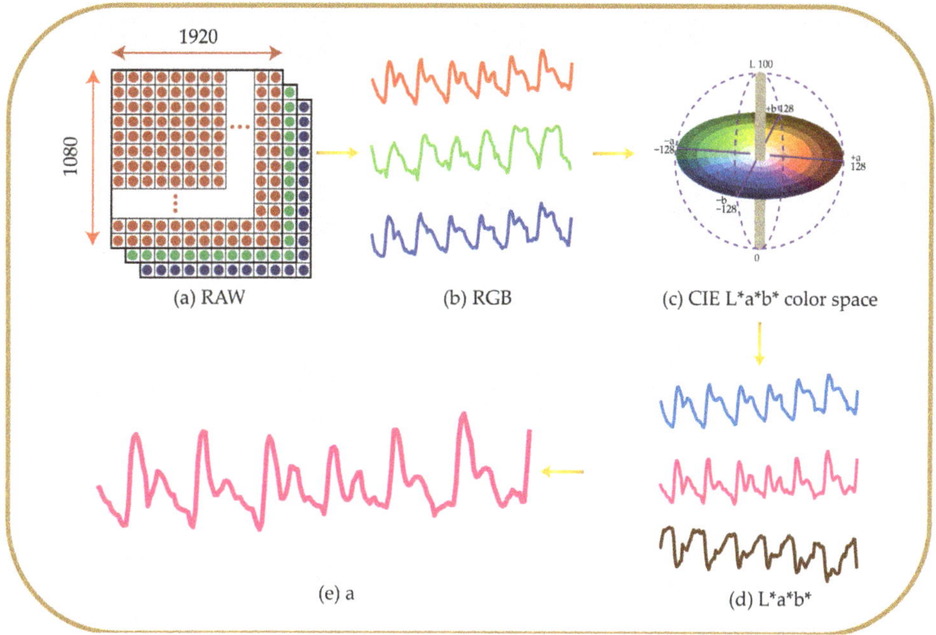

Figure 4. Smartphone collects the RGB channel information of the original image, converts it into the L*a*b* channel, and extracts the information of channel a for PPG calculations. (**a**) Schematic of a frame of image. (**b**) PPG signals of red, green, and blue channels. (**c**) CIE L*a*b* color space. (**d**) PPG signal of L*a*b* channels. (**e**) PPG signal of channel a.

2.5. Data Analysis

Figure 5a shows the PPG optical model of blood, which consists of tissue, venous blood, and arterial blood. Therefore, the model is simplified to only tissue and pulsating arterial blood, and then the modified Beer–Lambert law can be described as Equation (2),

$$A_\lambda = \ln(\frac{I_{out}}{I_{in}}) = (\varepsilon_{B_\lambda} \cdot c_B \cdot d_{B_\lambda} + \varepsilon_{T_\lambda} \cdot c_T \cdot d_{T_\lambda}) \cdot DPF + G. \quad (2)$$

where I_{out} represents the outgoing light intensity, I_{in} represents the incident light intensity, ε_{B_λ} represents the molar extinction coefficient of blood, c_B represents the concentration of blood, d_{B_λ} represents the light path of blood, ε_{T_λ} represents the molar extinction coefficient of tissue, c_T represents the concentration of tissue, d_{T_λ} represents the light path of tissue.

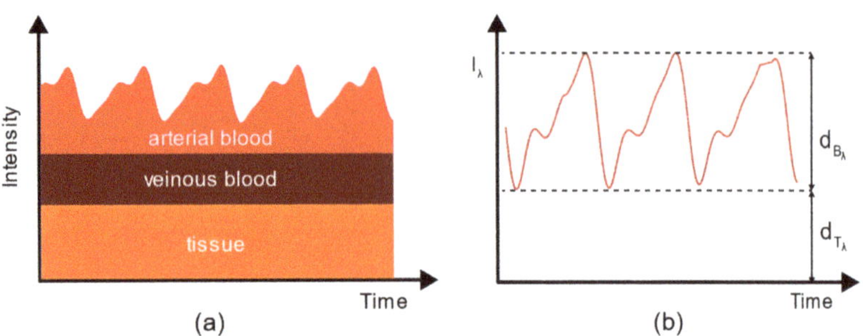

Figure 5. (**a**) PPG optical model of blood; (**b**) a characteristic PPG signal curve, where d_{T_λ} represents the optical path of tissue in finger, and d_{B_λ} represents the optical path of arterial blood in finger.

The absorption coefficient of the five wavelengths of light used in our system is different in human tissues. Under the same incident light intensity, the emitting light intensity of the strong absorbed wavelength will be relatively small, sometimes even similar to the intensity of the background noise. Therefore, the signal-to-noise ratio of such wavelength is low, meaning that it is necessary to strengthen the incident intensity of such a wavelength of light to increase the signal-to-noise ratio, that is, to increase the input current of the corresponding LED. For light with weak absorption wavelength in the tissue, a stronger incident light will lead to a stronger emitted light, which then saturates the signal of that wavelength, so its incident light intensity should be reduced. In summary, in order to achieve a good signal-to-noise ratio of light for all five wavelengths, the incident light intensity of each wavelength must be appropriately adjusted. To eliminate the effect of the current adjustment on the result, we use the AC/DC of each wavelength to obtain the characteristic parameter of R_λ as Equation (3),

$$R_\lambda = AC_\lambda / DC_\lambda = \frac{I_{in} \cdot e^{\varepsilon_{B_\lambda} \cdot c_B . d_{B_\lambda} \cdot DPF + G}}{I_{in} \cdot e^{\varepsilon_{T_\lambda} \cdot c_T \cdot d_{T_\lambda} \cdot DPF + G}} = e^{(\varepsilon_{B_\lambda} \cdot c_B \cdot d_{B_\lambda} - \varepsilon_{T_\lambda} \cdot c_T \cdot d_{T_\lambda}) \cdot DPF}. \quad (3)$$

To accurately obtain R_λ, we use a strict waveform screening tool and the skewness as a standard for waveform quality inspection. The waveforms that do not meet the requirements are not added to the calculation. Here, the parameter R_{1050} performs multiple linear regression fitting.

3. Results

3.1. Comparison of RGB and L*a*b* Color Spaces Results

We selected 50 waveforms in this section to calculate the R color channel of the RGB color space and channel a of the L*a*b color space according to Equation (3). The values were normalized to eliminate the error of the data scale. Figure 6a shows the R values of 660 nm, 810 nm, 900 nm, 970 nm, and 1050 nm, where the blue lines represent the R_a calculated by the "a" color channel, and the red lines represent the R_R calculated by the "R" color channel. It can be seen that for the same wavelength, the variation trends of R_a and R_R are similar for 660 nm, 810 nm, and 900 nm. To further show the distribution differences of R_a and R_R for each wavelength, we calculated the variance values to analyze their stability, as shown in Figure 6b. We can observe that the variance values of parameter R_a at 660 nm, 810 nm, 900 nm, and 1050 nm are smaller compared to those of parameter R_R, while the variance in R_a at 970 nm is larger than that in R_R. It can be concluded that for smartphone PPG predictions of hemoglobin concentration, the "a" color channel of L*a*b color space has a greater stability than the "R" color channel of RGB color space.

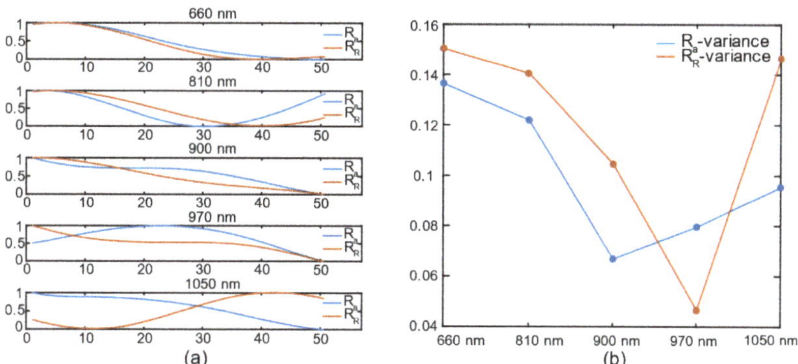

Figure 6. (**a**) Comparison of R_a and R_R at different wavelengths: 660 nm; 810 nm; 900 nm; 970 nm; 1050 nm; (**b**) variances in R_a and R_R at different wavelengths.

3.2. Prediction Results of Hemoglobin Concentration

Based on the method introduced in the above section, we collected 24 groups of effective waveforms of healthy people and patients, with the hemoglobin concentration distributed between 60 mg/dL and 170 mg/dL. We counted the five wavelengths of all volunteers. Due to the amplitude modulation of the pulse wave signal caused by respiration, the R of each wavelength fluctuates periodically with the breath. To eliminate this effect, we used wavelet changes to denoise it. Retaining only the trend term R to filter out the effects of respiration, we fed the five wavelengths and true hemoglobin concentration values of 24 volunteers into a multiple linear regressor, and the predicted results are shown in Figure 7, where the horizontal axis is the true value of the hemoglobin concentration, and the longitudinal axis is the predicted value of the hemoglobin concentration. Figure 7a,b show the prediction results obtained using the "R" parameter and "a" parameter, respectively. It can be seen in Figure 7 that both the parameters can provide good hemoglobin concentration predictions for the volunteers. It should be noted that, here, the "true values" were provided by the collaborating hospital. After the smartphone measurements, a nurse took the blood samples of the volunteers for hemoglobin concentration testing.

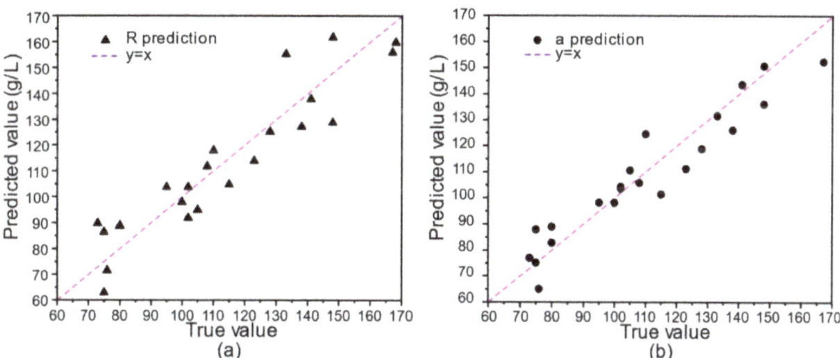

Figure 7. Prediction results of hemoglobin concentration using: (**a**) "R" channel and (**b**) "a" channel.

To further compare the prediction abilities of the "R" and "a" parameters, we used R^2, *RMSE* (Root Mean Square Error), and *MAPE* (Mean Absolute Percentage Error) for evaluation, as Equations (4)–(6) show,

$$R^2 = 1 - \frac{\sum_{i=1}^{n}(y_i - \hat{y}_i)^2}{\sum_{i=1}^{n}(y_i - \overline{y}_i)^2}, \in [0, 1] \quad (4)$$

$$RMSE = \sqrt{\frac{1}{n}\sum_{i=1}^{n}(y_i - \hat{y})^2}, \in [0, +\infty) \quad (5)$$

$$MAPE = \frac{100\%}{n}\sum_{i=1}^{n}\left|\frac{y_i - \hat{y}_i}{y_i}\right|, \in [0, +\infty) \quad (6)$$

Here, *MAPE* measures the average absolute value between the predicted and true values. The smaller the *MAPE* values of the model the better the prediction it can achieve. We compared the R^2, *RMSE*, *MAPE* metrics, and Durbin–Watson test for the "a" and "R" parameters, and the results are shown in Table 1.

Table 1. Comparison of the prediction results of "a" parameter and "R" parameter.

	Color Space	R^2	Adjusted R^2	*RMSE*	*MAPE*	Durbin–Watson Test
Model	a (L*a*b)	0.91	0.88	9.04	0.068	1.77
	R (RGB)	0.87	0.83	10.70	0.091	2.41

We can see from Table 1 that the adjusted R^2 of the multivariate linear regression model obtained by using the "a" parameter of the L*a*b color space reaches 0.88, which is larger than 0.83 achieved using the "R" parameter of the RGB color space. Moreover, the *RMSE* of the multiple linear regression model obtained by using the "a" parameter is 9.04, which is smaller than the value of 10.70 obtained by using the "R" parameter. The *MAPE* obtained by using the "a" parameter is 0.068, which is much smaller than the value of 0.091 obtained by using the "R" parameter. In summary, the R^2, *RMSE*, and *MAPE* of the regression results obtained by using the "a" parameter are better than those obtained by using the "R" parameter. In addition, the Durbin–Watson test was carried out to evaluate the autocorrelation problem of the independent variables. In general, the value of the Durbin–Watson test, shown in Table 1, confirms that there is no correlation between the residuals, and the subsequent improvement in the "a" parameter suggests that we should further increase the number of samples and expand the distribution range of hemoglobin concentration to further improve the prediction accuracy of the system.

4. Discussion

Most of the portable hemoglobin meters commercially available on the market are expensive. Although these devices can perform highly accurate hemoglobin measurements, they often require a finger prick to obtain the blood sample for testing, bringing additional infection risk for the users. In comparison to these devices, the smartphone-based biosensor proposed in this study provides a non-invasive and low-cost way of monitoring of total hemoglobin concentration. The testing results demonstrate that it has a satisfactory measurement accuracy, with irreplaceable portability and convenience. It is worth mentioning that there are other applicable methods for the direct monitoring of hemoglobin concentration using smartphone-based devices with high accuracy [36,37], whose R^2 values reached 0.9810 and 0.9583, respectively. Here, our device provides a multi-wavelength optical detection method, allowing one to achieve a satisfactory measurement accuracy non-invasively. It should be pointed out that, since the signal-to-noise ratio of the smartphone camera used to respond to infrared light needs to be improved, the data collected in this study have a certain error. For the wavelength of 1300 nm where water absorption is prominent, the response of the smartphone is limited. If the signal at 1300 nm can be added, the prediction result will be more accurate. Additionally, the response speed of the smartphone camera is relatively slow. It takes 2 s to stabilize the light intensity every time the LED is switched; hence, the time division multiplexing method cannot be used in this study. The waveform signals of five wavelengths in a very short time can be obtained. This method can further eliminate the interference from human movement and the modulation induced by breathing. In addition, only 24 volunteers were involved in this study, which may have led to certain errors in the regression coefficients calculation. As mentioned above, the prediction accuracy can be improved by increasing the number of samples.

It should be noted that there are issues of various smartphones having different types of cameras and lighting bias, which may result in different light intensity and color responses under the same illumination situation. As a possible solution for future development of this method and biosensor, a color calibration scheme could be adopted. More specifically, a color and intensity calibration algorithm could be designed based on the measurement results of standard samples and light sources when different smartphones are used to make sure that the light responses of different smartphone sensors are the same. As for the issue of different camera positions in different smartphones, a variable-position light source module can be designed and used.

5. Conclusions

In this paper, we proposed an accurate smartphone-based biosensor for detecting human total hemoglobin concentration. The sensor utilized the camera, memory, and computing power of the smartphone itself. We collected the data of 12 normal volunteers and 12 anemia patients and developed a new multi-wavelength light source system external to the smartphone. The smartphone can provide multi-spectral signals and used a 1050 nm wavelength LED light to provide information that is closely related to the absorption of water in the blood, thereby increasing the accuracy of the regression. Through the integrated LED light, the error of motion artifacts caused by switching lights of different wavelengths was effectively reduced. We designed a smartphone-fixing bracket to reduce the position error during the measurement. The "a" parameter of the L*a* b color space was used to predict the hemoglobin concentration when processing each picture frame. The experimental results demonstrate that, compared to the "R" parameter of the RGB color space, the "a" parameter has a better performance in predicting human hemoglobin concentration using PPG signals. The results show that the smartphone-based hardware and algorithm proposed in this study can meet the accuracy requirements of non-invasive detection of hemoglobin concentration in vivo.

Author Contributions: Conceptualization, Z.F., Y.Z., and H.H.; methodology, Z.F.; software, Z.F., Q.W. and H.Z.; formal analysis, Z.F.; resources, Y.Z.; data curation, Z.F.; writing—original draft preparation, Z.F.; writing—review and editing, Y.Z. and H.H. All authors have read and agreed to the published version of the manuscript.

Funding: This research was funded by Shenzhen Key Fundamental Research Project (No. JCYJ20210324120012035).

Institutional Review Board Statement: The study was conducted according to the guidelines of the Declaration of Helsinki and approved by the Ethics Committee of the Shenzhen International Graduate School, Tsinghua University (protocol code 2021-40, date of approval 31 May 2021).

Informed Consent Statement: Informed consent was obtained from all subjects involved in the study.

Data Availability Statement: The data presented in this study are available on request from the corresponding author.

Conflicts of Interest: The authors declare no conflict of interest.

References

1. Kamruzzaman, M.; Rabbani, M.G.; Saw, A.; Sayem, M.A.; Hossain, M.G. Differentials in the prevalence of anemia among non-pregnant, ever-married women in Bangladesh: Multilevel logistic regression analysis of data from the 2011 Bangladesh Demographic and Health Survey. *BMC Womens Health* **2015**, *15*, 54. [CrossRef] [PubMed]
2. Karakochuk, C.D.; Hess, S.Y.; Moorthy, D.; Namaste, S.; Parker, M.E.; Rappaport, A.I.; Wegmuller, R.; Dary, O.; Group, H.E.M.W. Measurement and interpretation of hemoglobin concentration in clinical and field settings: A narrative review. *Ann. N. Y. Acad. Sci.* **2019**, *1450*, 126–146. [CrossRef] [PubMed]
3. Causey, M.W.; Miller, S.; Foster, A.; Beekley, A.; Zenger, D.; Martin, M. Validation of noninvasive hemoglobin measurements using the Masimo Radical-7 SpHb Station. *Am. J. Surg.* **2011**, *201*, 592–598. [CrossRef] [PubMed]

4. Hiscock, R.; Kumar, D.; Simmons, S. Systematic review and meta-analysis of method comparison studies of Masimo pulse co-oximeters (Radical-7™ or Pronto-7™) and HemoCue® absorption spectrometers (B-Hemoglobin or 201+) with laboratory haemoglobin estimation. *Anaesth. Intensive Care* **2015**, *43*, 341–350. [CrossRef] [PubMed]
5. Park, Y.H.; Lee, J.H.; Song, H.G.; Byon, H.J.; Kim, H.S.; Kim, J.T. The accuracy of noninvasive hemoglobin monitoring using the radical-7 pulse CO-Oximeter in children undergoing neurosurgery. *Anesth. Analg.* **2012**, *115*, 1302–1307. [CrossRef] [PubMed]
6. Feiner, J.R.; Bickler, P.E.; Mannheimer, P.D. Accuracy of methemoglobin detection by pulse CO-oximetry during hypoxia. *Anesth. Analg.* **2010**, *111*, 143–148. [CrossRef]
7. Riess, M.L.; Pagel, P.S. Noninvasively Measured Hemoglobin Concentration Reflects Arterial Hemoglobin Concentration Before but Not after Cardiopulmonary Bypass in Patients Undergoing Coronary Artery or Valve Surgery. *J. Cardiothorac. Vasc. Anesth.* **2016**, *30*, 1167–1171. [CrossRef]
8. Song, J.; Pandian, V.; Mauk, M.G.; Bau, H.H.; Cherry, S.; Tisi, L.C.; Liu, C. Smartphone-Based Mobile Detection Platform for Molecular Diagnostics and Spatiotemporal Disease Mapping. *Anal. Chem.* **2018**, *90*, 4823–4831. [CrossRef]
9. Hussain, S.; Chen, X.; Wang, C.; Hao, Y.; Tian, X.; He, Y.; Li, J.; Shahid, M.; Iyer, P.K.; Gao, R. Aggregation and Binding-Directed FRET Modulation of Conjugated Polymer Materials for Selective and Point-of-Care Monitoring of Serum Albumins. *Anal. Chem.* **2022**, *94*, 10685–10694. [CrossRef]
10. Muthuraj, B.; Hussain, S.; Iyer, P.K. A rapid and sensitive detection of ferritin at a nanomolar level and disruption of amyloid β fibrils using fluorescent conjugated polymer. *Polym. Chem.* **2013**, *4*, 5096–5107. [CrossRef]
11. Pal, A.; Sinha, A.; Dutta Choudhury, A.; Chattopadyay, T.; Visvanathan, A. A robust heart rate detection using smart-phone video. In Proceedings of the 3rd ACM MobiHoc Workshop on Pervasive Wireless Healthcare, Bangalore, India, 29 July 2013; ACM: New York, NY, USA, 2013; pp. 43–48.
12. Pal, A.; Visvanathan, A.; Choudhury, A.D.; Sinha, A. Improved heart rate detection using smart phone. In Proceedings of the 29th Annual ACM Symposium on Applied Computing, Gyeongju, Korea, 24–28 March 2014; ACM: New York, NY, USA, 2014; pp. 8–13.
13. Gaoan, G.; Zhenmin, Z. Heart rate measurement via smart phone acceleration sensor. In Proceedings of the 2014 International Conference on Smart Computing, Hong Kong, China, 3–5 November 2014; pp. 295–300.
14. Bui, N.; Nguyen, A.; Nguyen, P.; Truong, H.; Ashok, A.; Dinh, T.; Deterding, R.; Vu, T. PhO2. In Proceedings of the 15th ACM Conference on Embedded Network Sensor Systems, Delft, The Netherlands, 6–8 November 2017; ACM: New York, NY, USA, 2017; pp. 1–14.
15. Fang, D.; Hu, J.; Wei, X.; Shao, H.; Luo, Y. A Smart Phone Healthcare Monitoring System for Oxygen Saturation and Heart Rate. In Proceedings of the 2014 International Conference on Cyber-Enabled Distributed Computing and Knowledge Discovery, Shanghai, China, 13–15 October 2014; pp. 245–247.
16. Nemcova, A.; Jordanova, I.; Varecka, M.; Smisek, R.; Marsanova, L.; Smital, L.; Vitek, M. Monitoring of heart rate, blood oxygen saturation, and blood pressure using a smartphone. *Biomed. Signal Process. Control* **2020**, *59*, 101928. [CrossRef]
17. Charlton, P.H.; Birrenkott, D.A.; Bonnici, T.; Pimentel, M.A.F.; Johnson, A.E.W.; Alastruey, J.; Tarassenko, L.; Watkinson, P.J.; Beale, R.; Clifton, D.A. Breathing Rate Estimation from the Electrocardiogram and Photoplethysmogram: A Review. *IEEE Rev. Biomed. Eng.* **2018**, *11*, 2–20. [CrossRef]
18. Karlen, W.; Garde, A.; Myers, D.; Scheffer, C.; Ansermino, J.M.; Dumont, G.A. Estimation of respiratory rate from photoplethysmographic imaging videos compared to pulse oximetry. *IEEE J. Biomed. Health Inform.* **2015**, *19*, 1331–1338. [CrossRef]
19. Batsis, J.A.; Boateng, G.G.; Seo, L.M.; Petersen, C.L.; Fortuna, K.L.; Wechsler, E.V.; Peterson, R.J.; Cook, S.B.; Pidgeon, D.; Dokko, R.S.; et al. Development and Usability Assessment of a Connected Resistance Exercise Band Application for Strength-Monitoring. *World Acad. Sci. Eng. Technol.* **2019**, *13*, 340–348. [CrossRef] [PubMed]
20. Wang, E.J.; Li, W.; Hawkins, D.; Gernsheimer, T.; Norby-Slycord, C.; Patel, S.N. HemaApp. In Proceedings of the 2016 ACM International Joint Conference on Pervasive and Ubiquitous Computing, Heidelberg, Germany, 12–16 September 2016; ACM: New York, NY, USA, 2016; pp. 593–604.
21. Ghatpande, N.S.; Apte, P.P.; Joshi, B.N.; Naik, S.S.; Bodas, D.; Sande, V.; Uttarwar, P.; Kulkarni, P.P. Development of a novel smartphone-based application for accurate and sensitive on-field hemoglobin measurement. *RSC Adv.* **2016**, *6*, 104067–104072. [CrossRef]
22. Swinehart, D.F. The beer-lambert law. *J. Chem. Educ.* **1962**, *39*, 333. [CrossRef]
23. Maikala, R.V. Modified Beer's Law—Historical perspectives and relevance in near-infrared monitoring of optical properties of human tissue. *Int. J. Ind. Ergon.* **2010**, *40*, 125–134. [CrossRef]
24. Pellicer, A.; Bravo Mdel, C. Near-infrared spectroscopy: A methodology-focused review. *Semin. Fetal Neonatal Med.* **2011**, *16*, 42–49. [CrossRef]
25. Duncan, A.; Meek, J.H.; Clemence, M.; Elwell, C.E.; Tyszczuk, L.; Cope, M.; Delpy, D. Optical pathlength measurements on adult head, calf and forearm and the head of the newborn infant using phase resolved optical spectroscopy. *Phys. Med. Biol.* **1995**, *40*, 295. [CrossRef]
26. Zee, P.; Cope, M.; Arridge, S.; Essenpreis, M.; Potter, L.; Edwards, A.; Wyatt, J.; McCormick, D.; Roth, S.; Reynolds, E. Experimentally measured optical pathlengths for the adult head, calf and forearm and the head of the newborn infant as a function of inter optode spacing. In *Oxygen Transport to Tissue XIII*; Springer: Berlin/Heidelberg, Germany, 1992; pp. 143–153.
27. Wray, S.; Cope, M.; Delpy, D.T.; Wyatt, J.S.; Reynolds, E.O.R. Characterization of the near infrared absorption spectra of cytochrome aa3 and haemoglobin for the non-invasive monitoring of cerebral oxygenation. *Biochim. Biophys. Acta (BBA)-Bioenerg.* **1988**, *933*, 184–192. [CrossRef]

28. Matcher, S.; Elwell, C.; Cooper, C.; Cope, M.; Delpy, D. Performance comparison of several published tissue near-infrared spectroscopy algorithms. *Anal. Biochem.* **1995**, *227*, 54–68. [CrossRef] [PubMed]
29. Timm, U.; Leen, G.; Lewis, E.; McGrath, D.; Kraitl, J.; Ewald, H. Non-invasive optical real-time measurement of total hemoglobin content. *Procedia Eng.* **2010**, *5*, 488–491. [CrossRef]
30. Jeon, K.J.; Kim, S.J.; Park, K.K.; Kim, J.W.; Yoon, G. Noninvasive total hemoglobin measurement. *J. Biomed. Opt.* **2002**, *7*, 45–50. [CrossRef]
31. Aziz, M.H.; Hasan, M.K.; Mahmood, A.; Love, R.R.; Ahamed, S.I. Automated Cardiac Pulse Cycle Analysis from Photoplethysmogram (PPG) Signals Generated from Fingertip Videos Captured Using a Smartphone to Measure Blood Hemoglobin Levels. *IEEE J. Biomed. Health Inform.* **2021**, *25*, 1385–1396. [CrossRef]
32. Yi, X.; Li, G.; Lin, L. Noninvasive hemoglobin measurement using dynamic spectrum. *Rev. Sci. Instrum.* **2017**, *88*, 083109. [CrossRef] [PubMed]
33. Kuru, K. Optimization and enhancement of H&E stained microscopical images by applying bilinear interpolation method on lab color mode. *Theor. Biol. Med. Model.* **2014**, *11*, 9. [CrossRef]
34. Stamatas, G.N.; Zmudzka, B.Z.; Kollias, N.; Beer, J.Z. Non-invasive measurements of skin pigmentation in situ. *Pigment Cell Res.* **2004**, *17*, 618–626. [CrossRef] [PubMed]
35. Kumar, M.R.; Mahadevappa, M.; Goswami, D. Low cost point of care estimation of Hemoglobin levels. In Proceedings of the 2014 International Conference on Medical Imaging, m-Health and Emerging Communication Systems (MedCom), Greater Noida, India, 7–8 November 2014; pp. 216–221.
36. Biswas, S.K.; Chatterjee, S.; Bandyopadhyay, S.; Kar, S.; Som, N.K.; Saha, S.; Chakraborty, S. Smartphone-Enabled Paper-Based Hemoglobin Sensor for Extreme Point-of-Care Diagnostics. *ACS Sens.* **2021**, *6*, 1077–1085. [CrossRef]
37. Lee, J.; Song, J.; Choi, J.H.; Kim, S.; Kim, U.; Nguyen, V.T.; Lee, J.S.; Joo, C. A Portable Smartphone-linked Device for Direct, Rapid and Chemical-Free Hemoglobin Assay. *Sci. Rep.* **2020**, *10*, 8606. [CrossRef]

 *biosensors* 

Review

Recent Advances in Field Effect Transistor Biosensors: Designing Strategies and Applications for Sensitive Assay

Ruisha Hao [1], Lei Liu [1], Jiangyan Yuan [1], Lingli Wu [2,*] and Shengbin Lei [1,*]

1 Tianjin Key Laboratory of Molecular Optoelectronic Science, Department of Chemistry, School of Science, Tianjin University, Tianjin 300072, China
2 Medical College, Northwest Minzu University, Lanzhou 730000, China
* Correspondence: 283112039@xbmu.edu.cn (L.W.); shengbin.lei@tju.edu.cn (S.L.)

Abstract: In comparison with traditional clinical diagnosis methods, field–effect transistor (FET)–based biosensors have the advantages of fast response, easy miniaturization and integration for high–throughput screening, which demonstrates their great technical potential in the biomarker detection platform. This mini review mainly summarizes recent advances in FET biosensors. Firstly, the review gives an overview of the design strategies of biosensors for sensitive assay, including the structures of devices, functionalization methods and semiconductor materials used. Having established this background, the review then focuses on the following aspects: immunoassay based on a single biosensor for disease diagnosis; the efficient integration of FET biosensors into a large–area array, where multiplexing provides valuable insights for high–throughput testing options; and the integration of FET biosensors into microfluidics, which contributes to the rapid development of lab–on–chip (LOC) sensing platforms and the integration of biosensors with other types of sensors for multifunctional applications. Finally, we summarize the long–term prospects for the commercialization of FET sensing systems.

Keywords: field effect transistor; biosensors; microfluidics; multiplexing; integration

Citation: Hao, R.; Liu, L.; Yuan, J.; Wu, L.; Lei, S. Recent Advances in Field Effect Transistor Biosensors: Designing Strategies and Applications for Sensitive Assay. *Biosensors* **2023**, *13*, 426. https://doi.org/10.3390/bios13040426

Received: 18 February 2023
Revised: 19 March 2023
Accepted: 23 March 2023
Published: 27 March 2023

1. Introduction

In the case of highly contagious and hidden viruses which spread recklessly around the world at an alarming rate, detecting and controlling an epidemic as early as possible can effectively reduce the harm caused to society by public health events to a large extent [1]. Therefore, early non–invasive diagnosis and the immediate detection of biomarkers has become a research hotspot. How to realize simple, rapid, sensitive and low–cost detection of biological target analytes such as viruses and various proteins has also become a major problem in the field of biosensors.

A biosensor is a device that is sensitive to biological substances and which can convert concentration signals into readable signals of light, electricity, and magnetism. It generally consists of biologically sensitive probes performing the identification of elements (enzymes, antibodies, antigens, nucleic acids and other biologically active substances), appropriate physical and chemical transducers (oxygen electrodes, field effect transistors etc.), and an analysis system composed of a signal amplification device [2]. Common biosensors include optical biosensors, thermal biosensors, resistive biosensors, and semiconductor biosensors. Particularly, FET biosensors have shown great technical potential in biomarker detection platforms due to their simple operation, high sensitivity, fast response speed, real–time signal amplification, easy miniaturization, and integration for high–throughput screening, which has caused them to become a promising candidate for various biosensing applications [3–6]. The main principle of FET biosensors for biological detection is that the bio–sensitive probe should specifically bind with the target analyte and generate charged ions, which will further induce the change of carriers in the channel material [7]. With the change of various electrical output parameters, such as mobility (μ), threshold voltage (V_{th}), on/off ratio (I_{on}/I_{off}) and source–drain currents (I_{ds}), the signals can also be effectively transmitted into electrical signals and amplified even in complex biological systems, thereby realizing the quantitative detection of biological substances [8]. Many FET biosensors have been successfully used for the sensitive detection of proteins, glucose, DNA, and cells, illustrating the rapid development of this exciting research field [9].

For FET biosensors, the realization of high efficiency signal transduction not only depends on optimizing the geometry of devices and the functionalization methods of devices, but also heavily relies on the development of semiconductor materials. Furthermore, detection for single analytes alone is far from sufficient to reach the required accuracy for early disease detection. Consequently, biosensor multiplexing has been developed to detect one analyte in multiple parallel channels or to detect multiple analytes simultaneously to improve accuracy and repeatability, and this multiplexing has been the key to the application of advanced FET biosensors in the practical medical field. FETs are small in size and compatible with traditional semiconductor microfabrication processes, so they could be integrated into microfluidic platforms. Integrating the microfluidics and immunoassays into lab–on–chip (LOC) devices can help detect biomarkers in a shorter analyzing time, with less reagent volume and lower power consumption automatically, which can contribute to developing handheld, miniaturized, medical diagnostic testing platforms. Therefore, we will discuss recent progress regarding the aspects mentioned above in this mini review and summarize the long–term prospects for the commercialization of FET sensing systems.

2. Biosensor Designing

2.1. Device Structures

FETs are mainly composed of three electrodes (gate, source and drain), an insulating layer and a semiconductor layer [10]. The device is "energized" only when the gate voltage reaches the "threshold voltage" (V_{th}). When it is above V_{th}, carriers flow along the channel between the source and drain. Therefore, the device state of "on" or "off" is related to the relative magnitude of the gate bias voltage (V_g) applied to the FET and V_{th}. According to the relative position of the electrode and the semiconductor layer, there are four basic structures of FET, as shown in Figure 1.

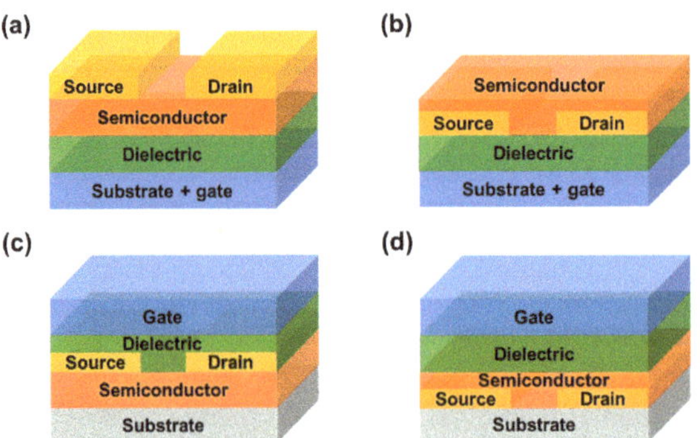

Figure 1. Schematic representation of four configurations of FETs. (**a**) Bottom–gate top contact (BGTC). (**b**) Bottom–gate bottom contact (BGBC). (**c**) Top–gate top contact (TGTC). (**d**) Top–gate bottom contact (TGBC).

In these device structures, when a metal and semiconductor are in contact, due to the difference in work function, free electrons will transfer from the metal to the semiconductor, or vice versa, forming a space charge region. Then, energy band edges in the semiconductor are shifted continuously because of an electric field generated by the charge transfer, which is called metal/semiconductor–contact–induced band bending. When an extra electric field is applied to the metal, an electric field is built between the metal and the semiconductor, and because of insufficient shielding by the charge carriers of low concentration, the electric field is penetrated into the near surface region of the semiconductor, causing field–effect–induced band bending [11]. In addition, charge transport within a device is also strongly influenced when charged molecules are adsorbed on a semiconductor surface. Specifically, when a molecule approaches the semiconductor surface, the potential energy gradient of electrons and holes in the near–surface region of the semiconductor is modified by adsorbed molecules, forming Helmholtz layers on the semiconductor surface and causing conduction and valence bands to bend. Therefore, due to the band bending effect, the efficiency of charge transfer from the semiconductor to the adsorbed molecule will be affected [12].

In recent years, in order to expand the application of FETs, researchers have replaced the traditional insulating layer materials with electrolytes, such as polymers or ionic liquids, and allowed contact with the gate electrodes to fabricate electrolyte–gated transistors (EGTs). Considering that electrochemical switching and field–effect modulation in semiconductor channels may often coexist, we will only discuss electrolyte–gate field–effect transistors (EGFETs) operating fully in field–effect mode here. Different from traditional FETs, the channel current of EGFETs is regulated by the gate electrode through the electrolyte solution, so that EGFETs show higher gate capacitance and lower operation voltage (less than 1V). In the EGFETs, depending on the position of the gate electrode relative to the semiconductor channel, there are several common geometric structures, as shown in Figure 2.

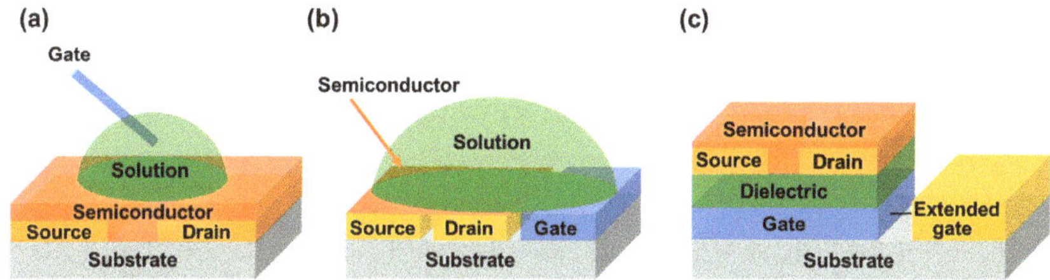

Figure 2. Schematic representation of three configurations of EGFETs. (**a**) Top–gate structure. (**b**) Side–gate architecture. (**c**) Extended–gate structure.

In the first structure (Figure 2a), the manually placed probe gate electrode is located above the semiconductor channel. For example, Horowitz's group used Au as the gate electrode and a simple water droplet as an insulating layer for the first time and fabricated a water–gate organic field–effect transistor (WGOFET) [13]. As water is the natural environment for livings, it is extremely suitable for detecting biological molecules. Following this, Kergoat et al. used WGOFETs for DNA testing. According to the formula calculation, the Debye length in PBS was 0.76 nm and a significant amount of negative charge of DNA was located outside of the Debye length, but it could be increased to 206 nm in deionized water at room temperature, which solved the problem of shielding DNA negative charge in high ion concentration solutions [14].

Because the position of a manually placed probe is arbitrary in the structure of WGOFETs, it is not easy to integrate such probe gate electrode structures into microfluidic channels. Consequently, side–gate architecture (Figure 2b) was proposed, in which the gate is on the same plane as the semiconductor channel. The main advantages of this structure were that the gate electrode position was highly controllable, the fabrication of devices was simplified greatly, and the source, drain, and gate electrodes could be simultaneously deposited by using a single pattern process. Kim et al. used liquid coplanar gate graphene FETs to detect and distinguish single strand (SS) and double strand (DS) DNA molecules [15].Compared with the traditional bottom–gate graphene field–effect transistors (GFETs), liquid coplanar–gate graphene FETs showed higher DNA detection sensitivity [16].

Considering that most of the research on biosensors is based on "bottom gate" or "solution gate" and the sensing region is placed on the semiconductor which is sensitive to factors such as water and oxygen, some researchers proposed the "extended gate" structure (Figure 2c) so as to protect semiconductors, which separated the sensing area from the transistor itself. Minamiki et al. achieved the label–free detection of phosphoproteins (α–casein) using Zn^{II}–DPA functionalized extended–gate electrodes; the detection of phosphoproteins can be applied in the fields of medicine and bioanalytical chemistry [17]. Zhang also reported an extended–gate organic FET sensing platform for exploiting the difference in weak steric interaction between cationic phenylcarbamoylated–CD and essential amino acids, which can be amplified strongly via organic field–effect transistors (OFETs), and it exhibited good chiral resolution for six essential amino acids [18]. This study provided a new direction for the molecular chirality study of natural amino acids.

2.2. Device Functionalization Methods

In addition to the rational design of the structure of devices, adopting suitable device functionalization methods was also important to achieve high sensitivity and selective detection of biological target analytes. The functionalization methods can be divided into two categories: physical functionalization methods and chemical functionalization methods [19].

2.2.1. Physical Functionalization Methods

The physical functionalization of semiconductors is to connect semiconductors and biological acceptors (which refers to any chemicals that have a recognition unit or reaction site with the target analyte) only through simple weak interaction such as van der Waals force and electrostatics interaction etc., instead of covalent bonds. One strategy is to blend them directly [20]. As shown in Figure 3a, Sun et al. chose glutaraldehyde (GA) as the dopant and achieved lower V_{th} and higher μ when adding 10% GA crosslinker to poly{2,2′–[(2,5–bis(2–octyldodecyl)–3,6–dioxo–2,3,5,6–tetrahydropyrrolo[3,4–c] pyr–role–1,4–diyl)] dithiophene–5,5′–diyl–alt–thieno[3,2–b] thiophen–2,5–diyl} (PDBT–Co–TT) solution. The main reasons for the obvious improvement in device performance were: (i) the gelation behavior of PDBT–co–TT polymer was effectively suppressed by the GA crosslinker, thus forming a better charge transport film; (ii) GA cross–linking agent acted as dopant and its strongly polar–CHO group facilitated the accumulation and transportation of charges, which contributed to improving the performance [21].

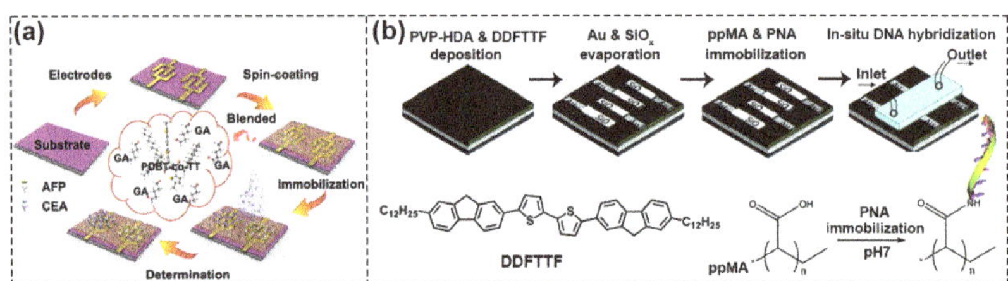

Figure 3. Schematic diagram of physical functionalization methods. (**a**) Flow chart of PDBT–co–TT/GA blend films. Reproduced with permission from [21]. Copyright 2021, American Chemical Society. (**b**) Process of deposition of a functional layer containing carboxyl groups on DDFTTF semiconductor surface by polymerization of MA monomer. Reproduced with permission from [22]. Copyright 2010, John Wiley and Sons.

Because blending often adversely affects the performance of FETs, researchers have tried to directly deposit the acceptor on the semiconductor to form a bilayer structure to physically functionalize the semiconductor, and the most commonly used method is Plasma Enhanced Chemical Vapor Deposition (PECVD). For example, Bao's group used this method to deposit maleic anhydride (MA) monomer in a plasma chamber onto the surface of 5,5′ –bis–(7–dodecyl–9H–fluoren–2–yl)–2,2′ –bithiophene (DDFTTF) semiconductor for DNA detection. MA was polymerized on the surface to form a 5 nm–thick ultrathin film containing carboxyl groups to allow for the covalent attachment of the peptide nucleic acid (PNA) strands (Figure 3b) [22,23]. As displayed in Figure 4a, Torsi's research group used ethylene and acrylic acid vapor as a precursor and used glow discharge in a plasma reactor to induce polymerization on the surface of P3HT [24]. Because the formed carboxyl functional layer was a hydrophilic layer, in order to reduce the possible influence of ions on the doping of semiconductors in the electrolyte solution, the researchers further modified the surface with immobilizing phospholipid (PL) molecules, where the deposited PL molecular layers were amphiphilic molecules with a non–polar nature, and the diffusion of ions through the membrane was minimized, ultimately limiting ion doping and maintaining good field–effect performance (Figure 4b) [25].

Mulla et al. used the spin–coating method to functionalize the PBTTT surface. A thin layer of polyacrylic acid (PAA) was spin–coated directly onto the PBTTT surface and then carboxyl functional group was generated by the UV–assisted cross–linking process to bind with biotinylated phospholipid (B–PL) containing membranes (Figure 4c) [26]. Sun et al. developed a novel material, 2,6–bis(4–formylphenyl)–anthracene (BFPA), to modify the poly{3,6–dithiophen–2–yl–2,5–di(2–octyldodecyl) pyrrolo [3,4–c] pyrrole–1,4–dione–alt–thienylenevinylene–2,5–yl} (PDVT–8) layer, as shown in Figure 4d, and achieved the ultrasensitive and reliable detection of AFP biomarkers in human serum with a sensitivity of up to femtomolar level. In this device, the BFPA layer played the dual roles of protection and functionalization [27]. In addition, depositing gold nanoparticles (AuNPs) on the semiconductor surface as a functional layer was also a common method. As an example, Figure 4e shows the block copolymer (BCP)–templated AuNP techniques used by Bao's group, in which hydrogen tetrachloroaurate ($HAuCl_4$) precursor was added to the poly(b4–vinylpryidine) (PS–b–P4VP) micelles and was then spin–coated on UV ozone–activated DDFTTF semiconductors; a large area of highly ordered AuNPs were deposited after the PS–b–P4VP was removed [28]. The AuNPs were subsequently functionalized to provide modular attachment points for DNA aptamer [29,30].

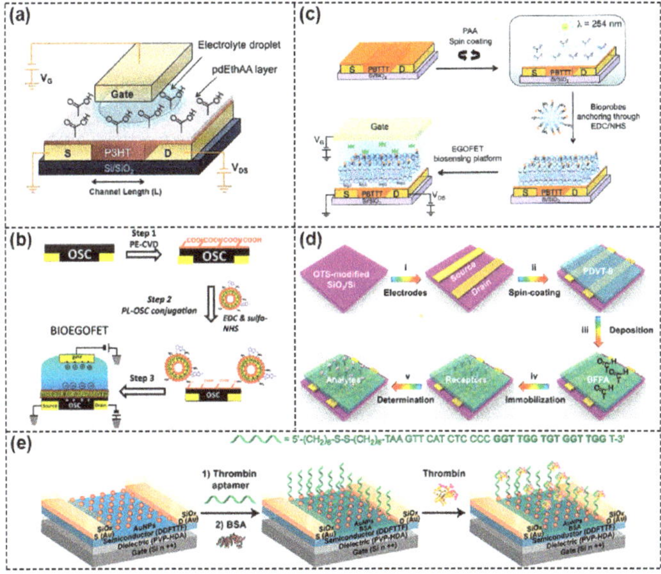

Figure 4. Schematic diagram of physical functionalization methods. (**a**) Schematic of a functional layer containing carboxyl groups deposited on the surface of P3HT semiconductor. Reproduced with

permission from [24]. Copyright 2013, WILEY–VCH Verlag GmbH & Co. KGaA, Weinheim. (**b**) Procedure of the PECVD method to introduce carboxyl functional layer onto the OSC surface and immobilization of phospholipid (PL) molecule for biological modification. Reproduced with permission from [25]. Copyright 2013, John Wiley and Sons. (**c**) Schematic diagram of introducing the carboxyl functional layer onto PBTTT surface, including the spin coating of PAA layer on PBTTT surface and subsequent UV–assisted cross–linking process. Reproduced with permission from [26]. Copyright 2015, Royal Society of Chemistry (**d**) Schematic diagram of BFPA layer prepared by spin–coating method as both protective layer and functional layer. Reproduced with permission from [27]. Copyright 2021, American Chemical Society. (**e**) Schematic of using deposited gold nanoparticles (AuNPs) on DDFTTF semiconductor surfaces as functional layers to provide binding sites for DNA aptamer. Reproduced with permission from [30]. Copyright 2013, American Chemical Society.

2.2.2. Chemical Functionalization Methods

One of the chemical functionalization methods is to introduce functional groups directly to semiconductors that have outstanding charge transport properties so that biological receptors can be immobilized onto them. Horowitz's group synthesized a new biotinylated polymer semiconductor consisting of biotin groups to detect avidin and streptavidin (Figure 5a) [31]. However, due to the introduction of functional groups, the molecular packing was changed and weakened π–π interaction among the molecules, which affected the transport path of charges and led to a sharp deterioration in device performance [32]. There were also some researchers using techniques such as ultraviolet (UV)ozone treatment and O_2 plasma treatment to generate a small number of defects on the semiconductor surface to serve as binding sites for biological receptors. For example, Zhu's group used the method of plasma–assisted–interface–grafting to introduce molecular antennas on the surface of semiconductors (Figure 5b). Minimized molecular gaps and reduced boundary interactions enhanced the interaction between the semiconductor active layer and adenosine triphosphate (ATP) in solution, reaching a low detection limit of 0.1 nM [33].

The O_2 plasma–generated oxygen–containing groups can be used to covalently tether the self–assembly membranes (SAMs), which can help to immobilize bio–sensitive probes in an efficient way [34,35]. As shown in Figure 5c, Lee et al. used O_2 plasma to treat mechanically exfoliated tungsten diselenide (WSe_2) flakes and then amino groups were introduced by using triaminopropyltriethoxysilane (APTES) as a silane coupling agent to immobilize bioreceptors.Compared with WSe_2 without O_2 plasma treatment, more surface defects were generated on the treated surface to serve as an additional binding site to hold APTES molecules. As a result of the additional binding sites of the biological receptor, sensitivity was further enhanced [36].

The other method involved using an Au gate as a sensor area, so that the SAMs layer was formed on the gold surface through the Au–S chemical bond. The bio–sensitive probes were fixed on the Au gate through the SAM layer for biological testing. Mulla et al. treated the gate region with 3–mercaptopropionic acid (3–MPA) solution to form a SAM layer, which realized the sensitive and quantitative detection of neutral enantiomers (Figure 5d) [37]. Biscarini's research team used cysteine to functionalize the Au gate and then Cys–protein G was adsorbed through chemical bonding onto the Au surface (Figure 5e). Because G protein could combine with the F_C region of the antibody specifically, the biosensor had a theoretical detection limit as low as 100 fM for anti–drug antibodies (ADA) detection [38]. Macchia et al. also utilized mixed alkyl mercaptan with carboxyl groups to link onto a gold surface to form Chem–SAM and then anti–human–Immunoglobulin–G (anti–IgG) was covalently connected with carboxyl groups to form the Bio–SAM on the gate at the same time (Figure 5f). In this way, single molecule detection of IgG was realized with a millimeter–sized transistor. The suggested sensing mechanism involved a work function change, which was assumed to propagate through the network of hydrogen bonds in the gating field [39,40].

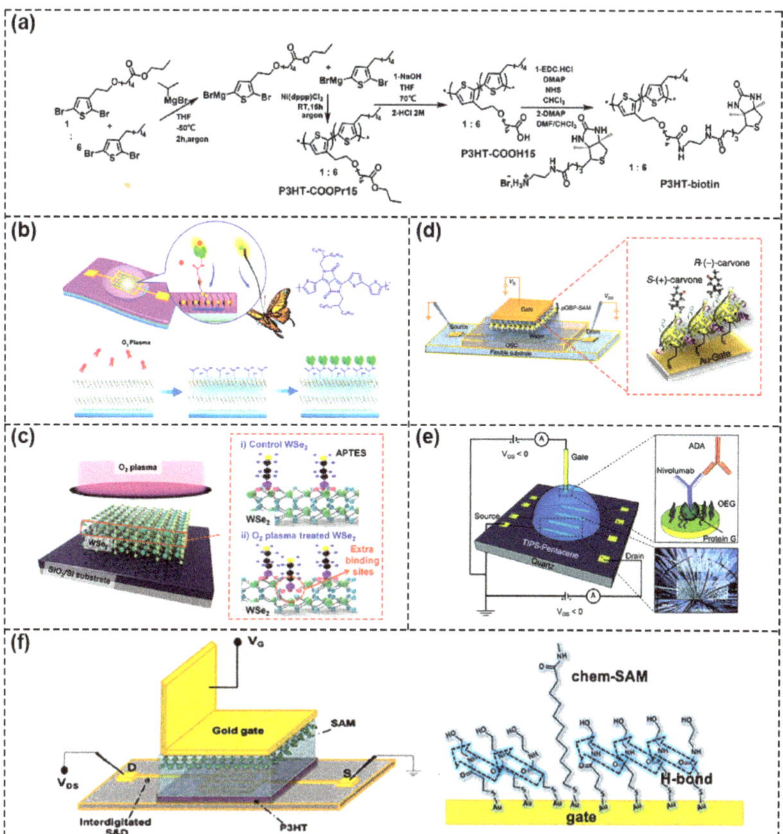

Figure 5. Schematic diagram of chemical functionalization methods. (**a**) The synthesis of a biotinylated polymer semiconductor consisting of carboxyl groups and biotin groups. Reproduced with permission from [31]. Copyright 2013, The Royal Society of Chemistry. (**b**) Schematic of the method of plasma–assisted–interface–grafting to introduce molecular antennas on the surface of semiconductor. Reproduced with permission from [33]. Copyright 2018, The Royal Society of Chemistry. (**c**) Schematic diagram of O_2 plasma treated WSe_2 flakes. Reproduced with permission from [36]. Copyright 2018 American Chemical Society. (**d**) Schematic representation of the SAMS layer on the gold surface by Au–S chemical bonds. Reproduced with permission from [37]. Copyright 2015, Macmillan Publishers Limited. (**e**) Schematic of the thiol groups of cysteine combined with Au gate to bind with Cys–protein G for detecting the F_C fragment of ADA. Reproduced with permission from [38]. Copyright 2021, The Royal Society of Chemistry. (**f**) Schematic diagram of the device structure using both Chem–SAM and Bio–SAM to modify the gate (**left**). Schematic of hydrogen bond network originated from Chem–SAM (**right**). Reproduced with permission from [40]. Copyright 2020, American Chemical Society.

In the case of functional steps and sensing detection, the long–term stability and high reproducibility of devices are very important for obtaining accurate and reliable detection results. For example, when FET sensors are immersed in a physiological environment, the surface of the silica insulation layer may be hydrolyzed by cationic electrolytes and thus destroyed, further reducing the reproducibility of the sensor response. Therefore, surface passivation is very important to achieve high stability and reproducibility in detecting target molecules [41,42].

The interfaces of OFETs, including OSC/electrode interface, OSC/insulation interface and OSC/air interface, largely determine the performance of devices. Due to defects such as traps and grain boundaries at these interfaces, charge would be trapped, which affects the charge transport, inevitably leading to deviation from the desired behavior of devices. In addition, the loose arrangement of organic molecules also makes it easier for water and oxygen in the air to be absorbed at the OSC/air interface. Charge injection and transfer will also be affected by these active impurities adsorbed at the interface, thus affecting the final performance of the devices [43]. In view of these interface problems, different solutions have been explored to improve device performance. For example, the formation of an organic monolayer on the surface of silicon–based sensors through a Si–C bond to achieve surface passivation and chemical functionalization has been discussed in a recent review by Justin Gooding [44]. Li et al. prepared high–stability devices through interface engineering and strain balance strategy [45]. Osaka et al. developed a simple surface coating technique and successfully achieved the long–term stability of FET biosensors in water environments by coating reduced GO to the surface of a silicon dioxide insulation layer, which effectively prevented cations in the electrolyte from invading the gate insulator of FETs [46].

In addition, another aspect is seldom accounted for by researchers working on FET sensing: the sensing surface will have a point of zero charge (PZC), where no excess charge is present at the electrode surface. A recent work by Darwish [47] has perfectly demonstrated that the kinetics of surface reactions depend on the surface PZC, and the adsorption and recognition of molecules on the surface can be controlled by applying potential, which will have a significant impact on the design and operation of the FET sensing interface. Furthermore, this may become a new issue for researchers to consider when functionalizing FET sensors in the future.

2.3. *Semiconductor Materials forActive Layers*

2.3.1. Two–Dimensional Materials (2D)

Since graphene was first introduced in 2004 [48], researchers have developed a wide variety of 2D materials. The thickness of 2D semiconductor materials is usually less than 5 nm and the carrier flow on the surface of the material is limited; this is conducive to achieving efficient signal acquisition and conversion because the 2D materials are directly exposed to the external environment. Because of these advantages, 2D materials have flourished in the field of FETs. Additionally, the large surface–volume ratio of the materials provides abundant modification sites for specific receptors, which is very important for FET–based biosensors [49].

- Two–dimensional layered materials;

Biosensors based on GFET have attracted much attention due to their high electron mobility, π–π stacking interactions with biomolecules and good stability. For example, Gao et al. fixed a DNA probe on the surface of the non–functionalized graphene only by using π–π interactions to achieve rapid and label–free miRNA detection within 20 min with detection limits of as low as 10 fM (Figure 6a) [50]. In order to enhance the interaction between graphene and biomolecules, some researchers have used 1–Pyrenebutanoic acid succinimidyl ester (PBASE) as a linker to treat graphene surface (Figure 6b) [51]. The pyrene group on one side of PBASE was bound to graphene through π–π interaction and the succinimide group on the other side was covalently bound to the DNA molecule. The edges and defect sites of graphene have high activity and the surface of oxidized graphene contains a large number of active epoxy groups and carboxyl groups [52], both can be used for functionalization. Therefore, Roberts et al. used 1–Ethyl–3–(3–dimethylaminopropyl) carbodiimide/N–hydroxysuccinimidesulfonate sodium salt (EDC/NHS) solution to functionalize graphene with carboxyl groups and monitored the resistance changes caused by antigen–antibody interaction in real time for the detection of Japanese encephalitis virus and avian influenza disease [53].

However, the lack of band gap in graphene results in a high leakage current of GFET biosensors, which reduces the sensors' dynamic range. The transition metal dichalcogenide

(TMD) material with X–M–X structure is composed of two atomic layers (X) and a transition metal layer (M) in between the two atomic layers (X) [54,55]. TMDs such as molybdenum disulfide (MoS_2) and WSe_2 exhibit a moderate band gap, which significantly reduces the leakage current in the FETs and improves detection sensitivity. Park et al. fabricated MoS_2 FET biosensors and made rigorous theoretical simulations, and the detection limit of prostate specific antigen (PSA) was as low as 100 fg mL^{-1} with standard errors below 9% [56]. WSe_2 FET biosensors were expected to show a good detection ability due to their high carrier mobility because high carrier mobility would affect several other performance indicators, such as current density and switching delay, in turn [57]. Hossain et al. developed a highly sensitive WSe_2 FET biosensor for PSA detection with a very low detection limit of 10 fg ml^{-1} [58]. Due to the absorption of H_2O and CO, the stability and detection capability of the original device would decrease and would probably lead to wrong signals. Zhang et al. used DNA tetrahedra and biotin–streptavidin (B–SA) to functionalize an MoS_2 FETs device, which provided a more stable anchoring system for antibody–antigen (Ab–Ag) binding, so it had an ultra–high sensitivity for PSA with a detection limit of 1 fg mL^{-1} (Figure 6c) [59].

- Two–dimensional organic materials

Two–dimensional organic materials such as 2D covalent organic framework (2D COFs) and metal organic frameworks (MOF) have the advantages of periodic planar network topology, good stability, good biocompatibility, ease of functionalization and they bear abundant modification sites, which enable them to anchor a large number of specific receptors favorable to be used in biosensors [60–64]. For instance, Wang et al. prepared Ni –Metal–Organic Framework (MOF)–based FETs using in situ grown $Ni_3(HITP)_2$ membrane as a channel material (Figure 6d). Tightly stacked MOF films with controllable thickness were prepared by adjusting the reaction time. Due to the tightly stacked sheet structure and bare surface, the material was conducive to carrier transmission and post modification. Following this, Ni–MOF was developed as a liquid–gated device with bipolar performance and excellent response to gluconic acid in the range of 10^{-6} to 10^{-3} g mL^{-1}, validating the potential of MOF–based FETs as biosensors [65].

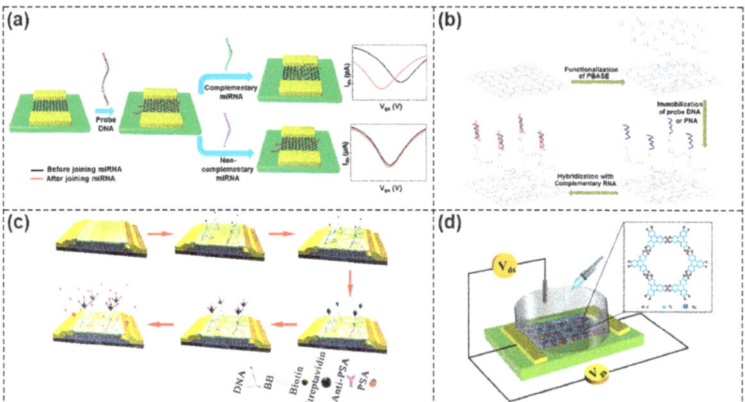

Figure 6. Representative FETs based on 2D materials. (**a**) Schematic of miRNA detection by GFET biosensors. Reproduced with permission from [50]. Copyright 2020, American Chemical Society. (**b**) Schematic of functionalization of PBASE as a linker on graphene surface. Reproduced with permission from [51]. Copyright 2020 Elsevier B.V. (**c**) Schematic of MoS_2 FET with DNA functionalization. Reproduced with permission from [59]. Copyright 2021, Elsevier B.V. (**d**) Schematic of Ni–MOF–FET as biosensors for gluconic acid detection. Reproduced with permission from [65]. Copyright 2019, American Chemical Society.

2.3.2. Polymer and Small Organic Molecule Materials

In comparison with inorganic semiconductor materials, organic semiconductor (OSC) materials have the following three advantages: (1) desired properties and functions can be obtained by simple chemical modification; (2) OSCs can be dissolved in common solvents to prepare devices by solution process methods such as spin coating and drop casting instead of the traditional vacuum deposition method, and it greatly simplifies the process of device preparation and decreases the cost; (3) there are many kinds of OSCs with good flexibility for integrating circuits and flexible displays. According to the molecular weight of the OSC materials, they can be divided into small molecule materials and polymer materials. The chemical structures of some typical OSC materials are shown in Figure 7.

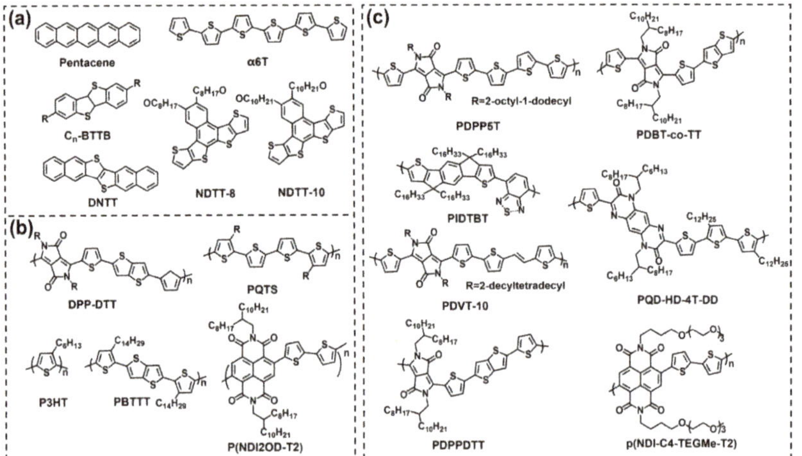

Figure 7. Chemical structures of organic small molecule semiconductors (**a**) and organic polymer semiconductors (**b**,**c**).

Typical small molecule materials include pentacene, α–sexithiophene (α6T), 2,7–dialkyl[1]benzothieno[3,2–b][1] benzothiophene (C_nBTBT), dinaphtho [2,3–b:20,30–f]thieno [3,2–b]thiophene (DNTT), and so on. For example, Song et al. prepared an extended gate OFET with pentacene as a semiconductor layer for the detection of glial fibrinous acidic protein [66]. Li et al. synthesized naphthodithieno [3, 2–b] –thiophene derivatives NDTT–8 and NDTT–10. They showed excellent water stability compared to pentacene and poly{3,6–dithiophen–2–yl–2,5–di(2–decyltetradecyl)–pyrrolo[3,4–c] pyr–role–1,4–dione–alt–thienylenevinylene–2,5–yl} (PDVT–10) polymers. After 90 days in water, the μ of the carrier remained above 50% (Figure 8a) [67]. However, because the performance of most small molecule OFETs degraded rapidly once they were exposed to moisture, they were not suitable for the detection of biomolecules in liquid environments [68]. In order to further improve the stability of devices, polymer semiconductor materials were applied. Typical polymer materials include poly[2,5–(2–octyldodecyl)–3,6–diketopyrrolopyrrole–alt–5,5–(2,5–di(thien–2–yl)thieno[3,2–b]thiophene)](DPP–DTT), poly(3–hexylthiophene) (P3HT), poly[2,5–bis(3–tetradecylthiophen–2–yl)thieno[3,2–b]thiophene](PBTTT), poly[[1,2,3,6,7,8–hexahydro–2,7–bis(2–octyldodecyl)–1,3,6,8–dioxobenzo[lmn][3,8]phenanthroline–4,9–diyl] [2,2′–bithiophene]–5,5′–diyl] [P(NDI2OD–T2)], diketopyrrolopyrrole-based π–conjugatedc-opolymer (PDPP5T) and so on. As shown in Figure 8b, Leong's group fabricated high–performance WGOFETs using PQD–HD–4T–DD polymer and the average μ was 9.76×10^{-3} $cm^2\ V^{-1}\ s^{-1}$, I_{on}/I_{off} was 4.41×10^4 [69]. Doumbia et al. synthesized two D–A polymers, (poly[2,5–(2–Octyldodecyl)–3, 6–Diketopyrrolopyrrole–alt–5,5–(2,5–di(thien–2–yl) thieno) [3,2–b] thiophene)] (PDPPDTT) and indacenodithiophene–co–benzothiadiazole (PIDTBT) for WGOFETs. The I_{on}/I_{off} were 3×10^3 (PDPPDTT) and 2×10^4 (PIDTBT), respectively.

The μ of PDPPDTT was 0.18 $cm^2\ V^{-1}\ s^{-1}$ and PIDTBT was 0.16 $cm^2\ V^{-1}\ s^{-1}$ (Figure 8c) [70]. Sun et al. synthesized π–conjugated polymer material PDBT–co–TT for WGOFETs with an average mobility of 0.22 $cm^2\ V^{-1}\ s^{-1}$ and a switching ratio of 5.13 × 10^3 (Figure 8d), which exceeded most of those reported WGOFETs to date [71]. Compared to P–type polymers, N–type polymers were affected heavily by air/water and had low performance, so they were not widely used in biosensors. Caironi et al. presented the first example of an N–type electrolyte–gated organic transistor based on an inkjet printing polymer, p(NDI–C4–TEGMe–T2) (Figure 8e).The device showed excellent working stability of more than 18 h and a switching ratio of more than 10^4 [72]. In terms of the material, they should have a suitable energy level and a good match with the work function of the source and drain to facilitate the effective injection and output of charge carriers, resulting in different detection performance (Table 1).

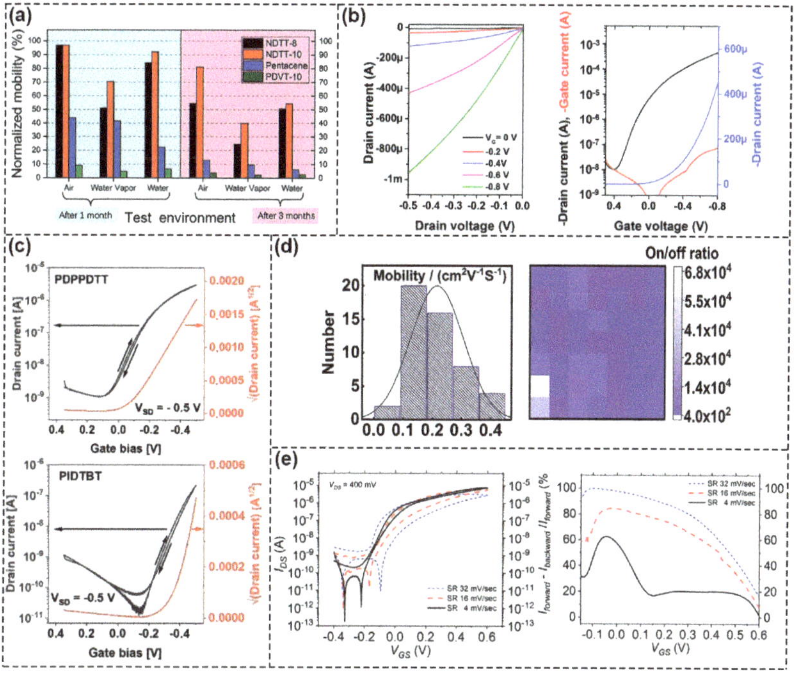

Figure 8. Performance of semiconductors used in FET Biosensors. (**a**) Stability testof NDTT–8 and NDTT–10 in water environment. Reproduced with permission from [67]. Copyright 2019, The Royal Society of Chemistry. (**b**) Characteristic I–V curves of PQD–HD–4T–DD polymer in water environments. Reproduced with permission from [69]. Copyright 2020, The Royal Society of Chemistry. (**c**) Representation of characteristic curves of PDPPDTT and PIDTBT transistors. Reproduced with permission from [70]. Copyright 2021, Wiley–VCH GmbH. (**d**) Saturation mobility and on/off ratio of PDBT–co–TT polymer transistors. Reproduced with permission from [71]. Copyright 2020, Elsevier B.V. I. (**e**) Characteristic I–V curves of N–type polymers, p(NDI–C4–TEGMe–T2). Reproduced with permission from [72]. Copyright 2022, Wiley–VCH GmbH.

Table 1. Semiconductor materials used for FET biosensors.

Characteristic	Semiconductor	Mobility ($cm^2V^{-1}s^{-1}$)	$I_{on/off}$	Analyte	Detection Limit	Times (min)	Sensitivity	Ref
① High surface–volume ratio, high theoretical carrier velocity (~10^6 m/s) and mobility; ② Zero band gap, large leakage current, reducing the dynamic range of the sensor, sensitive to external conditions, such as electric field and foreign doping impurities.	Graphene			SARS–CoV–2 antibody	10^{-18} M	<2	4%	1
				DNA	1 nM	minutes	N/A	15
				miRNA	10^{-15} M	20	5.99 mV/decade	50
				RNA	0.1 aM	minutes	14.8	51
		3.79 (hole) 3.78 (electron)	3100	Hg^{2+}	16 pM/L	minutes	N/A	52
				JEV/AIV	1 fM/10 fM	minutes	N/A	53
				SARS–CoV–2 antigen	2.42 × 10^2 copies/mL	>1	N/A	76
				SARS–CoV–2 Nucleic acid	0.03 copy/μL	~1	N/A	77
				SARS–CoV–2 protein	~8 fg/mL	minutes	12.8mV/decade	79
		>10,000 * (Room temperature)	N/A	K^+ Na^+ Ca^+	~100 μM	N/A	−54.7 ± 2.90 −56.8 ± 5.87 −30.1 ± 1.90 mV/decade	83
				DNA	1 nM	N/A	30.1mV/decade	86
				Nucleic acid	1.7 fM	~2.5	N/A	90
				IFN–γ	880 fM	minutes	N/A	97
① Adjustable intrinsic band gap, high carrier mobility, large switching ratio, low leakage current; ②Sensitive toexternal conditions;	WSe_2	133	~10^5	PSA	10 fg/mL	minutes	2.6	58
	WSe_2	N/A	N/A	Glucose	10 mM	N/A	2.87 × 10^5 A/A	36
	MoS_2	N/A	~10^6	PSA	100 fg/mL	minutes	N/A	56
	MoS_2	N/A	N/A	PSA	1 fg/mL	~4	0.05%	59
	MoS_2	19.4	~10^2	NMP22/CK8	0.027/0.019 aM	N/A	N/A	81
	MoS_2	83.5	~10^6	PSA	1 pg/mL	2~3	N/A	92
	MoS_2	9.18	~10^7	Glucose	N/A	N/A	N/A	93

Table 1. *Cont.*

Characteristic	Semiconductor	Mobility ($cm^2V^{-1}s^{-1}$)	$I_{on/off}$	Analyte	Detection Limit	Times (min)	Sensitivity	Ref
① Easy modification, adjustable energy band, high flexibility, easy solution processing, good hydrophobicity; ② Polymerization generally takes place at high temperature and consumes energy, low carrier mobility.	P3HT–COOH	0.5 ± 0.12	~10^3	DNA	N/A	minutes	N/A	14
	DDFTTF	~0.35	2 × 0^3	DNA	N/A	minutes	N/A	22
	P3HT	0.006	N/A	D–Phe	10^{-13} mol/L	N/A	N/A	18
	P3HT	10^{-3}	204 ± 91	SA	10 nM	45s	N/A	25
	PBTTT	N/A	N/A	α–casein	0.22 ppm	N/A	N/A	17
	PBTTT	N/A	N/A	BSA	6 × 10^{-13} M	<15	N/A	35
	PBTTT	~0.02	10^2–10^3	SA	10^{-11} M	minutes	N/A	26
	PBTTT–C14	(1.1 ± 0.2) × 10^{-1}	N/A	pOBP protein	50 pM	N/A	N/A	37
	PDVT–8	0.18	~10^5	AFP	4.5 fM	40	2.7%	27
	DDFTTF	0.25	2 × 10^3	Hg^{2+}	100 µM	N/A	N/A	29
	P3HT–biotin	~10^{-4}	~80	Streptavidin	N/A	minutes	2%	31
	PDPP3T	0.3~0.6	~10^3	ATP	0.1 nM	minutes	N/A	33
	PDBT–co–TT	0.22	5.13 × 10^3	AFP	0.15 ng/mL	45	N/A	71
	PDBT–co–TT	2.07	~10^6	AFP/ CEA	0.176 pM/ 65 fM	minutes	N/A	21
	PDBT–co–TT	~0.1	~10^3	AFP/CEA/ PSA	4.75 aM	N/A	N/A	82
① Clear structure, easy to purify, ② Poor film formation, not conducive to large area preparation. Sensitive to external conditions;	α 6T	4 × 10^{-2}	10^2–10^3	Penicillin	5 µM	minutes	50 µV/µM	34
	Pentacene	0.116	~10^6	BSA	N/A	N/A	N/A	23
	Pentacene	0.69 ± 0.07	26.0 ± 5.7	GFPA	1.0 ng/mL	minutes	N/A	66
	Pentacene	N/A	N/A	TNF α	3 pM	N/A	N/A	91
	TIPS–pentacene	N/A	N/A	ADAs	10^{-13} M	minutes	10^{11} M^{-1}	38

Notes: ① represents advantages and ② represents disadvantages of different materials. The asterisk (*) represents the theoretical value.

3. Application

3.1. Immunoassay Based on Single Biosensor

At present, serology and viral nucleic acid testing are two main diagnostic methods for COVID–19 [73–75], but they cannot meet the requirements of diagnostic accuracy and detection speed at the same time. It is becoming more and more important to develop biosensing devices with high sensitivity, fast detection speed and less volume, which is where researchers have concentrated a lot of effort.

Seo et al. reported on a FET biosensor for detecting SARS–CoV–2 virus in clinical samples, in which the SARS–CoV–2 spike antibody was coupled with a graphene sheet and used as sensing area (Figure 9a). It was able to detect SARS–CoV–2 spike protein in the clinical transport medium of 100 fg/mL [76]. Wei's group also developed a GFET biosensor modified with spike S1 protein (Figure 9b). Through the specific binding of SARS–CoV–2 antibody and S1, the conductance in graphene channels changed, and the ultra–low detection limit of SARS–CoV–2 antibody reached 2.6 aM [1]. The research group also tried to use DNA probes as recognition elements; however, conventional flexible SS DNA probes would aggregate and entangle at the sensing interface of conductive channels, leading to the inactivation of SS DNA probes, thus researchers used GFET and Y–shaped DNA dual probes (Y–dual probes) to detect SARS–CoV–2 nucleic acid. Due to the synergistic effect of probe sites targeting the ORF1ab and N gene regions, the biosensor had a high recognition rate for SARS–CoV–2 nucleic acid and reached a detection limit of three copies in 100 μL solution [77]. At present, most research on biological target analytes is focused on proteins including antigens, enzymes, etc., which are generally detected directly without an amplification process, leading to less accuracy than polymerase chain reaction (PCR). As shown in Figure 9c, Wei et al. demonstrated a multi–antibody FET sensor and successfully detected SARS–CoV–2 in artificial saliva with a detection limit of 3.5×10^{-17} g/mL and a detection limit of 0.173copies μL^{-1} in nasopharyngeal swabs [78]. In Figure 9d, Gao et al. fabricated biosensors using a van der Waals heterostructure of graphene and graphene oxide (GO) [79]. Compared with the GFET biosensor, the sensitivity for SARS–CoV–2 protein detection of the biosensors with GO/Gr heterostructure was increased threefold. This was mainly due to the fact that GO formed a uniform protective layer, which could prevent external ions from directly contacting the surface of graphene. At the same time, due to the formation of heterojunctions, the efficiency of electron exchange was improved through interface coupling and the charge mobility of the device was further improved. The advantage of 2D–layered materials is that they can be further integrated with other materials to form a special heterojunction at the atomic scale, which opens up new opportunities for constructing new biosensor components.

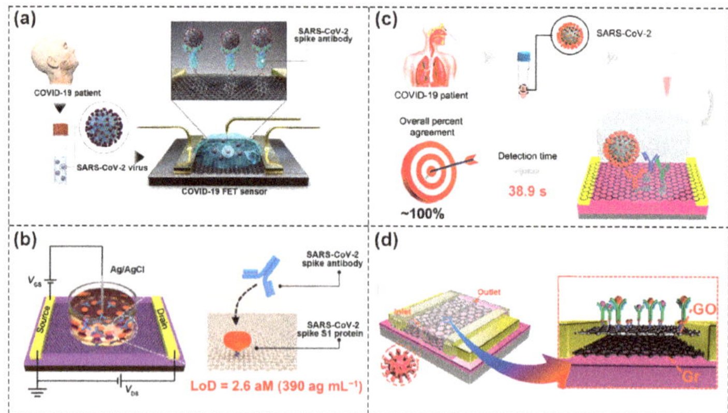

Figure 9. COVID–19 detection based on different FET biosensors. (**a**) Schematic of the SARS–CoV–2 spike antibody coupled to graphene sheet for detecting SARS–CoV–2 virus. Reproduced with permission

from [76]. Copyright 2020, American Chemical Society. (**b**) Schematic of a GFET biosensor modified with spike S1 protein for detecting SARS–CoV–2 spike antibody. Reproduced with permission from [1]. Copyright 2021, American Chemical Society. (**c**) Schematic of the multi–antibodies FET sensors for detecting SARS–CoV–2. Reproduced with permission from [78]. Copyright 2021, American Chemical Society. (**d**) Schematic of GO/Gr heterostructure biosensors for SARS–CoV–2 detection. Reproduced with permission from [79]. Copyright 2021, Elsevier B.V.

3.2. Integrated into Array for Multiplexing

The variability of devices due to uneven features during the process of material synthesis and device fabrication techniques is a critical concern in detecting single analytes, which may lead to certain errors. Li et al. constructed 120 silicon nanowires (SiNW) channels as the sensing area for sensitive detection of PIK3CA E542K ctDNA in parallel and the prepared SiNW FET sensors had good specificity and repeatability with an ultra–low detection limit of 10 aM [80]. The composition of real clinical samples is very complex and detecting a single analyte is far from meeting the need for early diagnosis of specific diseases. Therefore, the development of an efficient approach to simultaneously detect multiple markers and realize high–throughput screening is extremely necessary. With the rapid development of device miniaturization and integration, FET sensor arrays with multi–channel sensing units can be constructed to detect a variety of biomarkers so as to improve detection sensitivity and accuracy and to promote clinical application. As shown in Figure 10a, Yang et al. fabricated a FET biosensor composed of four sensing windows based on MoS_2 nanosheets, in which each module can be used to detect a single biomarker without interfering with the other. At the same time, each sensing window contained multiple parallel sensing units so as to achieve multi–channel detection. Bladder cancer biomarkers, nuclear matrix protein 22 (NMP22) and cytokeratin 8 (CK8), were detected simultaneously with detection limits of 0.027 and 0.019 aM, respectively, suggesting that properly designed multi–channel sensor arrays can be routinely used for detection with high sensitivity and accuracy [81]. Sun et al. integrated the prepared DMP [5]–COOH molecules as signal amplifiers with OFET devices and the sensing array was divided into different detection areas, which realized synchronous and immediate detection of three tumor markers with ultra–high sensitivity at aM level (Figure 10b) [82].Furthermore, as shown in Figure 10c, a graphene–based sensor array platform that consisted of more than 200 (16×16) integrated sensing units was constructed by Xue et al. The sensor chip was designed as three separate regions to enable the detection of potassium, sodium and calcium ions in complex solutions, such as artificial urine and artificial eccrine perspiration. The way to functionalize the graphene surface was by depositing three different ion–selective membranes (ISMs) using a 3D printing machine. Then, they further utilized the stochastic Forest algorithm model to demonstrate ion type classification, concentration prediction and disease diagnosis, thereby enhancing the reliability of the data. This also demonstrated the importance and effectiveness of combining experimental testing with machine model learning [83].In addition, the FET sensors could also be used in biomimetic human sensory systems. Kwon et al. reported on an artificial multiplex super bioelectronic nose (MSB–nose) using highly homogeneous graphene micropatterns (GMs) with two different human olfactory receptors attached to GMs as bio–probes [84]. It mimicked the human olfactory sensory system and had high performance in odor discrimination from mixtures. In addition, Ahn et al. developed GFET–based dual biological electronic tongues (DBTs) for the simultaneous detection of umami and sweet tastes, thus opening up new ways of mimicking human complex biomimetic systems and demonstrating the great potential of FET–based biosensors [85].

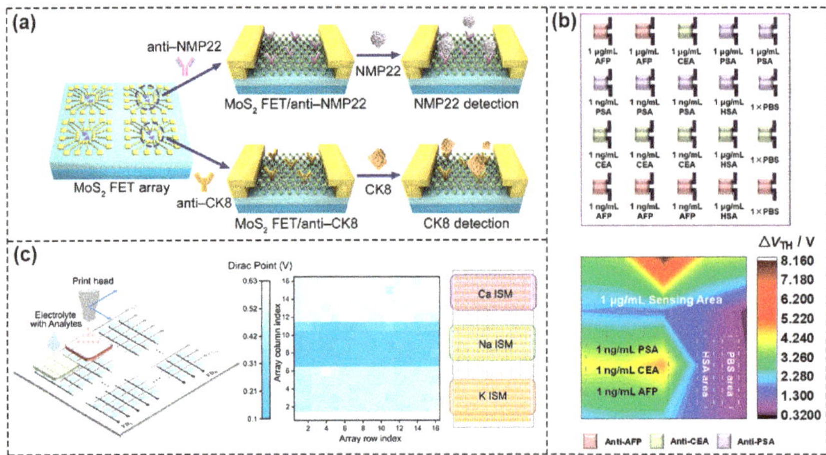

Figure 10. Integration of FET biosensors into array for multiplexing. (**a**) Schematic of FET sensor arrays based on MoS_2 nanosheets for simultaneous detection of multiple bladder cancer biomarkers. Reproduced with permission from [81]. Copyright 2020, Science China Press and Springer–Verlag GmbH Germany, part of Springer Nature.(**b**) Simultaneous determination of three biomarkers using a FET sensor array. Reproduced with permission from [82]. Copyright 2022, American Chemical Society. (**c**) Diagram of 16 × 16 sensor unit (**left**). Color map of Dirac points for three kinds of ion–sensing unit (**right**). Reproduced with permission from [83]. Copyright 2022, the author(s).

3.3. Integrated with Microfluidicsfor LAB–on–CHIP

Lab-on-chip (LOC) is a kind of device that integrates laboratory functions on a chip whose size is from a square millimeter to a few square centimeters. LOC has facilitated the development of handheld, miniaturized medical diagnostic test platforms. Integrating FET biosensors with microfluidic devices is an attractive direction in LOC [86].

Dai et al. realized the simultaneous detection of penicillin G and urea by designing urease–encoded and penicillinase–encoded polyethylene glycol hydrogels. The hydrogels were used as the biometric identification module to directly contact the graphene channel, in which they can be freely assembled and disassembled, which made the programmable sensing function of FET sensor chip systems possible [87]. Kim et al. combined the antibiotics conjugated graphene micropattern FET (ABX–GMFETs) with a microfluidic chip to detect dual bacterial Gram–positive bacteria (GPB) and Gram–negative bacteria (GNB) [88]. As shown in Figure 11a, Zhou et al. prepared an extended–gate FET biosensor chip modified with a supported lipid bilayer (SLB) and angiotensin–converting enzyme II (ACE2) receptor, where SARS–CoV–2 binding with ACE2 receptors infected host cells and SLB was used to provide the cell–simulated environment. The aim was to study the interaction between SARS–CoV–2 and cell membrane so as to facilitate the screening of effective anti–coronavirus drugs. The detection results showed that the presence of two different drugs had an effect on the interaction between coronavirus and the ACE2 receptor, with weak inhibition by hexapeptide and strong inhibition by HD5 peptide. The integrated system could translate the interaction between biological target analytes and receptors into real–time charge signal, so as to realize effective screening of therapeutic drugs [89]. Hajian et al. prepared CRISPR–Chip by modifying graphene surface with CRISPR–Cas9 complex. The chip could conveniently, rapidly, and selectively detect target sequences of CRISPR–Cas9′s gene and had the potential to extend the boundaries of digital genomics [90].

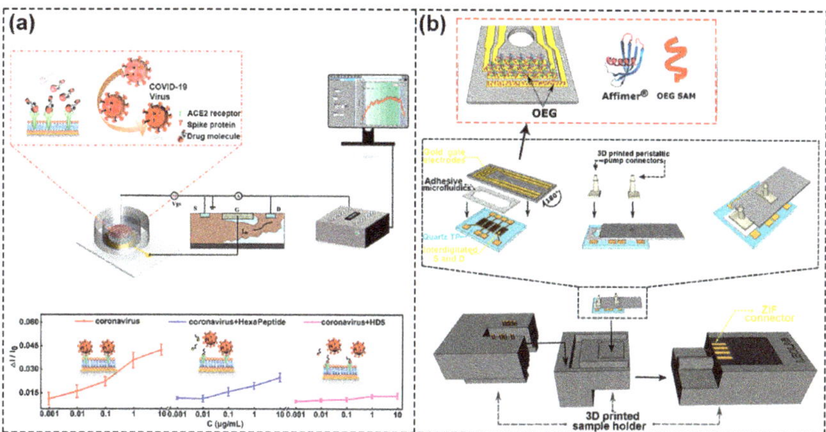

Figure 11. Schematic of integrated FET biosensors with microfluidic for lab–on–chip. (**a**) Schematic of biosensor chip modified by SLB (**top**) and the inhibitory response of two different drugs to the interaction between coronavirus and ACE2 receptor (**bottom**). Reproduced with permission from [89]. Copyright 2022, American Chemical Society. (**b**) Schematic of a lab–on–chip multi–gates organic transistor based on 3Dprinting and modified multi–gates in the red dotted box. Reproduced with permission from [91]. Copyright 2020, American Chemical Society.

In addition to the rapid detection of biomolecules, LOC can take advantage of a smaller sample volume and can conduct several sample tests simultaneously to assess the occurrence of non–specific interactions and minimize the chance of false positives. As shown in Figure 11b, Parkula et al. integrated multi–gates EGOFETs and a single reservoir microfluidic system in a 3D–printed sample box and detected binding events occurring at the gate–electrolyte interface in a 6.5 μL microfluidic channel with pM accuracy. To be specific, the proinflammatory cytokine tumor alpha (TNFα) samples were detected by three gates simultaneously, and the fourth electrode was used as a reference electrode to assess whether the detection response had to be attributed to the sensing event itself, which reduced the influence of non–specific adsorption [91]. It was a major step forward in the robustness and cost–effectiveness of detection, as it was able to increase the statistics of biomarker detection in the smallest sample volume and meet the trend of personalized medicine, which are guaranteed in biosensor applications at point of care (PoC).

3.4. Integrated with other Sensors for Multifunctional Applications

The integration of different sensors on the same chip allows multiple functions to be performed in a small volume. High integration means more functionalities in a smaller size with a lighter weight, which can meet the requirements of the next generation smart system. As shown in Figure 12a, Yoo et al. reported a flexible biochip within which a MoS_2 FET biosensor, readout circuit, and light–emitting diode (LED) were integrated. When 1 $\mu g \cdot mL^{-1}$ PSA was fixed on the MoS_2 surface, the corresponding off current increased and the output voltage amplified, which led to the lighting up of the LED indicator. Following this, when 100 $pg \cdot mL^{-1}$ PSA was bound to the immobilized antibody, the off current decreased, the output voltage dropped to 1.87 V, and the LED turned off, realizing the real–time and POC diagnosis of prostate cancer markers [92]. Stretchable and bendable devices integrated with multifunctional biosensors or devices implanted in the human body are able to sense physiological signals and environmental conditions in real time without affecting normal body movement. Guo et al. demonstrated a multifunctional smart contact lens sensor system based on ultrathin MoS_2 transistors including a photodetector to receive optical information, a glucose sensor to directly monitor glucose levels in tears, and a temperature sensor to diagnose underlying corneal diseases (Figure 12b) [93].

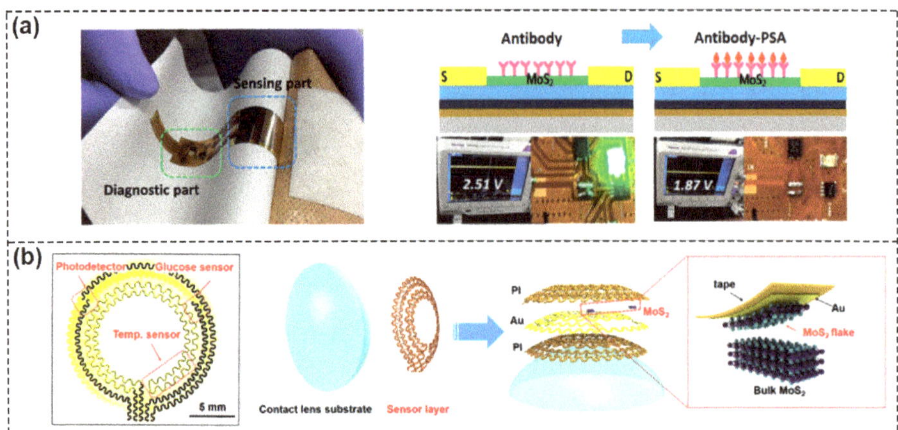

Figure 12. Schematic of multifunctional sensing systems. (**a**) The devices with system–level integration of flexible MoS_2 FET biosensors, read–out circuits and LEDs. Photograph of an epidermal skin–type MoS_2 biosensor system (**left**). Optical images of the LED indicator biochip for PSA detection (**right**). Reproduced with permission from [92]. Copyright 2017, Tsinghua University Press and Springer–Verlag Berlin Heidelberg. (**b**) Optical image of the serpentine mesh sensor system, including a photodetector, a temperature sensor and a glucose sensor, and schematic illustration of the different layers of smart contact lens structure attached to an eyeball. Reproduced with permission from [93]. Copyright 2021, Elsevier Inc.

4. Summary and Prospect

FET biosensors have made exciting progress in terms of device structure, material synthesis, device manufacturing, microfluidic industry–compatible technologies and multifunctional integrated applications. FET devices can detect a large variety of biomolecules/entities, from proteins to viruses, to bacteria, and cells in the body even at very low concentrations, thus opening up possible applications for almost any pathology and showing fresh vitality in wearable electronic devices and other fields [94,95].

Despite the fact that FET–based biosensors have the advantages of high sensitivity and fast detection speed, there are some aspects that still need to be improved and developed in the FET–based biosensors system. (1) Biomolecular immobilization technology: On the one hand, suitable methods to achieve stable and reliable immobilization of biomolecules on the sensor surface are still in high demand. On the other hand, methods to improve the density, the uniformity and orderly arrangement of the immobilized biomolecules on the sensing surface need to be developed to improve the sensing performance. (2) Selectivity and sensitivity: In addition to the target biomolecules, some non–target analytes also could be attached to the biosensor interface and will generate interference signals to the biosensors. Therefore, it is essential to develop methods to prevent the attachment of non–specific adsorbates, such as passivation of the excess functional groups by proper reagents. Designing masks and optimizing channel size can also play a role in improving sensitivity. (3) Reusability: Currently, most sensors are single use only, but the preparation of biosensors with a regenerative ability has a wider prospect in real–time applications. For example, Zhao et al. used Nafion solution to prepare a reproducible FET biosensor and realized the reusability of a single device [96]. (4) Microfluidic techniques for POC diagnosis have been shown to be effective in reducing sample size, testing cost, and time. Current leakage and power consumption problems must be considered in preparing microarrays, and integrating FET biosensors with microfluidic devices requires proper design of the FET structures, such as selecting dielectric layers with high k values to detect analytesat a low operating voltage(<1 V), etc. (5) Existing FET biosensors are mainly focused on in vitro detection of biological species, whereas bioelectronic devices are developing towards implantable, wearable and non–invasive measurement. Therefore, it is imperative to develop excellent biocompatible and flexible FET biosensors.

In addition, the transformation of this emerging technology from the laboratory to commercial production still requires the joint efforts of researchers and industrial circles. Developing and constructing FET biosensors of a small size, with low cost and commercial availability still presents great challenges, including: (1) Cost factor: researchers need to consider inexpensive methods and materials for mass production of standardized sensors. (2) Poor reliability: in addition to the cost factor, poor reliability is also a factor that cannot be ignored. In the process of commercialization, the inevitable quality problems in the large–scale manufacturing of devices must be taken into account. (3) Real–time communication capability: realizing real–time and remote data collection and processing for each individual through the Internet and to realize health monitoring and environmental testing, the balance of sensor performance and other parameters must be taken into account [97,98]. Furthermore, FET–based biosensors serve as an outstanding tool to bridge the worlds of electronics and biology, and further development of new sensing applications remains to be explored.

Author Contributions: Conceptualization, S.L., L.W., R.H.; investigation, R.H. and L.L.; writing—original draft preparation, R.H.; writing—review and editing, L.L., L.W. and S.L.; visualization, J.Y.; supervision, S.L. and L.W.; project administration, S.L.; funding acquisition, S.L. All authors have read and agreed to the published version of the manuscript.

Funding: This research was funded by the National Science Foundation of China, grant number 52073208.

Institutional Review Board Statement: Not applicable.

Informed Consent Statement: Not applicable.

Data Availability Statement: Not applicable.

Conflicts of Interest: The authors declare no conflict of interest.

References

1. Kang, H.; Wang, X.; Guo, M.; Dai, C.; Chen, R.; Yang, L.; Wu, Y.; Ying, T.; Zhu, Z.; Wei, D.; et al. Ultrasensitive Detection of SARS–CoV–2 Antibody by Graphene Field–Effect Transistors. *Nano Lett.* **2021**, *21*, 7897–7904. [CrossRef]
2. Li, P.; Lee, G.H.; Kim, S.Y.; Kwon, S.Y.; Kim, H.R.; Park, S. From Diagnosis to Treatment: Recent Advances in Patient–Friendly Biosensors and Implantable Devices. *ACS Nano* **2021**, *15*, 1960–2004. [CrossRef] [PubMed]
3. Rodrigues, D.; Barbosa, A.I.; Rebelo, R.; Kwon, I.K.; Reis, R.L.; Correlo, V.M. Skin–Integrated Wearable Systems and Implantable Biosensors: A Comprehensive Review. *Biosensors* **2020**, *10*, 79. [CrossRef] [PubMed]
4. Thriveni, G.; Ghosh, K. Advancement and Challenges of Biosensing Using Field Effect Transistors. *Biosensors* **2022**, *12*, 647. [CrossRef] [PubMed]
5. Tran, V.V. Conjugated Polymers–Based Biosensors for Virus Detection: Lessons from COVID–19. *Biosensors* **2022**, *12*, 748. [CrossRef] [PubMed]
6. Deng, Y.; Liu, L.; Li, J.; Gao, L. Sensors Based on the Carbon Nanotube Field–Effect Transistors for Chemical and Biological Analyses. *Biosensors* **2022**, *12*, 776. [CrossRef] [PubMed]
7. Li, H.; Shi, W.; Song, J.; Jang, H.J.; Dailey, J.; Yu, J.; Katz, H.E. Chemical and Biomolecule Sensing with Organic Field–Effect Transistors. *Chem. Rev.* **2019**, *119*, 3–35. [PubMed]
8. Shen, H.; Di, C.A.; Zhu, D. Organic transistor for bioelectronic applications. *Sci. China–Chem.* **2017**, *60*, 437–449. [CrossRef]
9. Sun, C.; Wang, X.; Auwalu, M.A.; Cheng, S.; Hu, W. Organic thin film transistors-based biosensors. *EcoMat* **2021**, *3*, e12094. [CrossRef]
10. Zaumseil, J.; Sirringhaus, H. Electron and ambipolar transport in organic field–effect transistors. *Chem. Rev.* **2007**, *107*, 1296–1323. [CrossRef]
11. Zhang, Z.; Yates, J.T., Jr. Band bending in semiconductors: Chemical and physical consequences at surfaces and interfaces. *Chem. Rev.* **2012**, *112*, 5520–5551. [CrossRef] [PubMed]
12. Zhang, Z.; Yates, J.T. Effect of Adsorbed Donor and Acceptor Molecules on Electron Stimulated Desorption: $O_2/TiO_2(110)$. *J. Phys. Chem. Lett.* **2010**, *1*, 2185–2188. [CrossRef]
13. Kergoat, L.; Herlogsson, L.; Braga, D.; Piro, B.; Pham, M.C.; Crispin, X.; Berggren, M.; Horowitz, G. A water–gate organic field–effect transistor. *Adv. Mater.* **2010**, *22*, 2565–2569. [CrossRef] [PubMed]

14. Kergoat, L.; Piro, B.; Berggren, M.; Pham, M.C.; Yassar, A.; Horowitz, G. DNA detection with a water–gated organic field–effect transistor. *Org. Electron.* **2012**, *13*, 1–6. [CrossRef]
15. Kim, J.W.; Jang, Y.H.; Ku, G.M.; Kim, S.; Lee, E.; Cho, K.; Lim, K.I.; Lee, W.H. Liquid coplanar–gate organic/graphene hybrid electronics for label–free detection of single and double–stranded DNA molecules. *Org. Electron.* **2018**, *62*, 163–167. [CrossRef]
16. Yaman, B.; Terkesli, I.; Turksoy, K.M.; Sanyal, A.; Mutlu, S. Fabrication of a planar water gated organic field effect transistor using a hydrophilic polythiophene for improved digital inverter performance. *Org. Electron.* **2014**, *15*, 646–653. [CrossRef]
17. Minamiki, T.; Minami, T.; Koutnik, P.; Anzenbacher, P., Jr.; Tokito, S. Antibody– and Label–Free Phosphoprotein Sensor Device Based on an Organic Transistor. *Anal. Chem.* **2016**, *88*, 1092–1095.
18. Zhang, J.J.; Wang, S.Y.; Zhang, P.; Fan, S.C.; Dai, H.T.; Xiao, Y.; Wang, Y. Engineering a cationic supramolecular charge switch for facile amino acids enantiodiscrimination based on extended–gate field effect transistors. *Chin. Chem. Lett.* **2022**, *33*, 3873–3878. [CrossRef]
19. Wang, Y.; Gong, Q.; Miao, Q. Structured and functionalized organic semiconductors for chemical and biological sensors based on organic field effect transistors. *Mater. Chem. Front.* **2020**, *4*, 3505–3520. [CrossRef]
20. Janasz, L.; Borkowski, M.; Blom, P.W.M.; Marszalek, T.; Pisula, W. Organic Semiconductor/Insulator Blends for Elastic Field-Effect Transistors and Sensors. *Adv. Funct. Mater.* **2022**, *32*, 2105456. [CrossRef]
21. Sun, C.; Vinayak, M.V.; Cheng, S.; Hu, W. Facile Functionalization Strategy for Ultrasensitive Organic Protein Biochips in Multi–Biomarker Determination. *Anal. Chem.* **2021**, *93*, 11305–11311. [CrossRef] [PubMed]
22. Khan, H.U.; Roberts, M.E.; Johnson, O.; Forch, R.; Knoll, W.; Bao, Z. In situ, label–free DNA detection using organic transistor sensors. *Adv. Mater.* **2010**, *22*, 4452–4456. [CrossRef]
23. Khan, H.U.; Jang, J.; Kim, J.J.; Knoll, W. In situ antibody detection and charge discrimination using aqueous stable pentacene transistor biosensors. *J. Am. Chem. Soc.* **2011**, *133*, 2170–2176. [CrossRef] [PubMed]
24. Magliulo, M.; Pistillo, B.R.; Mulla, M.Y.; Cotrone, S.; Ditaranto, N.; Cioffi, N.; Favia, P.; Torsi, L. PE–CVD of Hydrophilic–COOH Functionalized Coatings on Electrolyte Gated Field–Effect Transistor Electronic Layers. *Plasma Process. Polym.* **2013**, *10*, 102–109. [CrossRef]
25. Magliulo, M.; Mallardi, A.; Mulla, M.Y.; Cotrone, S.; Pistillo, B.R.; Favia, P.; Vikholm–Lundin, I.; Palazzo, G.; Torsi, L. Electrolyte–gated organic field–effect transistor sensors based on supported biotinylated phospholipid bilayer. *Adv. Mater.* **2013**, *25*, 2090–2094. [CrossRef] [PubMed]
26. Mulla, M.Y.; Seshadri, P.; Torsi, L.; Manoli, K.; Mallardi, A.; Ditaranto, N.; Santacroce, M.V.; DiFranco, C.; Scamarcio, G.; Magliulo, M. UV crosslinked poly (acrylic acid): A simple method to bio–functionalize electrolyte–gated OFET biosensors. *J. Mater. Chem. B* **2015**, *3*, 5049–5057. [CrossRef] [PubMed]
27. Sun, C.; Li, R.; Song, Y.; Jiang, X.; Zhang, C.; Cheng, S.; Hu, W. Ultrasensitive and Reliable Organic Field–Effect Transistor–Based Biosensors in Early Liver Cancer Diagnosis. *Anal. Chem.* **2021**, *93*, 6188–6194. [CrossRef] [PubMed]
28. Stoltenberg, R.M.; Liu, C.; Bao, Z. Selective surface chemistry using alumina nanoparticles generated from block copolymers. *Langmuir* **2011**, *27*, 445–451. [CrossRef]
29. Hammock, M.L.; Sokolov, A.N.; Stoltenberg, R.M.; Naab, B.D.; Bao, Z. Organic transistors with ordered nanoparticle arrays as a tailorable platform for selective, in situ detection. *ACS Nano* **2012**, *6*, 3100–3108. [CrossRef]
30. Hammock, M.L.; Knopfmacher, O.; Naab, B.D.; Tok, J.B.; Bao, Z. Investigation of protein detection parameters using nanofunctionalized organic field–effect transistors. *ACS Nano* **2013**, *7*, 3970–3980. [CrossRef] [PubMed]
31. Suspene, C.; Piro, B.; Reisberg, S.; Pham, M.C.; Toss, H.; Berggren, M.; Yassar, A.; Horowitz, G. Copolythiophene–based water–gated organic field–effect transistors for biosensing. *J. Mater. Chem. B* **2013**, *1*, 2090–2097. [CrossRef] [PubMed]
32. Zang, Y.; Huang, D.; Di, C.A.; Zhu, D. Device Engineered Organic Transistors for Flexible Sensing Applications. *Adv. Mater.* **2016**, *28*, 4549–4555. [CrossRef] [PubMed]
33. Shen, H.; Zou, Y.; Zang, Y.; Huang, D.; Jin, W.; Di, C.A.; Zhu, D. Molecular antenna tailored organic thin–film transistors for sensing application. *Mater. Horizons* **2018**, *5*, 240–247. [CrossRef]
34. Buth, F.; Donner, A.; Sachsenhauser, M.; Stutzmann, M.; Garrido, J.A. Biofunctional electrolyte–gated organic field–effect transistors. *Adv. Mater.* **2012**, *24*, 4511–4517. [CrossRef] [PubMed]
35. Minamiki, T.; Sasaki, Y.; Tokito, S.; Minami, T. Label–Free Direct Electrical Detection of a Histidine–Rich Protein with Sub–Femtomolar Sensitivity using an Organic Field–Effect Transistor. *ChemistryOpen* **2017**, *6*, 472–475. [CrossRef]
36. Lee, H.W.; Kang, D.H.; Cho, J.H.; Lee, S.; Jun, D.H.; Park, J.H. Highly Sensitive and Reusable Membraneless Field–Effect Transistor (FET)–Type Tungsten Diselenide (WSe_2) Biosensors. *ACS Appl. Mater. Interfaces* **2018**, *10*, 17639–17645. [CrossRef]
37. Mulla, M.Y.; Tuccori, E.; Magliulo, M.; Lattanzi, G.; Palazzo, G.; Persaud, K.; Torsi, L. Capacitance–modulated transistor detects odorant binding protein chiral interactions. *Nat. Commun.* **2015**, *6*, 6010. [CrossRef]
38. Sensi, M.; Berto, M.; Gentile, S.; Pinti, M.; Conti, A.; Pellacani, G.; Salvarani, C.; Cossarizza, A.; Bortolotti, C.A.; Biscarini, F. Anti–drug antibody detection with label–free electrolyte–gated organic field–effect transistors. *Chem. Commun.* **2021**, *57*, 367–370. [CrossRef]
39. Macchia, E.; Manoli, K.; Holzer, B.; Di Franco, C.; Ghittorelli, M.; Torricelli, F.; Alberga, D.; Mangiatordi, G.F.; Palazzo, G.; Scamarcio, G.; et al. Single–molecule detection with a millimetre–sized transistor. *Nat. Commun.* **2018**, *9*, 3223. [CrossRef]

40. Macchia, E.; Tiwari, A.; Manoli, K.; Holzer, B.; Ditaranto, N.; Picca, R.A.; Cioffi, N.; Di Franco, C.; Scamarcio, G.; Palazzo, G.; et al. Label–Free and Selective Single–Molecule Bioelectronic Sensing with a Millimeter–Wide Self–Assembled Monolayer of Anti–Immunoglobulins. *Chem. Mater.* **2019**, *31*, 6476–6483. [CrossRef]
41. Zhou, W.; Dai, X.; Fu, T.M.; Xie, C.; Liu, J.; Lieber, C.M. Long term stability of nanowire nanoelectronics in physiological environments. *Nano Lett.* **2014**, *14*, 1614–1619. [CrossRef] [PubMed]
42. Khanna, V.K. Remedial and adaptive solutions of ISFET non–ideal behaviour. *Sens. Rev.* **2013**, *33*, 228–237. [CrossRef]
43. Eswaran, M.; Chokkiah, B.; Pandit, S.; Rahimi, S.; Dhanusuraman, R.; Aleem, M.; Mijakovic, I. A Road Map toward Field–Effect Transistor Biosensor Technology for Early Stage Cancer Detection. *Small Methods* **2022**, *6*, e2200809. [CrossRef] [PubMed]
44. Ciampi, S.; Harper, J.B.; Gooding, J.J. Wet chemical routes to the assembly of organic monolayers on silicon surfaces via the formation of Si–C bonds: Surface preparation, passivation and functionalization. *Chem. Soc. Rev.* **2010**, *39*, 2158–2183. [CrossRef] [PubMed]
45. Wang, Z.; Lin, H.; Zhang, X.; Li, J.; Chen, X.; Wang, S.; Gong, W.; Yan, H.; Zhao, Q.; Lv, W.; et al. Revealing molecular conformation–induced stress at embedded interfaces of organic optoelectronic devices by sum frequency generation spectroscopy. *Sci. Adv.* **2021**, *7*, eabf8555. [CrossRef]
46. Hideshima, S.; Hayashi, H.; Takeuchi, R.; Wustoni, S.; Kuroiwa, S.; Nakanishi, T.; Momma, T.; Osaka, T. Improvement in long–term stability of field effect transistor biosensor in aqueous environments using a combination of silane and reduced graphene oxide coating. *Microelectron. Eng.* **2022**, *264*, 111859. [CrossRef]
47. Li, T.; Ciampi, S.; Darwish, N. The Surface Potential of Zero Charge Controls the Kinetics of Diazonium Salts Electropolymerization. *ChemElectroChem* **2022**, *9*, e202200255. [CrossRef]
48. Novoselov, K.S.; Geim, A.K.; Morozov, S.V.; Jiang, D.; Zhang, Y.; Dubonos, S.V.; Grigorieva, I.V.; Firsov, A.A. Electric field effect in atomically thin carbon films. *Science* **2004**, *306*, 666–669. [CrossRef]
49. Dai, C.; Liu, Y.; Wei, D. Two–Dimensional Field–Effect Transistor Sensors: The Road toward Commercialization. *Chem. Rev.* **2022**, *122*, 10319–10392. [CrossRef]
50. Gao, J.; Gao, Y.; Han, Y.; Pang, J.; Wang, C.; Wang, Y.; Liu, H.; Zhang, Y.; Han, L. Ultrasensitive Label–free MiRNA Sensing Based on a Flexible Graphene Field–Effect Transistor without Functionalization. *ACS Appl. Electron. Mater.* **2020**, *2*, 1090–1098. [CrossRef]
51. Tian, M.; Qiao, M.; Shen, C.; Meng, F.; Frank, L.A.; Krasitskaya, V.V.; Wang, T.; Zhang, X.; Song, R.; Li, Y.; et al. Highly–sensitive graphene field effect transistor biosensor using PNA and DNA probes for RNA detection. *Appl. Surf. Sci.* **2020**, *527*, 146839. [CrossRef]
52. Sun, M.; Zhang, C.; Chen, D.; Wang, J.; Ji, Y.; Liang, N.; Gao, H.; Cheng, S.; Liu, H. Ultrasensitive and stable all graphene field-effect transistor-based Hg^{2+} sensor constructed by using different covalently bonded RGO films assembled by different conjugate linking molecules. *SmartMat* **2021**, *2*, 213–225. [CrossRef]
53. Roberts, A.; Chauhan, N.; Islam, S.; Mahari, S.; Ghawri, B.; Gandham, R.K.; Majumdar, S.S.; Ghosh, A.; Gandhi, S. Graphene functionalized field–effect transistors for ultrasensitive detection of Japanese encephalitis and Avian influenza virus. *Sci. Rep.* **2020**, *10*, 14546. [CrossRef] [PubMed]
54. Li, J.; Yang, X.; Liu, Y.; Huang, B.; Wu, R.; Zhang, Z.; Zhao, B.; Ma, H.; Dang, W.; Wei, Z.; et al. General synthesis of two–dimensional van der Waals heterostructure arrays. *Nature* **2020**, *579*, 368–374. [CrossRef] [PubMed]
55. Zhao, Y.; Qiao, J.; Yu, Z.; Yu, P.; Xu, K.; Lau, S.P.; Zhou, W.; Liu, Z.; Wang, X.; Ji, W.; et al. High–Electron–Mobility and Air–Stable 2D Layered $PtSe_2$ FETs. *Adv. Mater.* **2017**, *29*, 1604230. [CrossRef]
56. Park, H.; Han, G.; Lee, S.W.; Lee, H.; Jeong, S.H.; Naqi, M.; AlMutairi, A.; Kim, Y.J.; Lee, J.; Kim, W.J.; et al. Label–Free and Recalibrated Multilayer MoS_2 Biosensor for Point–of–Care Diagnostics. *ACS Appl. Mater. Interfaces* **2017**, *9*, 43490–43497. [CrossRef]
57. Zhou, H.; Wang, C.; Shaw, J.C.; Cheng, R.; Chen, Y.; Huang, X.; Liu, Y.; Weiss, N.O.; Lin, Z.; Huang, Y.; et al. Large area growth and electrical properties of p–type WSe_2 atomic layers. *Nano Lett.* **2015**, *15*, 709–713. [CrossRef]
58. Hossain, M.M.; Shabbir, B.; Wu, Y.; Yu, W.; Krishnamurthi, V.; Uddin, H.; Mahmood, N.; Walia, S.; Bao, Q.; Alan, T.; et al. Ultrasensitive WSe_2 field–effect transistor–based biosensor for label–free detection of cancer in point–of–care applications. *2D Mater.* **2021**, *8*, 045005. [CrossRef]
59. Zhang, Y.; Feng, D.; Xu, Y.; Yin, Z.; Dou, W.; Habiba, U.E.; Pan, C.; Zhang, Z.; Mou, H.; Deng, H.; et al. DNA–based functionalization of two–dimensional MoS_2 FET biosensor for ultrasensitive detection of PSA. *Appl. Surf. Sci.* **2021**, *548*, 149169. [CrossRef]
60. Esrafili, A.; Wagner, A.; Inamdar, S.; Acharya, A.P. Covalent Organic Frameworks for Biomedical Applications. *Adv. Healthc. Mater.* **2021**, *10*, e2002090. [CrossRef]
61. Xing, C.; Mei, P.; Mu, Z.; Li, B.; Feng, X.; Zhang, Y.; Wang, B. Enhancing Enzyme Activity by the Modulation of Covalent Interactions in the Confined Channels of Covalent Organic Frameworks. *Angew. Chem.–Int. Edit.* **2022**, *61*, e202201378.
62. Li, W.; Yang, C.X.; Yan, X.P. A versatile covalent organic framework–based platform for sensing biomolecules. *Chem. Commun.* **2017**, *53*, 11469–11471. [CrossRef] [PubMed]
63. Wang, P.; Kang, M.; Sun, S.; Liu, Q.; Zhang, Z.; Fang, S. Imine–Linked Covalent Organic Framework on Surface for Biosensor. *Chin. J. Chem.* **2014**, *32*, 838–843. [CrossRef]

64. Lu, J.; Wang, M.; Han, Y.; Deng, Y.; Zeng, Y.; Li, C.; Yang, J.; Li, G. Functionalization of Covalent Organic Frameworks with DNA via Covalent Modification and the Application to Exosomes Detection. *Anal. Chem.* **2022**, *94*, 5055–5061. [CrossRef] [PubMed]
65. Wang, B.; Luo, Y.; Liu, B.; Duan, G. Field–Effect Transistor Based on an in Situ Grown Metal–Organic Framework Film as a Liquid–Gated Sensing Device. *ACS Appl. Mater. Interfaces* **2019**, *11*, 35935–35940. [CrossRef] [PubMed]
66. Song, J.; Dailey, J.; Li, H.; Jang, H.J.; Zhang, P.; Wang, J.T.; Everett, A.D.; Katz, H.E. Extended Solution Gate OFET–based Biosensor for Label–free Glial Fibrillary Acidic Protein Detection with Polyethylene Glycol–Containing Bioreceptor Layer. *Adv. Funct. Mater.* **2017**, *27*, 1606506. [CrossRef]
67. Li, C.; Zhang, J.; Li, Z.; Zhang, W.; Wong, M.S.; Yu, G. Water–stable organic field–effect transistors based on naphthodithieno[3,2–b] thiophene derivatives. *J. Mater. Chem. C* **2019**, *7*, 297–301. [CrossRef]
68. Lauro, M.D.; Berto, M.; Giordani, M.; Benaglia, S.; Schweicher, G.; Vuillaume, D.; Bortolotti, C.A.; Geerts, Y.H.; Biscarini, F. Liquid-Gated Organic Electronic Devices Based on High-Performance Solution-Processed Molecular Semiconductor. *Adv. Electron. Mater.* **2017**, *3*, 1700159. [CrossRef]
69. Ko, J.; Ng, C.K.; Arramel; Wee, A.T.S.; Tam, T.L.D.; Leong, W.L. Water robustness of organic thin–film transistors based on pyrazino[2,3–g] quinoxaline–dione conjugated polymer. *J. Mater. Chem. C* **2020**, *8*, 4157–4163. [CrossRef]
70. Doumbia, A.; Tong, J.; Wilson, R.J.; Turner, M.L. Investigation of the Performance of Donor–Acceptor Conjugated Polymers in Electrolyte-Gated Organic Field-Effect Transistors. *Adv. Electron. Mater.* **2021**, *7*, 2100071. [CrossRef]
71. Sun, C.; Wang, Y.X.; Sun, M.; Zou, Y.; Zhang, C.; Cheng, S.; Hu, W. Facile and cost–effective liver cancer diagnosis by water–gated organic field–effect transistors. *Biosens. Bioelectron.* **2020**, *164*, 112251. [CrossRef] [PubMed]
72. Viola, F.A.; Melloni, F.; Molazemhosseini, A.; Modena, F.; Sassi, M.; Beverina, L.; Caironi, M. A n-type, Stable Electrolyte Gated Organic Transistor Based on a Printed Polymer. *Adv. Electron. Mater.* **2022**, *9*, 2200573. [CrossRef]
73. Yuan, X.; Yang, C.; He, Q.; Chen, J.; Yu, D.; Li, J.; Zhai, S.; Qin, Z.; Du, K.; Chu, Z.; et al. Current and Perspective Diagnostic Techniques for COVID–19. *ACS Infect. Dis.* **2020**, *6*, 1998–2016. [CrossRef] [PubMed]
74. Chu, D.K.W.; Pan, Y.; Cheng, S.M.S.; Hui, K.P.Y.; Krishnan, P.; Liu, Y.; Ng, D.Y.M.; Wan, C.K.C.; Yang, P.; Wang, Q.; et al. Molecular Diagnosis of a Novel Coronavirus (2019–nCoV) Causing an Outbreak of Pneumonia. *Clin. Chem.* **2020**, *66*, 549–555. [CrossRef] [PubMed]
75. Zhang, F.; Abudayyeh, O.O.; Jonathan, S.G. A Protocol for Detection of COVID-19 Using CRISPR Diagnostics. Available online: https://www.broadinstitute.org/files/publications/special/COVID-19%20detection%20(updated).pdf (accessed on 1 May 2020).
76. Seo, G.; Lee, G.; Kim, M.J.; Baek, S.H.; Choi, M.; Ku, K.B.; Lee, C.S.; Jun, S.; Park, D.; Kim, H.G.; et al. Rapid Detection of COVID–19 Causative Virus (SARS–CoV–2) in Human Nasopharyngeal Swab Specimens Using Field–Effect Transistor–Based Biosensor. *ACS Nano* **2020**, *14*, 5135–5142. [CrossRef]
77. Kong, D.; Wang, X.; Gu, C.; Guo, M.; Wang, Y.; Ai, Z.; Zhang, S.; Chen, Y.; Liu, W.; Wu, Y.; et al. Direct SARS–CoV–2 Nucleic Acid Detection by Y-Shaped DNA Dual–Probe Transistor Assay. *J. Am. Chem. Soc.* **2021**, *143*, 17004–17014. [CrossRef]
78. Dai, C.; Guo, M.; Wu, Y.; Cao, B.P.; Wang, X.; Wu, Y.; Kang, H.; Kong, D.; Zhu, Z.; Ying, T.; et al. Ultraprecise Antigen 10–in–1 Pool Testing by Multiantibodies Transistor Assay. *J. Am. Chem. Soc.* **2021**, *143*, 19794–19801. [CrossRef]
79. Gao, J.; Wang, C.; Chu, Y.; Han, Y.; Gao, Y.; Wang, Y.; Wang, C.; Liu, H.; Han, L.; Zhang, Y. Graphene oxide–graphene Van der Waals heterostructure transistor biosensor for SARS–CoV–2 protein detection. *Talanta* **2022**, *240*, 123197. [CrossRef] [PubMed]
80. Li, D.; Chen, H.; Fan, K.; Labunov, V.; Lazarouk, S.; Yue, X.; Liu, C.; Yang, X.; Dong, L.; Wang, G. A supersensitive silicon nanowire array biosensor for quantitating tumor marker ctDNA. *Biosens. Bioelectron.* **2021**, *181*, 113147. [CrossRef]
81. Yang, Y.; Zeng, B.; Li, Y.; Liang, H.; Yang, Y.; Yuan, Q. Construction of MoS_2 field effect transistor sensor array for the detection of bladder cancer biomarkers. *Sci. China-Chem.* **2020**, *63*, 997–1003. [CrossRef]
82. Sun, C.; Feng, G.; Song, Y.; Cheng, S.; Lei, S.; Hu, W. Single Molecule Level and Label–Free Determination of Multibiomarkers with an Organic Field–Effect Transistor Platform in Early Cancer Diagnosis. *Anal. Chem.* **2022**, *94*, 6615–6620. [CrossRef] [PubMed]
83. Xue, M.; Mackin, C.; Weng, W.H.; Zhu, J.; Luo, Y.; Luo, S.L.; Lu, A.Y.; Hempel, M.; McVay, E.; Kong, J.; et al. Integrated biosensor platform based on graphene transistor arrays for real–time high–accuracy ion sensing. *Nat. Commun.* **2022**, *13*, 5064. [CrossRef] [PubMed]
84. Kwon, O.S.; Song, H.S.; Park, S.J.; Lee, S.H.; An, J.H.; Park, J.W.; Yang, H.; Yoon, H.; Bae, J.; Park, T.H.; et al. An Ultrasensitive, Selective, Multiplexed Superbioelectronic Nose That Mimics the Human Sense of Smell. *Nano Lett.* **2015**, *15*, 6559–6567. [CrossRef]
85. Ahn, S.R.; An, J.H.; Song, H.S.; Park, J.W.; Lee, S.H.; Kim, J.H.; Jang, J.; Park, T.H. Duplex Bioelectronic Tongue for Sensing Umami and Sweet Tastes Based on Human Taste Receptor Nanovesicles. *ACS Nano* **2016**, *10*, 7287–7296. [CrossRef]
86. Papamatthaiou, S.; Estrela, P.; Moschou, D. Printable graphene BioFETs for DNA quantification in Lab–on–PCB microsystems. *Sci. Rep.* **2021**, *11*, 9815. [CrossRef] [PubMed]
87. Dai, X.; Vo, R.; Hsu, H.-H.; Deng, P.; Zhang, Y.; Jiang, X. Modularized Field–Effect Transistor Biosensors. *Nano Lett.* **2019**, *19*, 6658–6664. [CrossRef] [PubMed]
88. Kim, K.H.; Park, S.J.; Park, C.S.; Seo, S.E.; Lee, J.; Kim, J.; Lee, S.H.; Lee, S.; Kim, J.S.; Ryu, C.M.; et al. High–performance portable graphene field–effect transistor device for detecting Gram–positive and –negative bacteria. *Biosens. Bioelectron.* **2020**, *167*, 112514. [CrossRef]
89. Zhou, F.; Pan, W.; Chang, Y.; Su, X.; Duan, X.; Xue, Q. A Supported Lipid Bilayer–Based Lab–on–a–Chip Biosensor for the Rapid Electrical Screening of Coronavirus Drugs. *ACS Sens.* **2022**, *7*, 2084–2092. [CrossRef]

90. Hajian, R.; Balderston, S.; Tran, T.; DeBoer, T.; Etienne, J.; Sandhu, M.; Wauford, N.A.; Chung, J.Y.; Nokes, J.; Athaiya, M.; et al. Detection of unamplified target genes via CRISPR–Cas9 immobilized on a graphene field–effect transistor. *Nat. Biomed. Eng.* **2019**, *3*, 427–437. [CrossRef]
91. Parkula, V.; Berto, M.; Diacci, C.; Patrahau, B.; Lauro, M.D.; Kovtun, A.; Liscio, A.; Sensi, M.; Samori, P.; Greco, P.; et al. Harnessing Selectivity and Sensitivity in Electronic Biosensing: A Novel Lab–on–Chip Multigate Organic Transistor. *Anal. Chem.* **2020**, *92*, 9330–9337. [CrossRef]
92. Yoo, G.; Park, H.; Kim, M.; Song, W.G.; Jeong, S.; Kim, M.H.; Lee, H.; Lee, S.W.; Hong, Y.K.; Lee, M.G.; et al. Real–time electrical detection of epidermal skin MoS_2 biosensor for point–of–care diagnostics. *Nano Res.* **2016**, *10*, 767–775. [CrossRef]
93. Guo, S.; Wu, K.; Li, C.; Wang, H.; Sun, Z.; Xi, D.; Zhang, S.; Ding, W.; Zaghloul, M.E.; Wang, C.; et al. Integrated contact lens sensor system based on multifunctional ultrathin MoS_2 transistors. *Matter* **2021**, *4*, 969–985. [CrossRef] [PubMed]
94. Ding, Q.; Wang, H.; Zhou, Z.; Wu, Z.; Tao, K.; Gui, X.; Liu, C.; Shi, W.; Wu, J. Stretchable, self-healable, and breathable biomimetic iontronics with superior humidity-sensing performance for wireless respiration monitoring. *SmartMat* **2023**, *4*, e1147. [CrossRef]
95. Li, J.; Li, H.; Xu, L.; Wang, L.; Hu, Z.; Liu, L.; Huang, Y.; Kotov, N.A. Biomimetic nanoporous aerogels from branched aramid nanofibers combining high heat insulation and compressive strength. *SmartMat* **2021**, *2*, 76–87. [CrossRef]
96. Wang, Z.; Hao, Z.; Wang, X.; Huang, C.; Lin, Q.; Zhao, X.; Pan, Y. A Flexible and Regenerative Aptameric Graphene–Nafion Biosensor for Cytokine Storm Biomarker Monitoring in Undiluted Biofluids toward Wearable Applications. *Adv. Funct. Mater.* **2020**, *31*, 2005958. [CrossRef]
97. Novodchuk, I.; Bajcsy, M.; Yavuz, M. Graphene–based field effect transistor biosensors for breast cancer detection: A review on biosensing strategies. *Carbon* **2021**, *172*, 431–453. [CrossRef]
98. Wadhera, T.; Kakkar, D.; Wadhwa, G.; Raj, B. Recent Advances and Progress in Development of the Field Effect Transistor Biosensor: A Review. *J. Electron. Mater.* **2019**, *48*, 7635–7646. [CrossRef]

Review

Label-Free Electrochemical Biosensor Platforms for Cancer Diagnosis: Recent Achievements and Challenges

Vildan Sanko [1] and Filiz Kuralay [2,*]

[1] Department of Chemistry, Gebze Technical University, 41400 Kocaeli, Turkey
[2] Department of Chemistry, Faculty of Science, Hacettepe University, 06800 Ankara, Turkey
* Correspondence: filizkur@hacettepe.edu.tr

Abstract: With its fatal effects, cancer is still one of the most important diseases of today's world. The underlying fact behind this scenario is most probably due to its late diagnosis. That is why the necessity for the detection of different cancer types is obvious. Cancer studies including cancer diagnosis and therapy have been one of the most laborious tasks. Since its early detection significantly affects the following therapy steps, cancer diagnosis is very important. Despite researchers' best efforts, the accurate and rapid diagnosis of cancer is still challenging and difficult to investigate. It is known that electrochemical techniques have been successfully adapted into the cancer diagnosis field. Electrochemical sensor platforms that are brought together with the excellent selectivity of biosensing elements, such as nucleic acids, aptamers or antibodies, have put forth very successful outputs. One of the remarkable achievements of these biomolecule-attached sensors is their lack of need for additional labeling steps, which bring extra burdens such as interference effects or demanding modification protocols. In this review, we aim to outline label-free cancer diagnosis platforms that use electrochemical methods to acquire signals. The classification of the sensing platforms is generally presented according to their recognition element, and the most recent achievements by using these attractive sensing substrates are described in detail. In addition, the current challenges are discussed.

Keywords: label-free electrochemical detection; electrochemical sensor; cancer diagnosis

Citation: Sanko, V.; Kuralay, F. Label-Free Electrochemical Biosensor Platforms for Cancer Diagnosis: Recent Achievements and Challenges. *Biosensors* **2023**, *13*, 333. https://doi.org/10.3390/bios13030333

Received: 1 December 2022
Revised: 17 February 2023
Accepted: 23 February 2023
Published: 1 March 2023

1. Introduction

Cancer, which causes premature death in almost all countries of the world, maintains its position at first place even if it is sometimes replaced by cardiac disease. In particular, due to demographic effects and the trends of these effects in cancer incidence in different locations, it is expected that instances of cancer will approximately double in the next 50 years globally. However, cancer does not affect the population of all countries at the same rate, and it is predicted that there will be a higher increase in countries that can be classified as low–middle income [1,2]. The Global Cancer Statistics 2020 report shows that the most common cancer in men is prostate cancer, followed by lung cancer, colorectal cancer and liver cancer, whereas breast cancer and cervical cancer are the most commonly diagnosed cancers in women. In addition, according to the same report, what is striking is that an estimated 19.3 million new cancer cases were detected worldwide and approximately 10.0 million deaths were calculated due to cancer only in 2020 [3].

Regardless of the type, the diagnosis and treatment of cancer at an early stage is very important to reduce both cancer incidence and mortality rates. As the traditional cancer detection method, enzyme-linked immunosorbent assay (ELISA), which detects cancer-specific protein biomarkers and is called the gold standard, is widely known [4]. Also, genomic- and proteomic-based molecular methods such as polymerase chain reaction (PCR), immunohistochemistry (IHC) and radioimmunoassay (RIA) are used for cancer diagnosis [5]. In addition, various clinical tools such as magnetic resonance imaging (MRI), positron emission tomography (PET), endoscopy, sonography, X-ray, computed tomography (CT) and biopsy are extensively utilized [5–7]. However, although the mentioned methods and technologies are efficient, most of them are expensive, time-consuming, invasive and limited to the laboratories of some hospitals. Especially with imaging methods, detecting cancer tumors below the millimeter size may be inconclusive. Similarly, invasive methods such as biopsy have the same problems and difficulties in diagnosing early-stage cancer tumors [6].

Early-stage cancer diagnosis increases the survival rate of the patient [8]. In addition, early diagnosis offers several advantages that lead to more appropriate treatment for the patient and even reduce the severity of the cancer [9]. One of the biggest problems limiting early diagnosis in cancer detection is the nonappearance of obvious symptoms in the early stages of cancer; the other is not detecting sufficiently sensitive biomarkers [10]. From an economic point of view, it is known that the costs used for cancer treatment are increasing rapidly, and this cost is expected to increase up to USD 246 billion by 2030. Therefore, detecting cancer at an early stage can reduce the potential economic burden for the patient and society [11]. There is a crucial need to develop low-cost, sensitive, non-invasive (bio)sensors for early-stage cancer diagnosis. In general, biological biomarkers show the genetic characteristics of cancer cells and diagnosis/monitoring of cancer with a biomarker-based biosensor is seen as one of the most promising approaches [12]. These biomarkers can be deoxyribonucleic acid (DNA), ribonucleic acid (RNA), hormones, protein, enzymes and specific cells that can be found in human bodily fluids such as urine, serum, plasma and blood [13,14]. For this aim, electrochemical sensors have been widely used in the field of cancer diagnosis. There are valuable studies in the literature that include various approaches to detect different cancer types such as breast cancer [15], ovarian cancer [16], prostate cancer [17], pancreatic cancer [18] and lung cancer [19]. Electrochemical sensors are prominent tools because they are sensitive, selective, fast, cost-effective, instrumentable and can be performed as on-site analysis [20,21]. Different electrochemical sensing methods such as potentiometric, amperometric, conductometric, impedimetric and voltammetric are used to convert the obtained signal into useful analytical data. Detection methods in biosensors can be grouped as labeled or unlabeled depending on the use of labels as electroactive molecules or nanomaterials. However, labeled systems are complex and expensive as they require an extra labeling process. Conversely, label-free biosensors have shorter analysis time and simplicity and they offer good advantages [22–24].

In this review, current label-free cancer diagnosis platforms in the literature, including the last three years, in which the electrochemical method is used as a signal converter, are detailed. Biorecognition elements and mechanisms used in biosensor design for cancer diagnosis are emphasized. In addition, the immobilization method and immobilization matrices, which are important parameters for the activity and stability of a biorecognition element, are also the subject of this study. Finally, current challenges and future perspectives are discussed.

2. Electrochemical Techniques as a Sensing Mechanism

Electroanalytical studies are included as a sub-discipline of analytical chemistry, which includes charge transfers in addition to oxidation–reduction reactions [25]. Biorecognition elements, which are one of the parameters that make up the biosensor, are important components for the analyte to be detected. This component needs to be used with a converter so that a meaningful signal can be generated according to the analyte concentration [26]. In an electrochemical transducer system, detectable signals such as current, potential, impedance and conductivity are obtained as a result of the interaction of samples with a bioreceptor. In connection with these signals, electrochemical biosensors are included in various classifications such as amperometric, potentiometric, impedimetric and conductometric. In addition, voltammetric techniques are important and sensitive techniques to help analyte determination [27,28]. Electrochemical detection systems, which provide analytical advantages such as low cost, simple design and portable features, are platforms that can make sensitive and selective detections even in body fluids with complex matrices such as serum [29,30].

Therefore, these detection systems have attracted great attention in biosensor technology owing to their unique properties.

Voltammetric techniques have been commonly utilized. For example, differential pulse voltammetry (DPV), where a pulse is applied to the electrode and provides current measurement. Before the pulse is applied and at the end of each pulse, the current is measured and the difference between the currents is calculated. This procedure effectively reduces the background current due to linear increase, thus resulting in a faradaic current with no capacitive current. The biggest advantage of DPV is a low capacitive current, which leads to high sensitivity. Small steps in DPV also lead to narrower voltammetric peaks, and therefore, DPV is often used to distinguish analytes with similar oxidation potentials. Thus, this technique is preferred in electrochemical cancer biosensors as it exhibits very sensitive properties against the reduction and oxidation of bio-electrochemical species [31,32]. Cyclic voltammetry (CV) is one of the most common methods to obtain information about redox potentials and to investigate the mechanisms and kinetic parameters involved in the reactions of electroactive analytes. In this method, the current between the working and counter electrodes is monitored, but changes in the potential of the working electrode due to the reference electrode are also controlled [33]. In the electrochemical impedance spectroscopy (EIS) technique, the impedance change in both faradaic and non-faradaic modes is measured. As an example, in the measurement system in the faradaic mode, the change in the electron transfer rate caused by the aptamer–analyte interaction is examined. In measurement systems taken in non-faradaic mode, the surface capacitance change due to the aptamer–analyte connection is detected [34]. In the amperometric technique, the working electrode is kept at a constant potential that is sufficient to reduce or oxidize the analyte of interest and the resulting current is monitored over time. Potential selection is critical as only one potential is applied in this technique. Due to the monitoring of current over time at a constant potential, all dynamic changes in the current can be observed [31]. On the other hand, in a potentiometric system based on potential measurements, the principle of changing the potential with the concentration of the analyte is used in the measuring system with the help of a reference electrode with a fixed electrode potential. Besides cancer diagnosis, electrochemical techniques are also highly preferred in routine laboratory analysis and clinical and environmental monitoring analysis [35].

When electrochemical techniques are compared with each other, it is observed that each of them can have limitations in different aspects. For example, the sensitivity of the potentiometric method depending on the environment and temperature is an important limitation. For the limitations of other methods, it can be said that redox elements are needed in the amperometric technique, whereas EIS is sensitive to the environment and requires theoretical stimulation for data analysis [36]. Voltammetric techniques show high selectivity and sensitivity due to the voltammetric peak potential applied to the analyte. However, one of the major problems encountered with these techniques is obtaining overlapping voltammetric responses due to very similar oxidation peak potentials. Various recently developed materials and protocols are used to overcome this problem [37]. Besides this, choosing an appropriate sensing technique for analyte detection can minimize the limitations. Additionally, parameters such as pretreatments applied to the working electrode and the biofunctionality of the electrodes can have a great impact on the precise and effective determination [34].

The electrochemical transformations occurring at the interface of the label-free sensing platform are determined by the affinity between the analyte and the biorecognition elements, regardless of the use of labels [29]. Thanks to the detectable signals obtained by electrochemistry, these techniques are widely preferred not only for cancer detection and follow-up but also for the accurate and sensitive detection of analytes in areas such as the detection of different diseases and environmental and food control [38–43]. In an electrochemical biosensor, two different reactions can be observed as a result of the interaction of the electrode surface and the analyte: the first is the positive read signal called "signal-on", and the other is the negative read signal called "signal-off" [44,45].

Label-free electrochemical biosensors are particularly interesting and important for studies in the biomedical field. In this type of electrochemical biosensor, the information in the reaction is converted into an electrical signal by the direct transfer of electrons between the electrode surface and the biorecognition elements as a result of the interaction between the biomolecule and the analyte [46]. Additionally, the surface characteristics of the electrodes significantly support improving the sensitivity of the biosensor. Therefore, surface modification is also important for good analytical performance. At this point, nanomaterials have been in the scope of scientists. The use of nanomaterials of different sizes, shapes and morphologies together with electrochemical transducers makes it possible to improve properties. Nanowires, metal/metal oxide nanoparticles, carbon nanotubes, graphene or graphene-like structures and conductive nanostructures such as polymers have provided more sensitive biosensors with high surface/volume ratios [47,48]. The scope of this study mainly covers the discussion of the technological developments and also problems/limitations in the development of label-free biosensors containing different biorecognition elements to serve cancer diagnosis.

Despite a lot of effort and good progress in the field of biosensors, it is seen as an inconsistency that only a few of them find a place in the commercial market. The first example of commercial biosensor is the enzymatic glucose biosensor, which is expected to have a market of USD 38 billion by 2027 [49]. This biosensor currently holds approximately 75% of the global biosensor market. There are still outstanding challenges, both to overcome the current constraints and to making the products available commercially. Firstly, understanding the mechanisms of biocatalytic work and charge transfers and also improvements in the properties of biorecognition elements that provide selectivity should be considered. In addition, the use of various nanoparticles and hydrogels has been reported to improve existing deficiencies, although not completely [50,51]. For this purpose, researchers are conducting detailed studies about the effects of parameters on biomolecule (such as enzymes) immobilization and the effect of these parameters on the performance of the biosensor platforms [52]. However, since the biomolecule redox reaction processes are still not fully known, in situ inspection techniques are used for evaluation [53]. Some of the obstacles in the transformation of biosensor studies from laboratory to commercial products are performance and nonspecific surface interaction problems in various body fluids, which have complex matrices [49].

Although electrochemical methods provide several advantages, each method may also have limitations. It is particularly important to focus on and discuss these limitations to put the developed technologies into clinical practice. Reducing or overcoming all the disadvantages could help to develop more accurate and sensitive electrochemical cancer biosensors. More effective platforms for early diagnosis can be created with a multidisciplinary study. In addition, the detection of new cancer biomarkers will greatly benefit the facilitation of early-stage diagnosis and thus the management and control of the cancer disease process. It is expected that the label-free electrochemical methods will increase in reliability after the difficulties we have mentioned have been overcome. As a result, they will find a regular use in the clinical field. To strengthen this reliability, novel and advanced electrochemical cancer biosensors with different perspectives need to be developed.

3. Importance of a Label-Free Electrochemical Sensing Platform

A typical electrochemical biosensor is expected to convert signals that are related to the presence of the analyte molecules into measurable quantities with the help of the biorecognition unit. In some cases, various markers/labels or tags are used for the detection of the analyte and the signal is obtained in conjunction with them. These biosensor systems are called label-based biosensors. The use of these labels, which are commonly classified as radioactive-, fluorescent- or chemiluminescence-based, is time consuming and laborious because it requires an extra process. More importantly, it is thought that in this case, the affinity between the biorecognition element and the analyte may be adversely affected. To eliminate these limiting factors, unlabeled detection systems have become highly preferred in recent years. If a direct measurement is made with the biorecognition system, this is called a label-free biosensor system [54].

In a typical label-free biosensor design, sensing can be performed by converting it to optical [55], mechanical [56] or electrical [57] signals and more accurate information can be provided as biorecognition systems are directly used. Within this classification, electrochemical label-free biosensors can be used actively in the field and can be also implanted in the body to detect biological analytes, increasing their future potential [58]. Various electrodes with different biorecognition elements and composite designs have been developed for analytes such as gliotoxin [59], microRNA (miRNA) [60], bacterial pathogens [61] and aflatoxin-B1 [62] in this biosensor group, which combines the advantages of both the electrochemical method and the label-free platform. For the continuation of the remarkable progress of the mentioned electrochemical label-free biosensors, a better understanding of the current working processes is required for the creation of sensitive and selective biosensing systems that find application in wider use. Based on this idea, we have detailed and discussed cancer studies classified on different biorecognition elements.

4. Biorecognition Elements for Label-Free Electrochemical Cancer Diagnosis

Basically, antibodies, aptamers, nucleic acids and cells are immobilized to surfaces/interfaces to achieve affinity and selective biorecognition. In this part, the classification of the label-free electrochemical cancer detection systems is divided into categories according to the type of the biorecognition element. Besides this classification, electrode material and the detection technique are also highlighted. Figure 1 demonstrates the schematic presentation of the label-free electrochemical cancer biosensors with successful electrode modifications, such as nanotechnology-based materials, biorecognition immobilization protocols and some of the powerful electrochemical detection techniques.

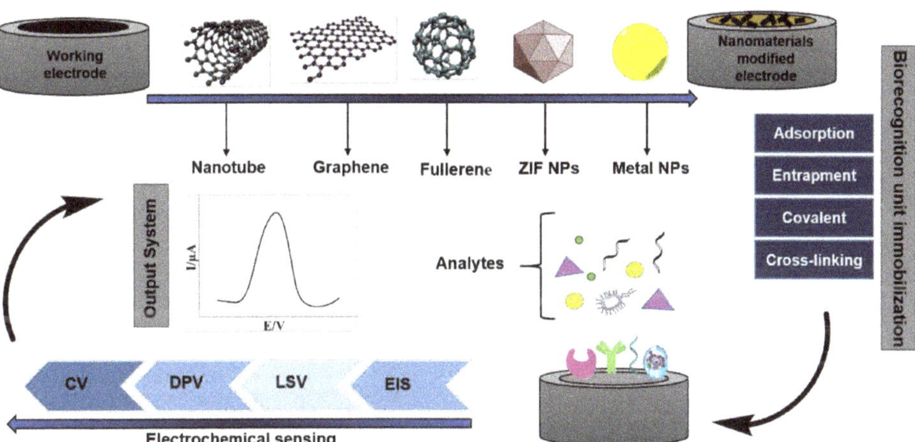

Figure 1. Label-free electrochemical cancer biosensors: electrode modifications such as nanotechnology-based materials, biorecognition immobilization protocols and some of the powerful electrochemical detection techniques.

4.1. Nucleic-Acid-Based Label-Free Cancer Biosensors

Nucleic acids are natural biopolymers that store genetic information in humans and almost all organisms [63]. Nucleic acids include DNA and RNA, which are composed of nucleotides. The well-known specific hybridization feature between nucleic acid chains also constitutes the main detection principle of DNA biosensors [64]. The development of biosensors for the detection of DNA sequences is important because of its application in gene identification, molecular diagnosis and drug screening [65]. Nucleic acids can be affected by environmental conditions such as temperature and pH [66]. Nevertheless, in many studies electrochemical signal amplification by means of nucleic acids has been successfully developed for cancer applications [67,68].

Studies in recent years show that excessive secretion of microRNAs is associated with malignancies that cause cancer [15,69–71]. In one study, Zhao et al. proposed MXene-molybdenum disulfide (MoS_2) constructs with thionine and gold nanoparticles for the label-free electrochemical detection of microRNA-21, which plays an important role in the emergence of cancer associated with proliferation/differentiation in cells. The modification of the prepared nanocomposite on glassy carbon electrode (GCE) was performed by drop casting. Then, the hairpin capture probe was dropped onto the modified electrode. The hybridization event was carried out in the presence of the target and a hairpin probe 2. The detection method was square wave voltammetry (SWV). Thanks to this structure, the capture probe immobilization was improved, the amplification of the electrochemical signal was achieved and microRNA-21 detection in the linear measurement range of 100 fM to 100 nM was obtained with a detection limit of 2 fM [72].

Pothipr et al. described a gold nanoparticle-dye/poly(3-aminobenzylamine)/two-dimensional molybdenum selenide ($MoSe_2$)-based electrochemical label-free biosensor for breast cancer diagnosis that could detect cancer antigen 15-3 and microRNA-21 simultaneously. Based on the complexity of the immune system in the human body and therefore the inadequacy of cancer assays using single biomarker systems, they introduced this bidirectional detection platform produced on a two-screen printed carbon electrode. DPV was used for the evaluation of the electrochemical performance of the biosensor and the detection limit was found to be 1.2 fM for microRNA-21 detection [73]. Jafari-Kashi et al. presented a DNA biosensor for the detection of cytokeratin 19 fragment 21-1, which is associated with lung cancer. They preferred DPV as an electrochemical technique to examine the interaction between the capture probe and target using GCE modified with reduced graphene oxide, polypyrrole, silver nanoparticles and single-stranded DNA (ssDNA). With this technique, no peak was detected before DNA hybridization, but a distinctive peak was obtained after hybridization according to the oxidation of guanine. They declared that the label-free DNA biosensor showed a good result for detection of cytokeratin 19 fragment 21-1, with a wide linear measurement range and a 2.14 fM limit of detection [74]. Avelino et al. presented a polypyrrole film containing DNA immobilized chitosan/zinc oxide nanoparticles for the diagnosis of myelocytic leukemia by BCR/ABL fusion gene detection. Oxidation and reduction steps were observed in line with the voltammetric measurements taken in 10 mM $[Fe(CN)_6]^{3-/4-}$. It is also stated that the biosensor was designed as a result of bioactivity tests and could be used as a new biosensing platform that enabled the identification of early-stage cancer [75].

4.2. Aptamer-Based Label-Free Cancer Biosensors

Aptamers are single-stranded DNA or RNA molecules that can usually be synthesized using an in vitro method. In fact, RNA-based aptamers were first found in 1990, followed by DNA-based aptamers, with the development of in vitro selection/amplification for the isolation of RNA sequences that could specifically bind to molecules [76]. In aptamer-based electrochemical sensors, it is necessary to be able to detect the conformational changes caused by the presence of the aptamer on the electrode surface for obtaining a signal [77]. Aptamers are widely used in the development of biosensors due to their high specificity, easy synthesis, simple modification and high chemical stability [78]. They offer the advantages of more cost-effective production, easy modification and thermal stability, especially when compared with monoclonal antibodies. After the aptamers are immobilized on a conductive matrix, their redox-active moieties allow the formation of aptamer–target complexes and thus the design of various electrochemical biosensors with the realized electron transfer properties [76]. The most important problem in this electrochemical process can be the generation of a determinable signal between the target analyte and the aptamer. In order to solve this problem, electrochemically active labeling units such as hemin [79], ferrocene [80] and methylene blue [81] have been introduced. However, labeling of aptamers introduces known disadvantages such as time consumption, poor affinity performance and cost [82].

In recent years, aptamers have attracted great interest in electrochemical label-free biosensor design, which has applications in the diagnosis and follow-up of various cancers. Label-free aptasensors also require an increased surface area to improve weak signal intensity. Nanomaterials contribute greatly to increasing the surface area because they act as electron-transfer tunnels, which increase the electrical communication between the redox regions of the aptamer and the electrode surface [83]. Zhang et al. developed a label-free aptasensor for the detection of cancer antigen 125 by immobilizing aptamer on the surface of nickel hexacyanoferrate nanocubes/polydopamine functionalized graphene. DPV was utilized for electroanalytical studies in this work, which was designed to provide a detectable electrochemical response with the help of increasing surface area and conductivity. Thanks to the insulating structure formed as a result of the combination of aptamer and cancer antigen 125 (CA125), or in other words aptamer–CA125 complex, the peak current value decreased as the analyte concentration increased. The linear measurement range and limit of detection were calculated as 0.10 pg mL^{-1}–1.0 μg mL^{-1} and 0.076 pg mL^{-1}, respectively. The measurements were carried out in phosphate buffer solution (PBS) [82]. In another study, a paper-based electrochemical label-free aptasensor was fabricated for the detection of epidermal growth factor receptors. Interestingly, the concept of origami as a valve for a paper-based biosensor was used in this study. As a result of the biochemical reaction, the data became an electrochemical response with the presence of the nanocomposites containing amino functionalized graphene/thionine/gold. This system in the form of origami was designed to increase the penetration of the liquid and shorten the time taken for flow, resulting in a shorter test time. The linear concentration range obtained with the sensor was from 0.05 ng mL^{-1} to 200 ng mL^{-1} and it had a detection limit of 5 pg mL^{-1} [84].

4.3. Antibody-Based Label-Free Cancer Biosensors

Antibodies are protective proteins produced by the immune system in response to the presence of antigens, including pathogens and toxic materials [78]. Biosensors that offer the advantages of high binding affinity and specificity and use antibodies for biorecognition take the advantage of the high affinity between antibodies and antigens for detection and are called immunosensors [85,86]. However, there are some parameters that limit their use. Apart from being adversely affected by environmental conditions and having difficulties for storage, it can be said that the production of polyclonal antibodies in animals is difficult and costly. Moreover, polyclonal antibodies may lack selectivity as they can have affinity for different epitopes [87]. With the help of the new and improved sensor interfaces developed in recent years, some disadvantages have been overcome and many antibody-based sensitive and selective label-free electrochemical biosensors have been designed. Also, these limitations pave the way for the development of new forms of biorecognition units that can replace antibodies, thus introducing new biosensor projections to the field.

Various electrochemical techniques have been used for antibody-based biosensors for gastric cancer [88], breast cancer [89–92], ovarian cancer [93–96], bladder cancer [97], colorectal cancer [98], lung cancer [99], prostate cancer [100–105], liver cancer [106] and more. In a study for a label-free electrochemical immunosensor developed for early-stage detection of prostate cancer, the surface of the indium tin oxide electrode was firstly coated with chitosan and reduced graphene oxide, and then the specific polyclonal anti-prostate-specific antigen (PSA) antibody as a recognition element was immobilized on the surface. It was determined that a linear decrease had been observed in the peak current values of the redox probe by using DPV with increasing concentrations of the antigen. It is reported that the linear measurement range determined for prostate-specific antigen detection was between 1 pg mL^{-1} and 5 ng mL^{-1}, and the limit of detection was 0.8 pg mL^{-1} [107].

CA125 was detected by DPV using a layer-by-layer assembly of ordered mesoporous carbon, gold nanoparticles and MgAl-layered double hydroxides containing ferrocene carboxylic acid composite. It is explained that the conductivity increased significantly with the addition of the ferrocene component to the composite. The electrochemical performance of the biosensor was determined based on the change of the peak current observed in the voltammogram at +0.27 V according to the ferrocene in the presence of different CA125 antigen concentrations. It is stated that the peak current value obtained with the increase in the CA125 concentration changed inversely, since the complex formed between the antigen and the antibody. The linear measuring range and limit of detection of the biosensor were described as 0.01 U mL^{-1}–1000 U mL^{-1} and 0.004 U mL^{-1}, respectively [108]. A label-free sandwich type biosensor was developed for the electrochemical detection of cytokeratin fragment antigen 21-1 (CYFRA 21-1), a lung cancer biomarker. An antibody–antigen–antibody sandwich structure was formed between the 4-(2-trimethylsilylethinyl)benzoic acid gold electrode used as a bridge and the poly(ε-caprolactone)-b-poly(ethylene oxide) copolymer. The linear concentration range and limit of detection for the sensor determined by electrochemical impedance spectroscopy were declared as 1.0 pg mL^{-1} to 10 ng mL^{-1} and 0.125 pg mL^{-1}, respectively. According to the impedance results, the electrochemical responses showed a linear response with the concentration of CYFRA 21-1 [109].

Liu et al. developed a gold nanoparticle/polyethyleneimine/reduced graphene oxide nanocomposite for the electrochemical detection of matrix metalloproteinase-1, a cancer biomarker, based on the knowledge that gold nanoparticles were supportive in maintaining the reversibility of redox reactions in electroanalytical reactions. They determined that the biosensor performance obtained by DPV had an operating range of 1 ng mL^{-1} to 50 ng mL^{-1}. In this work, the peak current value obtained from voltammetry decreased due to the increased antigen concentration blocking on the electrode surface. In the electrochemical measurements taken in 5 mM $Fe(CN)_6^{3-/4-}$ medium, it is stated that an insulating layer was formed due to the antigen–antibody complex, and therefore, a repulsive electrostatic interaction occurred between the antigen and $Fe(CN)_6^{3-/4-}$ [110]. Zhu et al. also developed a carbon-based nanocomposite to take advantage of its high surface area and good conductivity properties. The surface was used for the construction of an immunosensor for the detection of alpha-fetoprotein, which is a liver cancer biomarker. They calculated a linear measurement range of 0.10 ng mL^{-1} to 420 ng mL^{-1} and a limit of detection of 0.03 ng mL^{-1} using square wave voltammetry, a method that could suppress background current and provide sensitivity to the biosensor system [106].

4.4. Cell-Based Label-Free Cancer Biosensors

The use of cells as a biorecognition element dates back to the early 1970s and it is still preferred today. Cells offer an interesting alternative to other biorecognition units such as antibodies, enzymes and nucleic acids thanks to their relatively easy production and lower cost than antibodies and purified enzymes. As an example, since whole cells offer a multi-enzyme alternative, they can be preferred in the development of biosensors for the simultaneous determination of various analytes. In addition, cell-based biosensors enable in situ monitoring using suitable substrates [78,111,112]. However, some limitations such as maintenance and immobilization of cells can arise [113].

Human cervical carcinoma (HeLa) cells were used as a biorecognition unit in an electrochemical label-free cytosensor to evaluate the anticancer activity of pinoresinol, which had biological properties such as anticancer, anti-inflammatory and antifungal effects. HeLa cells were immobilized on a GCE surface modified with multi-walled carbon nanotubes and gold nanoparticles, and the performance of the biosensor was evaluated by electrochemical impedance spectroscopy with different pinoresinol concentrations. The limit of detection value for the biosensor, which showed a linear correlation with the pinoresinol concentration range of 10^2 to 10^6 cells mL^{-1}, was reported as 10^2 cells mL^{-1} [114]. Another cell-based label-free electrochemical biosensor was developed to investigate the interactions of cancer cells (HepG2 cells and A549 cells) with molecules and to screen anticancer drugs. Cancer cells were immobilized on the GCE coated with N-doped graphene–Pt nanoparticles–chitosan and polyaniline. It is stated that this electrode surface might be suitable for examining different cell lines by changing the targeted cells as a result of the electrochemical properties examined by DPV with its large surface area and catalytic properties [115].

Liu et al. carried out the detection of cell surface glycan that played an important role in processes such as cancer cell metastasis by means of a nano channel ion channel of porous anodic alumina hybrid combined with an electrochemical detector. Thus, the enhanced ionic current caused by the array nano channels along with the ionic current rectification gave a precise current response. The alumina was functionalized with amino-propyltriethoxysilane and glutaraldehyde to immobilize the cell surface glycan. The linear working range was obtained from 10 fM to 10 nM, and the limit of detection was calculated to be approximately 10.0 aM. It is stated that this biosensor was a promising alternative that could be used in cancer diagnosis and an important platform for label-free detection of cell surface glycan [116].

Despite the advantages of cell-based electrochemical biosensors, there are also various disadvantages faced by designers such as reproducibility and inability to selectively place cells at detection sites [117]. In addition, some difficulties in terms of electrochemical techniques such as amperometric and impedimetric have been reported in the literature. For example, the difficulties often observed in electrochemical impedance spectroscopy-based studies are that the measured electrochemical response is the total change produced by a set of cells and poor selectivity. Emerging technology, nanomaterial selection, new immobilization matrices, integration of different transducer mechanisms and advances in the control of the sensor interface are some of the promising approaches to overcome these challenges [105,106].

5. Immobilization Strategies of Biorecognition Elements

Biorecognition element immobilization or its integration is one of the important processes to be considered, since this step thoroughly affects the analytical performance of all types of biosensors. The efficient immobilization of the biorecognition element is a process applied to overcome the problems such as loss of activity and stability by integrating biomolecules into a suitable support material. The immobilization methods are classified as adsorption, covalent bonding, cross-linking, etc., according to the type of the biomolecule to be immobilized and the structure of the immobilization surface [118]. These methods are illustrated in Figure 2.

In Table 1, the immobilization methods used by some of the studies within the scope of this review are indicated. Some cancer detection studies in the literature for recent years, different biorecognition units, other biosensor components and the parameters used in these studies are listed. Metals, metal oxides, conductive polymers, biopolymers, carbon-based structures, quantum dots and their composites [93,100,107,109,119,120] have been used as the immobilization matrices for label-free electrochemical cancer biosensors. In general, electrostatic interactions can have negative effects on the stability of the biorecognition element or the repeatability of the biosensor [121,122]. However, these methods, which have very simple processes, are still actively used in the surface immobilization of many electrodes. The entrapment method also offers specific properties and contributes to the improvement of chemical and thermal stability. However, leakage and low biological activity limit this method. To overcome the leakage problem, crosslinkers are preferred in the immobilization step. However, at this stage, excessive chemical requirements are necessary [123].

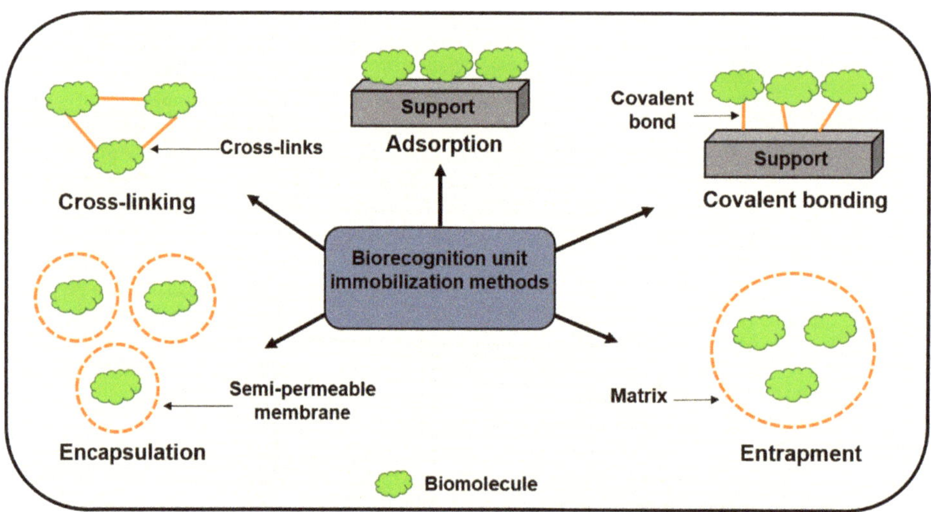

Figure 2. Various immobilization methods for the biorecognition elements.

In the study of Yaiwong et al., an immunosensor for label-free electrochemical cancer detection was developed. Electrostatic interaction was carried out for the immobilization of the anti-metalloproteinase-7 (MMP-7) capture antibody, which was used as a biorecognition element, on the surface of the screen-printed carbon electrode (SPCE) coated with two-dimensional (2D) MoS_2/graphene oxide [124]. More commonly, immobilization methods by covalent or cross-linking over carboxyl or amine groups are robust and reproducible ways to obtain an effective biosensor interface. Glutaraldehyde or carbodiimide structures that act as bridges in these binding reactions are preferred [121]. As an example, Yan et al. coated the surface of an indium tin oxide electrode with chitosan-modified reduced graphene oxide nanocomposite for prostate cancer detection. In order to detect prostate-specific antigens with this biosensor, they immobilized the recognition antibodies onto the electrode surface by covalent bonding. Chitosan naturally provided a large number of amino groups to the electrode surface, and glutaraldehyde, a bifunctional bridge, was used for covalent immobilization of the anti-PSA antibody with amino groups. Thus, a label-free electrochemical immunosensing platform based on antibody–antigen affinity was developed [107].

Echeverri et al. immobilized the anti-β-1,4-galactosyltransferase-V (β-1,4-GalT-V) antibody biorecognition element on the self-assembled monolayer (SAM)-coated SPCE by covalent bonding for the detection of colorectal cancer. The SAM provided a carboxylic acid group that allowed for antibody binding [98]. Generally, N-(3-dimethylaminopropyl)-N′-ethylcarbodiimide hydrochloride (EDC) and N-hydroxysuccinimide (NHS) pairs are used for this type of covalent bonding. In this way, a bridge is formed between the amine and carboxyl groups and a high binding efficiency is achieved [121]. Although covalent bonding seems to offer good efficiency and is an advantageous method, it can also have various disadvantages in some cases. For example, denaturation may occur due to the undesirable site orientation of the biorecognition element, and in addition, the bridging compounds are needed to use in the covalent bonding reaction. Therefore, there can be a decrease or disappearance of the biocatalytic effect expected from the biorecognition unit [125]. Moreover, covalent bonding, which causes a tight binding, can also restrict the movement of the biorecognition elements, which may also cause a loss of activity [126].

Table 1. Electrochemical-based label-free biosensors for cancer detection: biorecognition elements, sensor platforms, immobilization methods and electroanalytical performances.

Biorecognition Elements	Other Components	Immobilization Method	Cancer Type	Analyte	Electrochemical Technique	Limit of Detection	Linear Range	References
Biotinylated DNA probe	Multi-walled carbon nanotubes non-covalently functionalized with avidin	Covalent binding	Breast cancer	BRCA1	Electrochemical impedance spectroscopy	330 aM	1.0 fM–10 nM	[67]
Three-dimensional (3D) DNA walker	Au nanoparticles/ encapsulation of glucose oxidase in zeolitic imidazolate framework-8 (ZIF-8)	Electrostatic adsorption	Cancer	MicroRNA	Differential pulse voltammetry	29 pM	0.1 nM–10 μM	[68]
Hairpin probe (H1)	MXene/MoS_2/thionine/Au nanoparticles	Adsorption	Cancer	MicroRNA-21	Square wave voltammetry	26 fM	100 fM–100 nM	[72]
Anti-CA 15-3 antibodies	Gold nanoparticle-dye/poly(3-aminobenzylamine)/two dimensional $MoSe_2$/graphene oxide	Covalent binding	Breast cancer	CA 15-3	Differential pulse voltammetry	0.14 U mL^{-1}	0.0–500 U mL	[73]
DNA-21	Gold nanoparticle-dye/poly(3-aminobenzylamine)/two dimensional $MoSe_2$/graphene oxide	Adsorption	Breast cancer	MicroRNA-21	Differential pulse voltammetry	1.2 fM	0.0–1000 pM	[73]
DNA	Reduced-graphene oxide/polypyrrole/silver nanoparticles	Covalent binding	Lung Cancer	Cytokeratin 19 fragment 21–1 (CYFRA21-1)	Differential pulse voltammetry	2.4 fM	1.0×10^{-14}–1.0×10^{-6} M	[74]
DNA-aptamer probe	α-Fe_2O_3/Fe_3O_4@Au	Adsorption	Ovarian cancer	CA125	Differential pulse voltammetry	2.99 U mL^{-1}	5–125 U mL^{-1}	[77]
Aminoated CA 125 aptamers	Nickel hexacyanoferrate nanocubes/polydopamine functionalized graphene	Covalent binding	Ovarian cancer	CA125	Differential pulse voltammetry	0.076 pg mL^{-1}	0.10 pg mL^{-1}–1.0 μg mL^{-1}	[82]
Aptamer	Graphene oxide functionalized with aspartic acid	Covalent binding	Cancer	Cytochrome c	Differential pulse voltammetry	0.74 nM	10 nM–100 μM	[83]
Epidermal growth factor receptor (EGFR) aptamers	Amino-functionalized graphene/ thionine/gold particle nanocomposites	Covalent binding	Cancer	EGFR	Differential pulse voltammetry	5 pg mL^{-1}	0.05–200 ng mL^{-1}	[84]

Table 1. *Cont.*

Biorecognition Elements	Other Components	Immobilization Method	Cancer Type	Analyte	Electrochemical Technique	Limit of Detection	Linear Range	References
CA72-4 antibodies	Carbon nanotube–graphene oxide hybrid	Covalent binding	Gastric cancer	Antigen 72-4	Differential pulse voltammetry	0.4 U mL^{-1}	2.0–80.0 U mL^{-1}	[88]
Calreticulin antibody	Electrodeposited single-walled carbon nanotubes and polymerized oxiran-2-yl methyl 3-(1H-pyrrol-1-yl) propanoate monomer	Covalent binding	Breast cancer	Calreticulin	Electrochemical impedance spectroscopy	0.0046 pg mL^{-1}	0.015–60 pg mL^{-1}	[89]
Cancer antigen (CA 15-3) antibody	Ternary silver/titanium dioxide/reduced graphene oxides nanocomposite	Covalent binding	Breast cancer	CA 15-3 antigen	Amperometry	0.07 U mL^{-1}	0.1–300 U mL^{-1}	[90]
Anti-cancer antigen (CA125) antibody	Boron nitride nanosheets	Physical adsorption	Ovarian cancer	CA125	Differential pulse voltammetry	1.18 U mL^{-1}	5–100 U mL^{-1}	[94]
Anti-nuclear matrix protein 22 (NMP22)	Reduced graphene oxide/tetraethylene pentamine/Cu-based metal organic frameworks deposited silver nanoparticles	Covalent binding	Bladder cancer	NMP22	Differential pulse voltammetry	33.33 fg mL^{-1}	0.1 pg mL^{-1}–1000 ng mL^{-1}	[97]
Anti-β-1,4-galactosyltransferase-V (β-1,4-GalT-V) antibody	Self-assembled monolayer-coated screen-printed gold electrode	Covalent binding	Colorectal cancer	β-1,4-GalT-V glycoprotein	Electrochemical impedance spectroscopy	7 pM	5–150 pM	[98]
Prostate-specific membrane antibody (PSMA)	Cysteamine-modified gold nanoparticles	Cross-linking	Prostate cancer	PSMA protein	Differential pulse voltammetry	0.47 ng mL^{-1}	0–5 ng mL^{-1}	[100]
Anti-prostate-specific antigen (PSA)	Chitosan, graphene, ionic liquid and ferrocene cryogel	Chemical adsorption	Prostate cancer	PSA	Differential pulse voltammetry	4.8×10^{-8} ng mL^{-1}	1.0×10^{-7}–1.0×10^{-1} ng mL^{-1}	[103]
Anti-alpha-fetoprotein (AFP)	MnO_2 functionalized mesoporous carbon hollow sphere	Cross-linking	Liver cancer	AFP	Square wave voltammetry	0.03 ng mL^{-1}	0.10–420 ng mL^{-1}	[106]

Table 1. *Cont.*

Biorecognition Elements	Other Components	Immobilization Method	Cancer Type	Analyte	Electrochemical Technique	Limit of Detection	Linear Range	References
Polyclonal anti-PSA antibody	Chitosan–graphene-modified indium tin oxide electrode	Covalent binding	Prostate cancer	Prostate-specific antigen	Amperometry	0.8 pg mL^{-1}	1–5 ng mL^{-1}	[107]
Primary antibody (Ab1)	Linear poly(ε-caprolactone)-b-poly(ethylene oxide) copolymer	Cross-linking	Lung cancer	CYFRA 21-1	Electrochemical impedance spectroscopy	0.125 pg mL^{-1}	1 pg mL^{-1}–10 ng mL^{-1}	[109]
Human cervical carcinoma (HeLa) cells	Carboxylated multiwalled carbon nanotubes/gold nanoparticles	Adsorption	Cervical cancer	Pinoresinol	Electrochemical impedance spectroscopy	10^2 cells mL^{-1}	10^2–10^6 cells mL^{-1}	[114]
Anti-matrix metalloproteinase (MMP)-7 capture antibodies	Two-dimensional molybdenum disulfide/graphene oxide nanocomposite	Electrostatic interactions	Pancreatic and colorectal cancers	MMP-7	Differential pulse voltammetry	0.007 ng mL^{-1}	0.010–75 ng mL^{-1}	[124]
Antibodies specific to IL-8 (Anti-IL-8)	Silver molybdate nanoparticles	Covalent binding	Oral cancer	IL-8	Differential pulse voltammetry	90 pg mL^{-1}	1 fg mL^{-1}–40 ng mL^{-1}	[127]
Human epidermal growth factor receptor 2 (HER2) antibody	Fe_3O_4/TMU-21/multi-walled carbon nanotubes	Cross-linking	Breast cancer	HER2	Amperometry	0.3 pg mL^{-1}	1.0 pg mL^{-1}–100 ng mL^{-1}	[128]
Carcinoembryonic antigen (CEA) antibody	Fe_3O_4@Au nanoparticles	Adsorption	Cancer	CEA	Linear sweep voltammetry	0.10 pg mL^{-1}	0.001–100 ng mL^{-1}	[129]
DNA	Exo-III-assisted target recycling and dual enzymes	Covalent binding	Oral cancer	ORAOV1	Electrochemical impedance spectroscopy	0.019 fM	0.05 fM–20 pM	[130]
Capture strand	Au nanoparticles	Adsorption	Brain cancers	Cerebrospinal fluid microRNAs	Differential pulse voltammetry	56 fM	0.5–80 pM	[131]
CEA aptamer (AptGAC-P)	6-Mercapto-1-hexanol (MCH)/cpDNA2/gold	Adsorption	Cancer	CEA	Differential pulse voltammetry	0.24 ng mL^{-1}	2–45 ng mL^{-1}	[132]

Table 1. *Cont.*

Biorecognition Elements	Other Components	Immobilization Method	Cancer Type	Analyte	Electrochemical Technique	Limit of Detection	Linear Range	References
Carcinoembryonic antigen aptamer	Au nanoparticles	Self-assembly	Lung cancer	CEA	Electrochemical impedance spectroscopy	0.085 ng ml^{-1}	0.2–15.0 ng ml^{-1}	[133]
AS1411 aptamer	Reduced graphene oxide–chitosan–gold nanoparticle	Covalent binding	Breast cancer	MCF-7 cancer cells	Electrochemical impedance spectroscopy	4 cells mL^{-1}	1×10–1×10^6 cells mL^{-1}	[134]
DNA aptamer	Gold electrode	Covalent binding	Cancer	Cluster of differentiation-44 (CD44)	Electrochemical impedance spectroscopy	0.087 ng mL^{-1}	0.1–1000 ng mL^{-1}	[135]
PDGF-BB affinity aptamers	Carboxyl-functionalized photoresist-derived carbon	Covalent binding	Cancer	Platelet-derived growth factor-BB (PDGF-BB)	Cyclic voltammetry	7 pM	0.01–50 nM	[136]
DNA	Graphene oxide–chitosan/polyvinylpy rrolidone–gold nano urchin	Covalent binding	Lung Cancer	miR-141	Square wave voltammetry	0.94 fM	2.0–5.0×10^5 fM	[137]
DNA-21	Graphene/polypyrrole/gold nanoparticles	Electrostatic interaction	Cancer	miRNA-21	Differential pulse voltammetry	0.020 fM	1.0 fM to 1.0 nM	[138]
DNA	Chitosan-capped gold nanoparticles	Electrostatic interaction	Cervical cancer	HPV-16	Cyclic voltamme-try/square wave voltammetry	1.0 pM	1 pM–1 µM	[139]
ssDNA	L-cysteine functionalized ZnS quantum dots	Covalent binding	Ovarian cancer	miR-200a	Electrochemical impedance spectroscopy	8.4 fM	1.0×10^{-14}–1.0×10^{-6} M	[140]
ssDNA	Reduced graphene oxide/polyaniline nanofibers	Electrostatic interaction	Breast cancer	BRCA1	Differential pulse voltammetry	3.01×10^{-16} M	1.0×10^{-15}–1.0×10^{-7} M	[141]
acpcPNA-T9 probe	Ag@Au core–shell nanoparticles electrodeposited on graphene quantum dots	Adsorption	Cancer	miRNA-21	Chronoamperometry	5 pM	5 pM–5 mM	[142]

Although the immobilization of biorecognition elements on the surface of the biosensing platform is a very important step for the design of sensitive, selective and long operational lifetime biosensors, it is clear that each method has several advantages and disadvantages. Various factors such as the immobilization matrix and the charge or functional groups of the biorecognition units guide the selection of the appropriate method, and thus, effective interfaces are created.

6. Label-Free Electrochemical Cancer Biosensors for Point-of-Care Applications

Label-free electrochemical biosensors have a high capability of being adapted into point-of-care (POC) systems that can be used for outside the laboratory testing to minimize the need for healthcare services such as hospitals [14,143–145]. In POC testing particularly, microfluidic devices have attracted great attention lately for effective and accurate cancer diagnosis owing to their ability to separate analytes at a good resolution in a rapid reaction time and to minimize the handling errors and costs [143]. As a result, promising detection systems with high performances are acquired with the elimination of the need for trained personnel. Recently, in the study by Keyvani et al., a POC sensing device for the detection of cervical cancer was developed for whole blood. This system identified cancer circulating DNA with high purity by the help of a graphene oxide-dependent electrochemical sensor platform by using differential pulse voltammetry [146]. In another study, Ming et al. fabricated a cellulose-paper-based POC testing with the modification of amino redox graphene, thionine, streptavidin integrated gold nanoparticles and chitosan for the detection of biomarker 17β-estradiol, which may be associated with breast cancer. The detection strategy, realized with differential pulse voltammetry in phosphate buffer solution, was carried out via the interaction of the target biomarker and its biotin-modified aptamer on the surface of the paper. The linearity of the label-free sensor was between 10 pg mL^{-1} and 100 ng mL^{-1}, with a limit of detection value of 10 pg mL^{-1} [147].

Besides microfluidic devices, multiplex systems that can detect multiple analytes associated with cancer have several advantages in terms of label-free point-of-care testing. As an example, Kuntamung and his colleagues achieved simultaneous detection of breast cancer biomarkers: mucin1, cancer antigen 15-3 and human epidermal growth factor receptor 2 depending on the formed antibody and antigen interactions. For this purpose, redox species and antibody-conjugated polyethylenimine-modified gold nanoparticles were utilized as the modification elements of a SPCE. In addition to multiplex detection performance, the label-free biosensor kept 90% of its initial responses obtained via voltammetry [92]. In another approach that contained the fabrication of a flexible screen-printed electrode system, carcinoembryonic antigen was detected on graphene–ZnO nanorods deposited on a polyethylene terephthalate substrate with a screen-printed electrode by Chakraborty et al. ZnO nanorods were functionalized with aptamers and the resulting surface improved the mass transport through the electric field application. This system was integrated into smartphone interface technology and a handheld potentiostat. The linearity of the label-free sensor was between 0.001 pg mL^{-1} and 10 pg mL^{-1}, with a limit of the detection value of 1 fg mL^{-1} by using electrochemical impedance spectroscopy. The results were also validated using a commercial ELISA kit [148].

The use of label-free POC testing in cancer diagnosis is in increasing demand in recent years since POC systems yield rapid decisions, more frequent testing to monitor wellness, eliminate the need for trained staff and utilize small specimen volumes. In addition, they are cost-effective. Despite these advantages, they are still more open to false positives or negatives and incorrect interpretations. Also, these sensing platforms have a risk of external interference since the environment is not as well controlled as in laboratories. In some cases, the sampling procedure can be inconvenient, such as in cancer diagnosis protocols. Indeed, POC-based electrochemical cancer biosensors are not yet available on the market. One of the additional reasons for this issue could be the distance between physicians and electrochemical biosensor developers. It is believed that multidisciplinary studies between them will improve the quality of the developed platforms. Additionally, shelf-life and production control are important parameters to improve their commercialization capacity [149–152]. However, electrochemically based POC systems are promising tools for the accurate and fast detection of cancer with their overall characteristics.

7. Conclusions and Future Perspectives

In the current review, we have summarized the recent achievements and progresses around label-free electrochemical biosensors that are utilized for cancer detection. Since the type of biorecognition element is an important key parameter to enhance the selectivity of the detection, the classification of the biosensors is made according to the types of recognition elements. Besides the achievements, the current challenges are also outlined in detail. Label-free detection systems are in urgent demand owing to their properties, including reducing labored modification steps and interference effects.

The growing demand on clinical research and the medical industry for cancer studies has pushed scientists to perform early detection with practical analytical tools instead of time-consuming and back-breaking methods. In addition to detection, isolation of the cancer cells is also important to increase the survival rates and quality of life. The design and development of early-cancer diagnosis platforms has been one of the hot topics of the last decades. The recent advances in the field of cancer diagnosis show that electrochemical sensing methodologies have an important impact on the accurate, rapid and sensitive detection of cancer types. Particularly, label-free electrochemical biosensors maintain predominant features to obtain reliable, cost-effective and selective cancer diagnosis that can serve for future implementations. With the addition of advanced materials such as nanomaterials, not only sensitivity of the biosensors but also the selectivity of them can be significantly improved. Surface modification makes bare electrode substrates available and suitable for biorecognition element immobilization. Recent studies on label-free and electrochemical biosensing of cancers indicate how promising and operational these biosensors are. It is certain that their advantages will certify more powerful medical applications in the near future with the support of growing materials science technology.

Author Contributions: V.S.: Conceptualization, Writing—Original draft preparation, Review and Editing, Supervision; F.K.: Conceptualization, Writing—Original draft preparation, Review and Editing, Supervision, Project Administration. All authors have read and agreed to the published version of the manuscript.

Funding: This research received no external funding.

Institutional Review Board Statement: Not applicable.

Informed Consent Statement: Not applicable.

Data Availability Statement: Not applicable.

Acknowledgments: F.K. acknowledges Turkish Academy of Sciences as an associate member.

Conflicts of Interest: The authors declare no conflict of interest.

References

1. Soerjomataram, I.; Bray, F. Planning for tomorrow: Global cancer incidence and the role of prevention 2020–2070. *Nat. Rev. Clin. Oncol.* **2021**, *18*, 663–672. [CrossRef] [PubMed]
2. Bray, F.; Jemal, A.; Grey, N.; Ferlay, J.; Forman, D. Global cancer transitions according to the Human Development Index (2008–2030): A population-based study. *Lancet Oncol.* **2012**, *13*, 790–801. [CrossRef] [PubMed]
3. Sung, H.; Ferlay, J.; Siegel, R.L.; Laversanne, M.; Soerjomataram, I.; Jemal, A.; Bray, F. Global cancer statistics 2020: GLOBOCAN estimates of incidence and mortality worldwide for 36 cancers in 185 countries. *CA Cancer J. Clin.* **2021**, *71*, 209–249. [CrossRef] [PubMed]
4. Qian, L.; Li, Q.; Baryeh, K.; Qui, W.; Li, K.; Zhang, J.; Yu, Q.; Xu, D.; Liu, W.; Brand, R.E.; et al. Biosensors for early diagnosis of pancreatic cancer: A review. *Transl. Res.* **2019**, *213*, 67–89. [CrossRef] [PubMed]
5. Cui, F.; Zhou, Z.; Zhou, H.S. Measurement and analysis of cancer biomarkers based on electrochemical biosensors. *J. Electrochem. Soc.* **2019**, *167*, 037525. [CrossRef]

6. Khanmohammadi, A.; Aghaie, A.; Vahedi, E.; Qazvini, A.; Ghanei, M.; Afkhami, A.; Hajian, A.; Bagheri, H. Electrochemical biosensors for the detection of lung cancer biomarkers: A review. *Talanta* **2020**, *206*, 120251. [CrossRef]
7. Li, G.; Wu, J.; Qi, X.; Wan, X.; Liu, Y.; Chen, Y.; Xu, L. Molecularly imprinted polypyrrole film-coated poly (3,4-ethylenedioxythiophene):polystyrene sulfonate-functionalized black phosphorene for the selective and robust detection of norfloxacin. *Mater. Today Chem.* **2022**, *26*, 101043. [CrossRef]
8. Ahlquist, D.A. Universal cancer screening: Revolutionary, rational, and realizable. *NPJ Precis. Oncol.* **2018**, *2*, 23. [CrossRef]
9. Sadighbayan, D.; Sadighbayan, K.; Khosroushahi, A.Y.; Hasanzadeh, M. Recent advances on the DNA-based electrochemical biosensing of cancer biomarkers: Analytical approach. *TrAC Trends Anal. Chem.* **2019**, *119*, 115609. [CrossRef]
10. Wang, H.; Peng, R.; Wang, J.; Qin, Z.; Xue, L. Circulating microRNAs as potential cancer biomarkers: The advantage and disadvantage. *Clin. Epigenet.* **2018**, *10*, 59. [CrossRef]
11. Shih, Y.-C.T.; Sabik, L.M.; Stout, N.K.; Halpern, M.T.; Lipscomb, J.; Ramsey, S.; Ritzwoller, D.P. Health economics research in cancer screening: Research opportunities, challenges, and future directions. *JNCI Monogr.* **2022**, *59*, 42–50. [CrossRef]
12. Hasan, M.R.; Ahommed, M.S.; Daizy, M.; Bacchu, M.S.; Ali, M.R.; Al-Mamun, M.R.; Aly, M.A.S.; Khan, M.Z.H.; Hossain, S.I. Recent development in electrochemical biosensors for cancer biomarkers detection. *Biosens. Bioelectron. X* **2021**, *8*, 100075. [CrossRef]
13. Chen, Y.; Sun, L.; Qiao, X.; Zhang, Y.; Li, Y.; Ma, F. Signal-off/on electrogenerated chemiluminescence deoxyribosensors for assay of early lung cancer biomarker (NAP2) based on target-caused DNA charge transfer. *Anal. Chim. Acta* **2020**, *1103*, 67–74. [CrossRef]
14. Pacheco, J.G.; Silva, M.S.V.; Freitas, M.; Nouws, H.P.A.; Delerue-Matos, C. Molecularly imprinted electrochemical sensor for the point-of-care detection of a breast cancer biomarker (CA 15-3). *Sens. Actuators B* **2018**, *256*, 905–912. [CrossRef]
15. Pothipor, C.; Jakmunee, J.; Bamrungsap, S.; Ounnunka, K. An electrochemical biosensor for simultaneous detection of breast cancer clinically related microRNAs based on a gold nanoparticles/graphene quantum dots/graphene oxide film. *Analyst* **2021**, *146*, 4000–4009. [CrossRef]
16. Li, S.; Hu, C.; Chen, C.; Zhang, J.; Bai, Y.; Tan, C.S.; Ni, G.; He, F.; Li, W.; Ming, D. Molybdenum disulfide supported on metal–organic frameworks as an ultrasensitive layer for the electrochemical detection of the ovarian cancer biomarker CA125. *ACS Appl. Bio Mater.* **2021**, *4*, 5494–5502. [CrossRef]
17. Dou, Y.; Zhenhua, L.; Su, J.; Song, S. A portable biosensor based on Au nanoflower interface combined with electrochemical immunochromatography for POC detection of prostate-specific antigen. *Biosensors* **2022**, *12*, 259. [CrossRef]
18. Chen, S.-C.; Chen, K.-T.; Jou, A.F.-J. Polydopamine-gold composite-based electrochemical biosensor using dual-amplification strategy for detecting pancreatic cancer-associated microRNA. *Biosens. Bioelectron.* **2021**, *173*, 112815. [CrossRef]
19. Liu, Q.; Xie, H.; Liu, J.; Kong, J.; Zhang, X. A novel electrochemical biosensor for lung cancer-related gene detection based on copper ferrite-enhanced photoinitiated chain-growth amplification. *Anal. Chim. Acta* **2021**, *1179*, 338843. [CrossRef]
20. Jing, L.; Xie, C.; Li, Q.; Yang, M.; Li, S.; Li, H.; Xia, F. Electrochemical biosensors for the analysis of breast cancer biomarkers: From design to application. *Anal. Chem.* **2021**, *94*, 269–296. [CrossRef]
21. Kuralay, F.; Bayramlı, Y. Electrochemical determination of mitomycin C and its interaction with double-stranded DNA using a poly(*o*-phenylenediamine)-multi-walled carbon nanotube modified pencil graphite electrode. *Anal. Lett.* **2021**, *54*, 1295–1308. [CrossRef]
22. Li, G.; Qi, X.; Wu, J.; Xu, L.; Wan, X.; Liu, Y.; Chen, Y.; Li, Q. Ultrasensitive, label-free voltammetric determination of norfloxacin based on molecularly imprinted polymers and Au nanoparticle-functionalized black phosphorus nanosheet nanocomposite. *J. Hazard. Mater.* **2022**, *436*, 129107. [CrossRef]
23. Hai, X.; Li, Y.; Zhu, C.; Song, W.; Cao, J.; Bi, S. DNA-based label-free electrochemical biosensors: From principles to applications. *TrAC Trends Anal. Chem.* **2020**, *133*, 116098. [CrossRef]
24. Reta, N.; Saint, C.P.; Michelmore, A.; Prieto-Simon, B.; Voelcker, N.H. Nanostructured electrochemical biosensors for label-free detection of water-and food-borne pathogens. *ACS Appl. Mater. Interfaces* **2018**, *10*, 6055–6072. [CrossRef] [PubMed]
25. Jadon, N.; Jain, R.; Sharma, S.; Singh, K. Recent trends in electrochemical sensors for multianalyte detection—A review. *Talanta* **2016**, *161*, 894–916. [CrossRef] [PubMed]
26. Yahaya, M.L.; Noordin, R.; Razak, K.A. Advanced nanoparticle-based biosensors for diagnosing foodborne pathogens. In *Advanced Biosensors for Health Care Applications*; Inamuddin Mohammad, A., Khan, R., Asiri, A.M., Eds.; Elsevier: Amsterdam, The Netherlands, 2019; Chapter 1; pp. 1–43.
27. Sharma, H.; Mutharasan, R. Review of biosensors for foodborne pathogens and toxins. *Sens. Actuators B* **2013**, *183*, 535–549. [CrossRef]
28. Dükar, N.; Tunç, S.; Öztürk, K.; Demirci, S.; Dumangöz, M.; Sönmez Çelebi, M.; Kuralay, F. Highly sensitive and selective dopamine sensing in biological fluids with one-pot prepared graphene/poly(*o*-phenylenediamine) modified electrodes. *Mater. Chem. Phys.* **2019**, *228*, 357–362. [CrossRef]
29. Hassan, R.Y.A. Advances in electrochemical nano-biosensors for biomedical and environmental applications: From current work to future perspectives. *Sensors* **2022**, *22*, 7539. [CrossRef]
30. Sanko, V.; Şenocak, A.; Oğuz Tümay, S.; Orooji, Y.; Demirbas, E. An electrochemical sensor for detection of trace-level endocrine disruptor bisphenol A using $Mo_2Ti_2AlC_3$ MAX phase/MWCNT composite modified electrode. *Environ. Res.* **2022**, *212*, 113071. [CrossRef]

31. Patel, B.A. *Electrochemistry for Bioanalysis*; Elsevier: Amsterdam, The Netherlands, 2021.
32. Laborda, E.; González, J.; Molina, Á. Recent advances on the theory of pulse techniques: A mini review. *Electrochem. Commun.* **2014**, *43*, 25–30. [CrossRef]
33. Rezaei, B.; Irannejad, N. Electrochemical detection techniques in biosensor applications. In *Electrochemical Biosensors*; Ensafi, A.A., Ed.; Elsevier: Amsterdam, The Netherlands, 2019; Chapter 2; pp. 11–43.
34. Forouzanfar, S.; Alam, F.; Pala, N.; Wang, C. A review of electrochemical aptasensors for label-free cancer diagnosis. *J. Electrochem. Soc.* **2020**, *167*, 067511. [CrossRef]
35. Özbek, O.; Berkel, C.; Isildak, O.; Isildak, I. Potentiometric urea biosensors. *Clin. Chim. Acta* **2022**, *524*, 154–163. [CrossRef]
36. Sharifi, M.; Avadi, M.R.; Attar, F.; Dashtestani, F.; Ghorchian, H.; Rezayat, S.M.; Saboury, A.A.; Falahati, M. Cancer diagnosis using nanomaterials based electrochemical nanobiosensors. *Biosens. Bioelectron.* **2019**, *126*, 773–784. [CrossRef]
37. Chillawar, R.R.; Tadi, K.K.; Motghare, R.V. Voltammetric techniques at chemically modified electrodes. *J. Anal. Chem.* **2015**, *70*, 399–418. [CrossRef]
38. Kuralay, F.; Tunç, S.; Bozduman, F.; Oksuz, L.; Uygun Oksuz, A. Biosensing applications of titanium dioxide coated graphene modified disposable electrodes. *Talanta* **2016**, *160*, 325–331. [CrossRef]
39. Pourali, A.; Rashidi, M.R.; Barar, J.; Pavon-Djavid, G.; Omidi, Y. Voltammetric biosensors for analytical detection of cardiac troponin biomarkers in acute myocardial infarction. *TrAC Trends Anal. Chem.* **2021**, *134*, 116123. [CrossRef]
40. Walker, N.L.; Dick, J.E. Oxidase-loaded hydrogels for versatile potentiometric metabolite sensing. *Biosens. Bioelectron.* **2021**, *178*, 112997. [CrossRef]
41. Sciuto, E.L.; Petralia, S.; van der Meer, J.; Conoci, S. Miniaturized electrochemical biosensor based on whole-cell for heavy metal ions detection in water. *Biotechnol. Bioeng.* **2021**, *118*, 1456–1465. [CrossRef]
42. Hussein, H.A.; Kandeil, A.; Gomaa, M.; El Nashar, R.M.; El-Sherbiny, I.M.; Hassan, R.Y.A. SARS-CoV-2-impedimetric biosensor: Virus-imprinted chips for early and rapid diagnosis. *ACS Sens.* **2021**, *6*, 4098–4107. [CrossRef]
43. Sanko, V.; Şenocak, A.; Oğuz Tümay, S.; Çamurcu, T.; Demirbas, E. Core-shell hierarchical enzymatic biosensor based on hyaluronic acid capped copper ferrite nanoparticles for determination of endocrine-disrupting bisphenol A. *Electroanalysis* **2022**, *34*, 561–572. [CrossRef]
44. Yan, Y.; Qiao, Z.; Hai, X.; Song, W.; Bi, S. Versatile electrochemical biosensor based on bi-enzyme cascade biocatalysis spatially regulated by DNA architecture. *Biosens. Bioelectron.* **2021**, *174*, 112827. [CrossRef] [PubMed]
45. Mahshid, S.S.; Flynn, S.E.; Mahshid, S. The potential application of electrochemical biosensors in the COVID-19 pandemic: A perspective on the rapid diagnostics of SARS-CoV-2. *Biosens. Bioelectron.* **2021**, *176*, 112905. [CrossRef] [PubMed]
46. Vestergaard, M.; Kerman, K.; Tamiya, E. An overview of label-free electrochemical protein sensors. *Sensors* **2007**, *7*, 3442–3458. [CrossRef] [PubMed]
47. Grieshaber, D.; MacKenzie, R.; Vörös, J.; Reimhult, E. Electrochemical biosensors-sensor principles and architectures. *Sensors* **2008**, *8*, 1400–1458. [CrossRef]
48. Ozcelikay, G.; Kurbanoglu, S.; Yarman, A.; Scheller, F.W.; Ozkan, S.A. Au-Pt nanoparticles based molecularly imprinted nanosensor for electrochemical detection of the lipopeptide antibiotic drug Daptomycin. *Sens. Actuators B* **2020**, *320*, 128285. [CrossRef]
49. Colombo, R.N.P.; Sedenho, G.C.; Crespilho, F.N. Challenges in biomaterials science for electrochemical biosensing and bioenergy. *Chem. Mater.* **2022**, *34*, 10211–10222. [CrossRef]
50. George, S.M.; Tandon, S.; Kandasubramanian, B. Advancements in hydrogel-functionalized immunosensing platforms. *ACS Omega* **2020**, *5*, 2060–2068. [CrossRef]
51. Bhalla, N.; Pan, Y.; Yang, Z.; Payam, A.F. Opportunities and challenges for biosensors and nanoscale analytical tools for pandemics: COVID-19. *ACS Nano* **2020**, *14*, 7783–7807. [CrossRef]
52. Wandelt, K. *Encyclopedia of Interfacial Chemistry: Surface Science and Electrochemistry*; Elsevier: Amsterdam, The Netherlands, 2018.
53. Sedenho, G.C.; Hassan, A.; de Souza, J.C.P.; Crespilho, F.N. In situ and operando electrochemistry of redox enzymes. *Curr. Opin. Electrochem.* **2022**, *34*, 101015. [CrossRef]
54. Sang, S.; Wang, Y.; Feng, Q.; Wei, Y.; Ji, J.; Zhang, W. Progress of new label-free techniques for biosensors: A review. *Crit. Rev. Biotechnol.* **2016**, *36*, 465–481. [CrossRef]
55. Lee, S.-L.; Kim, J.; Choi, S.; Han, J.; Seo, G.; Lee, Y.W. Fiber-optic label-free biosensor for SARS-CoV-2 spike protein detection using biofunctionalized long-period fiber grating. *Talanta* **2021**, *235*, 122801. [CrossRef]
56. Yen, Y.-K.; Chiu, C.-Y. A CMOS MEMS-based membrane-bridge nanomechanical sensor for small molecule detection. *Sci. Rep.* **2020**, *10*, 2931. [CrossRef]
57. Faria, H.A.M.; Zucolotto, V. Label-free electrochemical DNA biosensor for zika virus identification. *Biosens. Bioelectron.* **2019**, *131*, 149–155. [CrossRef]
58. Syahir, A.; Usui, K.; Tomizaki, K.; Kajikawa, K.; Mihara, H. Label and label-free detection techniques for protein microarrays. *Microarrays* **2015**, *4*, 228–244. [CrossRef]
59. Wang, H.; Huang, Y.; Xiog, M.; Wang, F.; Li, C. A label-free electrochemical biosensor for highly sensitive detection of gliotoxin based on DNA nanostructure/MXene nanocomplexes. *Biosens. Bioelectron.* **2019**, *142*, 111531. [CrossRef]
60. Han, S.; Liu, W.; Yang, S.; Wang, R. Facile and label-free electrochemical biosensors for microRNA detection based on DNA origami nanostructures. *ACS Omega* **2019**, *4*, 11025–11031. [CrossRef]

61. Vu, Q.K.; Tran, Q.Y.; Vu, N.P.; Anh, T.-L.; Le Dnag, T.T.; Tonezzer, M.; Nguyen, T.H.H. A label-free electrochemical biosensor based on screen-printed electrodes modified with gold nanoparticles for quick detection of bacterial pathogens. *Mater. Today Commun.* **2021**, *26*, 101726. [CrossRef]
62. Singh, A.K.; Dhiman, T.K.; Lakshmi, G.B.V.S.; Solanki, P.R. Dimanganese trioxide (Mn_2O_3) based label-free electrochemical biosensor for detection of Aflatoxin-B1. *Bioelectrochemistry* **2021**, *137*, 107684. [CrossRef]
63. Karunakaran, C.; Rajkumar, R.; Bhargava, K. Introduction to biosensors. In *Biosensors and Bioelectronics*, 1st ed.; Karunakaran, C., Bhargava, K., Benjamin, R., Eds.; Elsevier: Amsterdam, The Netherlands, 2015; Chapter 1; pp. 1–68.
64. Du, Y.; Dong, S. Nucleic acid biosensors: Recent advances and perspectives. *Anal. Chem.* **2017**, *89*, 189–215. [CrossRef]
65. Kuralay, F.; Erdem, A. Gold nanoparticle/polymer nanocomposite for highly sensitive drug–DNA interaction. *Analyst* **2015**, *140*, 2876–2880.
66. Fu, Z.; Lu, Y.-C.; Lai, J.J. Recent advances in biosensors for nucleic acid and exosome detection. *Chonnam Med. J.* **2019**, *55*, 86–98. [CrossRef] [PubMed]
67. Mujica, M.L.; Rubianes, M.D.; Rivas, G. A multipurpose biocapture nanoplatform based on multiwalled-carbon nanotubes non-covalently functionalized with avidin: Analytical applications for the non-amplified and label-free impedimetric quantification of BRCA1. *Sens. Actuators B* **2022**, *357*, 131304. [CrossRef]
68. Kong, L.; Lv, S.; Qiao, Z.; Yan, Y.; Zhang, J.; Bi, S. Metal-organic framework nanoreactor-based electrochemical biosensor coupled with three-dimensional DNA walker for label-free detection of microRNA. *Biosens. Bioelectron.* **2022**, *207*, 114188. [CrossRef] [PubMed]
69. Gao, J.; Wang, C.; Chu, Y.; Wang, S.; Sun, M.Y.; Ji, H.; Gao, Y.; Wang, Y.; Han, Y.; Song, F.; et al. Poly-L-lysine-modified graphene field-effect transistor biosensors for ultrasensitive breast cancer miRNAs and SARS-CoV-2 RNA detection. *Anal. Chem.* **2022**, *94*, 1626–1636. [CrossRef]
70. Khodadoust, A.; Nasirizadeh, N.; Seyfati, S.M.; Taheri, R.A.; Ghanei, M.; Bagheri, H. High-performance strategy for the construction of electrochemical biosensor for simultaneous detection of miRNA-141 and miRNA-21 as lung cancer biomarkers. *Talanta* **2023**, *252*, 123863. [CrossRef]
71. Meng, F.; Yu, W.; Chen, C.; Guo, S.; Tian, X.; Miao, Y.; Ma, L.; Zhang, X.; Yu, Y.; Huang, L.; et al. A versatile electrochemical biosensor for the detection of circulating microRNA toward non-small cell lung cancer diagnosis. *Small* **2022**, *18*, 2200784. [CrossRef]
72. Zhao, J.; He, C.; Wu, W.; Yang, H.; Dong, J.; Wen, L.; Hu, Z.; Yang, M.; Hou, C.; Huo, D. MXene-MoS_2 heterostructure collaborated with catalyzed hairpin assembly for label-free electrochemical detection of microRNA-21. *Talanta* **2022**, *237*, 122927. [CrossRef]
73. Pothipor, C.; Bamrungsap, S.; Jakmunee, J.; Ounnunkad, K. A gold nanoparticle-dye/poly (3-aminobenzylamine)/two dimensional $MoSe_2$/graphene oxide electrode towards label-free electrochemical biosensor for simultaneous dual-mode detection of cancer antigen 15-3 and microRNA-21. *Colloids Surf. B* **2022**, *210*, 112260. [CrossRef]
74. Jafari-Kashi, A.; Rafiee-Pour, H.-A.; Shabani-Nooshabadi, M. A new strategy to design label-free electrochemical biosensor for ultrasensitive diagnosis of CYFRA 21–1 as a biomarker for detection of non-small cell lung cancer. *Chemosphere* **2022**, *301*, 134636. [CrossRef]
75. Avelino, K.Y.P.S.; Oliveira, L.S.; Santos, M.R.; Lucena-Silva, N.; Andrade, C.A.S.; Oliveira, M.D.L. Electrochemical DNA biosensor for chronic myelocytic leukemia based on hybrid nanostructure. *Bioelectrochemistry* **2022**, *147*, 108176. [CrossRef]
76. Song, S.; Wang, L.; Li, J.; Fan, C.; Zhao, J. Aptamer-based biosensors. *TrAC Trends Anal. Chem.* **2008**, *27*, 108–117. [CrossRef]
77. Ni, Y.; Ouyang, H.; Yu, L.; Ling, C.; Zhu, Z.; He, A.; Liu, R. Label-free electrochemical aptasensor based on magnetic α-Fe_2O_3/Fe_3O_4 heterogeneous hollow nanorods for the detection of cancer antigen 125. *Bioelectrochemistry* **2022**, *148*, 108255. [CrossRef]
78. Sawant, S.N. Development of biosensors from biopolymer composites. In *Biopolymer Composites in Electronics*, 1st ed.; Sadasivuni, K.K., Kim, J., AlMaadeed, M.A., Ponnamma, D., Cabibihan, J.-J., Eds.; Elsevier: Amsterdam, The Netherlands, 2017; Chapter 13; pp. 353–383.
79. Zhou, X.; Pu, Q.; Yu, H.; Peng, Y.; Li, J.; Yang, Y.; Chen, H.; Weng, Y.; Xie, G. An electrochemical biosensor based on hemin/G-quadruplex DNAzyme and PdRu/Pt heterostructures as signal amplifier for circulating tumor cells detection. *J. Colloid Interface Sci.* **2021**, *599*, 752–761. [CrossRef]
80. Nabok, A.; Abu-Ali, H.; Takita, S.; Smith, D.P. Electrochemical detection of prostate cancer biomarker PCA3 using specific RNA-based aptamer labelled with ferrocene. *Chemosensors* **2021**, *9*, 59. [CrossRef]
81. Wei, X.; Wang, S.; Zhan, Y.; Kai, T.; Ding, P. Sensitive identification of microcystin-LR via a reagent-free and reusable electrochemical biosensor using a methylene blue-labeled aptamer. *Biosensors* **2022**, *12*, 556. [CrossRef]
82. Zhang, F.; Fan, L.; Liu, Z.; Han, Y.; Guo, Y. A label-free electrochemical aptasensor for the detection of cancer antigen 125 based on nickel hexacyanoferrate nanocubes/polydopamine functionalized graphene. *J. Electroanal. Chem.* **2022**, *918*, 116424. [CrossRef]
83. Sadrabadi, E.A.; Benvidi, A.; Yazdanparast, S.; Amiri-zirtol, L. Fabrication of a label-free electrochemical aptasensor to detect cytochrome c in the early stage of cell apoptosis. *Microchim. Acta* **2022**, *189*, 279. [CrossRef]
84. Wang, Y.; Sun, S.; Luo, J.; Xiong, Y.; Ming, T.; Liu, J.; Ma, Y.; Yan, S.; Yang, Y.; Yang, Z.; et al. Low sample volume origami-paper-based graphene-modified aptasensors for label-free electrochemical detection of cancer biomarker-EGFR. *Microsyst. Nanoeng.* **2020**, *6*, 32. [CrossRef]
85. Holford, T.R.J.; Davis, F.; Higson, S.P.J. Recent trends in antibody based sensors. *Biosens. Bioelectron.* **2012**, *34*, 12–24. [CrossRef]

86. Liu, X.; Liu, J. Biosensors and sensors for dopamine detection. *View* **2021**, *2*, 20200102. [CrossRef]
87. Dover, J.E.; Hwang, G.M.; Mullen, E.H.; Prorok, B.C.; Suh, S.-J. Recent advances in peptide probe-based biosensors for detection of infectious agents. *J. Microbiol. Methods* **2009**, *78*, 10–19. [CrossRef] [PubMed]
88. Wei, C.; Xiao, J.; Liu, S.; Wang, Z.; Chen, L.; Teng, W. Simple and label-free electrochemical immuno determination of the gastric cancer biomarker carbohydrate antigen 72-4 with a carbon nanotube-graphene oxide hybrid as the sensing platform and ferrocyanide/ferricyanide as the probe. *Anal. Lett.* **2022**, *55*, 1306–1317. [CrossRef]
89. Aydın, E.B.; Aydın, M.; Sezgintürk, M.K. Impedimetric detection of calreticulin by a disposable immunosensor modified with a single-walled carbon nanotube-conducting polymer nanocomposite. *ACS Biomater. Sci. Eng.* **2022**, *8*, 3773–3784. [CrossRef] [PubMed]
90. Shawky, A.M.; El-Tohamy, M. Signal amplification strategy of label-free ultrasenstive electrochemical immunosensor based ternary Ag/TiO_2/rGO nanocomposites for detecting breast cancer biomarker CA 15-3. *Mater. Chem. Phys.* **2021**, *272*, 124983. [CrossRef]
91. Ortega, F.G.; Regiart, M.D.; Rodríguez-Martínez, A.; de Miguel-Pérez, D.; Serrano, M.J.; Lorente, J.A.; Tortella, G.; Rubilar, O.; Sapag, K.; Bertotti, M.; et al. Sandwich-type electrochemical paper-based immunosensor for claudin 7 and CD81 dual determination on extracellular vesicles from breast cancer patients. *Anal. Chem.* **2020**, *93*, 1143–1153. [CrossRef]
92. Kuntamung, K.; Jakmunee, J.; Ounnunkad, K. A label-free multiplex electrochemical biosensor for the detection of three breast cancer biomarker proteins employing dye/metal ion-loaded and antibody-conjugated polyethyleneimine-gold nanoparticles. *J. Mat. Chem. B* **2021**, *9*, 6576–6585. [CrossRef]
93. Chen, Z.; Li, B.; Liu, J.; Li, H.; Li, C.; Xuan, X.; Li, M. A label-free electrochemical immunosensor based on a gold–vertical graphene/TiO_2 nanotube electrode for CA125 detection in oxidation/reduction dual channels. *Microchim. Acta* **2022**, *189*, 257. [CrossRef]
94. Öndeş, B.; Evli, S.; Uygun, M.; Uygun, D.A. Boron nitride nanosheet modified label-free electrochemical immunosensor for cancer antigen 125 detection. *Biosens. Bioelectron.* **2021**, *191*, 113454. [CrossRef]
95. Biswas, S.; Lan, Q.; Xie, Y.; Sun, X.; Wang, Y. Label-free electrochemical immunosensor for ultrasensitive detection of carbohydrate antigen 125 based on antibody-immobilized biocompatible MOF-808/CNT. *ACS Appl. Mater. Interfaces* **2021**, *13*, 3295–3302. [CrossRef]
96. Rafique, S.; Tabassum, S.; Akram, R. Sensitive competitive label-free electrochemical immunosensor for primal detection of ovarian cancer. *Chem. Pap.* **2020**, *74*, 2591–2603. [CrossRef]
97. Rong, S.; Zou, L.; Zhu, L.; Zhang, Z.; Liu, H.; Zhang, Y.; Zhang, H.; Gao, H.; Guan, H.; Dong, J.; et al. 2D/3D material amplification strategy for disposable label-free electrochemical immunosensor based on rGO-TEPA@ Cu-MOFs@ SiO_2@AgNPs composites for NMP22 detection. *Microchem. J.* **2021**, *168*, 106410. [CrossRef]
98. Echeverri, D.; Orozco, J. β-1, 4-Galactosyltransferase-V colorectal cancer biomarker immunosensor with label-free electrochemical detection. *Talanta* **2022**, *243*, 123337. [CrossRef]
99. Kuntamung, K.; Sangthong, P.; Jakmunee, J.; Ounnunkad, K. A label-free immunosensor for the detection of a new lung cancer biomarker, GM2 activator protein, using a phosphomolybdic acid/polyethyleneimine coated gold nanoparticle composite. *Analyst* **2021**, *146*, 2203–2211. [CrossRef]
100. Kabay, G.; Yin, Y.; Singh, C.K.; Ahmad, N.; Gunasekaran, S.; Mutlu, M. Disposable electrochemical immunosensor for prostate cancer detection. *Sens. Actuators B* **2022**, *360*, 131667. [CrossRef]
101. Martínez-Rojas, F.; Castañeda, E.; Armijo, F. Conducting polymer applied in a label-free electrochemical immunosensor for the detection prostate-specific antigen using its redox response as an analytical signal. *J. Electroanal. Chem.* **2021**, *880*, 114877. [CrossRef]
102. Gui, J.-C.; Han, L.; Du, C.-X.; Yu, X.-N.; Hu, K.; Li, L.-H. An efficient label-free immunosensor based on ce-MoS_2/AgNR composites and screen-printed electrodes for PSA detection. *J. Solid State Electrochem.* **2021**, *25*, 973–982. [CrossRef]
103. Choosang, J.; Khumngern, S.; Thavarungkul, P.; Kanatharan, P.; Numnuam, A. An ultrasensitive label-free electrochemical immunosensor based on 3D porous chitosan–graphene–ionic liquid–ferrocene nanocomposite cryogel decorated with gold nanoparticles for prostate-specific antigen. *Talanta* **2021**, *224*, 121787. [CrossRef]
104. Chen, S.; Xu, L.; Sheng, K.; Zhou, Q.; Dong, B.; Bai, X.; Lu, G.; Song, H. A label-free electrochemical immunosensor based on facet-controlled Au nanorods/reduced graphene oxide composites for prostate specific antigen detection. *Sens. Actuators B* **2021**, *336*, 129748. [CrossRef]
105. Mishra, S.; Kim, E.-S.; Sharma, P.K.; Wang, Z.-J.; Yang, S.-Y.; Kaushik, A.K.; Wang, C.; Li, Y.; Kim, N.-Y. Tailored biofunctionalized biosensor for the label-free sensing of prostate-specific antigen. *ACS Appl. Bio Mater.* **2020**, *3*, 7821–7830. [CrossRef]
106. Zhu, X.; Dai, Y.; Sun, Y.; Liu, H.; Sun, W.; Lin, Y.; Gao, D.; Han, R. Rapid fabrication of electrode for the detection of alpha fetoprotein based on MnO_2 functionalized mesoporous carbon hollow sphere. *Mater. Sci. Eng. C* **2020**, *107*, 110206. [CrossRef]
107. Yan, L.; Zhang, C.; Xi, F. Disposable amperometric label-free immunosensor on chitosan–graphene-modified patterned ITO electrodes for prostate specific antigen. *Molecules* **2022**, *27*, 5895. [CrossRef] [PubMed]
108. Wu, M.; Liu, S.; Qi, F.; Qiu, R.; Feng, J.; Ren, X.; Rong, S.; Ma, H.; Chang, D.; Pan, H. A label-free electrochemical immunosensor for CA125 detection based on CMK-3 $(Au/Fc@MgAl\text{-}LDH)_n$ multilayer nanocomposites modification. *Talanta* **2022**, *241*, 123254. [CrossRef] [PubMed]

109. Zhang, Y.; Wang, X.; Fang, X.; Yuan, X.; Yang, H.; Kong, J. Label-free electrochemical immunoassay for detecting CYFRA 21-1 using poly (ε-caprolactone)-b-poly(ethylene oxide) block copolymer. *Microchem. J.* **2021**, *165*, 106119. [CrossRef]
110. Liu, X.; Lin, L.-Y.; Tseng, F.Y.; Tan, Y.-C.; Li, J.; Feng, L.; Song, L.; Lai, C.-F.; Li, X.; He, J.-H.; et al. Label-free electrochemical immunosensor based on gold nanoparticle/polyethyleneimine/reduced graphene oxide nanocomposites for the ultrasensitive detection of cancer biomarker matrix metalloproteinase-1. *Analyst* **2021**, *146*, 4066–4079. [CrossRef] [PubMed]
111. Liu, Q.; Wu, C.; Cai, H.; Hu, N.; Zhou, J.; Wang, P. Cell-based biosensors and their application in biomedicine. *Chem. Rev.* **2014**, *114*, 6423–6461. [CrossRef] [PubMed]
112. Gupta, N.; Wu, C.; Cai, H.; Hu, N.; Zhou, J.; Wang, P. Cell-based biosensors: Recent trends, challenges and future perspectives. *Biosens. Bioelectron.* **2019**, *141*, 111435. [CrossRef]
113. Lu, X.; Ye, Y.; Zhang, Y.; Sun, X. Current research progress of mammalian cell-based biosensors on the detection of foodborne pathogens and toxins. *Crit. Rev. Food Sci. Nutr.* **2021**, *61*, 3819–3835. [CrossRef]
114. Zhou, H.; Huang, R.; Su, T.; Li, B.; Zhou, H.; Ren, J.; Li, Z. A c-MWCNTs/AuNPs-based electrochemical cytosensor to evaluate the anticancer activity of pinoresinol from *Cinnamomum camphora* against HeLa cells. *Bioelectrochemistry* **2022**, *146*, 108133. [CrossRef]
115. Li, C.; Cui, Y.; Ren, J.; Zou, J.; Kuang, W.; Sun, X.; Hu, X.; Yan, Y.; Ling, X. Novel cells-based electrochemical sensor for investigating the interactions of cancer cells with molecules and screening multitarget anticancer drugs. *Anal. Chem.* **2020**, *93*, 1480–1488. [CrossRef]
116. Liu, F.-F.; Zhao, X.-P.; Liao, X.-W.; Liu, W.-Y.; Chen, Y.-M.; Wang, C. Ultrasensitive and label-free detection of cell surface glycan using nanochannel-ionchannel hybrid coupled with electrochemical detector. *Anal. Chem.* **2020**, *92*, 5509–5516. [CrossRef]
117. Ding, L.; Du, D.; Zhang, X.; Ju, H. Trends in cell-based electrochemical biosensors. *Curr. Med. Chem.* **2008**, *15*, 3160–3170. [CrossRef]
118. Sandhyarani, N. Surface modification methods for electrochemical biosensors. In *Electrochemical Biosensors*, 1st ed.; Ensafi, A.A., Ed.; Elsevier: Amsterdam, The Netherlands, 2019; Chapter 3; pp. 45–75.
119. Cruz-Pacheco, A.F.; Quinchia, J.; Orozco, J. Cerium oxide–doped PEDOT nanocomposite for label-free electrochemical immunosensing of anti-p53 autoantibodies. *Microchim. Acta* **2022**, *189*, 228. [CrossRef]
120. Ghanavati, M.; Tadayon, F.; Bagheri, H. A novel label-free impedimetric immunosensor for sensitive detection of prostate specific antigen using Au nanoparticles/MWCNTs-graphene quantum dots nanocomposite. *Microchem. J.* **2020**, *159*, 105301. [CrossRef]
121. Luo, X.; Davis, J.J. Electrical biosensors and the label free detection of protein disease biomarkers. *Chem. Soc. Rev.* **2013**, *42*, 5944–5962. [CrossRef]
122. Burcu Aydın, E.; Aydın, M.; Sezgintürk, M.K. Biosensors and the evaluation of food contaminant biosensors in terms of their performance criteria. *Int. J. Environ. Anal. Chem.* **2020**, *100*, 602–622. [CrossRef]
123. Agrahari, S.; Gautam, R.K.; Singh, A.K.; Tiwari, I. Nanoscale materials-based hybrid frameworks modified electrochemical biosensors for early cancer diagnostics: An overview of current trends and challenges. *Microchem. J.* **2022**, *172*, 106980. [CrossRef]
124. Yaiwong, P.; Semakul, N.; Bamrungsap, S.; Jakmunee, J.; Ounnunkad, K. Electrochemical detection of matrix metalloproteinase-7 using an immunoassay on a methylene blue/2D MoS_2/graphene oxide electrode. *Bioelectrochemistry* **2021**, *142*, 107944. [CrossRef]
125. Hartmann, M.; Kostrov, X. Immobilization of enzymes on porous silicas—Benefits and challenges. *Chem. Soc. Rev.* **2013**, *42*, 6277–6289. [CrossRef]
126. Eş, I.; Vieira, J.D.G.; Amaral, A.C. Principles, techniques, and applications of biocatalyst immobilization for industrial application. *Appl. Microbiol. Biotechnol.* **2015**, *99*, 2065–2082. [CrossRef]
127. Pachauri, N.; Lakshmi, G.B.V.S.; Sri, S.; Gupta, P.K.; Solanki, P.R. Silver molybdate nanoparticles based immunosensor for the non-invasive detection of Interleukin-8 biomarker. *Mater. Sci. Eng. C* **2020**, *113*, 110911. [CrossRef]
128. Ehzari, H.; Samimi, M.; Safari, M.; Gholivand, M.B. Label-free electrochemical immunosensor for sensitive HER2 biomarker detection using the core-shell magnetic metal-organic frameworks. *J. Electroanal. Chem.* **2020**, *877*, 114722. [CrossRef]
129. Butmee, P.; Tumcharern, G.; Thouand, G.; Kalcherd, K. An ultrasensitive immunosensor based on manganese dioxide-graphene nanoplatelets and core shell Fe_3O_4@Au nanoparticles for label-free detection of carcinoembryonic antigen. *Bioelectrochemistry* **2020**, *132*, 107452. [CrossRef] [PubMed]
130. Zhang, D.; Wang, Y.; Jin, X.; Xiao, Q.; Huang, S. A label-free and ultrasensitive electrochemical biosensor for oral cancer overexpressed 1 gene via exonuclease III-assisted target recycling and dual enzyme-assisted signal amplification strategies. *Analyst* **2022**, *147*, 2412–2424. [CrossRef] [PubMed]
131. Wang, C.; Liu, Y.; Chen, R.; Wang, X.; Wang, Y.; Wei, J.; Zhang, K.; Zhang, C. Electrochemical biosensing of circulating microRNA-21 in cerebrospinal fluid of medulloblastoma patients through target-induced redox signal amplification. *Microchim. Acta* **2022**, *189*, 105. [CrossRef]
132. Zhai, X.-J.; Wang, Q.-L.; Cui, H.-F.; Song, X.; Lv, Q.-Y.; Guo, Y. A DNAzyme-catalyzed label-free aptasensor based on multifunctional dendrimer-like DNA assembly for sensitive detection of carcinoembryonic antigen. *Biosens. Bioelectron.* **2021**, *194*, 113618. [CrossRef]
133. Wang, Y.; Chen, L.; Xuan, T.; Wang, J.; Wang, X. Label-free electrochemical impedance spectroscopy aptasensor for ultrasensitive detection of lung cancer biomarker carcinoembryonic antigen. *Front. Chem.* **2021**, *9*, 721008. [CrossRef]
134. Shafiei, F.; Saberi, R.S.; Mehrgardi, M.A. A label-free electrochemical aptasensor for breast cancer cell detection based on a reduced graphene oxide-chitosan-gold nanoparticle composite. *Bioelectrochemistry* **2021**, *140*, 107807. [CrossRef]

135. Zhou, J.; Cheng, K.; Chen, X.; Yang, R.; Lu, M.; Ming, L.; Chen, Y.; Lin, Z.; Chen, D. Determination of soluble CD44 in serum by using a label-free aptamer based electrochemical impedance biosensor. *Analyst* **2020**, *145*, 460–465. [CrossRef]
136. Forouzanfar, S.; Alam, F.; Pala, N.; Wang, C. Highly sensitive label-free electrochemical aptasensors based on photoresist derived carbon for cancer biomarker detection. *Biosens. Bioelectron.* **2020**, *170*, 112598. [CrossRef]
137. Khodadoust, A.; Nasirizadeh, N.; Taheri, R.A.; Dehghani, M.; Ghanei, M.; Bagheri, H. A ratiometric electrochemical DNA-biosensor for detection of miR-141. *Microchim. Acta* **2022**, *189*, 213. [CrossRef]
138. Pothipor, C.; Aroonyadet, N.; Bamrungsap, S.; Jakmunee, J.; Ounnunkad, K. A highly sensitive electrochemical microRNA-21 biosensor based on intercalating methylene blue signal amplification and a highly dispersed gold nanoparticles/graphene/polypyrrole composite. *Analyst* **2021**, *146*, 2679–2688. [CrossRef]
139. Pareek, S.; Jain, U.; Bharadwaj, M.; Chauhan, N. A label free nanosensing platform for the detection of cervical cancer through analysis of ultratrace DNA hybridization. *Sens. Bio-Sens. Res.* **2021**, *33*, 100444. [CrossRef]
140. Moazampour, M.; Zare, H.R.; Shekari, Z. Femtomolar determination of an ovarian cancer biomarker (miR-200a) in blood plasma using a label free electrochemical biosensor based on L-cysteine functionalized ZnS quantum dots. *Anal. Methods* **2021**, *13*, 2021–2029. [CrossRef]
141. Xia, Y.M.; Li, M.-Y.; Chen, C.-L.; Xia, M.; Zhang, W.; Gao, W.-W. Employing label-free electrochemical biosensor based on 3D-reduced graphene oxide and polyaniline nanofibers for ultrasensitive detection of breast cancer BRCA1 biomarker. *Electroanalysis* **2020**, *32*, 2045–2055. [CrossRef]
142. Farshchi, F.; Saadati, A.; Fathi, N.; Hasanzadeh, M.; Samiei, M. Flexible paper-based label-free electrochemical biosensor for the monitoring of miRNA-21 using core–shell Ag@Au/GQD nano-ink: A new platform for the accurate and rapid analysis by low cost lab-on-paper technology. *Anal. Methods* **2021**, *13*, 286–1294. [CrossRef]
143. Kaya, H.K.; Çağlayan, T.; Kuralay, F. Functionalized nanomaterial-based electrochemical sensors for point-of-care devices. In *Functionalized Nanomaterial-Based Electrochemical Sensors Principles, Fabrication Methods, and Applications*, 1st ed.; Hussain, C.M., Manjunatha, J.G., Eds.; Elsevier: Amsterdam, The Netherlands, 2022; Chapter 14; pp. 309–335.
144. Ebrahimi, G.; Samadi Pakchin, P.; Shamloo, A.; Mota, A.; de la Guardia, M.; Omidian, H.; Omid, Y. Label-free electrochemical microfluidic biosensors: Futuristic point-of-care analytical devices for monitoring diseases. *Microchim. Acta* **2021**, *189*, 252. [CrossRef]
145. Surucu, O.; Öztürk, E.; Kuralay, F. Nucleic acid integrated technologies for electrochemical point-of-care diagnostics: A comprehensive review. *Electroanalysis* **2021**, *34*, 148–160. [CrossRef]
146. Keyvani, F.; Debnath, N.; Ayman Saleh, M.; Poudineh, M. An integrated microfluidic electrochemical assay for cervical cancer detection at point-of-care testing. *Nanoscale* **2022**, *14*, 6761–6770. [CrossRef]
147. Ming, T.; Cheng, Y.; Xing, Y.; Luo, J.; Mao, M.; Liu, J.; Sun, S.; Kong, F.; Jin, H.; Cai, X. Electrochemical microfluidic paper-based aptasensor platform based on a biotin–streptavidin system for label-Free detection of biomarkers. *ACS Appl. Mater. Interfaces* **2021**, *13*, 46317–46324. [CrossRef]
148. Chakraborty, B.; Das, A.; Mandal, N.; Samanta, N.; Das, N.; Chaudkur, C.R. Label free, electric field mediated ultrasensitive electrochemical point-of-care device for CEA detection. *Sci. Rep.* **2021**, *11*, 2962. [CrossRef]
149. Laocharoensuk, R. Development of electrochemical immunosensors towards point-of-care cancer diagnostics: Clinically relevant studies. *Electroanalysis* **2016**, *28*, 1716–1729. [CrossRef]
150. Dai, Y.; Liu, C.C. Recent advances on electrochemical biosensing strategies toward universal point-of-care systems. *Angew. Chem. Int. Ed.* **2019**, *131*, 12483–12496. [CrossRef]
151. Syedmoradi, L.; Norton, M.L.; Omidfar, K. Point-of-care cancer diagnostic devices: From academic research to clinical translation. *Talanta* **2021**, *225*, 122002. [CrossRef] [PubMed]
152. Lopes, L.C.; Santos, A.; Bueno, P.R. An outlook on electrochemical approaches for molecular diagnostics assays and discussions on the limitations of miniaturized technologies for point-of-care devices. *Sens. Actuators Rep.* **2022**, *4*, 100087. [CrossRef]

Review

Biofunctionalization of Multiplexed Silicon Photonic Biosensors

Lauren S. Puumala [1,2,*], Samantha M. Grist [1,2,3], Jennifer M. Morales [4], Justin R. Bickford [4], Lukas Chrostowski [3,5,6], Sudip Shekhar [3,5] and Karen C. Cheung [1,2,5,*]

1 School of Biomedical Engineering, University of British Columbia, 2222 Health Sciences Mall, Vancouver, BC V6T 1Z3, Canada
2 Centre for Blood Research, University of British Columbia, 2350 Health Sciences Mall, Vancouver, BC V6T 1Z3, Canada
3 Dream Photonics Inc., Vancouver, BC V6T 0A7, Canada
4 Army Research Laboratory, US Army Combat Capabilities Development Command, 2800 Powder Mill Rd., Adelphi, MD 20783, USA
5 Department of Electrical and Computer Engineering, University of British Columbia, 2332 Main Mall, Vancouver, BC V6T 1Z4, Canada
6 Stewart Blusson Quantum Matter Institute, University of British Columbia, 2355 East Mall, Vancouver, BC V6T 1Z4, Canada
* Correspondence: lpuumala@student.ubc.ca (L.S.P.); kcheung@ece.ubc.ca (K.C.C.)

Abstract: Silicon photonic (SiP) sensors offer a promising platform for robust and low-cost decentralized diagnostics due to their high scalability, low limit of detection, and ability to integrate multiple sensors for multiplexed analyte detection. Their CMOS-compatible fabrication enables chip-scale miniaturization, high scalability, and low-cost mass production. Sensitive, specific detection with silicon photonic sensors is afforded through biofunctionalization of the sensor surface; consequently, this functionalization chemistry is inextricably linked to sensor performance. In this review, we first highlight the biofunctionalization needs for SiP biosensors, including sensitivity, specificity, cost, shelf-stability, and replicability and establish a set of performance criteria. We then benchmark biofunctionalization strategies for SiP biosensors against these criteria, organizing the review around three key aspects: bioreceptor selection, immobilization strategies, and patterning techniques. First, we evaluate bioreceptors, including antibodies, aptamers, nucleic acid probes, molecularly imprinted polymers, peptides, glycans, and lectins. We then compare adsorption, bioaffinity, and covalent chemistries for immobilizing bioreceptors on SiP surfaces. Finally, we compare biopatterning techniques for spatially controlling and multiplexing the biofunctionalization of SiP sensors, including microcontact printing, pin- and pipette-based spotting, microfluidic patterning in channels, inkjet printing, and microfluidic probes.

Keywords: silicon photonics; evanescent field biosensor; SOI biosensor; biofunctionalization; functionalization; bioreceptor; immobilization chemistry; biopatterning; microfluidics

Citation: Puumala, L.S.; Grist, S.M.; Morales, J.M.; Bickford, J.R.; Chrostowski, L.; Shekhar, S.; Cheung, K.C. Biofunctionalization of Multiplexed Silicon Photonic Biosensors. *Biosensors* **2023**, *13*, 53. https://doi.org/10.3390/bios13010053

Received: 8 November 2022
Revised: 10 December 2022
Accepted: 23 December 2022
Published: 29 December 2022

1. Introduction

Biosensors, which comprise a transducer and biorecognition element, aim to meet increasing demands for medical diagnostics by permitting rapid testing, guiding personalized care, and reducing healthcare costs in decentralized and low-resource settings [1–3]. Silicon photonic (SiP) sensors are one class of optical refractometric sensors with promise as sensitive, rapid, and inexpensive transducers for point-of-care (POC) biosensing [4]. Compared to other types of transducers employed for biosensing, such as electrochemical [5], piezoelectric [6], and mechanical (e.g., microcantilever) [7] sensors, some advantages of SiP sensors are their high sensitivity, wide dynamic range, compatibility with label-free operation, mechanical stability, and insensitivity to electromagnetic interferences [8]. SiP devices can be patterned with wafer-scale semiconductor fabrication techniques, allowing for reproducible, inexpensive, and highly scalable production [1,9,10]. These devices consist of nanoscale patterned silicon or silicon nitride structures that can guide and manipulate light, owing to the high refractive index contrast between the structures themselves and the surrounding media [1,11]. In SiP sensors, near-infrared light is confined in silicon or silicon nitride waveguides [1,12]. A portion of the light's electric field, known as the evanescent field, extends outside the waveguide and interacts with the surrounding medium to create a refractive index-sensitive region (Figure 1a) [1]. A change in the refractive index within this region due to analyte capture on the waveguide surface, for example, perturbs the evanescent field and changes the effective refractive index, n_{eff}, of the guided optical mode [1,4]. This translates to a shift in the optical phase, and in the case of resonant circuit architectures, leads to a resonance wavelength shift that is proportional to the amount of bound analyte, yielding a quantifiable change in the device's optical spectrum [1,4,12]. This change is typically read out using benchtop-scale optical inputs (e.g., broadband optical source or tunable laser) and outputs (e.g., spectrum analyzer or photodetector) [12–16].

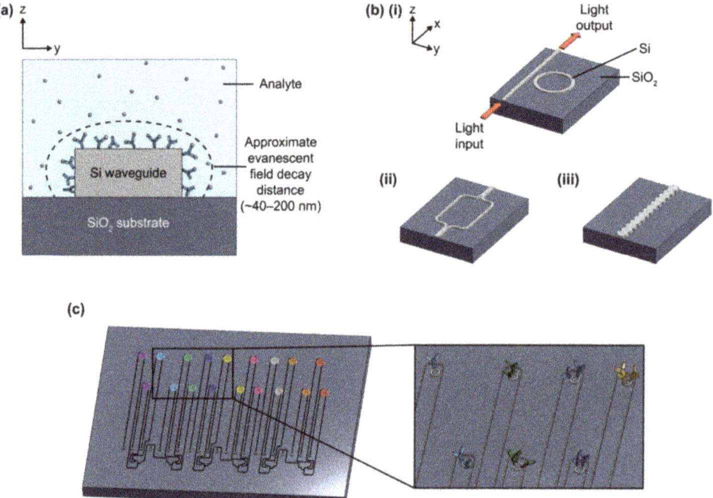

Figure 1. (**a**) Illustration of cross-section of silicon photonic (SiP) sensor, showing the SiO_2 substrate, Si strip waveguide (height: 220 nm, width: 500 nm), and approximate evanescent field decay distance (~40–200 nm, depending on waveguide geometry and light polarization). (**b**) Illustration of four different SiP sensing architectures, including (i) microring resonator (MRR), (ii) Mach-Zehnder interferometer (MZI), and (iii) Bragg grating sensor. (**c**) Visual depiction of a multiplexed SiP MRR sensor chip, showing different rings functionalized with different antibodies (different antibodies are represented by different colors). Antibodies in (**a**,**c**) are not to scale.

Interferometers, microring resonators (MRR), and Bragg gratings (Figure 1b) are among the SiP biosensing architectures that have been demonstrated for disease biomarker detection at concentrations down to the pg/mL scale [17,18]. Readers are directed elsewhere [1] for a detailed description of the principles of operation of each of these sensing architectures. Porous silicon sensors, which are fabricated with electrochemically etched crystalline silicon, have also been widely used in Bragg reflector and PhC configurations for biosensing since the late 1990s and are compatible with many of the same functionalization approaches [19]. This review, however, will mainly focus on planar SiP sensors, which permit greater optical confinement and guidance. Dozens of these individually addressable planar SiP sensors can be fabricated on a single millimeter-scale chip [10]. This permits multiplexed sensing, which is the simultaneous detection of multiple analytes from a single sample. Some benefits afforded by multiplexed biosensing are (1) the opportunity to diagnose multiple conditions/diseases from the same sample, (2) more selective and reliable diagnosis of a single condition by using multiple biomarkers to inform decision-making [20–22], and (3) the opportunity to include controls and reference sensors (e.g., to control for temperature fluctuations) to improve measurement accuracy [23–26]. In addition to these benefits afforded by multiplexed functionalization with different bioreceptors, multiple sensors on the same chip with identical functionalization offer the benefit of replicate measurements to improve accuracy and replicability (e.g., serving as technical replicates allowing for exclusion of failed measurements and averaging out the effects of sensor-to-sensor variability and some assay issues) [27].

The process of functionalizing the sensor surface with biorecognition elements (also called bioreceptors) that selectively bind target analytes is essential to accurate SiP biosensing. The performance characteristics of the biosensor, such as sensitivity, reproducibility, and stability, are inextricably linked to the biofunctionalization chemistry [28]. Here, we broadly characterize biofunctionalization in terms of bioreceptor selection, bioreceptor immobilization strategy (attachment to the sensor surface), and biopatterning technique. Designing antifouling surface modifications is also often included in biofunctionalization procedures to prevent non-specific binding. However, this topic has been reviewed in detail elsewhere [29] and will not be a major focus of the current review.

Many different biofunctionalization strategies are available and should be carefully chosen and optimized to suit the application and sensor architecture. In general, the selected bioreceptor should have good selectivity toward the target analyte to ensure low cross-reactivity with non-target molecules in the sample, high affinity toward the target to achieve fast, sensitive detection, good stability to retain consistent binding activity over time, and reproducible production to ensure predictable and replicable sensor performance across batches/lots of reagents [30]. The strategy used to immobilize bioreceptors on the sensor must not damage the sensor surface or the bioreceptors, and it should be compatible with any system-level integration required for the sensor chips (e.g., chip-mounted lasers and detectors, photonic wire bonds, etc.). It should also allow for oriented bioreceptor immobilization to optimize target accessibility and binding activity, permit uniform bioreceptor coverage on the sensor surface to ensure predictable and consistent target binding across all active sensing areas, have good stability to prevent bioreceptor detachment, and be reproducible [29,31]. The patterning strategy refers to the method by which bioreceptors are deposited on specific locations of the sensor surface (Figure 1c). This is required for multiplexed sensing and to confine bioreceptors to active sensing areas, thus preventing target depletion from dilute samples during sensing [32–34]. The selected patterning technique should not damage the sensor surface or bioreceptors. It should also have sufficient resolution for the selected application, be multiplexable so multiple different bioreceptors can be handled and deposited on a single substrate, produce uniform patterns with good spot-to-spot reproducibility, be compatible with the immobilization protocol (e.g., patterning under conditions that preserve functional groups on the silicon surface), and have low reagent consumption to conserve costly and precious reagents.

In addition to the general biosensor functionalization needs outlined in the previous paragraph, SiP devices have unique needs that distinguish them from other biosensors. Many immobilization techniques (e.g., covalent crosslinking) and bioreceptor types (e.g., antibodies, aptamers, etc.) [35] are shared across an array of sensing technologies including lateral flow assays [36], electrochemical probes [37], piezoelectric sensors [38] and other optical sensors like SPR [39]. While these sensing technology applications can provide valuable insight to inform functionalization strategies for SiP devices, only some of the findings are relevant because they utilize a variety of surfaces including glass, paper, polymers, specialized membranes (nitrocellulose), quartz, nanomaterials, alloys, metals (gold), and ceramics. Here, we focus specifically on immobilization techniques for silicon, silicon nitride, and other like materials.

Among these other transducer types, SiP sensors likely share the most similarities with SPR sensors, which employ a similar evanescent field-based detection principle. Nevertheless, SiP and SPR sensors exhibit differences in their surface chemistries, evanescent field propagation distances, miniaturizability, and multiplexability, as summarized in Table 1 [4,11,12,40–53]. Due to these differences, SiP devices have unique biofunctionalization needs, providing the main motivation for this review. For example, as SiP surfaces typically consist of 90–220 nm-thick silicon or silicon nitride nanostructures patterned on a silicon dioxide substrate [11,40], while SPR sensors typically have gold surfaces, the efficient thiol self-assembled monolayer-based strategies often used to modify metallic SPR biosensor surfaces are not suitable for SiP devices; instead silane-based chemistries are typically used [31,41]. Another unique consideration is the evanescent field penetration depth. For SiP devices, this is ~40–200 nm, depending on waveguide geometry and polarization (Figure 1a) [4,12]. Consequently, SiP sensors require a very thin biofunctional layer that brings target analytes within ~40–200 nm of the sensor surface. The size of this refractive index-sensitive region must be considered when choosing both the biorecognition element and the immobilization chemistry.

Table 1. Comparison of SiP and surface plasmon resonance (SPR) sensors, including SPR imaging (SPRi) and localized SPR (LSPR) devices.

	SiP	SPR
Surface [11,12,42]	Si or Si_3N_4, typically coated with native SiO_2 film	Au
Approx. evanescent field decay distance [4,12,43–46]	~40–200 nm; depends on waveguide geometry and polarization	Typically ~200 nm; up to 600 nm when using an infrared laser or long-range surface plasmons
Miniaturization [12,13,47–49]	Very compact: chip-level integration with microfluidics, electronics, and optical inputs/outputs possible	Moderate: portable instrumentation demonstrated with dimensions ~10–20 cm
Sensor size [4,50–52]	Total sensor chip dimensions ~1–10 mm; active sensing spot dimensions ~10–100 µm	Total sensor chip dimensions ~10 mm; active sensing spot dimensions ~100 µm–1000 µm for multiplexed SPRi and LSPR devices
Cost [12,53]	Low (at high volume)	High
Multiplexing [53]	Multiplexable	Not possible with conventional SPR; multiplexable with SPRi and LSPR

Broadly, the more well-established field of SPR sensing offers a few advantages over SiP sensing. For example, SPR permits the use of simple thiol-based self-assembled monolayer functionalization strategies [29,31]. SPR variants (e.g., SPRi and LSPR) are also compatible with excitation via direct illumination and simple colorimetric readout, which are attractive for portable sensing [42,53,54]. Multimodal SPR-SERS (surface-enhanced Raman scattering) sensing is also possible for highly sensitive and reliable analyte detection [55–57], while multi-modal sensing strategies based on SiP still require further research and development [58,59]. Nevertheless, large-scale and low-cost production remains a challenge for widespread use of SPR-based sensors outside of the laboratory environment [12,53].

SiP biosensor chips, themselves, are uniquely suited to reliable point-of-care (POC) use owing to their ease of miniaturization, low cost, and ease of multiplexing [1,12]. POC biosensing not only permits accessible diagnosis in decentralized and resource-limited settings, but also facilitates treatment decision-making in situations like stroke and sepsis where rapid confirmation of clinical findings is required and conventional lab-based assays may be too time consuming [20,60]. Further, wearable sensors that can be interfaced with flexible electronics may permit real-time and noninvasive monitoring of physiologically relevant analytes (e.g., in sweat) [61,62]. However, one major challenge associated with the translation of SiP biosensors to POC applications is that SiP devices are typically operated with expensive benchtop-scale fluidics and optical readout systems [13]. Miniaturized system-level integration is possible in principle, though, and work to integrate SiP sensors with microfluidics, CMOS electronics, and on-chip lasers and detectors via photonic wire bonds is underway to produce low-cost and portable complete-system PCB-mounted sensors [13,63–66]. Another major challenge with this translation is biofunctionalization.

Given the potential of SiP devices for POC biosensing, a major focus of this work is benchmarking SiP biofunctionalization strategies against needs pertaining to their multiplexed use at the POC. Some of these needs include good environmental and temporal stability to ensure predictable performance after transport and storage at ambient conditions, scalability and manufacturability to permit large-scale deployment, low cost to ensure accessibility, compatibility with easy-to-collect biological samples, such as whole blood, urine, and saliva, and biopatterning resolution on the order of 10 μm to complement the sensor miniaturization afforded by SiP technologies [2]. Reusability is another desirable feature for POC devices that could further reduce sensing costs and improve the accessibility of diagnostic tests in remote and low-resource settings [30]. Chip-level integration of SiP sensors introduces additional biofunctionalization needs. Not only must the biofunctionalization workflow be compatible with the SiP chip architecture, but it also must be compatible with attached optical inputs/outputs and electronics. For example, the immobilization chemistry and patterning technique must not damage electrical or photonic wire bonds, chip-mounted lasers, or PCB materials. Additionally, the immobilized bioreceptors need to be stable through any processing and packaging that needs to be done after immobilization.

To date, numerous existing reviews provide an overview of SiP biosensing technologies, focusing largely on the transduction techniques [1,12,14,19,42,67], with limited discussion about surface biofunctionalization. Others have focused on a single class of bioreceptors for biosensing applications (e.g., antibodies [68,69], nucleic acid probes [70,71], and molecularly imprinted polymers [72]), often including discussion about immobilization chemistries specific to that bioreceptor, and others have focused solely on the comparison of multiple bioreceptor classes for biosensing [30,73]. Several reviews have provided detailed discussion about bioreceptor immobilization chemistries for SiP sensors [31,69] and other biosensing technologies [43,74–76]. A number of works have explored different patterning techniques for the preparation of microarrays and the multiplexed functionalization of biosensors [32,77,78]. Finally, some reviews have discussed at least two of the three key aspects of biofunctionalization (bioreceptor selection, bioreceptor immobilization strategy, and biopatterning technique) for SiP [29] and other sensor technologies (e.g., SPR [41,79,80] and electrochemical sensors [80,81]). Distinct from these existing works, the current review (1) focuses on the unique functionalization needs and strategies of multiplexed SiP biosensors, (2) discusses all three key aspects of biofunctionalization (bioreceptor selection, immobilization chemistry, and patterning technique) and how they are interrelated, and (3) includes a review of biofunctionalization strategies that have been previously implemented on SiP biosensors. To our knowledge, our review is the first contribution to comprehensively summarize and categorize the biofunctionalization strategies previously demonstrated for SiP biosensors (from 2005 to present) as well as present a critical analysis of the various existing (demonstrated on SiP) and potential (demonstrated on similar sensor types) strategies towards the goal of meeting the performance criteria most relevant to SiP biosensors.

Here, we benchmark biofunctionalization strategies against the needs outlined in Table 2, with specific focus placed on biosensor design for multiplexed POC use [82,83]. First, we critically discuss several bioreceptor classes as biorecognition elements for SiP biosensors. Examples of SiP biosensors employing these bioreceptors are highlighted, including their demonstrated sensing performance and assay format. Strategies for bioreceptor immobilization on SiP platforms are discussed along with their advantages and limitations, with particular focus on gold standard silane-mediated covalent chemistries. Finally, contact and contact-free techniques for patterning bioreceptors on SiP sensors are identified and their performance characteristics are discussed. This review aims to present a balanced discussion of the tradeoffs of a range of biofunctionalization strategies to help guide those designing SiP biosensors in selecting a biofunctionalization approach that meets the unique needs of their intended application.

Table 2. Biofunctionalization needs for SiP biosensors. Please note that the performance metrics included in this table are general guidelines and designers should tailor these metrics based on their application. Interdependencies between the different columns of this table should also be considered (e.g., more expensive bioreceptors may still be suitable when combined with patterning techniques that permit very low reagent consumption).

Bioreceptor	Immobilization Chemistry	Patterning Technique
• High affinity (K_D~nM or lower) • Selective (in the ideal case, signal change due to non-specific binding is less than the system limit of detection) • Stable (can be stored at ambient conditions with minimal activity loss for times scales on the order of weeks; stable in biological analytes for several hours) • Available as validated commercial products • Scalable and reproducible production (in the ideal case, variations in target capture due to lot-to-lot variability is less than the system limit of detection) • Regenerable or reversible (<5% signal loss between regeneration cycles and >10 consecutive regenerations possible) [82] • Small (much smaller than evanescent field decay distance; ~10 nm or less) • Low cost (~CAD 1–10/mg)	• Compatible with Si or Si_3N_4 surfaces (or native SiO_2) • Stable (can be stored at ambient conditions for time scales on the order of weeks; stable in biological analytes for several hours) • Thin (a few nm or less) • Does not introduce a reduction in bioreceptor affinity due to denaturation or random orientation • Replicable and uniform (<1 nm intra- and inter-chip variation in immobilization layer thickness) • Compatible with system-level sensor integration (e.g., must not damage photonic wire bonds, chip-mounted lasers, or PCB materials) • Scalable and simple (does not require highly skilled operators) • Mild (no damage to sensor surface or bioreceptors)	• Resolution ~10 μm or less • Multiplexable (multiple reagents can be patterned on different regions of one surface) • Uniform spots (<10% spot-to-spot variation; <10% intra-spot variation in bioreceptor loading density) [83] • Reproducible (<10% run-to-run variation in spot size, shape, and bioreceptor loading density) • High throughput (~10 spots per second or more) • Low reagent consumption (minimal reagent waste) • Simple (does not require highly skilled operators) • Compatible with system-level sensor integration (e.g., must not damage photonic wire bonds, chip-mounted lasers, or PCB materials) • No damage to sensor surface or bioreceptors • Available as cost-effective commercial products or services

K_D: dissociation constant, PCB: printed circuit board.

2. Bioreceptors

In this section, we introduce several classes of bioreceptors that have been used for SiP sensor functionalization and benchmark them against performance criteria outlined in Table 2. A high-level comparison of these bioreceptors is provided in Table 3. We have included subsections for each bioreceptor class to provide details about the opportunities and tradeoffs associated with each of these bioreceptors. For each bioreceptor class, tables summarizing their key advantages and limitations, and categorizing their use in SiP sensor functionalization approaches demonstrated in the previous literature are provided. Because strategies to improve sensitivity, specificity, stability, and other performance metrics are in many cases dependent on the bioreceptor class, within each subsection we have outlined strategies for these types of improvements as well as provided comparisons with other classes where relevant and available. Where appropriate, comparisons between bioreceptor subtypes are also tabulated according to these performance metrics.

Table 3. Comparison of different bioreceptor classes based on biofunctionalization needs for SiP biosensors.

Bioreceptor	Targets	Affinity	Specificity	Stability	Availability	Reproducibility of Production	Regenerability	Size	Cost *
Antibody	Antigens: diverse range of small molecules, complex macro-molecules, viruses, and bacteria; exclude many toxins and non-immunogenic targets [29,33,84,85]	Very high; dissociation constant (K_D) values typically in the low-nM to pM range [85,86]	Very high [87–90]	Poor [24,91]	High [92,93] • Commercial products available from many vendors (e.g., Abcam plc, GenScript, Thermo Fisher Scientific Inc., etc.) † [94,95]	Poor to moderate [87–89]	Moderate; may experience activity loss [96]	• Molecular weight (MW): ~150 kDa • Diameter: ~20 nm [30,87,89]	High; ~USD 500/100 µg [89,97–99]
Aptamer	Very wide range including ions, small inorganic molecules, peptides, proteins, toxins, viral particles, and cells [85,89,100]	Very high; K_D values typically in the low-nM to pM range [85,100,101]	Very high [89,100]	• High physical and thermal stability • Long shelf life • Unmodified versions susceptible to nuclease degradation in biological fluids [89,100, 102]	Moderate • Often require custom design and synthesis [89] • Design/discovery available via companies such as Aptagen, Base Pair Biotechnologies, Aptamer Sciences (AptaSci), etc.† [89] • Synthesis of designed aptamers widely available via companies such as Integrated DNA Technologies (IDT), Thermo Fisher Scientific, and Twist Bioscience, etc.† [103,104]	High due to chemical synthesis [89]	Good [96,105]	• MW: 5–30 kDa (for aptamers consisting of nucleic acid sequences ~15–100 bases in length) • Diameter: ~2 nm [85,89,102]	• DNA aptamer: low; USD 0.07–5.40/mg for large scale synthesis [89] • RNA aptamer: moderate; USD 20–67/mg for large scale synthesis [106]

Table 3. *Cont.*

Bioreceptor	Targets	Affinity	Specificity	Stability	Availability	Reproducibility of Production	Regenerability	Size	Cost *
Nucleic acid probe	Nucleic acids	High to very high, depending on type of nucleic acid; K_D values demonstrated in the low-nM range for DNA-DNA duplexes [107]	High to very high [29,73,81,108, 109]	• High physical and thermal stability • Long shelf life • Unmodified versions susceptible to nuclease degradation in biological fluids • Peptide nucleic acids (PNA), locked nucleic acids (LNA), and morpholinos have improved enzymatic stability [73,81,101,108,110]	Moderate-high • Easy to design once target sequence is known • Usually require custom synthesis, which is widely available via companies such as IDT, Thermo Fisher Scientific, and Twist Bioscience, etc.† [103,104]	High due to chemical synthesis [111, 112]	Good [80]	• MW: 1.5–30 kDa (for nucleic acid sequences ~5–100 bases in length) [113] • Length: ~0.33 nm/base [114]	• DNA: low; USD 0.07–5.40/mg for large scale synthesis [89] • RNA: moderate; USD 20–67/mg for large scale synthesis [106] • PNA, LNA, and morpholinos: high [115–117]
Molecularly imprinted polymer (MIP)	Various small molecules; template molecules must be able to survive in organic solvents	Wide range; K_D values reported in the 10^{-15}–10^{-5} M range [72]	Moderate	High mechanical and chemical stability [118]	Low • Custom synthesis available from some companies including Affinisep, MIP Diagnostics, MIP Technologies, etc.† [72,119]	High	Good	Thin film (sub-nanometer to micrometers) [72]	Dependent on development
Synthetic peptides [91, 120]	A range of small molecules to proteins	Very high	High	• High temperature stability up to 100 °C • Resistant to enzymatic degradation	High • Commercially available synthesis (e.g., Custom Peptide synthesis via Thermo Fisher, Millipore Sigma, AnaSpec, GenScript) †	Very high	Good	• 2–14 kDa; 89–204 Da per amino acid • Protein-catalyzed capture agents (PCCs) are 2–4 kDa	Moderate; CAD 10–100 for custom synthesis dependent on amount and purity for <0.5 g

Table 3. *Cont.*

Bioreceptor	Targets	Affinity	Specificity	Stability	Availability	Reproducibility of Production	Regenerability	Size	Cost *
					High				
Native peptides [120]	A range of small molecules to proteins	Very high	High	Poor	• Commercially available (e.g., via Thermo Fisher, Millipore Sigma, ABclonal, RayBiotech) †	Very high	Moderate	2–14 kDa; 89–204 Da per amino acid	High; CAD 100–1000; dependent on peptide, purity, and quantity
Glycan and lectin	• Glycan: lectins, toxins, and viruses • Lectin: gly-cans and glyco-conju-gates	Low; K_D in the µM–mM range [121,122]	Low [121,123]	• Glycan: good shelf life and stability under dry and ambient conditions [124–126] • Lectin: poor [121]	• Glycan: low-moderate; some commercial products (e.g., from Sigma Aldrich, Biosynth, etc.) †; custom chemical synthesis possible (e.g., via Creative Biolabs, Asparia Glycomics, Glycan Therapeutics, etc.) †, but challenging [127–129] • Lectin: moderate-high; some commercial products (e.g., from Sigma Aldrich, Vector Laboratories, Medicago AB, etc.) †; custom synthesis possible [130]	• Glycan: high due to chemical synthesis [127,128] • Lectin: moder-ate [121]	• Glycan: good regenera-bility using concen-trated salt and high/low pH solu-tions [126] • Lectin: moder-ate; may experi-ence activity loss [121]	• Glycan: highly variable [128] • Lectin: ~10–140 kDa [131–134]	• Glycan: very high; ~CAD 200–1200/10 µg for commercial products [135] • Lectin: moderate; ~CAD 1–300/mg [136]

Table 3. *Cont.*

Bioreceptor	Targets	Affinity	Specificity	Stability	Availability	Reproducibility of Production	Regenerability	Size	Cost *
High contrast cleavage detection (HCCD)	Nucleic acids	High	Very high [137,138]	• Nucleic acid-conjugated reporters: good thermal and physical stability; susceptible to nuclease degradation in biological fluids • CRISPR-Cas enzymes: poor [33]	Moderate • Most reagents commercially available (e.g., via IDT, Abcam, Sigma Aldrich, etc.) †	Moderate-high	Not regenerable	• Reporters: ~15–20 nm diameter • Single-stranded DNA (ssDNA) anchor: ~12 kDa, ~13.2 nm long (~40 bases) [137,139]	High • Guide RNA: CAD 180/10 nmol [140] • Cas12a nuclease: CAD 1250/500 μg (3.2 nmol) [140] • Reporters: ~CAD 400–600/100 mL for gold nanoparticle reporters [141]; ~CAD 200–900/10 mg for quantum dot reporters [142]
CRISPR-dCas9-mediated detection	Nucleic acids	High	Very high [143]	• Nucleic acid probes: good thermal and physical stability; susceptible to nuclease degradation in biological fluids • Cas9 and recombinase polymerase amplification (RPA) enzymes: poor	Moderate • Most reagents commercially available (e.g., via IDT, Thermo-Fisher Scientific, TwistDX, etc.) †	Moderate-high	Good	• Probe MW: 1.5–30 kDa (for nucleic acid sequences ~5–100 bases in length) [113] • Probe length: ~0.33 nm/base [114]	High • Probe ssDNA: CAD 0.07–5.40 for large scale synthesis [89] • Guide RNA: CAD 180/10 nmol [144] • Cas9 nuclease: CAD 1250/500 μg [144] • RPA reagents: ~CAD 5.00/reaction [145–147]

Table 3. *Cont.*

Bioreceptor	Targets	Affinity	Specificity	Stability	Availability	Reproducibility of Production	Regenerability	Size	Cost *
Lipid nanodisc	Proteins	Moderate; K_D in nM–μM range for phospholipid nanodiscs [148,149]	Variable; depends on membrane protein content [148–150]	Poor due to protein content	Low-moderate • Few commercial products available (e.g., Cube Biotech) † • Typically require custom synthesis in laboratory [151]	Moderate	Good [148,149]	8–16 nm diameter [150,151]	High • ~CAD 770–1011/25 nmol (~$770–$1011/500 μg) [152] • ~CAD 0.19–0.54/chip (based on chip functionalization procedure in [148])

* Prices are listed in Canadian dollars (CAD) and are based on commercial products available as of July 2022. † Vendors listed are based on an exploratory search and are not endorsed or suggested by the authors.

2.1. Antibodies

Antibodies (Figure 2) are the most commonly used bioreceptors for diagnostic assays [91,153]. Antibodies are Y-shaped proteins of ~150 kDa in size, which consist of two identical Fab regions (fragment, antigen-binding), and a single Fc region (fragment, crystallizable) [30,33,87]. The Fab regions specifically bind with high affinity to target molecules called antigens via binding sites called epitopes on the antigen surface. Antigens comprise a diverse range of biological molecules including simple sugars, hormones and lipids, complex macromolecules like proteins, nucleic acids, phospholipids and carbohydrates, and even viruses and bacteria [29,33,84]. On the other hand, the Fc region typically interacts with effector molecules and cells in biological systems and may be targeted for antibody immobilization on a solid substrate in biosensing applications [33,87,154]. Millions of antibodies have been validated for tens of thousands of antigen targets, making them a widely-available and flexible bioreceptor option for many different use cases [92–95]. Antibody production starts by immunizing animals against an antigen to stimulate the production of antigen-specific antibodies by the animals' B cells [88,155]. Then, the antibodies can be obtained directly from the animal immune-sera. Alternatively, antibody-producing B cells can be immortalized by fusion with hybridoma cells for long-term production.

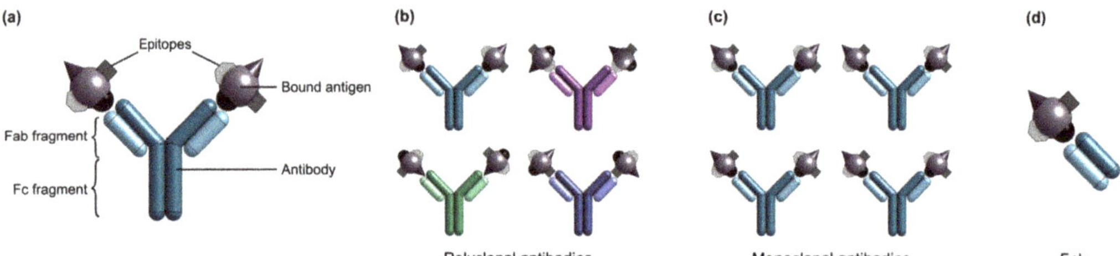

Figure 2. (**a**) Illustration of an antibody and bound antigens. Illustrations of different antibody subtypes, including (**b**) polyclonal antibodies, (**c**) monoclonal antibodies, and (**d**) a Fab fragment. Note that polyclonal antibodies are produced as heterogeneous mixtures in which different antibodies may bind to different epitopes of the same antigen. Monoclonal antibodies are produced as homogeneous samples in which all antibodies bind to the same epitope.

There are two major classes of antibodies: polyclonal and monoclonal. Polyclonal antibodies are produced as heterogeneous mixtures from animal serum and individual antibodies in a serum sample may bind to various epitopes on a single antigen [87]. Polyclonal antibodies exhibit significant batch-to-batch variability, partly owing to their animal origin [156]. Antibody quality can vary from animal-to-animal and even throughout an individual animal's lifetime [156]. Conversely, monoclonal antibodies are produced from immortalized cell lines, are homogeneous in nature, and bind to a single epitope on the target antigen surface [88,156]. Monoclonal antibodies offer excellent specificity and reduced cross-reactivity and variability compared to their polyclonal counterparts; as a result, monoclonal antibodies have been widely used in diagnostic assay applications [86–88,90]. More recently, molecular engineering has also been used to generate shorter antibody variants including Fabs, single chain variable fragments, and single domain antibodies that can be produced more easily in vitro and used for applications that solely require epitope binding [29,75,157]. A comparison of polyclonal antibodies, monoclonal antibodies, and Fab fragments as bioreceptors for SiP biosensors is provided in Table 4.

Table 4. Comparison of antibody subtypes as bioreceptors for SiP biosensors.

	Affinity [86–89]	Specificity [87–90]	Availability [92–95,158,159]	Reproducibility of Production [87–89]	Size [30,87,89]	Cost * [89,97–99]
Polyclonal antibody	Very high, but with significant variability within a sample	High • Greater risk of cross-reactivity than monoclonal antibodies • Target multiple antigen epitopes and are less sensitive to small changes in epitope structure	High; many vendors available (key companies include Abcam plc, GenScript, Thermo Fisher Scientific Inc., etc.) †	Poor	• MW: ~150 kDa • Diameter: ~15 nm	High ~CAD 500/100 μg
Monoclonal antibody	Very high with K_D values as low as 10 pM	Very high • Target a single antigen epitope and are very sensitive to small changes in epitope structure • Low risk of cross-reactivity	High; many vendors available (key companies include Abcam plc, GenScript, Thermo Fisher Scientific Inc., etc.) †	Moderate	• MW: ~150 kDa • Diameter: ~15 nm	High ~CAD 500/100 μg
Fab fragment	Very high	High to very high, depending on origin	High; can be prepared from available antibodies or purchased commercially (major antibody-fragment producing companies include AbbVie, Amgen, Novartis AG, etc.) ‡	Poor to moderate, depending on origin	MW: ~50 kDa	High

* Prices are listed in CAD and are based on commercially available products as of July 2022. † The named vendors comprise a small subset of the major competitors in the global research antibodies market. Readers are directed elsewhere [94,95] for a more comprehensive list of major vendors. Vendors listed are not endorsed or suggested by the authors. ‡ The named vendors comprise a small subset of the major competitors in the global antibody fragments market. Readers are directed elsewhere [158,159] for a more comprehensive list of major vendors. Vendors listed are not endorsed or suggested by the authors.

Numerous SiP biosensing platforms using antibodies as bioreceptors have been reported in the literature. Conventional ELISAs are typically done in sandwich or competitive assay formats, requiring labeled secondary antibodies or labeled analyte molecules, respectively [160]. SiP platforms, however, permit label-free assays [1]. In the label-free format, binding of a target analyte to surface-bound antibodies is directly monitored, offering the advantages of real-time detection and simple sample preparation [14,161]. Nevertheless, sandwich formats using an unlabeled secondary antibody [18] or labeled antibody combined with subsequent enzymatic amplification [17,162] or protein-based multilayer signal enhancement [163] have been used to achieve more sensitive and specific detection for low-concentration and low-molecular weight analytes. To tether the capture antibodies to the sensor, these antibody-based SiP platforms typically rely on randomly oriented covalent immobilization strategies that target abundant amine or carboxyl groups on the antibody surface [75]. However, other covalent and non-covalent immobilization strategies have also been used [75].

SiP biosensors using antibodies as bioreceptors (Table 5) have been proposed for the biomarker-based diagnosis of cancer [17,18,22,161,163], cardiac disorders [164,165], inflammation [166], and viral infection [167], in addition to the detection of toxins [25,168], viral particles [169–171], and bacteria [172]. Such antibody-based SiP platforms have achieved LoDs as low as the pg/mL range using enzymatically or layer-by-layer-enhanced sandwich assay formats [17,163]. Other antibody-based SiP platforms have achieved label-free analyte detection with LoDs in the low-ng/mL range [161,169]. While most of the aforementioned examples employ whole polyclonal or monoclonal antibodies, Chalyan et al. [25] functionalized thiolated silicon oxynitride microring resonators with Fab fragments obtained from protease digestion of polyclonal antibodies for the detection of a carcinogenic mycotoxin, Aflatoxin M1, with a LoD of ~5 nM. The functionalization strategy used in this work targeted sulfhydryl (–SH) groups present on the Fab surface that were liberated from splitting the intact antibody; since these sulfhydryl groups are located opposite to the antigen-binding sites, this strategy ensures highly oriented bioreceptor immobilization, making it an attractive alternative to amine- and carboxyl-targeting strategies [75,173]. Shia and Bailey [168] functionalized silicon microring resonators with recombinantly derived single domain antibodies for the detection of ricin, a lethal protein toxin. The single domain antibodies exhibited improved specificity and lower cross-reactivity compared to a commercial polyclonal anti-ricin antibody.

Despite their excellent sensitivity and specificity, antibody-based biosensors present notable challenges regarding POC sensing. Namely, antibody discovery is achieved by months-long in vivo screening processes, which are expensive and laborious [89]. Antibody production largely relies on mammalian cell lines, which means that these bioreceptors are costly and require highly trained personnel to produce, precluding their use in highly scalable and low-cost sensors [2,97–99,157]. Moreover, among antibody vendors, there is a lack of consistency in the context-specific validation and reporting of antibody specificity and reproducibility for different applications [92,156,178]. The use of animals and cell colonies in antibody production makes these bioreceptors susceptible to sample contamination [89]. This means that choosing successful antibodies for biosensors is often an expensive and time-consuming task involving troubleshooting and returning failed antibodies to suppliers [156,178]. Antibodies are also susceptible to denaturation and require carefully controlled storage conditions, which may be difficult to maintain in POC settings [24,91]. Further, antibody immobilization on a solid substrate is known to reduce antibody binding activity, making the optimization of immobilization strategies using mild chemistries a particular challenge in the design of highly sensitive biosensors [75]. The key advantages and limitations of antibodies as bioreceptors are highlighted in Table 6. Given the limitations of antibodies discussed here, several classes of synthetic affinity reagents have been developed as alternatives to antibodies and have been demonstrated as bioreceptors on SiP platforms [2].

Table 5. Demonstrations of SiP biosensors using antibodies or antibody fragments as the biorecognition element and their sensing performance.

Bioreceptor Description	Sensor Type	Target	Detection Performance		Assay Format	Refs.
			Figure of Merit	Value		
Monoclonal antibody	Si MRR	Thrombin	Min. detected concentration	500 pM	Label-free	[174]
Monoclonal antibodies	Si MRR	Carcinoembryonic antigen	Min. detected concentration	10 ng/mL	Label-free	[22]
		Prostate-specific antigen				
		α-fetoprotein				
		Interleukin-8				
		Tumor necrosis factor-α				
Monoclonal antibody	Si MRR	Monocyte chemotactic protein 1	Limit of detection (LoD)	0.5 pg/mL	Amplification with secondary antibody and enzymatic enhancement	[17]
Monoclonal antibody	Si MRR	Carcinoembryonic antigen	LoD	2 ng/mL in buffer, 25 ng/mL in serum	Label-free	[161]
Monoclonal antibodies	Si MRR	Interleukin-2	LoD	100 pg/mL	Sandwich immunoassay	[18]
		Interleukin-8	LoD	100 pg/mL		
Monoclonal antibody	Si MRR	Bean pod mottle virus	LoD	10 ng/mL	Label-free	[169]
Monoclonal antibody	Si MRR	C-reactive protein	-	-	Label-free	[166]
Monoclonal antibodies	Si MRR	α-fetoprotein	Working range	0.3–20.6 ng/mL	Amplification with secondary antibody and protein-based multilayer signal enhancement	[163]
		Activated leukocyte cell adhesion molecule	Working range	1.0–43.7 ng/mL		
		Cancer antigen 15-3	Working range	2.0–91.5 units/mL		
		Cancer antigen 19-9	Working range	2.5–96.6 units/mL		
		Cancer antigen-125	Working range	2.4–95.6 units/mL		
		Carcinoembryonic antigen	Working range	0.16–20.2 ng/mL		
		Osteopontin	Working range	4.3–50.3 ng/mL		
		Prostate specific antigen	Working range	0.054–4.7 ng/mL		

Table 5. *Cont.*

Bioreceptor Description	Sensor Type	Target	Detection Performance		Assay Format	Refs.
			Figure of Merit	Value		
Monoclonal antibodies	Si MRR	Interleukin-2	LoD	1 pg/mL	Amplification with secondary antibody and enzymatic enhancement	[162]
		Interleukin-6	LoD	1 pg/mL		
		Interleukin-8	LoD	0.5 pg/mL		
Monoclonal antibody	Hydex MRR	*E. coli* O157:H7 bacterial cells	LoD	10^5 CFU/mL	Label-free	[172]
Monoclonal antibody	Si PhC	Human Papillomavirus virus-like particles	LoD	1.5 nM	Label-free	[170]
Antibody	Si PhC	Cardiac myoglobin	Min. detected concentration	70 ng/mL	Label-free	[165]
Monoclonal antibodies	Si_3N_4 planar waveguide interferometer	Hemagglutinin (H7N2 and H7N3)	Min. detected concentration	0.05 hemagglutination (HA) units/mL	Label-free	[175]
Polyclonal antibodies	Si_3N_4 planar waveguide interferometer	Hemagglutinin (H7N2)	Min. detected concentration	0.0005 HA units/mL		
		Hemagglutinin (H7N3)	Min. detected concentration	0.005 HA units/mL		
Polyclonal antibody	Porous Si sensor using reflectometric interference spectroscopy	Insulin	LoD	4.3 µg/mL	Label-free	[176]
Antibody	Si PhC total internal reflection	Cardiac troponin I	LoD	0.01 ng/mL	Label-free	[164]
Antibody	Si_3N_4/SiO_2 slot-waveguide MRR	Bovine serum albumin	LoD	16 pg/mm^2	Label-free	[177]
Antigen-binding fragment (Fab) from protease digestion of polyclonal IgG	SiO_xN_y MRR	Aflatoxin M1	LoD	5 nM	Label-free	[25]
Single domain antibodies	Si MRR	Ricin	LoD	200 pM	Label-free	[168]

Si: silicon, LoD: limit of detection, Si_3N_4: silicon nitride, SiO_2: silicon dioxide, SiO_xN_y: silicon oxynitride.

Table 6. Advantages and limitations of antibodies as bioreceptors.

Advantages	Limitations
• Diverse targets including small molecules, complex macromolecules, viruses, and bacteria [29,33,84] • High affinity and specificity [87–89] • Widely available [92,93] • Regeneration possible for multiple binding cycles [18]	• Poor stability [24,91] • Batch-to-batch and vendor-to-vendor variability [92,156,178] • Expensive [89,157] • Potential activity loss from immobilization and regeneration procedures [75,96] • Time-consuming and laborious discovery and production [30,89] • Susceptible to sample contamination [89]

2.2. Aptamers

Aptamers, which have been referred to as "synthetic antibodies", are short, single-stranded DNA or RNA molecules that are systematically selected to bind to a given target molecule (Figure 3) [87,89]. These single-stranded oligonucleotides fold into unique sequence-specific three-dimensional structures that bind to targets with high specificity and affinity via non-covalent effects, including electrostatic interactions, van der Waals, and hydrogen bonding [89,100]. Aptamers are generated using an in vitro process called SELEX (systematic evolution of ligands by exponential enrichment), which allows for the selection of unique target-binding DNA or RNA molecules from a large library (Figure 3c) [100]. The SELEX process begins with a library of around 10^{15} single-stranded oligonucleotides, each containing a different random sequence of 20–60 nucleotides, flanked by fixed sequences on the 3′ and 5′ ends [89,100]. This library is amplified by the polymerase chain reaction (PCR), then strand-separated to yield ssDNA or transcribed to yield RNA, depending on whether a DNA or RNA aptamer is desired [100,102,179]. These amplified products are then incubated with target molecules and target-bound DNA or RNA are separated from unbound sequences, followed by elution of the bound species. The amplification and target-binding stages of this process are repeated with the enriched pool of target-binding sequences. The process is repeated for a total of 8–20 cycles during which competitive binding causes high-affinity binding sequences to outcompete lower-affinity ones, eventually yielding a pool dominated by sequences with the strongest affinity to the target [100–102,179]. An additional negative selection step can also be included in the SELEX process to reduce cross-reactivity of aptamers to structurally similar targets, thus enhancing selectivity [102]. The selected oligonucleotides can subsequently be sequenced and synthesized for analysis and use [100]. The resulting aptamers can achieve comparable, or even better, affinity to their targets when compared to monoclonal antibodies, with typical dissociation constants (K_D) in the low nanomolar to picomolar range [85,100,101].

Since their discovery three decades ago, aptamers have been generated against inorganic ions, metabolites, dyes, drugs, amino acids, peptides, proteins, cells, and even tissues [89,100,101,105]. Because the production of antibodies relies on the immune response, antibodies can only be generated for immunogenic and non-toxic targets [89,100]. Conversely, the in vitro SELEX process theoretically allows for the generation of aptamers against any target. Further, given the small size of aptamers (5–30 kDa) compared to antibodies (150–180 kDa), aptamers can be designed against small molecule targets that are inaccessible to antibodies [89]. In evanescent field-based sensing applications, the smaller size of aptamers can allow for greater surface immobilization density and can bring captured analytes closer to the sensor surface, potentially improving sensitivity [113,114]. The selection environment (e.g., buffer type, ionic strength, pH, temperature, etc.) during aptamer generation can also be tailored to the binding conditions required for the intended use case [89,100,180]. This is contrasted to antibodies which are limited to target recognition under physiological conditions.

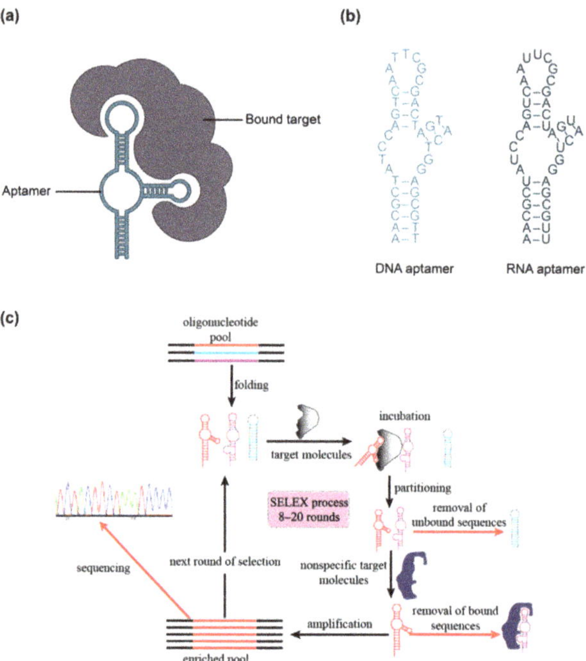

Figure 3. (**a**) Illustration of aptamer and bound target. (**b**) Visual representation of aptamer subtypes: DNA and RNA aptamers. (**c**) Illustration of SELEX (systematic evolution of ligands by exponential enrichment) process to design aptamers against a target. In (**c**), different colors in the oligonucleotide pool represent different nucleic acid sequences, while different colors in the sequencing step represent different nucleic acid bases identified by Sanger sequencing or high-throughput sequencing methods. Part (**c**) is reprinted from Ref. [101] in accordance with the Creative Commons Attribution 4.0 International (CC BY 4.0) license.

Other advantages conferred to aptamers by the nature of the SELEX discovery process include fast discovery and low batch-to-batch variability [89]. While antibody discovery requires upward of 6 months, the SELEX process can be completed in a matter of days if high-throughput automated methods are used [89,102]. Additionally, since antibody synthesis relies on animals or cell cultures, batch-to-batch variability can be high; this variability is avoided in aptamer samples because they are generated via chemical synthesis procedures with a low risk of contamination [89]. Aptamers also exhibit better environmental stability, especially thermal stability, and long shelf lives compared to antibodies [89,105]. Namely, aptamers are resistant to high temperatures up to 95 °C and cycles of denaturation and renaturation, while they can also be lyophilized and stored at room temperature [89]. This makes aptamers attractive bioreceptors for point of care devices and opens opportunities for surface regeneration and reusable sensors [96,105]. Finally, aptamer discovery and manufacture are generally lower cost than for antibodies. For example, CamBio offers custom aptamer discovery down to USD 5000 per target [181]. After the aptamer has been selected and sequenced, it can be manufactured at low cost using common oligonucleotide synthesis techniques. For example, Aptagen offers aptamer manufacture at USD 1–4 per milligram for microgram-scale synthesis, US $300 per gram for milligram-scale synthesis, and USD 50 per gram for gram-scale synthesis, while IDT offers DNA oligonucleotide synthesis at CAD 1.40–2.40 per base for 1 µmol quantities for sequences of 5–100 bases in length [89,182]. However, the manufacture of RNA sequences, especially those exceeding 60 bases can be more costly. For example, Bio-Synthesis, Inc. manufactures RNA sequences of 10–30 bases in length for USD 14.50–50 per milligram for 50–1000 mg-scale synthesis, while IDT manufactures RNA sequences of 5–60 bases in length for CAD 24.00 per base at 1 µmol quantities and RNA sequences of 60–120 bases in length for CAD 23.00 per base at 80 nmol quantities [89,106,182]. Table 7 provides a high-level comparison of DNA and RNA aptamers for SiP biosensing.

Table 8 summarizes aptamer-functionalized SiP biosensors that have been demonstrated in the literature. In all of these aptamer-based SiP sensors, label-free sensing formats were used. Park et al. [24] demonstrated IgE and thrombin detection on an aptamer-functionalized silicon microring resonator and demonstrated reproducible surface regenerations for up to 10 cycles after IgE and thrombin binding using a NaOH solution. Byeon and Bailey [174] compared thrombin binding on aptamer-functionalized silicon microring resonators to antibody-functionalized resonators and demonstrated aptamer-functionalized surface regeneration using proteinase K. The authors found that the aptamer had a lower affinity toward thrombin (K_D = 8.2 nM) compared to the antibody (K_D = 3.3 nM), suggesting a poorer limit of detection for sensing applications relying on steady-state binding. However, the aptamer-functionalized sensors demonstrated faster thrombin-binding kinetics, which could produce a theoretically lower LoD for the aptamer-based sensor in applications that leverage binding kinetics measurements to generate a calibration curve (e.g., by linearly fitting the initial slope of the binding kinetics curve to quantify analyte concentration [161,184]). Christenson et al. [164] presented a comparative study in which aptamer- and antibody-functionalized Photonic Crystal-Total Internal Reflection biosensors were investigated for the detection of cardiac troponin I. The aptamer- and antibody-functionalized sensors achieved detection limits of 0.1 ng/mL and 0.01 ng/mL, respectively. While the aptamer-functionalized sensor demonstrated poorer sensitivity, both sensors achieved clinically relevant limits of detection, and the aptamer sensor was lower cost and did not require refrigeration during storage. Chalyan et al. [25] compared the performance of aptamer- and Fab-functionalized silicon oxynitride microring resonator biosensors for the detection of Aflatoxin M1. A limit of detection of 5 nM was reported for both the aptamer- and Fab-functionalized sensors, though the Fab-functionalized sensor was deemed preferable due to its superior reproducibility. Both Chalyan et al. [25] and Guider et al. [185] reported effective sensor regeneration after Aflatoxin M1 binding using glycine solutions.

While aptamers offer notable advantages over antibodies in the context of POC diagnostics, they still face challenges such as degradation in biological fluids, low SELEX success rates, lower availability, and highly variable costs. Firstly, aptamers, especially RNA aptamers, are susceptible to nuclease degradation in biological fluids [100,102]. For example, in human serum, the half-life of an unmodified aptamer is about one minute [180]. This limits the use of unmodified aptamers as bioreceptors in diagnostic devices using blood or serum samples. RNA aptamers are also more susceptible to hydrolysis than DNA aptamers at pH > 6 [183]. However, chemical modifications, such as the incorporation of 2′-fluoro or 2′-amino-modified nucleotides, are often introduced to aptamers either at the beginning of SELEX or during chemical synthesis to improve their resistance to nuclease degradation [89,186]. These types of modifications can increase an aptamer's half-life in biological fluids to multiple days [180], but modifications introduced during and after SELEX can add complexity to the SELEX process or change the folding structure and binding properties of the aptamer, respectively [89]. As such, careful optimization is required to achieve effective nuclease resistance without compromising binding performance.

Table 7. Comparison of aptamer subtypes as bioreceptors for SiP biosensors.

	Affinity [89,100]	Specificity [89,100]	Stability [89,100,102,183]	Availability [89]	Reproducibility of Production [89]	Attachment Chemistry	Size * [85,89,102]	Cost † [89,106,182]
DNA aptamer	High	High	• High physical and thermal stability • Unmodified versions susceptible to nuclease degradation	Moderate • May require custom design via SELEX (e.g., via Aptagen, Base Pair Biotechnologies, Aptamer Sciences (AptaSci), etc.) ‡ • Custom synthesis of already-designed aptamers widely available (e.g., via IDT, Thermo Fisher Scientific, and Twist Bioscience, etc.) §	High due to chemical synthesis	Typically, covalent immobilization via terminal functional groups (e.g., 5′ amines)	• MW: 5–30 kDa • Diameter: ~2 nm	Low • Research scale synthesis: CAD 1.40–2.40/base for 1 µmol quantities for sequences of 5–100 bases in length from IDT • Large scale synthesis: USD 0.05–4/mg
RNA aptamer	Very high due to more diverse 3D conformations	Very high due to more diverse 3D conformations	• High physical and thermal stability • Unmodified versions more susceptible to more susceptible to nuclease degradation and hydrolysis at pH > 6 than DNA aptamers • Stronger RNA-RNA intra-strand interactions than in DNA aptamers	Moderate • May require custom design via SELEX (e.g., via Aptagen, Base Pair Biotechnologies, Aptamer Sciences (AptaSci), etc.) ‡ • Custom synthesis of already-designed aptamers widely available (e.g., via IDT, Thermo Fisher Scientific, and Twist Bioscience, etc.) §	High due to chemical synthesis	Typically, covalent immobilization via terminal functional groups (e.g., 5′ amines)	• MW: 5–30 kDa • Diameter: ~2 nm	Moderate • Research scale synthesis: CAD 24.00/base at 1 µmol quantities for sequences of 5–60 bases in length and CAD 23.00/base at 80 nmol quantities for sequences of 60–120 bases in length from IDT • Large scale synthesis: USD 14.50–50/mg

* Molecular weight and size for aptamers consisting of 15–100 nucleotides. † Prices are listed in CAD for products available as of July 2022, unless otherwise specified. ‡ Readers are directed elsewhere [89] for a more comprehensive list of key companies in the global aptamers market. Vendors listed are not endorsed or suggested by the authors. § Readers are directed elsewhere [103,104] for a more comprehensive list of key companies in the global oligonucleotide synthesis market. Vendors listed are not endorsed or suggested by the authors.

Table 8. Demonstrations of SiP biosensors using DNA aptamers as the biorecognition element and their sensing performance. All demonstrations tabulated here used label-free assay formats.

Sensor Type	Target	Detection Performance		Refs.
		Figure of Merit	Value	
Si MRR	IgE	LoD	33 pM	[24]
	Thrombin	LoD	1.4 nM	
Si MRR	Thrombin	Min. detected concentration	500 pM	[174]
SiO_xN_y MRR	Aflatoxin M1	LoD	5 nM	[25]
SiO_xN_y MRR	Aflatoxin M1	Min. detected concentration	1.58 nM	[185]
Si PhC total internal reflection	Cardiac troponin I	LoD	0.1 ng/mL	[164]
Porous Si reflectometric interference spectroscopy	Insulin	LoD	1.9 μg/mL	[176]

Next, the success rate of SELEX aptamer generation is lower than in vivo antibody generation, likely due to the lower structural diversity of nucleotides compared to amino acids and the small size of aptamers [101,180]. This increases the time and resources required to optimize aptamers for new targets. However, this <30% SELEX success rate could be improved through the use of specialized SELEX technology variants, personalized protocols, optimized oligonucleotide libraries, and quality control measures [180,187]. The target-binding performance of an aptamer depends on its structural conformation, which can be influenced by pH, ionic strength, and temperature [180]. Therefore, to ensure predictable binding, aptamer selection must be carried out in buffer systems similar to those used in the final application. However, this may also mean that an aptamer that performs well in solutions of a purified target in buffer may not perform as well in complex biological samples. Lastly, aptamers lack the type of extensive commercial infrastructure and investment seen in the antibody market and usually must be custom-synthesized by a handful of companies [89]. A summary of the key advantages and limitations of aptamers as bioreceptors is provided in Table 9.

2.3. Nucleic Acid Probes (Hybridization-Based Sensing)

Short, single-stranded nucleic acid probes have been widely used for the detection of nucleic acid targets via hybridization-based SiP sensing (Figure 4) [80,107,188]. Both ssDNA and RNA sequences can be immobilized on a biosensor surface, where they bind complementary nucleic acid target sequences through hydrogen bond formation, yielding DNA-DNA, DNA-RNA, or RNA-RNA duplexes [33,70,189]. Such biosensors are often called genosensors [81]. Compared to aptamers, which can be designed to bind many different types of target molecules, nucleic acid probes can only bind other nucleic acids [30]. Additionally, the function of nucleic acid probes depends primarily on their nucleotide sequence, not on their three dimensional structure: once the target gene sequence is known, the complementary probe can be designed directly [30]. This means that nucleic acid probes can be designed against a new target very quickly compared to antibodies and aptamers. Short nucleic acid probes of 100 nucleotides or less can be synthesized using well-characterized phosphoramidite chemistry [103,104,111,112]. This synthetic method of nucleic acid synthesis is highly reproducible, allows for the incorporation of functional groups like thiols and amines to aid in probe immobilization on solid substrates, and is typically low-cost [81,111,112,190]. Another key advantage of nucleic acid probe-based biosensors is that they can be thermally or chemically regenerated with good reproducibility between sensing cycles [80].

Table 9. Advantages and limitations of aptamers as bioreceptors.

Advantages	Limitations
• High affinity and specificity [89,100] • Can be designed for theoretically any target, including toxins and non-immunogenic species [85,89,100] • Small size [89] • Relatively low-cost and rapid discovery via SELEX process [89,102] • Produced via chemical synthesis yielding low batch-to-batch variability and allowing for the introduction of chemical modifications for improved functionality [85,89] • Good stability and long shelf life [89,105] • Good regenerability: regeneration can be achieved using temperature or concentrated salt, acidic, basic, chaotropic agent, surfactant, and chelating agent solutions [96,105] • Low production costs for DNA aptamers [89]	• Nuclease susceptibility of unmodified aptamers [100,102,180] • High production costs for RNA aptamers [89] • Low SELEX success rates [101,180] • Structural conformation and binding are sensitive to pH, ionic strength, and temperature [180] • Less widely available than antibodies and usually require custom synthesis [89]

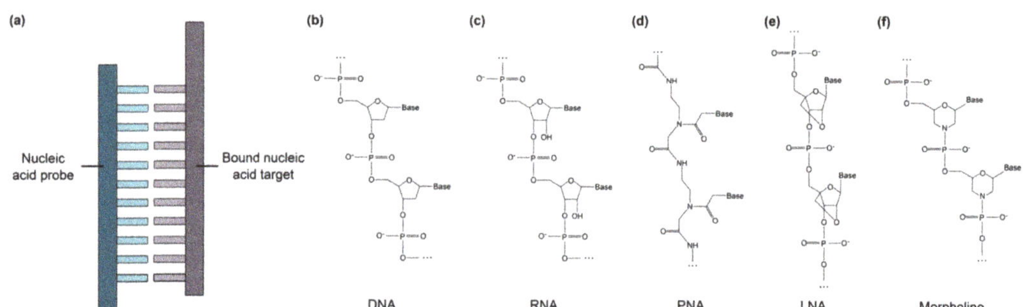

Figure 4. (**a**) Illustration of nucleic acid bioreceptor and bound nucleic acid target. Comparisons of the chemical structures of different nucleic acid subtypes, including (**b**) DNA, (**c**) RNA, (**d**) PNA, (**e**) LNA, and (**f**) morpholino, shown as line structures informed by Refs. [73,191].

In addition to conventional ssDNA and RNA probes, synthetic nucleic acid analogues with functional chemical modifications to improve binding performance and biostability have recently been explored for biosensing applications. These include peptide nucleic acids (PNAs), locked nucleic acids (LNAs), and morpholinos [23,30,73,81,110,115–117,191,192]. PNAs (Figure 4d) are synthetic DNA mimics that can hybridize to complementary DNA and RNA, but have a backbone consisting of *N*-(2-aminoethyl)-glycine units linked by peptide bonds, rather than the sugar-phosphate backbone usually found in DNA [81]. Unlike natural nucleic acids, PNAs are uncharged, giving them improved hybridization stability [73]. Their hybridization stability is also impacted to a greater extent by single base mismatches than DNA-DNA hybridization, making PNAs more selective than DNA probes and a good choice for detecting single nucleotide polymorphisms [193]. PNAs also exhibit ionic insensitivity and improved pH, thermal, and enzymatic stability [73]. LNAs (Figure 4e) are another class of synthetic DNA mimics in which the ribose is locked in the 3′-endo conformation, resulting in reduced conformational flexibility, improved biostability, and enhanced binding affinity toward the target sequence [30,81,101]. Morpholinos (Figure 4f) are synthetic nucleic acid analogues in which the sugar-phosphate backbone is replaced by alternating morpholine rings, connected by phosphoramidite groups [110]. Morpholinos are uncharged and possess many of the same characteristics as PNAs, but morpholinos exhibit improved solubility, poorer stability at low pH, and improved flexibility of synthesis regarding sequence length, offering the opportunity to bind longer DNA and RNA target sequences, compared to PNAs [108]. Table 10 provides a comparison between these nucleic acid subtypes and benchmarks them against functionalization performance criteria for SiP biosensing.

Numerous SiP sensing platforms have been demonstrated in the literature using nucleic acids or nucleic acid analogues as biorecognition elements for the detection of ssDNA and RNA biomarkers with applications in the detection of cancer [23,192,194–197] and bacteria [198,199] (Table 11). Often, a label-free assay format is used on these sensing platforms. For example, Sepúlveda et al. [200] demonstrated label-free detection of short ssDNA targets down to 300 pM using a silicon nitride Mach-Zehnder interferometer sensor functionalized with ssDNA probes, while Shin et al. [197] demonstrated specific and label-free detection of longer ssDNA targets (>100 nucleotides) on ssDNA-functionalized silicon microring resonators down to 400 fmol, which corresponds to 16 μL of a 25 nM sample. A silicon nitride slot waveguide Mach-Zehnder interferometer functionalized with methylated ssDNA probes was demonstrated by Liu et al. [192] to quantify the methylation density of a DNA-based cancer biomarker at sample concentrations down to 1 fmol/μL or 1 nM. Nucleic acid-functionalized SiP sensors have also been used for microRNA detection, as demonstrated by Qavi and Bailey [194], who used a ssDNA-functionalized silicon MRR sensor for the rapid and label-free quantification of microRNAs. In this work, the authors reported a limit of detection of 150 fmol, which represented the minimum quantity of microRNA that could be reasonably detected in solution with the reported biosensor. Based on the supporting information provided for this work, this detection limit corresponded to a 75 μL analysis volume of 2 nM microRNA. Synthetic nucleic acid analogues have been demonstrated as receptors and targets for SiP sensors. Yousuf et al. [110] recently demonstrated the detection of short ssDNA targets on morpholino-functionalized suspended silicon microrings down to 250 pM, while Hu et al. [201] demonstrated PNA detection using ssDNA-functionalized planar SiP sensors.

In contrast to these label-free methods, Qavi et al. [109] amplified the detection of microRNA on a ssDNA-functionalized silicon microring resonator sensor using S9.6 anti-DNA:RNA antibodies. The S9.6 antibody selectively binds to DNA-RNA heteroduplexes and was shown here to effectively amplify the signal after microRNA hybridization, achieving a limit of detection of 350 amol, corresponding to 35 μL of a 10 pM microRNA sample. This was a 3-fold improvement compared to label-free microRNA detection on the same sensor. This work also demonstrated preliminary results demonstrating that LNA probes could be used to capture the microRNA targets, followed by successful, albeit slightly less effective, amplification with the S9.6 antibody. Kindt and Bailey [196] improved the limit of detection of a ssDNA-functionalized silicon microring resonator sensor for the detection of mRNA using streptavidin-coated beads. This bead-based amplification improved the sensor's limit of detection to 512 amol, compared to 32 fmol without bead-based amplification.

To date, most nucleic acid hybridization-based biosensors have been demonstrated for the detection of short target sequences due to the tendency of longer sequences to fold and obtain secondary structures [70,198]. These secondary structures significantly slow down binding kinetics, thus increasing sensing times. This challenge can be mitigated by pre-treating the targets via thermal denaturation, fragmentation, or the use of short nucleic acid chaperones which disrupt the nucleic acid target's secondary structure [196,198]. In one work [198], the folded structures of long transfer-messenger RNA (tmRNA) targets were modified using one of the three following strategies prior to detection: (1) chemical fragmentation, (2) thermal denaturation, or (3) thermal denaturation in the presence of chaperone probes. Subsequently, the treated tmRNA targets were detected in a label-free format on ssDNA-functionalized silicon microring resonators. Chemical fragmentation was found to be the most effective RNA pre-treatment strategy for increasing the binding kinetics and magnitude of the sensor response. In another work [196], short DNA chaperone molecules were used to disrupt the secondary structure of full length mRNA transcripts prior to detection on ssDNA-functionalized silicon microring resonators. This effectively improved the sensing assay's binding kinetics.

Table 10. Comparison of nucleic acid subtypes as bioreceptors for SiP biosensors.

	Affinity [29,30,81,108]	Specificity [29,73,81,108,109]	Stability [73,81,101,108,110,183,189]	Availability	Reproducibility of Production [89]	Attachment Chemistry	Cost * [115–117,182]
DNA	High	High	• High physical and thermal stability • Susceptible to nuclease degradation	Moderate • Custom synthesis required • Many vendors available (e.g., IDT, Thermo Fisher Scientific, Twist Bioscience, etc.) ‡	High due to chemical synthesis	Typically, covalent immobilization via terminal functional groups (e.g., 5′ amines)	Low • CAD 2.40/base for 5–100 bases at 1 µmol production scale from IDT • 25-base custom DNA oligo: CAD 60 at 1 µmol production scale
RNA	High	High	• High physical and thermal stability • More susceptible to nuclease degradation and hydrolysis at pH > 6 than DNA aptamers • RNA-RNA duplexes have higher thermal stability than DNA-DNA duplexes of the same sequence	Moderate • Custom synthesis required • Many vendors available (e.g., IDT, Thermo Fisher Scientific, Twist Bioscience, etc.) ‡	High due to chemical synthesis	Typically, covalent immobilization via terminal functional groups (e.g., 5′ amines)	Medium • CAD 24.00/base for 5–60 bases at 1 µmol production scale from IDT • 25-base custom RNA oligo: CAD 600 at 1 µmol production scale
PNA	Higher than DNA/RNA due to neutral charge	Very high	Very high physical, thermal, and enzymatic stability	Low • Custom synthesis required • Few vendors available (e.g., biomers.net, PNA Bio Inc., Panagene, etc.) §	High due to chemical synthesis	Typically, covalent immobilization via terminal functional groups (e.g., 5′ amines)	High • 25-base custom PNA oligo: ~CAD 8100 at 1 µmol production scale from biomers.net
LNA	Higher than DNA/RNA due to decreased configurational flexibility	Very high	Very high physical, thermal, and enzymatic stability	Low • Custom synthesis required • Few vendors available (e.g., IDT Affinity Plus™ products, Qiagen, LGC Biosearch Technologies, etc.) §	High due to chemical synthesis	Typically, covalent immobilization via terminal functional groups (e.g., 5′ amines)	High • CAD 52.00/base for 5–100 bases at 1 µmol production scale from IDT † • 25-base custom LNA-containing oligo: CAD 1300 at 1 µmol production scale

Table 10. *Cont.*

	Affinity [29,30,81,108]	Specificity [29,73,81,108,109]	Stability [73,81,101,108,110,183,189]	Availability	Reproducibility of Production [89]	Attachment Chemistry	Cost * [115–117,182]
Morpholino	Higher than DNA/RNA due to neutral charge, but lower affinity than PNA	Very high	• Very high physical, thermal, and enzymatic stability • Unstable at low pH	Low • Custom synthesis required • Few vendors available (e.g., Gene Tools, E-nnovation Life Sciences) §	High due to chemical synthesis	Typically, covalent immobilization via terminal functional groups (e.g., 5′ amines)	High • 25-base custom morpholino oligo: ~CAD 1300 at 1 µmol production scale from Gene Tools

* All prices are listed in CAD for oligonucleotides with no chemical modifications. Prices are provided for µmol-scale production for comparative purposes. Note that lower prices are available for large scale synthesis. † Price provided for DNA oligonucleotides containing 1–20 LNA bases. ‡ Readers are directed elsewhere [103,104] for a more comprehensive list of key companies in the global DNA and RNA oligonucleotide synthesis market. Vendors listed are not endorsed or suggested by the authors. § These are not exhaustive lists of PNA, LNA, and morpholino vendors. Vendors listed are based on an exploratory search and are not endorsed or suggested by the authors.

Table 11. Demonstrations of SiP biosensors using nucleic acid probes as the biorecognition element and their sensing performance.

Bioreceptor	Sensor Type	Target	Detection Performance		Assay Format *	Refs.
			Figure of Merit	Value		
ssDNA	Si MRR	microRNA	LoD	150 fmol (i.e., 75 µL of 2 nM microRNA solution)	Label-free	[194]
ssDNA	Si MRR	Complementary DNA generated from targeted microRNAs	-	-	Label-free	[195]
ssDNA	Si MRR	Full-length mRNA transcripts	LoD	32 fmol for label-free detection; 512 amol with bead-based amplification	Label-free and with streptavidin-coated bead-based amplification	[196]
ssDNA	Si MRR	microRNA	LoD	10 pM (i.e., 350 amol in a 35 µL sample)	Amplification with anti-DNA:RNA antibodies	[109]
ssDNA	Si MRR	ssDNA	LoD	400 fmol (i.e., 16 µL of 25 nM ssDNA solution)	Label-free	[197]
ssDNA	Si MRR	Bacterial transfer-messenger RNA (tmRNA)	LoD	52.4 fmol (i.e., 100 µL of 524 pM tmRNA solution)	Label-free	[198]
ssDNA	Si MRR	Methylated DNA	-	-	Label-free	[23]
ssDNA	Cascaded Si MRRs	IS6110 ssDNA biomarker	LoD	1 fg (corresponds to 10 µL of 0.1 pg/mL ssDNA solution)	Label-free	[199]
		IS1081 ssDNA biomarker	LoD	10 fg (i.e., 10 µL of 1 pg/mL ssDNA solution)		
ssDNA	Si_3N_4 MRR	ssDNA	-	-	Label-free	[202]
ssDNA	N-doped Si MRR electrophotonic sensor	ssDNA	-	-	label-free	[59]
ssDNA	Si_3N_4 MZI	ssDNA	LoD	300 pM	Label-free	[200]
ssDNA	Planar Si PhC waveguide	ssDNA	LoD	19.8 nM	Label-free	[203]
ssDNA (directly conjugated)	Si MRR and Si PhC	ssDNA	LoD	50 nM	Label-free	[201]
		ssPNA	-	-		
ssDNA (synthesized in situ)	Si MRR and Silicon PhC	ssDNA	LoD	10 nM		
		ssPNA	-	-		
Methylated ssDNA	Si_3N_4 slot waveguide MZI	Methylated ssDNA	Min. detected concentration	1 fmol/µL (1nM)	Label-free	[192]
Morpholino	Suspended Si MRR	ssDNA	Min. detected concentration	250 pM	Label-free	[110]

* Does not include PCR amplification prior to introduction to sensor surface.

Indeed, the greatest limitation of nucleic acid-based bioreceptors is their limited applicability: they are only suitable for applications requiring nucleic acid targets [30]. Further, nucleic acid targets usually require significant sample preparation prior to detection [188]. For DNA targets, the sample usually must undergo fragmentation to ensure that the target sequence is accessible to the capture probes, followed by denaturation to yield single-stranded sequences. Depending on the abundance of the target, it may also require amplification through PCR or isothermal strategies prior to detection [188,195,199]. For RNA targets, sample preparation may be simpler, but still typically requires a fragmentation step [188]. Finally, DNA and RNA carry an inherent negative charge, making them susceptible to non-specific binding due to electrostatic interactions with non-target molecules [30]. This also poses challenges regarding nucleic acid probe immobilization. For example, nucleic acid probes are repelled by an unmodified SiP sensor's negatively charged native oxide surface, which means that the SiP surface must be modified with a cationic film should passive adsorption be used for probe immobilization [204]. When covalent immobilization strategies are used, this negative charge increases steric hindrance between adjacent nucleic acid probes, which affects the maximum density of probes that can be immobilized on the sensor surface and the number of available binding sites for targets, potentially limiting sensor sensitivity [201]. This effect, however, can be reduced by employing in situ synthesis of nucleic acid probes on the SiP surface. Hu et al. [201] demonstrated a greater than 5-fold increase in ssDNA probe surface coverage and a greater than 5-fold increase in detection sensitivity for SiP microring resonators and photonic crystal sensors when functionalized via in situ probe synthesis, compared to the covalent immobilization of full ssDNA sequences. Conversely, if the immobilization strategy is optimized and the density of immobilized nucleic probes on the surface becomes too high, hybridization of targets to the surface-bound probes is hindered by steric crowding and electrostatic repulsion, also limiting sensor sensitivity [71]. As such, careful tuning of the spacing between immobilized probes is required for optimal performance. Some of these limitations can be mitigated by the use of uncharged synthetic DNA analogues including PNAs or morpholinos [30]. For example, in a study investigating DNA- and PNA-functionalized electrochemical sensors for the capture of DNA targets, the PNA-functionalized sensors exhibited stronger target capture and demonstrated optimal sensing performance at higher probe surface density than the DNA-functionalized sensors, likely due to reduced steric and electrostatic effects [205]. This contributed, in part, to a greater sensitivity for the PNA-functionalized sensor, which had a very wide dynamic range from pM to μM and a LoD that was 370 times lower than that achieved when using DNA probes. However, the lack of electrostatic repulsion between uncharged DNA analogues can lead to local clustering on the sensor surface, creating a heterogeneous layer of these uncharged probes, thus hindering the reproducibility of the functionalization strategy [193,206]. A summary of the key advantages and limitations of nucleic acid probes for SiP biosensing is provided in Table 12.

2.4. Molecularly Imprinted Polymers (MIPs)

Molecularly Imprinted Polymers (MIPs) are a type of label-free synthetic receptor for binding a broad spectrum of analytes from small molecules and viruses to larger proteins and cell membrane structures (Figure 5) [72]. The first imprinted polymers were developed in the early 1990s and demonstrated the ability to change impedance in response to target binding. Later, more developed MIP films exhibited changes in refractive index upon binding, making them ideal for optical sensors.

Table 12. Advantages and limitations of nucleic acid probes as bioreceptors.

Advantages	Limitations
• Simple to design once target sequence is known [30] • Reproducible chemical synthesis [111,112] • Chemical modifications can be introduced [81,111,112,190] • Amenable to thermal or chemical regeneration [80] • Synthetic DNA analogs (PNA, LNA, and morpholinos) available to enhance affinity, specificity, and stability [30,73,81,110]	• Limited to nucleic acid targets [30] • Challenging to capture long targets due to secondary structures [70,198] • Targets often require significant sample preparation [188] • DNA and RNA susceptible to non-specific interactions due to negative charge [30] • Steric hindrance effects may limit probe immobilization density and binding capacity [71,201] • Low molecular weight nucleic acid targets are challenging to detect without amplification [138]

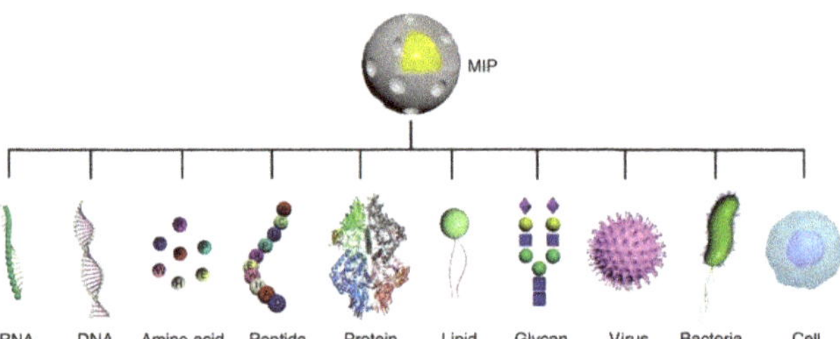

Figure 5. MIPs can be templated with an array of targets including: RNA, DNA, amino acids, peptides, proteins, lipids, glycans, viruses, and bacterial or cell epitopes. Reproduced from Ref. [207] in accordance with the Creative Commons Attribution 4.0 International license (CC BY 4.0).

Several strategies for MIP preparation on SiP platforms and representative surfaces have been summarized in Table 13. MIPs are created via template assisted synthesis where an analyte is cured within a polymer making a 3D impression in the form of a binding pocket (Figure 6a) [119,208]. There are two main methods of MIP polymerization or "templating" for optical sensors: solution based (Figure 6b) or surface stamping (Figure 6c) [119,209]. In solution-based MIPs, a target, or template, is solvated in organic solvents with precursors, initiators, and monomers [72]. Smaller molecules are primarily used directly as a template, whereas larger targets (proteins, peptides, etc.) use a smaller binding epitope for imprinting. These formulations are specific to the template and form complexes of reversible covalent or noncovalent interactions with the template's chemical structure. Next the solution is deposited on a surface and cured by ultraviolet (UV) or thermal polymerization. Solution-based MIPs can be templated onto many shapes such as coatings, thin films, and nanoparticles [210]. This is advantageous since they can conform to many different fiber and waveguide topologies. MIP films can be grown on a variety of photonic sensor designs, dipped on optical fibers, or developed in solution on microspheres [211]. Following MIP synthesis, the template molecules must be extracted, which is often achieved by washing or soaking in solution [211–215] or by plasma-treatment [211], though physically assisted solvent extraction (e.g., microwave- or ultrasound-assisted extraction) and extraction using supercritical or subcritical fluids have also been used [216]. This produces a distribution of exposed binding site geometries due to the template's random orientation on the polymer surface (Figure 6b). Surface stamping using support molds was the first method of casting [209,217]. Template molecules are crosslinked to a surface mold and pressed onto the polymer surface over the sensor prior to curing. Removal of the mold leaves imprinted binding sites stamped on the surface of the polymer. This method produces more regular pockets in comparison to solution-based MIPs due to the added control over the depth of imprinted binding sites and the opportunity to control template orientation on the surface mold (Figure 6c) [218].

Table 13. Strategies demonstrated for the preparation of MIPs on SiP sensors and representative surfaces.

MIP Type	Film Constituents	Template	Film Deposition Technique	Template Extraction	Sensor Type	Refs.
Polymer synthesis	Methacrylic acid (functional monomer) and ethylene glycol dimethacrylate (crosslinking agent)	Testosterone	Casting, followed by thermopoly-merization	Acetic acid and ethanol-based chemical extraction	Si MRR	[212]
Polymer synthesis	Methacrylic acid (functional monomer) and ethylene glycol dimethacrylate (crosslinking agent)	Progesterone	Coating, followed by UV photopoly-merization	-	Cascaded Si MRRs	[219]
Sol-gel	Bis(trimethoxysilylethyl)benzene and 2-(2-pyridylethyl)trimethoxysilane, prepared in tetrahydrofuran	Carbamate (used to create trinitrotoluene binding sites)	Airspray coating or electrospray ionization	HCl and chloroform-based chemical extraction	Si MRR	[213]
Sol-gel	Ethanol, methyltrimethoxysilane, aminopropyltriethoxysilane, and HCl, prepared in dimethyl sulfoxide	Fluorescein isothiocyanate	Dip coating	Oxygen plasma degradation or chemical extraction with solutions of ethanol, acetic acid, and chloroform or acetonitrile.	SiO_2 microsphere whispering gallery mode resonators	[211]
Sol-gel	Tetraethoxysilane (TEOS), water, ethanol, and HCl or methyltriethoxysilane (C1-TriEOS), ethanol, HCl, and MPTMS.	Cortisol	Spin coating	Ethanol-based chemical extraction.	Si chips	[214, 215]

These methods produce specific binding pockets on the polymer surface that match the three-dimensional molecular structure of the template. Targets primarily bind via hydrogen bonding, electrostatic interactions, and Van der Waals forces. Reversible covalent bonding is less common since it is dependent on the template's molecular structure, available specialized monomers, and more complex synthesis [220]. Direct adsorption of analytes into the binding pockets produces a change in refractive index or electrochemical (impedance) signal that can be read out by optical and amperometric sensors, respectively.

MIPs are considered an alternative to antibodies since they are highly sensitive, reversible and have both chemical and mechanical stability. They are synthetic making them robust, scalable, low-cost, and shelf-stable [118]. They have been shown to be stable over months in a large temperature range (up to 150 °C) with over 50 adsorption/desorption cycles in organic solvents, acids, and bases [221]. Divinylbenzene MIP bases are twice as robust (up to 100 cycles) in comparison to methacrylate- or acrylamide-based polymers over a larger pH range. Although MIPs are an excellent synthetic method of producing a non-refrigerated product with a long shelf life, there are several limitations to the technology. Currently, synthesis is developed for one target at a time and requires computational studies to downselect polymer precursors and benchtop chemistry to optimize formulation. Computational studies include quantum mechanics/molecular mechanics (QM/MM) calculations (ab initio, molecular dynamics, etc.) between possible precursors and the template molecule [222]. These calculations determine which reagents interact with the chemical structure of the template molecule. Then, MIPs are formulated based on the set and ratios of precursors are empirically tested. The final MIP formulation is selected to maximize sensitivity and specificity, based on these empirical data.

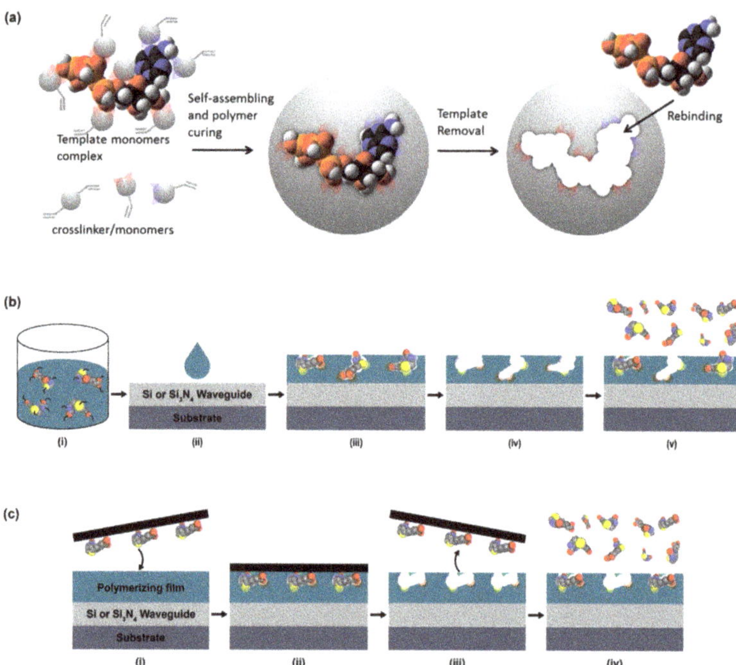

Figure 6. Illustration of MIP templating approaches. (**a**) MIP templating begins with a template mixed with polymer precursors followed by curing and the template removal. (**b**,**c**) Illustration of molecularly imprinted polymer (MIP) showing the random and oriented nature of template orientation on the surface of solution based (**b**) and stamped (**c**) MIPs, respectively. In solution-based MIP preparation (**b**): templates are first solvated in organic solvents with precursors, initiators, and functional monomers (i), followed by deposition on the sensor surface (ii), curing (iii), and template extraction (iv), after which the MIP can be used for target capture (v). Note that the small pieces of color left behind in the binding sites after template extraction, as seen in (iv,v), represent sites where the functional monomers formed non-covalent or covalent bonds with the template. In surface stamping based MIP preparation (**c**): templates are immobilized on a surface mold (i) and pressed into a polymer film on the sensor surface (i,ii) prior to curing. After curing, the surface mold is removed (iii), leaving imprinted binding sites on the sensor surface, which can be subsequently used for target capture (iv). Part (**a**) is reproduced with permission from Ref. [210]. Copyright 2016, American Chemical Society.

MIPs have limited specificity in complex solutions due to the imprinted nature of the polymers, which include an array of heterogeneous binding pocket orientations [30]. Smaller or like molecules can fill the binding pockets, producing a background signal or affecting the MIP's affinity toward its target [223]. Formulations thus need to be thoroughly optimized for the template (as described above) and tested against non-imprinted polymers (NIPs) [224]. NIPs are the same composition as the MIPs, only formulated without the template. They are used as a control to determine the sensitivity of the MIPs against nonspecific adsorption. Further studies testing MIPs in real bioanalyte samples are essential to validate their specificity [30]. A summary of the advantages and limitations of MIPs as bioreceptors is provided in Table 14.

Table 14. Advantages and limitations of MIPs as bioreceptors.

Advantages	Limitations
• Diverse targets [72] • Amenable to many (>50) cycles of reversible binding [221] • Excellent chemical, mechanical, and temporal stability [221] • Scalable and low-cost production due to synthetic nature	• Poor specificity [223] • Heterogeneous binding sites • Low availability and complicated optimization of formulations

The use of refractive index sensing with MIPs in silicon photonics (Table 15) is limited, although they have been well demonstrated with SPR-based sensors [209]. MIPs can be drop-cast, spray-coated, spin-coated and inkjet printed on the sensor surface. Chen et al. [212] demonstrated thermally polymerized, drop-cast ultrathin film MIPs on a passive SOI microring resonator sensor for testosterone. This method is highly sensitive for sensing ultralow concentrations with a sensitivity of 4.803 nm/(ng·mL). First the template solution is premixed to promote self-assembly between the template and monomers specific to its chemical structure. This produces a pre-polymerized layer surrounding the template in solution that is further complexed with the addition of carboxyl-terminated monomers. This matrix is then drop-cast on the sensor's surface and thermally treated for 12 h. The combination of the pre-polymerized matrix and dilute solution results in an ultrathin assembled monolayer of MIPs on the surface with a limit of detection of 48.7 pg/mL. Multiple cycles of MIP regeneration (using a 1:1 acetic acid-ethanol rinse) and sensing with a solution of 1 ng/mL testosterone were tested on this platform to assess reproducibility. There was a drift in the sensor response and corresponding decrease in sensitivity as the number of regenerations increased, which the authors attributed to damage to the MIP during testing. Selectivity was also assessed by introducing the small molecule toxin, microcystin-LR, to the sensor, which produced a negligible response.

Table 15. Demonstrations of SiP biosensors using MIPs as the biorecognition element and their sensing performance.

Bioreceptor	Sensor Type	Target	Detection Performance		Assay Format	Refs.
			Figure of Merit	Value		
MIP film	Si MRR	Testosterone	LoD	48.7 pg/mL	Label-free	[212]
Sol-gel MIP	Si racetrack resonators	Trinitrotoluene vapor	Min. detected concentration	5 ppb	Label-free	[213]
MIP film	Cascaded Si MRRs	Progesterone	LoD	83.5 fg/mL	Label-free	[219]
MIP film	SiO_2 microsphere whispering gallery mode resonator	Fluorescein isothiocyanate	-	Fluorescence intensity	-	[211]
MIP film	SiO_xN_y dual polarization interferometer	Hemoglobin	LoD	2 µg/mL	Label-free	[225]

Photopolymerization can be achieved all at once by direct UV polymerization or in stages by pre-polymerizing in a dilute crosslinking solution followed by the addition of a UV initiator for a final cure. Xie et al. [219] used this process with cascaded microring resonators for sensing progesterone. They used an SU-8 cladding and a slightly larger ring diameter to match the free spectral range of the reference ring to the MIP-coated sensing ring. The MIP is prepared by pre-polymerizing acetic and methacrylic acid with progesterone for 3 h followed by adding UV crosslinkers in a specialized tank for UV curing. This produces a thin self-assembled film on the sensor surface. Their results showed a limit of detection of 83.5 fg/mL which is approximately 3 orders of magnitude lower than enzyme-linked immunosorbent assays (ELISA). The sensor shows good selectivity to progesterone with little to no response with testosterone and the NIP. Eisner et al. [213] used

MIP sol-gels to compare airbrush versus electrospray ionization deposition techniques. These sol-gels are formed by hydrolysis and polycondensation of a colloidal liquid into a gel at low temperatures. The colloid includes metal oxides, salts, or alkoxides suspended in solvents. This ceramic-based MIP was designed for the detection of trinitrotoluene (TNT) vapor and was coated on passive silicon racetrack resonators with thicknesses of 500–700 nm to minimize resonant wavelength shift artifacts due to changes in the bulk refractive index surrounding the MIP. The results showed a ~10× increase in response and sensitivity in the electrospray MIP in comparison to airbrushing. The MIP-coated sensors showed a nonspecific response to other nitro-based explosives (2,4-dinitrotoluene (DNT) and 1,3-dinitrobenzene (DNB)); however, the device's sensitivity was about an order of magnitude greater for TNT than for DNT and DNB.

Hydrogel-based MIP thin films are less successful since they expand and contract based on water content and salinity, producing unwanted effects. Reddy et al. [225] sensed hemoglobin on silicon oxynitride waveguides for dual polarization interferometry. The gels initially increased in thickness and mass upon injection of a control solution, but the response was transient suggesting adsorption and desorption of the control on the hydrogel surface. In contrast, the target, hemoglobin, produced a continuous signal and remained selective in solutions containing <1% pooled bovine serum.

2.5. Peptides and Protein-Catalyzed Capture Agents

Synthetic and native peptides are an attractive method of capture for chemical and biological targets in SiP due to their small size in comparison to antibodies, aptamers, and other larger components (Figure 7) [226–228]. Peptides are differentiated from proteins by their size (2–70 amino acids) and flexible structure. There are two main types of peptides for attachment: native and synthetic [120]. Native peptides are small binding epitopes or ligands found in nature that selectively bind to a specific site on the target of interest. They are primarily recombinant and produced by cloning the peptide in an organism. The peptide sequence is inserted into a plasmid, expressed in bacteria, insect or mammalian cells and purified for processing [229].

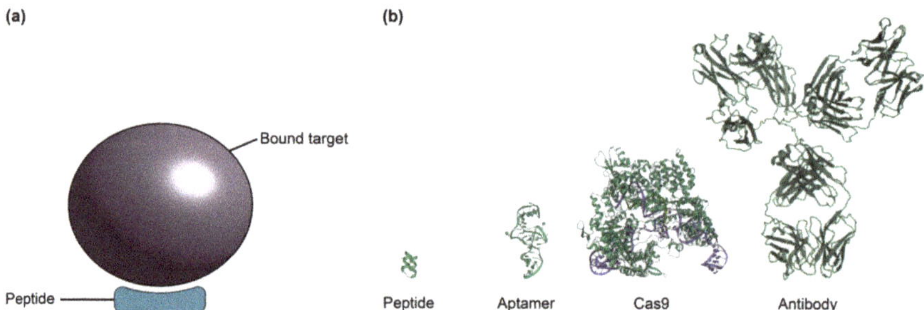

Figure 7. (**a**) Illustration of peptide bound target. (**b**) Comparison of peptide, aptamer, Cas9 enzyme and antibody relative sizes, informed by protein data bank crystal structures 2AU4, 4OO8 and 1IGY [226–228]. Peptides are smaller than aptamers, antibodies, and many other bioreceptor classes discussed here, offering potential improvement in SiP biosensor sensitivity by bringing the binding interaction into a region of the evanescent field with higher field intensity.

Synthetic peptides are chemically synthesized using solid phase peptide synthesis (SPPS) or solution-based synthesis (SPS) [230]. Synthetic peptides are made using D-amino acids instead of the more naturally occurring L-amino acids seen in native peptides. D- and L-amino acids are enantiomers, or the same amino acid sequence with a mirror image structure. This change in configuration makes D-amino acids less susceptible to enzyme degradation without changing their biological function. SPS was the first synthesis method, developed in 1901, where a chain of amino acids is grown one residue at a time in solution [231]. SPPS followed in 1963 and uses a solid support for anchoring the peptide chain that enables washing steps between the addition of successive amino acids. Both methods start from a primary amino acid using selective protecting groups (FMOC, BOC) where successive amino acids are added in a step-by-step fashion to form a chain [232]. Generally, SPPS is the most common method since it is a well-established commercially available process and contributed to the Nobel Prize in Chemistry in 1984 [231]. Its use of a support and wash cycles results in a higher production of correctly formed peptides, removes reaction byproducts, as well as decreases the tendency of aggregation and incomplete reactions. However, SPS is still used since the lack of a support enables more challenging structures (cyclic), nonstandard components, and a larger array of coupling conditions (acidic, oxidative) [233].

Protein-catalyzed capture (PCC) agents are specialized, short (20 amino acid), synthetic peptides optimized to capture a target of interest [234]. They are considered "synthetic antibodies" due to their comparable high specificity and affinity for a target without the temperature sensitivity or stability issues common in enzymes, aptamers, and antibodies [235]. PCCs are highly selective since they are computationally designed based on the binding sites of proteins and other targets. Screening of chemical peptide libraries, such as one-bead one-compound (OBOC), identifies peptide components with high specificity and selectivity to the target of interest [236,237]. Due to this design, their affinity can be tailored to the specific dynamic range needed for sensing. Agnew et al. [234] evaluated the epitope binding sites and affinity of PCCs to those of monoclonal antibodies of the same target using principal component analysis. Their analysis covered 14 different protein targets as well as considered their physicochemical properties and molecular binding interactions. The results showed that PCCs are able to match and surpass antibody affinities with the majority of the binding driven by electrostatic interactions and hydrogen bonding.

In the literature, peptides and PCCs have been demonstrated as bioreceptors against antibodies [238], cancer cells [239], viral proteins [91], and streptavidin [240] on SiP platforms (Table 16). Angelopoulou et al. compared recombinant SARS-CoV-2 spike protein peptide on silicon nitride MZI sensors to conventional ELISA assays [238]. Silicon nitride MZI sensors were crosslinked via glutaraldehyde to the spike peptide against SARS-CoV-2 in a manner that selectively attached the peptides to only the silicon nitride waveguides and not the surrounding silicon dioxide. The reference was blocked with bovine serum albumin as a control for non-specific binding. The label-free peptide MZI showed a 80 ng/mL limit of detection and correlated with the ELISA results of 37 diluted serum samples. The addition of an antibody as a label improved the limit of detection to 20 ng/mL.

Table 16. Demonstrations of SiP biosensors using peptides as the biorecognition element and their sensing performance. All tabulated demonstrations used a label-free assay format.

Bioreceptor	Sensor Type	Target	Detection Performance		Refs.
			Figure of Merit	Value	
Peptide	Si_3N_4 MZI	SARS-CoV-2 antibodies	LoD	80 ng/mL	[238]
Peptide	Planar Si and porous Si microcavity	A20 lymphoma cancer cells	Coverage efficiency after 2 h incubation with 50,000 A20 cells	~85% and ~4% for planar and porous functionalized surfaces, respectively	[239]
PCC	Porous Si microcavity	Chikungunya virus E2 protein	Resonance shift after 3 h incubation with 1 μM E2 protein	1.7 ± 0.3 nm	[91]
PCC	Porous Si microcavity	Streptavidin	Resonance redshift after 1 h exposure to 5 μM streptavidin	12.9 nm	[240]

Martucci et al. [239] used idiotype peptides to determine the surface capture efficiency of tumor cells on silicon surfaces. Idiotype peptides are ligands from the binding site of receptors on the surface of immune cells that bind to antigens on the surface of lymphoma cells. They are specific to a subset of B-cells and can specifically identify lymphoma cells. The authors functionalized the surface of porous silicon microcavities by submerging in a 5% amino-terminated silane solution, crosslinking with a double N-succinimidyl terminated linker to crosslink to a primary amine on the peptide. The authors moved away from crosslinking antibodies to silicon surfaces since antibodies are known to have problems assembling monolayers in the same orientation due to their large size and multiple crosslinking sites [241,242]. Their results showed that the idiotype peptide covered 85% of the sensor's surface with a uniform, oriented layer and had a detection efficiency of 8.5×10^{-3} cells/μm^2.

PCCs are starting to become a more well-known method for biological sensing using silicon photonics due to their good stability and long shelf life. They are temperature stable, showing little to no change in affinity after heating to 90 °C, and resistance against protease degradation [91,243]. Layouni et al. [91] showed a PCC specific to Chikungunya virus E2 protein on porous silicon microcavities and with positive detection in response to 1 μM E2 viral protein. In addition, their results showed no statistical significance in sensor response between previously heated (90 °C, 1 h) and unheated PCCs. This stability was further confirmed by PCCs for vascular endothelial growth factor maintaining 81% of their affinity after 1 h using standard ELISA assays. Another work with porous silicon microcavities for PCC sensing of streptavidin showed detection of 5 μM streptavidin using PCCs immobilized via click chemistry crosslinking [240]. A summary of the advantages and limitations of peptides and PCCs as bioreceptors is provided in Table 17.

Table 17. Advantages and limitations of peptides and PCCs as bioreceptors.

Advantages	Limitations
• Small size • Synthetic production is available for scalable and flexible production [230] • Good specificity and selectivity [235] • Good temperature stability and resistance to protease degradation [91,235,243]	• Limited availability • Limited data available to assess performance (particularly on SiP platforms)

2.6. Glycans and Lectins

Both glycans and lectins have been employed as biosensor recognition elements on SiP devices (Figure 8). Glycans are carbohydrates which are covalently conjugated to proteins (glycoproteins) and lipids (glycolipids) [122,244]. In biological systems, glycoconjugates are typically found on cell surfaces, in the extracellular matrix, or in cellular secretions, and participate in intermolecular and cell–cell recognition events. Glycans consist of monosaccharides linked together in linear or branched structures by glycosidic bonds [244]. The diversity of their constituent monosaccharide residues and the position and configuration of their glycosidic bonds give glycans significant structural variability [128,244]. Lectins are non-immune proteins that recognize and bind glycoconjugates and non-conjugated glycans via carbohydrate recognition domains (CRD) [121,122,134]. Specific lectin-glycan binding is affinity-based and facilitated by hydrogen bonding, metal coordination, van der Waals and hydrophobic interactions [121]. The CRDs of lectins may target monosaccharide residues or they may show poor affinity toward monosaccharides and, instead, preferentially bind oligosaccharides based on their glycosidic linkages [121,122,244]. The affinity of individual CRD-glycan interactions are weak, with dissociation constants in the micromolar to millimolar range [121,122]. Multivalent binding between lectins and glycans, however, allows for higher-avidity interactions, with dissociation constants that are multiple orders of magnitude lower [122,123]. Namely, some lectins possess multiple CRDs that bind to multiple monosaccharide residues on a polysaccharide or to multiple proximal carbohydrates immobilized on a densely-coated solid substrate [121–123]; moreover, lectins can recognize homogeneous carbohydrate-coated surfaces or mixed glycan patches. Conversely, in the case of lectins with only one CRD, higher-avidity binding may be achieved by the clustering of many lectin molecules [122]. While many lectins have been identified and their glycan-binding characteristics have been characterized, these only encompass a small fraction of the diverse set of glycans that are found in nature [123]. Compared to proteins and nucleic acids, the functional study of glycans lags far behind [129].

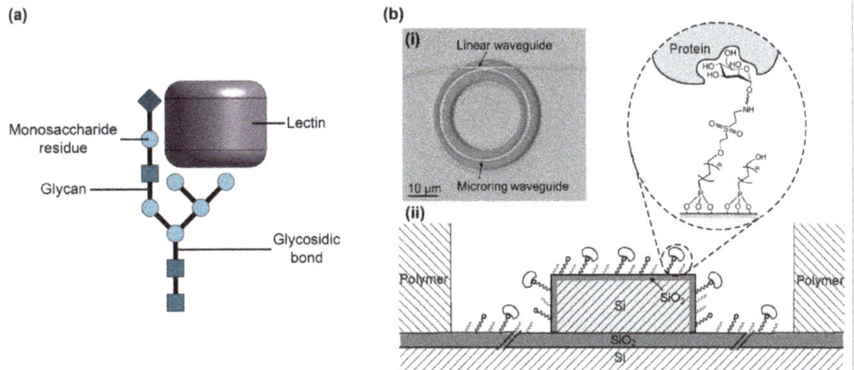

Figure 8. (**a**) Illustration of a glycan and bound lectin. (**b**) (i) SEM image of a microring resonator and (ii) cross-section of microring resonator waveguide using glycans as bioreceptors. The glycans are immobilized using an organophosphonate linking strategy and used for lectin (protein) capture. Part (**b**) is adapted with permission from Ref. [126]. Copyright 2012 American Chemical Society.

Glycans can be immobilized easily on biosensor surfaces in an oriented manner; for example, the terminal amine group of a glycan derivative can be targeted for site-directed covalent amine coupling to a surface [244]. In comparison, lectins possess more complex structures, making oriented immobilization more challenging.

Homogeneous glycan samples for biosensing applications cannot be synthesized easily in large quantities using biological systems, making chemical and chemoenzymatic synthesis the preferred routes of production for structurally defined glycans and glycoconjugates [127,245]. Multi-milligram quantities of polysaccharides up to 50 mers in length can be rapidly and reproducibly synthesized and optionally conjugated to nonglycan entities, like proteins, to yield glycoconjugates [127,128]. Nevertheless, chemical glycan synthesis is in its infancy and is inherently more challenging than oligonucleotide and oligopeptide synthesis because glycans are often highly branched and their biosynthesis is not template-driven [129]. Chemical glycan synthesis requires the modification of one monosaccharide hydroxyl group at a time in the presence of many others and the careful control of glycosidic linkage positions [127]. Currently, the synthesis of complex and highly branched glycan structures remains a major challenge [129].

Lectins may be purified from various organisms, though yields, especially for animal-derived lectins, are often too low for practical use [130]. Consequently, recombinant techniques are usually required for the production of lectins in multi-gram quantities [130]. Notably, anti-carbohydrate antibodies can be generated for glycan capture, but, due to the poor immunogenicity of carbohydrates, these antibodies typically have poor affinities toward their targets and limited versatility, making lectins preferable for carbohydrate detection [121]. In comparison to antibodies, the cost of lectin production is also lower. However, similarly to antibodies, the commercial synthesis of lectins is cell-based, and samples may vary in purity, properties, availability, and activity within and between vendors [121]. An overall comparison of glycans and lectins as bioreceptors for SiP biosensors is detailed in Table 18.

Table 18. Comparison of glycans and lectins as bioreceptors for SiP biosensors.

	Compatible Targets [121,122]	Affinity [121,122]	Specificity [121,123]	Stability [121,124–126]	Availability [127–130]	Reproducibility of Production [121,127,128]	Attachment Chemistry [244]	Size [121,128,131–134]	Cost *[135,136]
Glycan	Lectins, toxins, and viruses	Low; K_D in μM–mM range	Low	Good shelf life and stable under dry and ambient conditions	Low–moderate • Some commercial products (e.g., via Sigma Aldrich, Biosynth, etc.) † • Custom chemical synthesis possible for small glycans, but challenging for complex structures (e.g., synthesis available via Creative Biolabs, Asparia Glycomics, Glycan Therapeutics, etc.) †	High due to chemical synthesis	Site-directed immobilization via terminal amine group	Diverse (polysaccharides containing a few monosaccharide units to thousands)	Very high; ~CAD 200–CAD 1200/10 μg for commercial products
Lectin	Carbohydrates, including glycans and glycoconjugates	Low; K_D in μM–mM range	Low	Low due to susceptibility to permanent denaturation	Moderate–high • Some commercial products available (e.g., via Sigma Aldrich, Vector Laboratories, Medicago AB, etc.) † • Custom synthesis via recombinant techniques available	Moderate due to variability introduced by cell-based synthesis techniques	Challenging to achieve oriented immobilization due to complex structure	Diverse, but typically smaller than antibodies (~10–140 kDa)	Moderate; ~CAD 1–300/mg

* Prices are listed in CAD and are based on commercially available products as of July 2022. † These are not exhaustive lists of glycan or lectin vendors. Vendors listed are based on an exploratory search and are not endorsed or suggested by the authors.

Glycan-coupled SiP biosensors can be used for lectin capture and have applications in toxin [132] and virus [126] detection. For example, Ghasemi et al. [132] covalently immobilized GM1 ganglioside glycans on the surface of a TM mode silicon nitride microring resonator sensor for label-free detection of Cholera Toxin subunit B. The authors reported an absolute limit of detection of 400 ag, which corresponds to a surface coverage of 8 pg/mm^2. Shang et al. [126] used an organophosphonate strategy to tether glycans and glycoproteins to silicon microring resonators for label-free detection of various lectins and norovirus-like particles. The authors reported a limit of detection of 250 ng/mL for the norovirus-like particles. The functionalized sensors also demonstrated excellent stability, retaining strong binding performance after one month of storage at ambient conditions and after multiple cycles of surface regenerations with high-salt and high- and low-pH solutions. Indeed, the good chemical stability of glycans, even at ambient and dry conditions for prolonged periods of time, is an attractive characteristic of glycan-conjugated biosensors [124,125]. Other publications have demonstrated glycan- and glycoconjugate-functionalized SiP sensors for the label-free detection of common lectins, with limits of detection down to the ng/mL range [133,246].

Given that various diseases, such as cancer, autoimmune diseases, infections, and chronic inflammatory diseases are associated with glycan aberrations, glycans are valuable disease biomarkers [121,129]. Lectin-coupled biosensors have, therefore, been proposed for glycan biomarker-based disease diagnosis [121,129]. While lectin-coupled SiP sensors have seldom been reported in the literature, Yaghoubi et al. [131] reported a lectin-coupled porous silicon sensor using reflectometric interference Fourier transform spectroscopy for label-free detection of bacteria. The authors functionalized sensors with three different lectins, concanavalin A (Con A), wheat germ agglutinin (WGA), and ulex europaeus agglutinin (UEA), and found that the Con A- and WGA-coupled sensors demonstrated the greatest binding affinities for *E. coli* and *S. aureus*, respectively and demonstrated limits of detection of approximately 10^3 cells/mL. Table 19 provides a summary of SiP biosensors demonstrated in the literature that use glycans or lectins as bioreceptors.

Table 19. Demonstrations of SiP biosensors using glycans or lectins as the biorecognition element and their sensing performance. All tabulated demonstrations used label-free assay formats.

Bioreceptor	Sensor Type	Target	Detection Performance		Refs.
			Figure of Merit	Value	
Lacto-N-fucopentaose III-human serum albumin (LNFPIII-HSA) glycoprotein	Si MRR	Norovirus-like particles	LoD	250 ng/mL	[126]
BSA-mannose, BSA-lactose, BSA-galactose and RNase B glycoconjugates	Si MRR	Lectins: concanavalin A, griffithsin, and ricin	-	-	[246]
GM1 ganglioside glycan	Si_3N_4 MRR	Cholera toxin subunit B	LoD	400 ag (corresponds to 8 pg/mm^2)	[132]
3-fucosyl lactose glycan	Si_3N_4 MRR	Aleuria Aurantia Lectin	LoD	0.5 ng/mL (7 pM)	[133]
α2,6-disialylated biantennary N-glycan	Si_3N_4 MRR	Sambucus Nigra Lectin	LoD	12 ng/mL (86 pM)	
Concanavalin A lectin	Porous Si	*Escherichia coli*	LoD	10^3 cells/mL	[131]
Wheat germ agglutinin lectin	Porous Si	*Staphylococcus aureus*	LoD	10^3 cells/mL	

The greatest limitation of biosensors using glycan-lectin binding is their specificity. Unlike antibodies, lectins often bind to more than one glycostructure and demonstrate broader specificity, thus requiring extensive selectivity and cross-reactivity characterization prior to use [121,123]. The poorer selectivity of glycan-lectin interactions complicates their detection in complex biological samples and makes it difficult to detect small aberrations in the target structure [121,244]. Moreover, the avidity of glycan-lectin interactions is highly variable and depends not only on the structure of the biomolecules, but also on their multivalency and packing density on the sensor surface [122]. While glycans offer simple oriented conjugation to sensor surfaces and improved stability compared to antibodies, the discovery and production of biologically relevant glycans, especially complex and highly branched ones, is limited by current structural characterization and synthesis techniques [129]. Commercially available glycans are very expensive, at roughly CAD 200–1200/10 μg [135], while custom glycan synthesis is also costly [247]. On the other hand, lectins can be characterized and produced using mature and cost-effective techniques, but these proteins suffer from the same batch and vendor variability and pH-, temperature-, and buffer-sensitivity issues as antibodies [121,136]. While lectin regeneration is possible, it is likely to result in activity loss [121]. Table 20 highlights the key advantages and limitations of glycans and lectins as bioreceptors for SIP biosensors.

Table 20. Advantages and limitations of glycans and lectins as bioreceptors.

	Advantages	Limitations
Glycan	• Easy to achieve oriented immobilization [244] • Highly reproducible chemical synthesis [127,128] • Good stability [124–126] • Good regenerability [126]	• Poor affinity and specificity [121–123] • Functional study of glycans is less advanced than proteins and nucleic acids [123,129] • Chemical synthesis is challenging for long and branched structures [129] • Expensive [247]
Lectin	• Well-understood and relatively inexpensive synthesis by recombinant techniques [121]	• Poor affinity and specificity [121–123] • Challenging to achieve oriented immobilization [244] • Poor reproducibility due to cell-based synthesis [121] • Limited regenerability [121]

2.7. Other

2.7.1. High Contrast Cleavage Detection (i.e., CRISPR Cleavage Detection)

Recently, high contrast cleavage detection (HCCD) has been proposed as a detection mechanism for optical biosensing employing CRISPR-associated proteins as the biorecognition elements [137–139]. This is a clustered regularly interspaced short palindromic repeats (CRISPR)-based biosensing approach that can be used for sensitive detection of nucleic acid (DNA or RNA) targets. CRISPR systems contain CRISPR-associated (Cas) proteins, which possess endonuclease activity to cleave targets via guide RNA [140,248]. Most reported CRISPR based biosensors use Cas9, Cas12, or Cas13 effectors, which demonstrate different cleavage activities [249]. Namely, CRISPR-Cas9 cleaves target dsDNA based on guidance from single guide RNA [249]. CRISPR-Cas12 captures target DNA that is complementary to its guide RNA, activating non-specific collateral cleavage (or trans-cleavage) of nearby ssDNA [250]. Similarly, CRISPR-Cas13 captures target RNA that is complementary to its guide RNA, activating non-specific collateral cleavage of nearby ssRNA [250]. In the HCCD technique, Cas12 or Cas13 effectors can be used [138].

Most SiP biosensors rely on affinity-based detection, whereby low-index bioanalytes are captured on the sensor surface upon introduction of the target analyte. The HCCD method, however, adopts a different architecture and relies on the removal of high-index contrast reporters from the sensor surface upon introduction of the target analyte (Figure 9). In HCCD, the sensor surface is first decorated with high-index contrast reporters, such as silicon nanoparticles, gold nanoparticles or quantum dots, tethered to the surface by single-stranded oligonucleotides [137–139,141,142,251]. Then, the analyte is combined with CRISPR-Cas12 or CRISPR-Cas13, which have guide RNA complementary to the target [137]. Once activated, these CRISPR-Cas complexes cleave the reporters from the surface, leading to a change in the local refractive index that can be transduced by the SiP device.

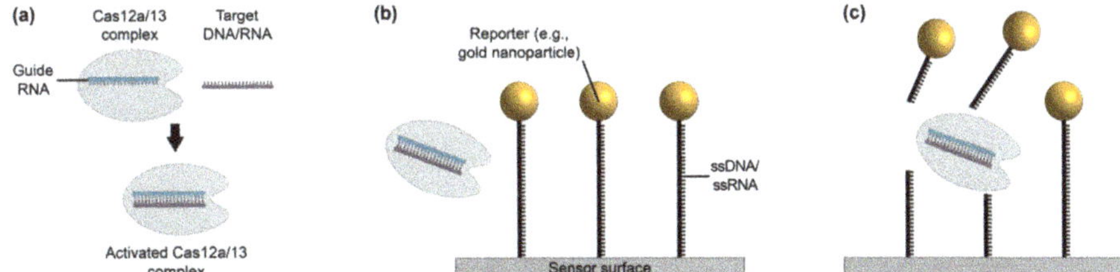

Figure 9. Illustration of HCCD, showing (**a**) activation of the CRISPR-Cas12a/13 effector by the target nucleic acid sample, (**b**) high index contrast reporters (e.g., gold nanoparticles) tethered to the sensor surface by single-stranded DNA or RNA prior to cleavage by the activated CRISPR-Cas12a/13 complex, and (**c**) non-specific collateral cleavage of single-stranded DNA or RNA by the activated CRISPR-Cas12a/13 complex, leading to the removal of reporters from the sensor surface.

The first experimental implementation of this method was reported in 2021 by Layouni et al. [137] on a porous silicon interferometer platform. This was a proof-of-concept study in which the sensor surface was decorated with nucleic-acid-conjugated quantum dot reporters, then exposed to a DNase solution, which cleaved reporters from the surface. While this work did not report specific analyte detection, it demonstrated the ability to detect a large shift in the sensor's reflectance peak upon enzyme-mediated removal of reporters from the porous silicon surface. This work paved the way for another preliminary study in which Liu et al. [139] demonstrated the detection of SARS-CoV-2 target DNA on a silicon microring resonator chip using HCCD (Table 21). The authors reported a ~8 nm blue shift in the resonance wavelength upon cleavage of gold nanoparticle reporters from the sensor surface by CRISPR-Cas12a activated by a 1 nM sample of target DNA in buffer solution. To our knowledge, SiP sensors using HCCD have yet to be demonstrated for the detection of nucleic acid targets in complex biological samples like whole blood, serum, and plasma. Chung et al. [251] proposed an inverse-designed waveguide-based integrated silicon photonic biosensor for HCCD-mediated sensing. However, this biosensing architecture has not yet been demonstrated experimentally.

Table 21. Demonstration of SiP biosensor using HCCD its sensing performance.

Bioreceptor	Sensor Type	Target	Detection Performance		Assay Format	Ref.
			Figure of Merit	Value		
CRISPR-Cas12a with guide RNA complementary to target	Si MRR	SARS-CoV-2 ssDNA	Resonance shift after exposure to 1 nM of target DNA	~8 nm	Labeled: gold nanoparticle reporters tethered to sensor surface by ssDNA were cleaved by activated Cas12a effector; amplification via collateral cleavage	[139]

The HCCD technique touts several advantages compared to traditional hybridization-based nucleic acid sensing. On a SiP sensor using hybridization-based sensing, signal generation relies on the small difference in refractive index between the sample buffer and the target nucleic acids. Typically, to achieve a detectable signal, the nucleic acid sample needs to be PCR amplified prior to detection or a secondary amplification molecule must be used after hybridization [138]. In HCCD, the refractive index contrast between the high-index reporters and background fluid is greater, leading to a greater signal change upon reporter removal compared to the binding of unlabeled targets [137,138]. Each activated CRISPR-Cas complex can perform up to 10^4 non-specific probe cleavages after activation, leading to multiplicative signal amplification, thus enhancing sensitivity [138,251]. Further, since HCCD relies on the removal of reporters from the surface, the SiP sensor experiences a blue shift in resonant frequency for a positive result; this is in contrast to affinity-based sensing in which a positive result causes a red shift. This means that HCCD is less susceptible to false positives caused by non-specific adsorption of biomolecules to the sensor surface [137]. Another beneficial feature of the HCCD method is that it derives its specificity from the CRISPR-Cas12 or -Cas13 complexes, which are activated in a highly specific manner by their nucleic acid targets. Since specificity is conferred by the CRISPR-Cas complexes rather than biomolecules immobilized on the sensor surface, there is an opportunity to develop universal reporter-functionalized SiP sensors which can be used with application-specific CRISPR-Cas reagents, thus reducing the costs of sensor development and production [138].

While the sensitivity of this detection strategy is bolstered by the collateral cleavage of the activated CRISPR-Cas complexes, this non-specific cleavage also makes multiplexing challenging. To the best of our knowledge, multiplexed nucleic acid detection based on HCCD has not yet been demonstrated. Another limitation of HCCD is that the irreversible cleavage of reporters from the SiP surface prevents facile regeneration of the functionalized sensor for repeated use. Further, HCCD is only suitable for the detection of nucleic acid targets, limiting its versatility. Finally, while the nucleic-acid-based reporter-modified surface has improved storage stability compared to antibody-functionalized surfaces, Cas enzyme activity is sensitive to storage conditions, complicating POC use [33]. This could potentially be addressed by lyophilizing the assay reagents [145]. Overall, while HCCD remains in its infancy and is yet to be validated for detection in complex media, this method addresses some of the limitations of hybridization-based nucleic acid detection schemes and offers potential as a highly sensitive and specific strategy for SiP sensing. Table 22 highlights the key advantages and limitations of HCCD.

Table 22. Advantages and limitations of HCCD.

Advantages	Limitations
• Very high sensitivity • Signal enhancement due to high index contrast reporters [137,138] • Multiplicative signal enhancement due to collateral reporter cleavage by Cas12/Cas13 effectors [138,251] • Insensitive to non-specific interactions [137] • Universal reporters for simple sensor functionalization [138]	• Challenging to multiplex • Not regenerable • Limited to nucleic acid targets • Poor stability of Cas enzymes [33] • Limited precedent for use and not yet demonstrated for sensing in complex biological fluids

2.7.2. CRISPR-dCas9-Mediated Sensing

CRISPR-associated proteins have also been used as a biorecognition element for signal amplification in silicon photonic sensors, in combination with nucleic acid probes. In 2018, Koo et al. [143] proposed a CRISPR-dCas9-mediated SiP biosensor for highly specific and sensitive detection of pathogenic DNA and RNA fragments for the diagnosis of tick-borne diseases. Broadly, this sensing method relies on twofold signal enhancement. Firstly, recombinase polymerase amplification (RPA) is used to amplify nucleic acid targets. RPA is a rapid enzyme-mediated DNA amplification technique that can be completed isothermally at mild temperatures [147,252]. This isothermal strategy obviates the need for power-intensive thermal cycling, which is required for conventional DNA amplification via PCR [147]. As such, RPA has been identified as an attractive alternative for POC use [145]. Additionally, reverse transcriptase (RT) can be added to the RPA reagents to facilitate isothermal amplification of RNA targets and reverse transcription of cDNA from RNA [143,147]. Secondly, nuclease-deactivated Cas9 (dCas9) is used in this sensing method as a labeling molecule. Like its active form, dCas9 binds to target dsDNA based on guidance from a target-specific single guide RNA (sgRNA) sequence [143,249]. Unlike its active form, dCas9 cannot cleave target sequences.

Koo et al. demonstrated this sensing method on SiP microring resonator sensors for the detection of pathogenic DNA and RNA sequences for scrub typhus (ST) and severe fever with thrombocytopenia syndrome (SFTS), respectively (Table 23) [143]. The sensor surface was first functionalized with single-stranded nucleic acid probes, complementary to the target sequences (Figure 10a) [143]. RPA or RT-RPA reagents were prepared, then added to the pathogenic DNA or RNA samples, along with dCas9 effectors and sgRNA. This mixture was incubated on the sensor chip in acrylic wells at 38 °C (for DNA targets) or 43 °C (for RNA targets). During this on-chip incubation, three key events took place: (1) the target DNA or RNA was amplified via RPA or RT-RPA, respectively, (2) amplified targets bound to complementary probes immobilized on the sensor surface (Figure 10b) and (3) dCas9 effectors bound to the hybridized targets to increase the refractive index change associated with each bound target (Figure 10c).

Table 23. Demonstrations of SiP biosensors using CRISPR-dCas9-mediated sensing and their performance.

Bioreceptor	Sensor Type	Target	Detection Performance		Assay Format	Ref.
			Figure of Merit	Value		
ssDNA probes	Si MRR	Scrub typhus viral DNA	LoD	0.54 aM	Isothermal pre-amplification of targets; target-specific CRISPR-dCas9 signal amplification	[143]
		Severe fever with thrombocytopenia viral RNA	LoD	0.63 aM		

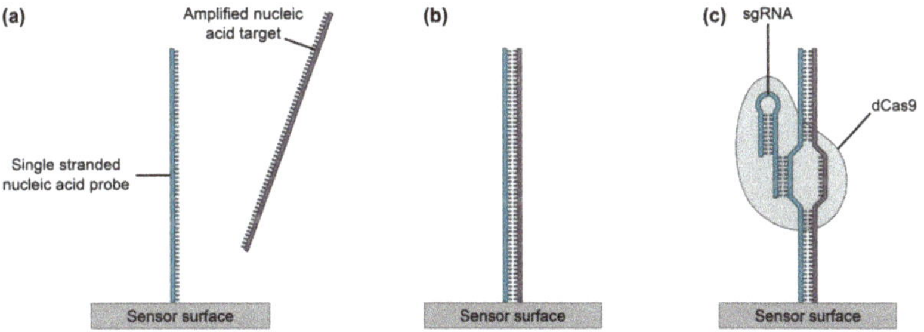

Figure 10. Illustration of CRISPR-dCas9-mediated sensing. (**a**) Single-stranded nucleic acid probes are immobilized on the sensor surface and the nucleic acid targets (amplified by recombinase polymerase amplification) are introduced to the sensor surface. (**b**) The nucleic acid targets hybridize to the surface-bound probes. (**c**) Deactivated Cas9 (dCas9), guided by single guide RNA (sgRNA) specifically binds to the nucleic acid duplex to amplify the signal, without cleaving the nucleic acid duplex.

The authors reported the detection of pathogenic DNA for ST with a detection limit of 0.54 aM and the detection of pathogenic RNA for SFTS with a detection limit of 0.63 aM [143]. The platform effectively discriminated between ST and SFTS in clinical blood serum samples in just 20 min. Indeed, this platform allows for exceptional sensitivity as a result of the aforementioned twofold signal enhancement. Further, specificity is ensured in three ways. Firstly, the nucleic acid probes immobilized on the sensor surface facilitate selective hybridization of complementary targets. Secondly, RPA or RT-RPA nucleic acid amplification is guided by primers to selectively amplify target sequences in the sample [252]. Thirdly, dCas9 solely binds to double-stranded target sequences based on sgRNA guidance, so dCas9 signal enhancement can only occur after targets have hybridized to complementary probes on the sensor surface. As such, this is a promising method for applications requiring highly sensitive detection of nucleic acid targets in complex samples.

To our knowledge, this is the only example of CRISPR-dCas9-mediated biosensing on a SiP platform in the literature. Because this sensing method uses dCas9, which does not demonstrate collateral cleavage, it may offer more straight-forward multiplexing compared to the HCCD technique, but at the cost of increased assay complexity [248,253]. Multiplexing may be possible if multiple microrings on a single chip are functionalized with different target-specific nucleic acid probes in a spatially defined manner, and multiple target-specific RPA primers and dCas9/sgRNA complexes are used [254].

Regarding costs, the short synthetic nucleic acid probes and CRISPR-Cas reagents required for this detection method can be produced at moderate cost, but the RPA reagents are more expensive [144,145,248,249,255]. For example, a single CRISPR-based diagnostics reaction involving RPA pre-amplification costs an estimated USD 0.61–5.00 in a laboratory setting, with RPA reagents making up the majority of this price [145–147]. Nevertheless, given the microlitre-scale reagent and sample volume requirements of SiP-based assays, these costs are unlikely to be prohibitive for POC use.

In this detection method, the sensor surface is prepared similarly to conventional nucleic acid hybridization-based biosensors, as described in Section 2.3, which allows for superior sensor stability compared to antibody-functionalized devices. However, one key limitation of this method is the requirement for many different assay reagents, including RPA enzymes and primers, dCas9 enzymes, and sgRNA. This increases the complexity of the assay preparation and requires environmentally controlled storage of the assay reagents, especially the enzymes, making POC use less feasible. However, lyophilization of environmentally sensitive reagents for transport and storage before use is a potential solution to this challenge [145]. The use of such a platform in a POC setting is further complicated by the need to implement careful thermal control over the RPA reaction. Finally, as is the case with classic nucleic acid hybridization-based biosensors, this platform only allows for the detection of nucleic acid targets, limiting its breadth of applications. A summary of the advantages and limitations of CRISPR-dCas9-mediated sensing as a biodetection technique for SiP biosensors is provided in Table 24.

Table 24. Advantages and limitations of CRISPR-dCas9-mediated sensing.

Advantages	Limitations
• Very high sensitivity and specificity [143] • Multiplexable • Good nucleic acid probe stability	• Limited to nucleic acid targets • Requires many assay reagents • Expensive reagents • Limited precedent for use

2.7.3. Lipid Nanodiscs

Lipid nanodisc-functionalized SiP sensors have been proposed to study signaling and interactions at cell membranes [148–150]. Lipid nanodiscs are 8–16 nm scale discoidal lipid bilayers, held together and made soluble by two encircling amphipathic protein belts, called membrane scaffold proteins (Figure 11) [150,151]. These nanodisc structures recapitulate the native cell membrane environment and allow for the precise control of lipid composition. This permits the study of biochemical processes that occur at cell membranes, and which require specific lipid compositions for full functionality [256]. Lipid nanodiscs also solubilize and stabilize membrane proteins, which typically demonstrate loss of activity and function outside of the phospholipid membrane environment [151]. Given that membrane proteins are involved in vital regulatory cell functions and are often the target of therapeutic drugs, lipid nanodiscs are a valuable tool for studying cell membrane interactions involving these proteins. Compared to other structures, such as liposomes and detergent-stabilized micelles, which are used to mimic the cell membrane environment, nanodiscs offer improved consistency, monodispersity, production yield, and control over lipid and protein composition [150,151].

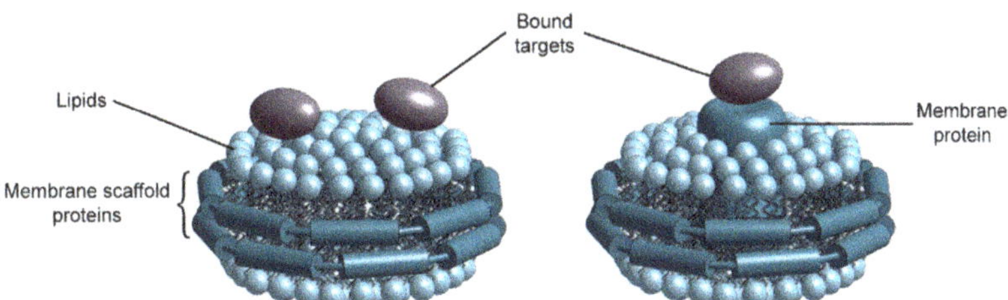

Figure 11. Illustrations of lipid nanodiscs with bound targets. The nanodiscs consist of lipid bilayers, held together by two encircling membrane scaffold proteins. The nanodiscs may be prepared without (**left**) or with (**right**) embedded membrane proteins.

SiP sensors are an appealing platform on which to investigate interactions between lipid nanodiscs and other biomolecules. The multiplexability of SiP sensors permits high-throughput screening of cell membrane interactions. Further, membrane proteins are challenging to produce and typically have low yields, making the low reagent volume requirements of SiP sensors particularly attractive [148]. Finally, nanodiscs physisorb directly onto silicon dioxide, permitting their facile immobilization onto the native oxide surfaces of silicon and silicon nitride waveguides [150].

In the literature, lipid nanodisc-functionalized silicon microring resonator sensors have been used to probe interactions between soluble proteins and lipids, glycolipids, and membrane proteins embedded in nanodiscs (Table 25) [148–150]. In a study by Sloan et al. [150], lipid nanodiscs prepared with varying compositions of the phospholipids, 1-palmitoyl-2-oleoyl-sn-glycero-3-phosphocholine (POPC) and 1-palitoyl-2-oleoyl-sn-glycero-3-[phospho-L-serine] (POPS), were used to probe the binding of annexin V, a lipid-binding protein. Nanodiscs prepared with glycolipids (1,2-dimyristoyl-sn-glycero-3-phosphocholine/monosialotetrahexosyl ganglioside, GM1), biotinylated lipids (*N*-(biotinoyl)-1,2-dipalmitoyl-sn-glycer-3-phoaphoethanolamine, biotin-DPPE), and enzymes (cytochrome P450 3A4, CYP3A4) were also used to probe binding interactions with cholera toxin subunit B, streptavidin, and anti-CYP3A4, respectively. A 4-plex assay was prepared by microfluidically patterning the sensor chip with POPS, GM1, biotin-DPPE, and CYP3A4 nanodiscs, then exposing the whole sensor surface to annexin V, CTB, streptavidin, and anti-CYP3A4 solutions in sequence. This multiplexed assay demonstrated effective binding with minimal cross-reactivity for each specific protein-nanodisc combination.

Table 25. Demonstrations of SiP biosensors using lipid nanodiscs as the biorecognition element and their sensing performance. All tabulated demonstrations used a label-free assay format.

Bioreceptor	Sensor Type	Target	Detection Performance		Ref.
			Figure of Merit	Value	
Lipid nanodiscs containing PC, four binary compositions of PC and PS, and two binary combinations of PS and PA	Si MRR	Blood clotting proteins: pro-thrombin, factor X, activated factor VII, and activated protein C	-	-	[149]
Lipid nanodiscs containing POPC and POPC/POPS	Si MRR	Annexin V	-	-	[150]
Lipid nanodiscs containing GM1	Si MRR	Cholera Toxin Subunit B	-	-	
Lipid nanodiscs containing biotin-DPPE	Si MRR	Streptavidin	-	-	
Lipid nanodiscs containing CYP3A4	Si MRR	Anti-CYP3A4 antibody	-	-	
Lipid nanodiscs with 9 different compositions containing PS, PE, and PC.	Si MRR	Protein clotting factors: prothrombin, activated factor VII, factor IX, factor X, activated protein C, protein S, and protein Z	-	-	[148]

PC: phosphatidylcholine, PS: phosphatidylserine, PA: phosphatidic acid, POPC: 1-palmitoyl-2-oleoyl-sn-glycero-3-phosphocholine, POPS: 1-palmitoyl-2-oleoyl-sn-glycero-3-(phospho-L-serine), GM1: monosialotetrahexosyl ganglioside, biotin-DPPE: *N*-(biotinoyl)-1,2-dipalmitoyl-sn-glycer-3-phosphoethanolamine, CYP3A4: cytochrome P450 3A4, and PE: phosphatidylethanolamine.

In another work by Muehl et al. [149], a SiP microring resonator platform was used to investigate interactions between four different blood clotting proteins (pro-thrombin, factor X, activated factor VII, and activated protein C) and lipid nanodiscs prepared with seven different binary lipid combinations of phosphatidylcholine (PC), phosphatidylserine (PS), and phosphatidic acid (PA). A 7-plex sensor was demonstrated using these seven nanodisc preparations to obtain dissociation constants for binding between the coagulation proteins and lipid surfaces. All of the coagulation proteins studied in this work bind in a Ca^{2+} manner, so the nanodisc-functionalized surfaces were regenerated with good replicability using a Ca^{2+}-free buffer after protein binding. In a subsequent work, Medfisch et al. [148] used a SiP microring resonator platform to study the binding interactions of seven different protein clotting factors (prothrombin, activated factor VII, factor IX, factor X, activated protein C, protein S, and protein Z) and lipid nanodiscs prepared with nine different phospholipid compositions involving PS, phosphatidylethanolamine (PE), and PC. The effect of PE-PS lipid synergy on the membrane binding of clotting factors was investigated. Again, surface regeneration after binding events was achieved using Ca^{2+}-free buffer.

So far, SiP sensors functionalized with lipid nanodiscs have demonstrated value in the study of binding interactions at cell membranes. This is in contrast with other classes of bioreceptors discussed in this review, which have primarily been proposed for toxin and pathogen detection and/or diagnostic applications. The nanodisc-protein interactions demonstrated by Muehl et al. [149] and Medfisch et al. [148] have limited specificity, with all of the investigated clotting proteins binding, albeit to different extents, to the lipid nanodisc-functionalized surfaces. Hence, these nanodisc-functionalized sensors are likely unsuitable for selective discrimination between multiple targets. The incorporation of embedded membrane proteins or glycolipids into the nanodiscs, however, may offer more selective detection of soluble proteins, as demonstrated by Sloan et al. [150]. Lipid nanodiscs are typically custom-synthesized in the laboratory setting, allowing for the precise control of lipid composition and membrane protein content; while this leverages the flexibility of lipid nanodiscs, it limits their accessibility for assay-development and widespread use [151]. Overall, nanodisc-functionalized SiP sensors offer an excellent opportunity for high-throughput laboratory-based cell membrane interaction studies, but their potential in

POC diagnostics may be limited. A summary of the advantages and limitations of lipid nanodiscs as bioreceptors for SiP biosensors is provided in Table 26.

Table 26. Advantages and limitations of lipid nanodiscs as bioreceptors.

Advantages	Limitations
• Solubilize and stabilize membrane proteins for studying cell membrane interactions [151] • Better consistency, production, monodispersity, production yield, and control over lipid composition than other cell membrane mimics [150,151] • Easy immobilization by physisorption [150] • Regenerable [148,149]	• Poor selectivity, but can be improved by incorporation of membrane proteins [148–150] • Limited availability; usually custom synthesized in lab [151] • Biosensing applications are mainly limited to the study of cell membrane interactions

2.8. Summary and Future Directions

Given the myriad of potential applications for SiP biosensors and the complex trade-offs of each bioreceptor class, there is no simple formula for selecting an optimal bioreceptor as each class has its own set of advantages and limitations that must be balanced with the needs of the application. For studies specifically probing carbohydrate-protein or cell membrane interactions, the choice is simple, with glycans/lectins or lipid nanodiscs typically being the most appropriate options, respectively. For other applications, the choice of bioreceptor can initially be narrowed down based on compatibility with the target of interest (see Table 3). Beyond this, the specific functionalization needs for the application of interest must be identified and used to guide further bioreceptor selection. For example, for non-nucleic acid targets, one must choose between antibodies, aptamers, MIPs, PCCs, and peptides. For non-POC applications where stability, regenerability, and cost are less important, monoclonal antibodies may be a suitable option due to their widespread availability and good binding affinity and selectivity. For POC applications, antibodies may not be suitable, and the choice between aptamers, MIPs, PCCs, and peptides will likely depend on the availability of pre-designed and validated products for the target of interest, or access to the relevant expertise and resources to design a custom bioreceptor for the target of interest. Trade-offs between affinity, selectivity, and stability should also be considered as relevant to the desired application. For nucleic acid targets, nucleic acid probes may be the best option for applications where assay simplicity, cost, stability, and/or multiplexing are the most important considerations. The opportunity to choose between different nucleic acid analogues (e.g., DNA, RNA, PNA, LNA, morpholinos) and chemical modifications can be used to tailor the stability and affinity of the nucleic acid probes for the application of interest. Applications requiring exceptional sensitivity and selectivity may benefit from the use of the more complex and early-stage HCCD or CRISPR-dCas9-mediated sensing strategies.

Regarding future directions, further research and development are required to improve the availability of pre-designed synthetic antibody analogues (e.g., aptamers, MIPs, PCCs) against various biomarkers. The availability of successful MIP formulations may be enhanced by increased use of computational methods. Such computational methods can aid in the development and optimization of MIP formulations for targets of interest and reduce experimental effort by guiding researchers toward promising systems [119]. Future SiP biosensing studies should focus on biomarker detection in complex biological fluids to quantify bioreceptor selectivity and to ensure reliable detection performance when using clinically relevant samples. For instance, the validation of aptamers for target detection in complex biological samples is essential for their translation to real-world sensing applications due to the sensitivity of their three-dimensional conformation and binding affinity to the ionic strength and pH of the sample [180]. HCCD and CRISPR-dCas9-mediated sensing are in their infancy and future studies should focus on validating these strategies for sensing in complex biological samples. Moreover, future work should focus on multiplexing these CRISPR-based methods to enable simultaneous detection of multiple targets.

3. Bioreceptor Immobilization Strategies

The surface of unmodified SiP sensors consists of a native silicon dioxide layer, which grows on silicon and silicon nitride upon exposure to air and moisture [31,257]. This oxide surface is hydrophilic in character [258,259] and has a negative surface charge density above pH 3.9 [260]. Strategies for immobilizing bioreceptors on SiP devices generally rely on non-covalent interactions between bioreceptors and the native oxide surface or target surface silanol groups for covalent attachment. In this section we discuss bioreceptor immobilization strategies for SiP biofunctionalization, focusing on passive adsorption, bioaffinity binding, and covalent immobilization (Figure 12). We discuss methods relevant to antibody, aptamer, nucleic acid probe, peptide, PCC, glycan, lectin, and lipid nanodisc immobilization and present tables categorizing bioreceptor immobilization strategies that have been used in functionalization approaches in the previous literature. It should be noted that strategies discussed in the following subsections generally are not relevant to MIP-based bioreceptors, which are immobilized on SiP surfaces during synthesis via casting and/or in situ polymerization; as such, MIPs are not discussed in detail here. Table 27 provides a summary of bioreceptor immobilization strategies that have been employed on SiP devices in the literature and benchmarks these strategies against biofunctionalization needs for SiP biosensors.

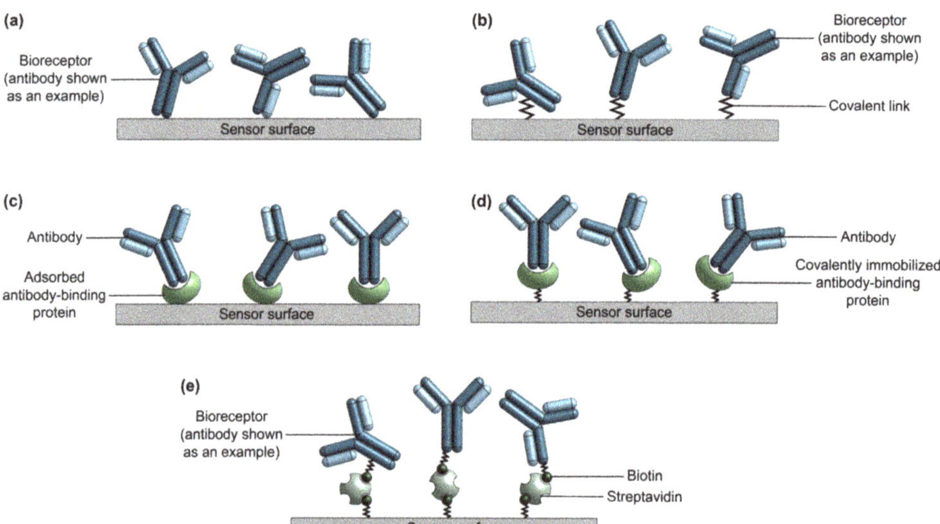

Figure 12. Illustrations of different strategies for immobilizing bioreceptors (antibodies are shown as an example) on SiP sensor surfaces. The depicted immobilization strategies include (**a**) non-covalent passive adsorption, (**b**) covalent attachment, (**c**) bioaffinity-based oriented immobilization using antibody-binding proteins adsorbed to the surface, (**d**) bioaffinity-based oriented immobilization using antibody-binding proteins covalently linked to the surface, and (**e**) bioaffinity-based immobilization in which the surface and bioreceptor are covalently conjugated with biotin and streptavidin is used as a linking molecule.

Table 27. Comparison of different bioreceptor immobilization chemistries based on SiP biofunctionalization needs.

Immobilization Chemistry	Compatible Bioreceptors	Surface Modification	Bioreceptor Modification	Required Linkers	Stability	Thickness	Oriented Bioreceptor Immobilization?	Impact on Bioreceptor Function	Typical Replicability and Uniformity	Compatibility with System-Level Sensor Integration	Process Scalability
Passive adsorption											
Passive adsorption	All	Not required	Not required	None	Typically, poor [31,69,261]	No added thickness	Typically, no	Likely to reduce bioreceptor binding activity [29,69,262,263]	Poor	Good	Very good
Bioaffinity											
Antibody (Ab)-binding protein	Antibodies	Ab-binding protein is typically passively adsorbed on sensor	None	None	Moderate [29,75]	3–4 nm for adsorbed PrA [259,264]	Yes	Preserves bioreceptor binding activity	Depends on Ab-binding protein immobilization strategy	Depends on Ab-binding protein immobilization strategy	Depends on Ab-binding protein immobilization strategy
Biotin/(strept)avidin	Antibodies, aptamers, nucleic acid probes, peptides, PCCs, glycans/lectins	Often silanization	Biotinylation	(Strept)avidin acts as bioaffinity linker	Good [74,265]	~6–7 nm plus thickness of chemical layer used to immobilize (strept)avidin [266]	Possible	Preserves bioreceptor binding activity [263]	Depends on (strept)avidin immobilization strategy	Depends on (strept)avidin immobilization strategy	Poor due to complexity
Covalent											
Silane chemistry	Antibodies, aptamers, nucleic acid probes, peptides, PCCs, glycans/lectins	Silanization	Aptamers and nucleic acids require modification with terminal functional groups (e.g., amine, carboxyl, thiol). S-4FB conjugation required for SoluLink chemistry.	Often required; popular options include GA, BS^3, and EDC/NHS	Good [31]	<1 nm for silane monolayer; multilayer films may exceed 10 nm [267]	Possible	Typically preserves bioreceptor binding activity; strategies requiring antibody disulfide bond reduction may reduce antibody binding activity [69,75]	Variable for solution-phase silanization; good for vapor-phase silanization [261,267–269]	Poor for solution-phase silanization; good for vapor-phase silanization	Fair-good, depending on reaction conditions and linker requirements
Organophosphonate chemistry	Antibodies, aptamers, nucleic acid probes, peptides, PCCs, glycans/lectins	UDPA deposition	Aptamers and nucleic acids require modification with terminal functional groups	Required; DVS and 3-maleimidopropionic-acid-*N*-hydroxysuccinimidester have been used [126,270]	Very good [126]	~1 nm [270]	Possible	Preserves bioreceptor binding activity [126,270]	Very good	Poor due to solution-phase UDPA deposition	Fair; multistep process

Table 27. *Cont.*

Immobilization Chemistry	Compatible Bioreceptors	Surface Modification	Bioreceptor Modification	Required Linkers	Stability	Thickness	Oriented Bioreceptor Immobilization?	Impact on Bioreceptor Function	Typical Replicability and Uniformity	Compatibility with System-Level Sensor Integration	Process Scalability
Click chemistry	Antibodies, aptamers, nucleic acid probes, peptides, PCCs, glycans/lectins	Azide/alkyne derivatization	Modification with azide/alkyne moieties, DBCO, or tetrazine	Azide, alkyne, DBCO, or tetrazine terminations required	Good [91]	<1 nm	Possible	Preserves bioreceptor binding activity	Good; insensitive to oxygen and moisture [271]	Poor due to solution-phase chemistry and aggressive surface pre-treatment strategies, which may damage sensor [59,91,240]	Fair; multistep process
UV-crosslinking	Aptamers [272] and nucleic acids [273]	None required	Modification with poly(T) or poly(TC) tags	None	Good [272,273]	Thickness added by poly(T) or poly(TC) tag; depends on tag length	Yes	Preserves bioreceptor binding activity [272,273]	Good due to process simplicity [272]	Very good	Very good

S-4FB: sulfo succinimidyl 4-formylbenzoate, GA: glutaraldehyde, BS^3: bis(sulfosuccinimidyl)suberate, EDC/NHS: 1-ethyl-3-[3-(dimethylamino)propyl]-carbodiimide/*N*-hydroxysuccinimide, UDPA: 11-hydroxyundecylphosphonic acid, DVS: divinyl sulfone, DBCO: dibenzoazacyclooctyne or aza-dibenzocyclooctyne, poy(T)/poly(TC): poly(thymine)/poly(thymine cytosine).

3.1. Passive Adsorption

Adsorption (Figure 12a) is the fastest and simplest method by which bioreceptors can be immobilized on a biosensor surface [29,68,69,80]. Adsorption-based bioreceptor immobilization has been widely used, especially in preliminary demonstrations of novel sensing architectures [31]. Bioreceptors may adsorb to a bare or modified SiP surface due to electrostatic, hydrophobic, polar-polar or Van der Waals interactions, or some combination of these non-covalent interactions [29,262]. Nevertheless, covalent and affinity-based strategies are typically preferred to adsorption-based immobilization.

One major disadvantage of adsorption is that it provides little control over the orientation of immobilized bioreceptors [31,69,204,263,274]. This may render binding sites unavailable for target capture, reducing the target binding capacity and, therefore, the sensitivity of the sensor. This random orientation, combined with intermolecular interactions, may also lead to poor bioreceptor loading density on the sensor surface [263]. Adsorption-based immobilization may lead to reduced bioreceptor activity due to folding or denaturation. This is especially relevant for protein-based bioreceptors, like antibodies, which are known to denature when adsorbed to surfaces, potentially changing the structure of their Fab fragments and diminishing their antigen-binding capacity [29,69,262,263]. Further, adsorbed bioreceptors are susceptible to desorption, leading to poor surface stability [31,69,261]. This is particularly relevant when the sensor is operated in flow conditions or when surface regeneration involving the release of targets from the sensor for multiple cycles of reproducible binding is desired. For example, Jönsson et al. [261] demonstrated that antibodies physisorbed onto chemically modified silicon dioxide surfaces were unstable toward changes in the surrounding medium, demonstrating significant desorption upon exposure to low pH, low surface tension, detergent, urea, and high ionic strength solutions. Finally, surfaces allowing for strong adsorption of bioreceptors may also be amenable to the adsorption of other biomolecules present in a complex biological sample, such as blood, leading to non-specific adsorption and high background signals [33]. Similarly, if other proteins possessing higher adsorption affinities to the sensor surface are present in the fluid, the bioreceptors may leach off the sensor [75]. This, in turn, compromises the selectivity of the sensor.

Despite the numerous limitations of adsorption-based functionalization, SiP sensors functionalized with lipid nanodiscs have demonstrated good stability and selectivity using adsorptive immobilization [148–150]. Indeed, lipid nanodiscs are especially amenable to adsorption-based immobilization because, like lipid bilayers, they are known to adsorb well to silicon dioxide surfaces [275,276]. This allows for simple nanodisc immobilization without the need to chemically modify the SiP surface or the nanodiscs. These nanodisc-functionalized sensors were regenerated after target binding with good reproducibility and no appreciable nanodisc desorption using Ca^{2+}-free buffer, indicating stable immobilization [148,149]. These sensors were also used for multiplexed detection of soluble proteins with minimal non-specific binding [150]. However, the native silicon dioxide surface of SiP sensors is negatively charged at physiological pHs, such as the buffered systems employed in these nanodisc studies, meaning that nanodiscs with lipid compositions containing a high percentage of anionic lipids show a lower affinity for the sensor surface, leading to poorer surface coverage [148,149]. Fortunately, this reduced affinity is predictable and could be counteracted, at least in part, by using higher spotting concentrations [149].

3.2. Bioaffinity-Based Immobilization

Bioaffinity-based receptor immobilization involves the creation of multiple non-covalent interactions between a bioreceptor and biomolecule(s) acting as linker to the substrate [41]. The sum of many weak interactions yields a strong link between the bioreceptor and the surface. Two of the most common bioaffinity-based receptor immobilization strategies used for SiP sensor functionalization involve antibody-binding proteins and the biotin-avidin system, both of which can achieve oriented bioreceptor immobilization [29,69,74,75].

Antibody-binding proteins, including Protein A, Protein G, Protein A/G and Protein L, have been widely used for the oriented capture of antibodies on biosensor surfaces [29,75,263,277]. Protein A is derived from *Staphylococcus aureus*, while Protein G is derived from *Streptococcus* species, and Protein L is derived from *Peptostreptococcus magnus* [75,277]. Both Proteins A and G reversibly bind the Fc region of antibodies, binding a maximum of two antibodies at a time, and have variable antibody-binding affinities that depend on the immunoglobulin (Ig) subclass and the species of origin [75]. Protein A can capture mammalian IgGs with dissociation constants as low as the 1–10 nM range, while Protein G can achieve slightly higher affinity capture of mouse and human IgGs with dissociation constants as low at the 0.1–1 nM range [278,279]. The oriented capture of antibodies by these proteins ensures that the antibody's Fab fragments are accessible for antigen capture, significantly enhancing the functionalized surface's antigen binding activity [277,278]. For example, Ikeda et al. [278] demonstrated a 4- to 5-fold increase in antigen binding capacity for antibodies immobilized on silicon wafers using Protein A, compared to antibodies immobilized via physisorption alone. This was attributed to the improved steric accessibility of the antigen binding sites of the well-oriented Protein A-immobilized antibodies. In addition to its Fc-binding regions, native Protein G has additional sites for albumin and cell surface binding; however, recombinant Protein G, containing only Fc-binding domains, has been produced using *E. coli* to prevent this nonspecific binding [74,75]. Protein A/G is a recombinant protein that contains the Fc-binding domains from both Protein A and G [74]. Similarly to Proteins A and G, Protein L also binds antibodies in an oriented manner, but instead of binding to the Fc region, Protein L binds to antibodies' κ-light chains outside of the antigen-binding site with dissociation constants as low as ~10 nM [277,279]. As a result, Protein L can bind any class of antibody, in addition to Fab fragments, which lack an Fc region, though its binding affinity is species-specific [277,280]. Indeed, a significant challenge associated with antibody-binding protein-directed bioreceptor immobilization is this Ig subclass and/or species-based variation in antibody-binding affinity; further this technique cannot be used to immobilize any bioreceptors aside from antibodies.

While Proteins A, G, and L allow for optimal orientation of immobilized capture antibodies, oriented immobilization of these antibody-binding proteins remains a challenge [75]. Fortunately, these antibody-binding proteins have several high affinity binding sites for antibodies, making their orientation on sensor surfaces less critical [41]. In the literature, antibody-functionalized SiP microring resonator sensors have been prepared using Protein A physisorbed on the sensor surface (Figure 12c) [1,264]. It has been reported that Protein A adsorbs onto silicon surfaces in a two-step process to yield a ~3–4 nm-thick adlayer [259,264]. First, a monolayer of Protein A is rapidly adsorbed on the surface; this first monolayer is denatured due to very strong non-covalent binding to the surface. Next, a second and third monolayer of Protein A are slowly adsorbed on the surface; these layers consist of non-denatured proteins which retain their ability to effectively bind the Fc region of antibodies. This strategy of passive Protein A adsorption followed by oriented antibody capture was used on sub-wavelength grating SiP microring resonators by Flueckiger et al. [264] and Luan et al. [1] to immobilize anti-streptavidin for model streptavidin-binding assays.

Others have tagged antibody-binding proteins with small molecules or other proteins to achieve higher-affinity binding to SiP sensor substrates. For example, Ikeda et al. [278] constructed a fusion of Protein A and bacterial ribosomal protein L2, which is termed "Si-tag" and binds strongly to silicon dioxide surfaces [281]. The authors demonstrated that the fusion protein was strongly immobilized on silicon dioxide surfaces in an oriented manner with a dissociation constant of 0.31 nM. The Si-tagged protein A also strongly bound mouse IgGs with a dissociation constant of 3.8 nM. The fusion protein immobilized 30–70% more IgG compared to physisorption of IgGs on bare silicon dioxide surface. Further, the fusion protein-immobilized IgGs demonstrated a 4- to 5-fold increase in antigen binding performance compared to the physisorbed IgGs. This functionalization strategy was subsequently demonstrated on a SiP microring resonator platform [282]. Christenson et al. [164]

leveraged the strong bioaffinity interaction between biotin and streptavidin to immobilize recombinant Protein G on silicon photonic crystal-total internal reflection sensors. In this work, the sensor surface was modified with silane-PEG-biotin molecules, followed by streptavidin, then biotinylated recombinant Protein G. Antibodies were immobilized on this surface and used to detect cardiac troponin I. Covalent immobilization of antibody-binding proteins to silicon-based substrates (Figure 12d) may also be facilitated via methods such as surface modification with silane and a crosslinker, followed by Protein A or G attachment, as demonstrated by Anderson et al. [283] or via click chemistry, as demonstrated by Seo et al. [277] on glass substrates.

Interactions between antibodies and antibody-binding proteins are reversible and can be disrupted by variations in pH [29,75]. This limits biosensor stability and complicates sensor regeneration because antigens cannot be easily eluted from the sensor surface without also removing the capture antibodies. In this way, sensor regeneration is possible, but requires that both the antigen and capture antibody be eluted from the antibody-binding protein-functionalized surface, followed by reapplication of the capture antibody for another round of detection [75,263,283]. For example, Seo et al. [277] covalently bound Protein A onto glass slides, followed by the immobilization of receptor antibodies (rabbit anti-goat IgGs). These antibody-functionalized slides were used to capture target antibodies (goat anti-human IgGs) and were then treated with a low pH glycine-HCl buffer to remove the receptor and target antibodies. After this wash step, only the covalently bound Protein A remained. The surfaces were then successfully regenerated for a second round of binding by reapplying the receptor antibodies. Similarly, Anderson et al. [283] covalently bound Protein A to silicon dioxide optical fibers, followed by the immobilization of capture antibodies (rabbit anti-goat IgGs). Then, the functionalized fibers were used to capture fluorescently labeled targets (Cy5.5-goat IgG). The surfaces were regenerated using a pH 2.5 glycine-HCl, 2% acetic acid solution, followed by re-application of the capture antibody. Four cycles of regeneration were performed successfully with no appreciable reduction in Protein A's Fc-binding capacity. However, the authors also reported unsuccessful regeneration of Protein A and G for an assay detecting plague F1 antigen, showing that regeneration of antibody-binding proteins may depend on the selected capture antibody and antigen. Here, the necessary reapplication of the receptor antibody also increases the cost and complexity of sensor reuse compared to strategies in which the functionalized surface can be regenerated solely by the removal of the target.

Another common bioaffinity interaction coupling method used in biosensor functionalization is based on the biotin-avidin/streptavidin complex, whereby the sensor surface is coated with avidin or streptavidin and used to immobilized biotinylated receptors (Figure 12e) [41]. Biotin is a small vitamin and avidin is a glycoprotein found in egg whites, which contains four biotin binding sites [265]. The biotin-avidin interaction is one of the highest affinity non-covalent interactions known in biology, with a dissociation constant on the order of 10^{-15} M [75,80,265]. This nearly irreversible non-covalent interaction is extremely resistant to variations in temperature, buffer salt, pH, and the presence of denaturants and detergents [74,265]. Streptavidin is a biotin-binding protein, derived from *Streptomyces avidinii*, which shows similar biotin-binding activity to avidin [265]. Streptavidin, however, has a pI of 5, while avidin has a pI of 10.5; as such, streptavidin is less susceptible to nonspecific interactions at physiologic pH, often making it the preferred choice [80,265].

The high-affinity nature of the biotin-streptavidin interaction means that biosensor regeneration via target removal can be achieved without disrupting the link between the receptor and the surface [284]. This means that the sensor can be used for multiple cycles of target binding without reapplying receptors. For example, Choi et al. [285] functionalized silicon nitride chips for reflectometric interference spectroscopy by covalently linking biotin to the surface, followed by avidin, and biotinylated concanavalin A. This lectin-coupled chip was used to reproducibly capture ovalbumin, a glycoprotein, over multiple binding cycles by regenerating the surface with a 10 mM glycine-HCl (pH 1.5) solution, which removed captured glycoproteins, while leaving the lectin-functionalized surface intact.

In another work [284], SPR surfaces were functionalized with a biotin analogue, desthiobiotin, followed by streptavidin, and biotinylated IgGs. The authors reported that the functionalized surface was stable throughout multiple cycles of regeneration with solutions commonly used for target removal from bioreceptors, including HCl, Na_2CO_3, glycine buffer, and SDS solutions. Lü et al. [286] functionalized optical fiber probes by covalently linking streptavidin to the exposed silicon dioxide core, followed by the oriented immobilization of 5′-biotinylated DNA probes. These surfaces were used to bind complementary DNA targets, followed by thermal regeneration via washing for 2 min in hybridization buffer at 70 °C, or chemical regeneration via washing in 4 M urea solution. The surfaces demonstrated no appreciable loss in hybridization ability over six cycles of thermal or chemical regeneration. Efforts have also been made to break biotin-streptavidin interactions for complete surface regeneration whereby receptors are completely removed from the surface [266,284]. This has been achieved using a pH 7 chemical buffer solution [266], sequential rinsing with free biotin, guanidinium thiocyanate, pepsin, and sodium dodecyl sulfate [284], and washing with water at 70 °C [287]. These strategies require the reapplication of biotinylated receptors and sometimes streptavidin/avidin between binding cycles, but also open the possibility for sensors to be reused with different bioreceptors for each cycle.

The biotin-avidin/streptavidin-based immobilization strategy is more flexible than antibody-binding proteins, in that many different classes of receptors can be tagged with biotin and immobilized on avidin/streptavidin-coated sensor surfaces. On SiP platforms, this biotin-avidin/streptavidin bioaffinity functionalization strategy has been used to immobilize antibodies [288,289] and nucleic acid probes for both hybridization sensing [203] and CRISPR-Cas-modulated high contrast cleavage detection [137,139]. Similarly, it has been used to immobilize lectins on a silicon nitride sensor using reflectometric interference spectroscopy as the transduction technique [285]. This strategy can achieve unoriented or oriented receptor immobilization. For antibodies, amine, carboxyl, sulfhydryl, and carbohydrate groups can all be targeted for biotinylation, depending on the choice of biotin derivative; this can lead to unoriented antibody capture in the case of amine and carboxyl targeting or oriented capture in the case of sulfhydryl or carbohydrate targeting [29,69,265]. Optimally oriented nucleic acid probe immobilization has been achieved through biotinylation at terminal groups [137,139,286].

3.3. Covalent Immobilization

Covalent strategies are the gold standard for bioreceptor immobilization on SiP biosensors. Covalent immobilization (Figure 12b) is versatile, robust, and can be used to tether many different types of bioreceptors to SiP surfaces, yielding irreversibly bound functional layers [29,31]. This irreversible immobilization is beneficial for stable sensor performance under flow conditions and across multiple cycles of regeneration. Covalent methods may yield a higher density of immobilized bioreceptors compared to physisorption and bioaffinity techniques, which may, in turn, increase sensitivity [263]. Designing and optimizing a suitable covalent immobilization chemistry, however, can complicate assay development and preparation. Surface pre-treatments, reagents, and reaction conditions must be carefully chosen to yield reproducible and homogeneously thin surface modifications, while avoiding damage to the biosensor surface and bioreceptors [31]. For the design of POC sensors, further considerations may include selecting a scalable chemistry and designing a workflow that is suitable for SiP chips integrated with electronics and optical inputs and outputs.

3.3.1. Silane-Mediated Immobilization

Most covalent immobilization strategies for SiP sensors involve silanization with organosilanes. Organosilane-based methods have been used for antibody, aptamer, nucleic acid probe, glycan, and lectin immobilization on SiP devices. Silanes consist of a silicon atom bonded to four other constituents [290]. Organosilanes include silane reactive groups and at least one functional organic group. The silane reactive groups covalently couple to the sensor's native oxide surface by forming siloxane linkages with surface hydroxyls (Figure 13) [31,290]. A surface pre-treatment step is typically performed prior to silane deposition to remove organic contaminants and increase the number of surface hydroxyl groups available for silane grafting (Figure 13a) [31,74]. This pre-treatment step, which often involves oxidation via piranha, UV radiation and ozone, or plasma treatment, is essential to improving the silane grafting density and reproducibility of silanization. The silanization reaction can be performed using solution- or vapor-phase processes, with solution-phase processes being more widely used on SiP devices. However, no consensus on optimal reagents or reaction conditions exists, with significant variations in solvent choice, reagent concentrations, reaction time, and reaction temperature existing in the literature. After the silane is attached to the sensor surface, the silane's organic groups can react with other organic molecules to facilitate bioreceptor attachment. While it is possible to directly attach bioreceptors to the organosilane surface [25,172,185,291], it is more common to attach bioreceptors using a bifunctional crosslinker that is highly reactive toward both the silane and the bioreceptors, as the most commonly used organosilanes lack sufficient reactivity toward bioreceptors [29].

When attaching the bioreceptor, native reactive groups or non-native reactive groups introduced during synthesis are targeted for immobilization. These may include amine, carboxyl, thiol, or carbohydrate groups. The choice of targeted functional group affects the orientation of the immobilized bioreceptor. Antibodies, for example, possess native amines in their lysine residues and native carboxyls in their aspartate and glutamate residues [69]. These residues are abundant on the antibody surface, so targeting amines or carboxyls leads to unoriented antibody immobilization. Conversely, thiol groups present in cysteine residues of the hinge region can be targeted for site-directed antibody immobilization [69,75]. However, creating reactive thiol groups to target requires reduction of the hinge disulfide bonds, which may lead to undesired reduction of other disulfide bonds, potentially reducing the antibody's activity toward its target [75]. Native carbohydrate moieties present in the Fc region of antibodies can also be targeted for oriented capture [69,75]. Synthetic bioreceptors, including aptamers, nucleic acid probes, and glycans can be immobilized on silanized SiP surfaces by targeting terminal amine or thiol groups introduced during synthesis; this allows for oriented immobilization.

The most commonly used silanes for SiP functionalization are aminosilanes, particularly 3-aminopropyltriethoxysilane (APTES) (Figure 13). Aminosilanes contain organic groups that terminate in a primary amine, which can be targeted by amine-reactive crosslinkers for bioreceptor conjugation [290]. In order to initiate the reaction between the silane reactive groups of APTES and the hydroxyl groups present on the SiP surface, APTES must be hydrolyzed by moisture or water (Figure 13b) [74,290]. In the literature, APTES silanization of SiP sensors has been performed in anhydrous solvents such as toluene [201,239,292], acetone [17,195,293,294], and ethanol [132,202]. In these reactions, APTES hydrolysis is initiated by trace amounts of moisture present in the solvent [74]. APTES silanization has also been performed on SiP sensors using aqueous reaction solutions that contain a small quantity of water (e.g., ~5%) to catalyze APTES hydrolysis, combined with an organic solvent, typically ethanol [23,24,161,166,192,194,197,199,295]. These aqueous reactions are simpler than anhydrous ones, as they typically do not require drying the solvent or carrying out the reaction in a rigorously controlled inert atmosphere and/or under reflux [290,296]. However, in aqueous solutions, APTES is susceptible to copolymerization in the liquid phase prior to attachment to the solid substrate [258,268,297]. This can lead to the formation of thick and uneven films containing large silane aggregates (Figure 13d). Consequently, using an anhydrous solvent or maintaining low water content (~0.1%) in the reaction solution may yield thinner and more uniform silane layers [261,297,298]. Aside from solvent choice, an APTES concentration of 1–5% is typically used for solution-phase deposition [166,195,292], while reaction times vary significantly from several minutes [166,195] to overnight [202]. An alternative approach is vapor-phase silanization, which has been used to create uniform monolayer aminosilane films on silicon substrates [261,267,268,298]. In vapor-phase techniques, APTES is hydrolyzed by atmospheric moisture [74]. Compared to solution-phase reactions, vapor-phase aminosilane deposition has been reported as more reproducible, less sensitive to reagent purity and atmospheric conditions, and less likely to deposit polymeric silane particles [261,267–269]. Further, vapor-phase silanization may be more suitable than solution-phase methods when functionalizing SiP chips integrated with chip-mounted electronics and optical inputs/outputs, as vapor-phase processes do not require solvents that may degrade PCB or photonic wire bond materials. The final step of APTES silanization is typically a curing step at elevated temperature, which aids in the removal of moisture and the formation of siloxane bonds between the silane and surface [192,197,202,290].

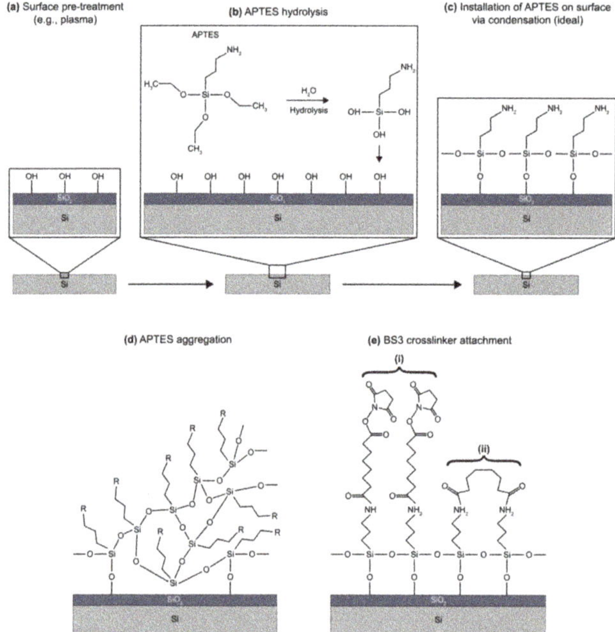

Figure 13. Silanization of SiP surface using 3-aminopropyltriethoxysilane (APTES). (**a**) The native oxide surface of the Si waveguide is pre-treated to remove organic contaminants and activate the surface hydroxyl groups, (**b**) APTES is hydrolyzed to form reactive silanols, and (**c**) adjacent APTES molecules are covalently linked together via silanol condensation and APTES is covalently bound to the surface. This yields a covalently bound APTES monolayer presenting functional amine groups for linker or bioreceptor immobilization. (**d**) Undesirable formation of large silane aggregates on the surface. (**e**) Attachment of a homobifunctional crosslinker to the aminosilane-coated surface, showing (i) ideal homobifunctional crosslinker attachment whereby one reactive group reacts with the silanized surface and the other remains available for conjugation with the bioreceptor, and (ii) undesirable crosslinker-mediated bridging whereby both ends of the homobifunctional crosslinker react with functional groups on the silanized surface, becoming unavailable for bioreceptor immobilization. BS^3 is used here as an example of a homobifunctional crosslinker.

Once the SiP surface has been modified with an aminosilane, bioreceptors are covalently linked to the surface via functional linkers such as glutaraldehyde (GA), bis(sulfosuccinimidyl)suberate (BS^3), or 1-ethyl-3-[3-(dimethylamino)propyl]-carbodiimide/N-hydroxysuccinimide (EDC/NHS) [74]. GA and BS^3 are homobifunctional linkers, which crosslink amine groups on the silanized substrate to amine groups on the bioreceptor. GA contains aldehyde groups which form imine bonds with amines via the formation of Schiff bases [31,74]. GA has been used to link antibodies [170], amine-terminated aptamers [24], amine-terminated DNA probes [23,197,199], and amine-terminated morpholinos [110] to aminosilane-modified SiP sensors.

GA linking has also been combined with SiP surface modification strategies whereby hydrofluoric acid (HF) is used to produce primary amines on silicon nitride waveguide surfaces [177]. These HF crosslinking approaches are particularly attractive for use with silicon nitride waveguides since they can be designed so that the amines are only produced on the nitride and not on the surrounding oxide [299]. This method uses basic cleaning methods followed by a HF dip to produce primary amines on the waveguide surface without the need of an additional aminosilane surface coating step. Next the sensor is immersed in a 2.5% GA crosslinker solution and washed. Bañuls et al. [299] developed this process to increase and localize biotarget capture to waveguide surfaces. The authors hypothesized that oxide comprised 98% of their sensor surface area with only 2% of the surface belonging to the silicon nitride sensing waveguides. This suggested that non-selective bioreceptor immobilization would lead to the majority of the target being captured by bioreceptors immobilized outside the sensing region. To show selective attachment to silicon nitride, slot waveguide ring resonator biosensors were modified with BSA and anti-BSA using the HF/GA procedure. Their results showed a detection limit of 28 pg/mm^2 for anti-BSA antibody immobilization on the surface and 16 pg/mm^2 for BSA. A similar procedure was used by Angelopoulou et al. [238] who modified MZI sensors with HF and GA, then spotted mouse IgG on individual sensors using an inkjet printer for multiplexing, followed by incubation with fluorescently labeled goat anti-mouse IgG antibodies and washing steps. The authors tested this direct attachment method in comparison to physical adsorption of the bioreceptors on amine-terminated silane (APTES) coated waveguides. The silane protocol yielded fluorescently tagged antibodies attached to both the waveguides and the surrounding oxide, whereas the HF procedure only functionalized the silicon nitride waveguides (Figure 14). Next, both sensors were spotted with a peptide, Receptor Binding Domain (RBD) of SARS-CoV-2 Spike 1 protein, and a BSA blocking protein on the sensing and reference waveguides, respectively. The HF method produced well-coated waveguides with the response of the reference sensor showing little change compared to the baseline signal upon exposure to anti-RBD antibodies. In comparison, the APTES modified reference sensor response could be clearly distinguished from the baseline signal. This suggests that BSA did not fully coat the APTES coated waveguides.

BS^3 consists of sulfo-NHS esters at either end of an 8-carbon spacer arm [300]. The NHS esters react with primary amines to form stable amide bonds. BS^3 has been used to conjugate antibodies [17,166], amine-terminated DNA probes [195], and peptides [239] to APTES-modified SiP sensors. When applied to antibody immobilization, GA and BS^3 target native amine functional groups that are abundant on the antibody surface, leading to random antibody orientation. Moreover, as these immobilization strategies target functional groups that are abundant on the antibody surface, they may result in the formation of multiple bonds between the antibody and the surface [74]. This may lead to conformational changes of the antibody and render binding sites inaccessible for target capture. As such, spacer molecules or hydrophilic polymers can be incorporated into the linking chemistry to reduce steric hindrance and the risk of bioreceptor denaturation. The hydrophilic polymer, oligo(ethylene glycol), which can be used for this purpose, has also shown antifouling properties with short chains ($\leq$7 repeats), which create a less ordered surface and decrease non-specific adsorption [301]. When applied to bioreceptors modified with terminal amine groups, such as 5$'$ amine-modified aptamers or nucleic acid probes, linking strategies using GA and BS^3 permit site-directed immobilization. Another notable limitation of homobifunctional crosslinkers like these is that they may form bridged structures where both reactive ends are linked to the substrate, limiting the number of binding sites available for bioreceptor attachment and thus reducing bioreceptor density (Figure 13e) [302]. This can be avoided with heterobifunctional crosslinkers. EDC/NHS is a heterobifunctional crosslinker combination using carbodiimide chemistry, which links carboxyl groups on the bioreceptor to amine groups on the silanized substrate via the formation of stable amide bonds [74,176]. This linker chemistry has been used to covalently attach antibodies to APTES-modified MRRs [166] and silicon photonic crystals [165]. Since this strategy targets abundant carboxyl groups, which are also abundant on antibody surfaces, it results in unoriented antibody immobilization and may cause conformational changes, as described above. A similar carbodiimide chemistry was used by Peserico et al. [202] in which an APTES-modified MRR chip was carboxylated with succinic anhydride, then EDC was used to covalently link 5′ amine-modified DNA probes to the carboxyl-presenting surface, this time in an oriented manner.

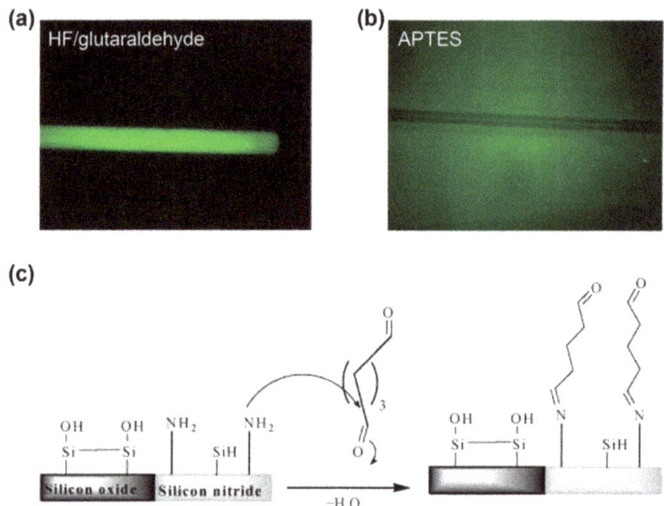

Figure 14. Silicon nitride waveguides from a MZI sensing window with fluorescently tagged (Alexa Fluor 488) antibodies (goat anti-mouse IgG) attached by (**a**) covalent HF/glutaraldehyde-based immobilization and (**b**) APTES functionalization followed by passive adsorption [238]. Parts (**a**,**b**) are adapted with permission from Ref. [238]. Copyright 2022 Elsevier. (**c**) Chemical reaction mechanism for selective silicon nitride functionalization by HF and glutaraldehyde crosslinking [299]. In (**c**), the subscript of "3" on the glutaraldehyde structure indicates that only one of the three carbons between the formyl end groups has been drawn for brevity. Part (**c**) is adapted with permission from Ref. [299]. Copyright 2010 Elsevier.

Despite their popularity, GA, BS^3, and EDC/NHS linker chemistries pose reproducibility challenges. GA polymerizes in aqueous solutions and the extent and nature of this polymerization depends on the age of the solution and can be difficult to control and reproduce [300]. BS^3 and EDC/NHS linker chemistries both involve NHS ester groups which rapidly hydrolyze in aqueous solutions [31,74]. This rapid hydrolysis competes with biomolecule conjugation and is highly sensitive to reaction conditions, hindering reproducibility and limiting the yield of the conjugation reaction.

Bioreceptor conjugation using SoluLink chemistry is another silane-based strategy which offers good reproducibility and has been extensively used on SiP devices, namely the commercial Genalyte MRR platform [31,295]. In the literature, this chemistry has been used to covalently immobilize antibodies [18,22,161,168,169,174,295], 5′ amine-modified aptamers [174], and 5′ amine-modified DNA probes [109,194,196,198,303] on MRRs. Using this strategy, the bioreceptor is reacted with succinimidyl-4-formylbenzamide (S-4FB), which targets primary amines via succinimide coupling. The substrate surface is either modified with an aminosilane, followed by reaction with 6-hydrazinonicotinamide (S-HyNic) [161,194], or the bare SiP surface is directly reacted with HyNic-silane [18,22,109,168,169,174,196,198,295,303]. The 4FB-conjugated bioreceptors are introduced to the HyNic-modified surface, leading to bioreceptor immobilization through hydrazone bond formation. This reaction proceeds slowly, but aniline can be used as a catalyst to increase the rate of reaction, improve bioreceptor loading on the substrate, and allow for lower reagent consumption [295]. Despite its good reproducibility, chemically modifying bioreceptors with 4FB prior to immobilization adds time and complexity to this technique. More recent demonstrations on the Genalyte platform have instead used APTES silanization and BS^3 to immobilize unmodified amine-containing bioreceptors for simple and flexible assay design [17,27,195,304,305].

Others have used 3-mercaptopropyltrimethoxysilane (MPTMS) to install thiol groups on SiP sensor surfaces to mediate bioreceptor immobilization. Thiolated bioreceptors can be directly conjugated to MPTMS-modified surfaces without an intermediate crosslinker through the formation of disulfide bonds [31]. For example, Chalyan et al. [25] directly immobilized Fab fragments on a MPTMS-modified SiP sensor. In this work, the Fab fragments were generated from protease digestion of polyclonal antibodies, followed by the reduction of hinge disulfide bonds to generate reactive thiol groups [25,69]. A similar strategy omitting the protease digestion step can also be used for site-directed antibody capture on MPTMS-modified surfaces [306]. However, covalent immobilization via thiol-bearing cysteine residues, which are usually internal to the antibody structure, and the unintentional reduction of non-target disulfide bonds may disrupt antibody conformation and binding affinity [69,75]. In addition to antibodies, this thiol-directed covalent strategy has been used for nucleic acid probe immobilization. Sepúlveda et al. [200] modified silicon nitride Mach-Zehnder interferometer sensors with MPTMS, followed by covalent and oriented immobilization of 5′ thiol-modified ssDNA probes.

Bioreceptors that lack reactive thiols can also be conjugated to MPTMS-modified surfaces using maleimide linkers. For example, Xu et al. [175] covalently immobilized antibodies on a MPTMS-modified planar silicon nitride optical waveguide interferometric biosensor using m-maleimidobenzoyl-N-hydroxysuccinimide ester as a thiol-to-amine crosslinker. Ghasemi et al. [133] covalently attached amine-derivatized glycans to MPTMS-modified silicon nitride MRRs using a SM(PEG)12 linker. SM(PRG)12 contains a polyethylene glycol (PEG) chain terminated by NHS ester and maleimide reactive groups. As such, it acted as a heterobifunctional linker between the thiolated surface and amine-derivatized glycans, while the PEG chain prevented nonspecific interactions between non-target molecules and the sensor surface.

3-Glycidoxypropyltrimethoxysilane (GPTMS) is another silane that can mediate direct covalent immobilization of bioreceptors on SiP sensors. GPTMS installs epoxy groups on silicon surfaces, which are reactive toward amine, thiol, or hydroxyl groups [31,290]. Ramachandran et al. [172] conjugated monoclonal antibodies and 5′ amine-modified ssDNA probes to GPTMS-modified glass (Hydex) MRRs. Using this strategy, the bioreceptors were covalently linked to the surface via amine reactive groups, resulting in unoriented and oriented antibody and ssDNA probe capture, respectively. Chalyan et al. [25] and Guider et al. [185] covalently immobilized amine-terminated aptamers on GPTMS-modified silicon oxynitride MRRs in an oriented manner.

3.3.2. Organophosphonate-Mediated Immobilization

Organophosphonate chemistry presents a promising alternative to silane chemistry. Compared to silanes, phosphonate films can achieve greater monolayer density, surface coverage, and stability, and have a lower tendency to form multilayered structures [270,307]. Shang et al. [126] demonstrated covalent immobilization of amine-bearing glycan and glycoprotein bioreceptors on silicon MRRs using an organophosphonate surface coating and an amine-vinyl sulfone linker (Figure 15). After treating the surface with piranha solution to increase the number of available surface hydroxyl groups for organophosphonate grafting, the sensor surface was coated with a monolayer of 11-hydroxyundecylphosphonic acid (UDPA). This was achieved using the "T-BAG" method whereby UDPA is adsorbed onto the substrate, then heated to 120–140 °C to activate the formation of covalent linkages [126,307]. After the sensors were modified with UDPA, divinyl sulfone (DVS) was used to link the hydroxyl-terminated organophosphonate film to the amine-bearing bioreceptors [126]. In this work, the MRRs demonstrated excellent stability and reproducibility across multiple cycles of chemical regeneration and long-term storage at ambient conditions. A similar strategy was used to functionalize the surface of silicon nanowires with cysteine-modified PNA oligonucleotides [270]. Here, 3-maleimidopropionic-acid-N-hydroxysuccinimidester was used instead of DVS as a heterobifunctional linker to attach the thiol-containing PNA oligonucleotides to the UDPA-modified nanowires.

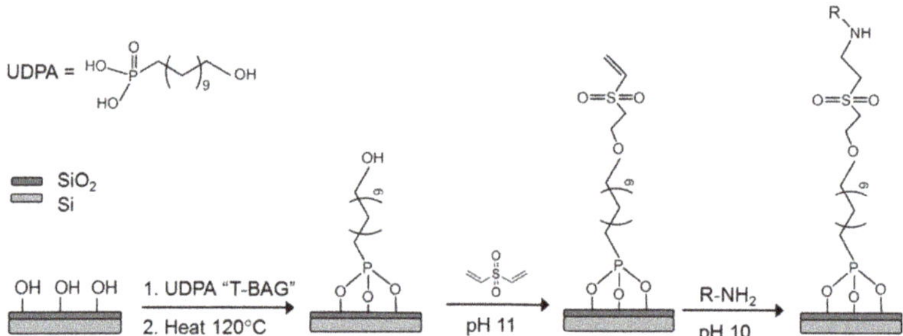

Figure 15. Organophosphonate-based surface functionalization scheme whereby the surface is coated with a film of UDPA using the T-BAG method and a DVS linking strategy is used for the immobilization of aminated bioreceptors. Reprinted with permission from Ref. [126]. Copyright 2012 American Chemical Society.

3.3.3. Click Chemistry

Click chemistry is a widely used crosslinking technique for simple, fast, and selective attachments with high efficiency. This method is attractive for biorecognition components since it uses physiological reaction conditions (neutral pH, buffered solution). Briefly, click chemistry involves linking molecules via heteroatom links (C–X–C) [271]. There are three main click procedures based on Cu(I) catalyzed azide-alkyne, strain promoted azide-alkyne, and tetrazine-alkene ligation reactions. The reaction is simple, more efficient than EDC/NHS chemistry, selective to only click reagents, has many commercially available modular components, and is not sensitive to oxygen or water [271].

This method has been used to immobilize ssDNA probes [59] and PCCs [91,240] on the surfaces of silicon-based optical biosensors. Juan-Colás et al. [59] demonstrated a novel silicon electrophotonic biosensor consisting of silicon MRRs fabricated with a thin n-doped layer at their surface to combine high-Q-factor photonic ring resonance with electrochemical sensing (Figure 16). In this work, the MRRs were covalently functionalized with ssDNA probes using the popular copper-catalyzed azide-alkyne click reaction. Firstly, two electrophotonic MRRs fabricated on a single chip were modified by electrografting azidoaniline or ethynylaniline onto the rings to install azide or alkyne groups, respectively (Figure 16a). The two electrophotonic MRRs were individually addressable, allowing for site-directed electrografting of azide groups on one ring and alkyne groups on the other. Next, the copper-catalyzed azide-alkyne click reaction was performed to conjugate azide-modified ssDNA probes to the alkyne-modified ring and alkyne-modified ssDNA probes to the azide-modified ring (Figure 16b,c). This unique strategy permits high-density multiplexed functionalization with submicrometer- to micrometer-scale precision, though it is not suitable for traditional SiP sensors that lack electrochemical control. Click chemistry was also used by Cao et al. [240] and Layouni et al. [91] to covalently link PCCs to porous silicon surfaces. In these works, the surfaces were modified with alkyne moieties by thermal hydrolyzation with 1,8-nonadiyne, followed by copper-catalyzed azide alkyne cycloaddition to attach azide-modified PCCs. This method requires removal of the substrate's native oxide layer by exposure to HF prior to hydrolyzation. Consequently, this method may not be suitable for SiP devices patterned with extremely fragile silicon structures like sub-wavelength gratings, which may be partially etched or delaminated upon exposure to HF. Overall, some of the key advantages of click chemistry compared to silane-mediated strategies are its insensitivity to oxygen and water and its chemoselectivity, which prevents side reactions with other bioreceptor functional groups and preserves bioreceptor activity [29,91,308]. However, a limitation is the requirement for prior surface and bioreceptor modification with functional tags, like azide and alkyne groups, which adds complexity to the functionalization process [29,234].

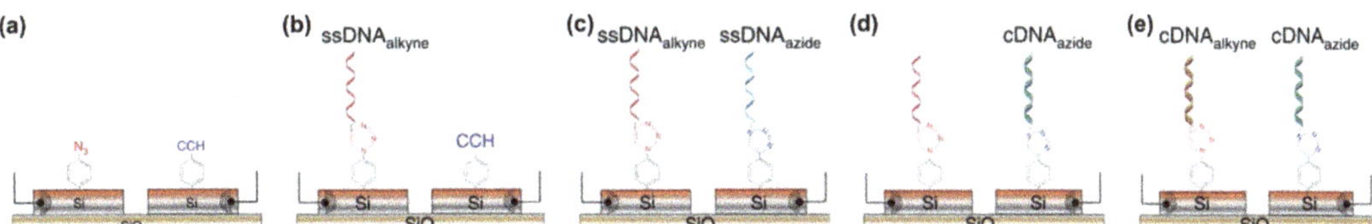

Figure 16. Click chemistry-mediated immobilization of nucleic acid probes on an electrophotonic ring resonator. (**a**) Two different diazonium salts (azidoaniline and 4-ethynylbenzene) are electrografted on the electrophotonic rings, which are electrically isolated. The individual microrings are then functionalized with alkyne- and azide-modified DNA probes using copper-catalyzed azide-alkyne click reaction. (**b**) First, the azide-modified sensor is functionalized with the alkyne-modified single-stranded DNA probe ($ssDNA_{alkyne}$). (**c**) Next, the alkyne-modified sensor is functionalized with the azide-modified single-stranded DNA probe ($ssDNA_{azide}$). The target sequences complementary to $ssDNA_{azide}$ (**d**) $ssDNA_{alkyne}$ (**e**) (labeled $cDNA_{azide}$ and $cDNA_{alkyne}$, respectively) are introduced and hybridize to the functionalized sensors. Adapted from Ref. [59] in accordance with the Creative Commons Attribution 4.0 International (CC BY 4.0) license.

3.3.4. UV-Crosslinking

Direct UV-crosslinking of nucleic-acid-based bioreceptors has been demonstrated on planar glass and silicon dioxide wafers. This is a simple and inexpensive method that could be extended to SiP biosensors. Gudnason et al. [273] linked poly(T)10-poly(C)10-tagged ssDNA probes to unmodified glass surfaces using UV light irradiation. The immobilized probes demonstrated similar hybridization efficiency when compared to ssDNA probes immobilized on an amino-silane surface via traditional chemical crosslinking. The UV-linked probes showed no appreciable decrease in hybridization performance after incubation in water at 100 °C for 20 min, demonstrating strong thermal stability. In this work, the hybridization assay was performed in PerfectHyb Plus buffer to obviate the need for a surface blocking step. A similar strategy was used by Chen et al. [272] to covalently link thrombin-binding DNA aptamers with poly(T)20 tails to unmodified glass and silicon dioxide wafer surfaces using UV irradiation, while maintaining strong target affinity. Note that in this work, thrombin binding was performed in the presence of BSA and Tween-20 surfactant to reduce non-specific binding of the target to unmodified regions of the substrates. This UV-linking strategy is both simple and rapid because it requires no prior chemical modification of the substrate. Additionally, the nucleic acid-based bioreceptors do not require chemical modifications with reactive functional groups, lowering synthesis costs. However, to our knowledge, this strategy has not yet been demonstrated on patterned SiP sensor surfaces.

Tables 28, 29, 31 and 33 and Section 3.4 outline strategies that have been demonstrated on SiP sensors and representative surfaces for the immobilization of antibodies (Table 28), aptamers (Table 29), nucleic acid probes (Table ??), peptides and PCCs (Table 31), glycans and lectins (Table ??), HCCD reporters (Table 33), CRISPR-dCas9-mediated sensing probes (Table ??), and lipid nanodiscs (Table ??) in the previous literature.

3.4. Summary and Future Directions

We have discussed adsorption, bioaffinity, and covalent strategies for immobilizing bioreceptors on SiP surfaces. While adsorption-based strategies offer excellent simplicity, their poor stability and lack of control over bioreceptor orientation limit their suitability for SiP biosensing applications. However, novel polymeric coating materials, such as PAcrAm™ and AziGrip4™ from SuSoS AG, are available and replicably self-assemble as stable monolayers on silicon substrates by adsorption from solution [309–311]. These polymeric coatings have customizable functional binding groups and allow for covalent and electrostatic capture of bioreceptors on the adsorbed coating [309–311]. This may create the opportunity for bioreceptor immobilization with similar simplicity to passive adsorption, but with improved stability and more controllable bioreceptor orientation, making this a potentially valuable future research direction. To the best of our knowledge, such functionalization techniques have not yet been demonstrated on SiP platforms.

Bioaffinity and covalent strategies typically offer improved stability and control over bioreceptor orientation compared to adsorption, but at the cost of increased complexity [41]. Bioaffinity strategies involving antibody-binding proteins permit controlled antibody orientation, but have limited stability compared to biotin-based and covalent methods [265,278,279]. Covalent strategies, especially those using silanization, have been widely used on SiP platforms, as they can permit very stable and tailorable bioreceptor immobilization [29,31]. When designing a covalent immobilization protocol, surface pre-treatment must be carefully considered to ensure that the sensor surface is free of organic contaminants prior to applying the immobilization chemistry, and to activate surface functional groups (e.g., hydroxyls) that will be targeted by the immobilization chemistry [31,74]. Such pre-treatments improve grafting density on the sensor surface, while also improving the reproducibility of the immobilization protocol [74]. Pre-treatment approaches that have been used in SiP bioreceptor immobilization protocols, such as piranha, UV radiation and ozone, plasma, and HF treatments, have been comprehensively summarized in Tables 28, 29, 31 and 33 and Section 3.4. Future work should focus on optimizing standardized silanization protocols that can be used for highly replicable, scalable, and robust surface modifications with limited silane aggregation. In parallel with future work focusing on the system-level integration of SiP sensors for POC use, immobilization chemistries that are compatible with these integrated sensor architectures should be designed and tested. For example, translating solution-phase surface modification protocols to vapor-phase ones may reduce the risk of damage to the sensing system during functionalization, while improving scalability, reproducibility, and film uniformity [261,267–269]. Immobilization strategies using UV-crosslinking of bioreceptors directly to unmodified surfaces should also be explored on SiP sensors as a potentially simple, low cost, and scalable immobilization technique [272,273].

In designing immobilization protocols, potential steric crowding effects should also be considered in the context of bioreceptor immobilization and target capture. For example, crowding of bioaffinity linkers on the sensor surface may hinder subsequent bioreceptor immobilization [289]. These steric effects can be counteracted by using a higher bioreceptor concentration in the immobilization protocol or by using long linking molecules to increase the distance between the sensor surface and bioreceptors, providing more flexibility for the receptors to optimize steric crowding. When using these longer linking molecules, however, the potential sensitivity trade-offs associated with moving the binding reaction farther away from the sensor surface should also be considered. Immobilization approaches using these longer linking molecules may be most suitable for SiP architectures with greater evanescent field penetration depths (e.g., those based on ultra-thin [40] or sub-wavelength grating [16,46,264] waveguides). Similarly, dense receptor packing on the sensor surface may not always enhance target binding. Steric hindrance effects due to target molecule binding can reduce the rate of the forward binding reaction for neighboring receptors and affect the dynamic range of the sensor [312]. Thus, these steric effects should be accounted for when optimizing bioreceptor immobilization protocols.

Table 28. Strategies demonstrated for the immobilization of antibodies on SiP sensors.

Antibody Immobilization						
Immobilization Strategy	**Bioreceptor Subtype**	**Surface Pre-Treatment**	**Chemical Surface Modification**	**Linking Strategy**	**Sensor Type**	**Refs.**
Covalent	Fab fragment	Argon plasma treatment	MPTMS silanization	Direct conjugation to silanized surface	SiO_xN_y MRR	[25]
Covalent	Monoclonal antibodies	NaOH and ethanol/water cleaning	GPTMS silanization	Direct conjugation to silanized surface	Hydex MRR	[172]
Covalent	Monoclonal [18,22,109,169,174,295] and single-domain antibodies [168]	Piranha treatment	HyNic silane surface modification	Conjugation of antibody with S-4FB for hydrazone bond formation with modified surface	Si MRR	[18,22, 162,168, 169,174, 295]
Covalent	Monoclonal antibody	Piranha treatment	APTES silanization + S-HyNic surface modification	Conjugation of antibody with S-4FB for hydrazone bond formation with modified surface	Si MRR	[161]
Covalent	Monoclonal antibodies	Acetone/isopropanol cleaning [17,27] or oxygen plasma [166]	APTES silanization	BS^3 crosslinker	Si MRR	[17,27, 166]
Covalent	Monoclonal antibody	Piranha treatment	APTES silanization	*N*,*N*-diisopropylethylamine and *N*,*N*′-disuccinimidyl carbonate linker	Si MRR	[295]
Covalent	Monoclonal antibody	Oxygen plasma	APTES silanization	EDC/NHS activation	Si MRR	[166]
Covalent	Antibody	Piranha treatment	APTES silanization	EDC/NHS activation	Si PhC	[165]
Covalent	Monoclonal antibody	Piranha treatment	APDMES silanization	Glutaraldehyde linker	Si PhC	[170]
Covalent	Polyclonal antibody	Oxidization	CTES silanization	EDC/NHS activation	Si_3N_4 MRR	[171]
Covalent	Polyclonal antibody	-	Thermal hydrolyzation with undecylenic acid	EDC/NHS activation	Porous Si sensor	[176]
Covalent	Antibody	Piranha treatment	Native oxide removal with HF	Glutaraldehyde linker	Si_3N_4/SiO_2 MRR	[177]
Covalent	Monoclonal and polyclonal antibodies	Cleaning with Micro-90 solution and chromic acid	MPTMS silanization	*m*-maleimidobenzoyl-*N*-hydroxysuccinimide ester linker	Si_3N_4 planar waveguide interferometer	[175]

Table 28. *Cont.*

Antibody Immobilization						
Immobilization Strategy	**Bioreceptor Subtype**	**Surface Pre-Treatment**	**Chemical Surface Modification**	**Linking Strategy**	**Sensor Type**	**Refs.**
Bioaffinity	Antibody	-	-	Protein A adsorbed on surface	Si MRR (sub-wavelength grating, SWG)	[264]
Bioaffinity	Antibody	-	-	Protein A adsorbed on surface	Si MRR (multibox SWG)	[1]
Bioaffinity	Antibodies	-	-	Antibody-binding fusion protein consisting of Si-tag and Protein A	Si wafer	[278]
Bioaffinity	Antibodies	-	-	Antibody-binding fusion protein consisting of Si-tag and Protein A	Si MRR	[282]
Covalent + bioaffinity	Monoclonal antibodies (oligonucleotide-conjugated)	Piranha treatment	HyNic silane surface modification	Intermediate oligonucleotides conjugated with S-4FB for hydrazone bond formation with modified surface, then used as a bioaffinity linker	Si MRR	[163]
Covalent + bioaffinity	Antibody	Plasma treatment	Silane-PEG-biotin surface modification	Streptavidin used as a bioaffinity linker to immobilize biotinylated Protein G	Si PhC	[164]
Covalent + bioaffinity	Antibodies (biotinylated)	Dry thermal oxidation	APTMS silanization	Sulfo-NHS-LC-LC-biotin linker + streptavidin used as bioaffinity linkers	Porous Si sensor	[288,289]
Covalent + bioaffinity	Monoclonal and polyclonal antibodies	HF treatment	MPTMS silanization	*N*-succinimidyl-4-malemidobutyrate crosslinker + Protein A or G	Multimode optical fibers	[283]

MPTMS: (3-mercaptopropyl)trimethoxysilane, GPTMS: 3-glycidoxypropyltrimethoxysilane, HyNic silane: 3-*N*-((6-(*N*′-Isopropylidene-hydrazino))nicotinamide)propyltriethyoxysilane, S-HyNic: succinimidyl 6-hydra- zinonicotinamide acetone hydrazone, APTES: 3-aminopropyltriethoxysilane, APTMS: 3-aminopropyltrimethoxysilane, Sulfo-NHS-LC-LC-biotin: sulfosuccinimidyl-6-(biotinamido)-6-hexanamido hexanoate, HF: hydrofluoric acid.

Table 29. Strategies demonstrated for the immobilization of aptamers on SiP sensors and representative surfaces. All works listed in this table used DNA aptamers.

Aptamer Immobilization						
Immobilization Strategy	**Surface Pre-Treatment**	**Chemical Surface Modification**	**Linking Strategy**	**Targeted Aptamer Terminal Group**	**Sensor Type**	**Refs.**
Covalent	Argon plasma [25] or piranha treatment [185]	GPTMS silanization	Direct conjugation to silanized surface	Amine	SiO_xN_y MRR	[25,185]
Covalent	Oxygen plasma	APTES silanization	Glutaraldehyde linker	Amine	Si MRR	[24]
Covalent	Piranha treatment	HyNic silane surface modification	Conjugation of aptamer with S-4FB for hydrazone bond formation with modified surface	Amine	Si MRR	[174]
Covalent	Plasma	Silane-PEG-COOH surface modification	EDC/NHS activation	Amine	Si PhC	[164]
Covalent	-	Thermal hydrolyzation with undecylenic acid	EDC/NHS activation	Amine	Porous Si sensor	[176]
Covalent	-	-	Direct UV crosslinking on surfaces	poly(T) and poly(TC) tags	Glass slides and SiO_2 wafers	[272]

Silane-PEG-COOH: slane-polyethylene glycol-carboxyl.

Table 30. Strategies demonstrated for the immobilization of nucleic acid probes on SiP sensors and representative surfaces.

Immobilization of Nucleic Acid Probes for Hybridization Sensing							
Immobilization Strategy Type	**Bioreceptor Subtype**	**Surface Pre-Treatment**	**Chemical Surface Modification**	**Linking Strategy**	**Targeted Probe Terminal Group**	**Sensor Type**	**Refs.**
Covalent	DNA	Piranha treatment	ICPTS silanization	Direct conjugation to silanized surface	Amine	Si wafers and nanostructured Si	[291]
Covalent	DNA	NaOH and ethanol/water cleaning	GPTMS silanization	Direct conjugation to silanized surface	Amine	Hydex MRR	[172]
Covalent	DNA	Piranha treatment and thermal oxidation	APTES silanization	Sulfosuccinimidyl-4-(*N*-maleimidomethyl)cyclohexane-1-carboxylate (Sulfo-SMCC) linker	Thiol	Si MRR and Si PhC	[201]
Covalent	DNA	Piranha treatment and thermal oxidation	TEOS-HBA silanization	Base-by-base in situ ssDNA probe synthesis via phosphoramidite method	-	Si MRR and Si PhC	[201]

Table 30. *Cont.*

Immobilization of Nucleic Acid Probes for Hybridization Sensing							
Immobilization Strategy Type	Bioreceptor Subtype	Surface Pre-Treatment	Chemical Surface Modification	Linking Strategy	Targeted Probe Terminal Group	Sensor Type	Refs.
Covalent	DNA	Piranha treatment	APTES silanization + S-HyNic surface modification	Conjugation of DNA probe with S-4FB for hydrazone bond formation with modified surface	Amine	Si MRR	[194]
Covalent	DNA	Nitric acid wash	APTES silanization	Succinic anhydride and EDC linker	Amine	Si_3N_4 MRR	[202]
Covalent	PNA	Oxygen plasma	11-hydroxyundecylphosphonate surface modification	3-maleimidopropionic-acid-*N*-hydroxysuccinimide linker	Cysteine tag	Si nanowires	[270]
Covalent	DNA	Ethanol and water rinse	-	Direct UV crosslinking on surfaces	No tag and poly(TC) tag	Glass slides (unmodified and GAPS II™ aminosilane coated)	[273]
Covalent	DNA	Piranha treatment	APTES silanization	BS^3 linker	Amine	Si MRR	[195]
Covalent	DNA	Piranha and oxygen plasma treatment	Electrografting of alkyne- and azide-presenting diazonium salts	Cu-catalyzed azide-alkyne click reaction	Azide and alkyne	N-doped Si MRR electrophotonic sensor	[59]
Covalent	DNA	Piranha treatment	HyNic silane surface modification	Conjugation of DNA probe with S-4FB for hydrazone bond formation with modified surface	Amine	Si MRR	[109, 196, 198, 303]
Covalent	DNA	Oxygen plasma and nitric acid treatment	MPTMS silanization	Direct conjugation via disulphide bond formation	Thiol	Si_3N_4 Mach-Zehnder interferometer	[200]
Covalent	DNA	Oxygen plasma	APTES silanization	Glutaraldehyde linker	Amine	Si MRR	[23, 197, 199]
Covalent	Morpholino	Piranha treatment	APTMS silanization	Glutaraldehyde linker	Amine	Suspended Si photonic microring resonator	[110]
Covalent + bioaffinity	DNA (biotinylated)	-	3-isocyanatepropyl thriethoxysilane vapor silanization	Streptavidin conjugated to silanized surface and used as a bioaffinity linker	Biotin tag	Planar photonic crystal-waveguide-based optical sensor	[203]

ICPTS: 3-isocyanatepropyl triethoxysilane, TEOS-HBA: *N*-(3-triethoxysilylpropyl)-4-hydroxybutyramide.

Table 31. Strategies demonstrated for the immobilization of peptides and PCCs on SiP sensors.

Peptide and PCC Immobilization						
Immobilization Strategy	**Bioreceptor Subtype**	**Surface Pre-Treatment**	**Chemical Surface Modification**	**Linking Strategy**	**Sensor Type**	**Refs.**
Covalent	Peptide	Acetone and isopropanol cleaning	HF treatment to create secondary amines	Glutaraldehyde linker	Si_3N_4 Mach-Zehnder interferometer	[238]
Passive adsorption	Peptide	Acetone and isopropanol cleaning + Piranha treatment	APTES silanization	-	Si MRR	[174]
Covalent	Peptide	Piranha treatment	APTES silanization	BS^3	Porous Si microcavity	[239]
Covalent	PCC	HF treatment	Thermal hydrolyzation with 1,8-nonadiyne	PCC attachment via click chemistry with copper(I)-catalyzed azide alkyne cycloaddition	Porous Si microcavity sensor	[91, 240]

Table 32. Strategies demonstrated for the immobilization of glycans and lectins on SiP sensors and representative surfaces.

Glycan and Lectin Immobilization						
Immobilization Strategy	**Bioreceptor Subtype**	**Surface Pre-Treatment**	**Chemical Surface Modification**	**Linking Strategy**	**Sensor Type**	**Refs.**
Covalent	Glycan	Piranha treatment	UDPA organophosphonate surface modification	DVS activation	Si MRR	[126]
Covalent	Glycan	Piranha and UV/ozone plasma treatment	MPTMS silanization	SM(PEG)12 linker	Si_3N_4 MRR	[133]
Covalent	Glycan	Piranha treatment	APTES silanization	BS(PEG)9 linker	Si_3N_4 MRR	[132]
Covalent	Lectin	Hydrogen peroxide and thermal treatment	APTES silanization	Glutaraldehyde linker	Porous Si sensor	[131]
Non-covalent	Glycoproteins and neoglycoconjugates	-	-	Passive adsorption	Si MRR	[246]
Covalent + bioaffinity	Lectin (biotinylated)	UV/ozone clean	APTMS silanization	NHS-PEG$_4$-biotin linker + avidin	Si_3N_4 reflectometric interference spectroscopy sensor	[285]

SM(PEG)12: succinimidyl-([*N*-maleimidopropionamido]-dodecaethyleneglycol) ester, BS(PEG)9: bis-*N*-succinimidyl-(nonaethylene glycol) ester, NHS-PEG$_4$-biotin: *N*-hydroxysuccinimide ester-polyethylene glycol-biotin.

Table 33. Strategies demonstrated for the immobilization of HCCD reporters on SiP sensors.

HCCD Reporter Immobilization						
Immobilization Strategy	**Reporter Type**	**Surface Pre-Treatment**	**Chemical Surface Modification**	**Linking Strategy**	**Sensor Type**	**Refs.**
Covalent + bioaffinity	Biotinylated dsDNA-quantum dot reporters	Thermal oxidation	APTES silanization	Glutaraldehyde linker + streptavidin used as a bioaffinity linker	Porous Si sensor	[137]
Covalent + bioaffinity	Biotinylated ssDNA-gold nanoparticle reporters	Plasma treatment	Silane-PEG-biotin surface modification	Streptavidin used as a bioaffinity linker	Si MRR	[139]

Table 34. Strategies demonstrated for the immobilization of nucleic acid probes for CRISPR-dCas9-mediated sensing on SiP sensors.

CRISPR-Ca9-Mediated Sensing					
Immobilization Strategy	**Surface Pre-Treatment**	**Chemical Surface Modification**	**Nucleic Acid Probe Linking Strategy**	**Sensor Type**	**Refs.**
Covalent	Oxygen plasma	APTES silanization	Glutaraldehyde linker	Si MRR	[143]

Table 35. Strategies demonstrated for the immobilization of lipid nanodiscs on SiP sensors.

Lipid Nanodisc Immobilization					
Immobilization Strategy	**Surface Pre-Treatment**	**Chemical Surface Modification**	**Linking Strategy**	**Sensor Type**	**Refs.**
Non-covalent	Piranha treatment	-	Passive adsorption	Si MRR	[149]
Non-covalent	Piranha treatment	-	Passive adsorption	Si MRR	[150]
Non-covalent	Acetone and isopropanol wash	-	Passive adsorption	Si MRR	[148]

4. Patterning Techniques

In this section, we introduce several patterning techniques that can be used for SiP sensor functionalization and benchmark them against the critical patterning performance criteria relevant to SiP biosensing, as outlined in Table 2. A high-level comparison of these patterning techniques is provided in Table 36. The subsequent subsections provide further details about each patterning technique, outline their opportunities and limitations for multiplexed SiP biofunctionalization, and highlight demonstrations from the previous literature in which these patterning techniques have been used to deposit bioreceptors on SiP biosensors. For each patterning technique, tables categorizing these demonstrations from the previous literature are provided.

4.1. Microcontact Printing

Microcontact printing (μCP), also called microstamping, is a soft lithography method whereby geometrically defined 2D patterns of biomolecules are transferred to a substrate using an elastomeric stamp (Figure 17a) [320,321]. This technique has been used to prepare patterns of bioreceptors like antibodies [32,322], DNA [323–325], MIPs [326], and carbohydrates [327] on solid substrates.

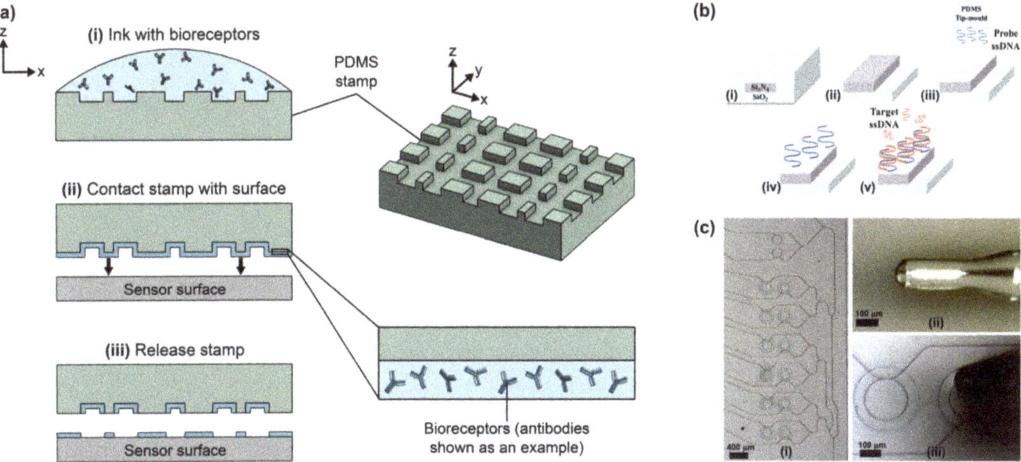

Figure 17. (**a**) Illustration of the process to pattern a surface using μCP. (i) First, the elastomeric stamp is inked with a bioreceptor solution whereby bioreceptors adsorb to the stamp surface. Inking may be achieved using soak, spray-on, or robotic feature-feature ink transfer methods. Subsequently, the stamp can be rinsed and dried or used wet for stamping. (ii) The stamp is contacted with the sensor surface and gentle pressure is applied to transfer the bioreceptors from the stamp to the surface at the regions of contact. (iii) The stamp is released to reveal the bioreceptor-patterned surface. (**b**) Graphical representation of the functionalization of a SiP waveguide using tip-mold microcontact printing, showing the (i) waveguide cross-section of a reference microring, (ii) waveguide cross-section of a control microring, (iii,iv) application of ssDNA probes on the waveguide surface using a PDMS tip-mold μCP tool, and (v) hybridization of ssDNA targets to the immobilized ssDNA probes on the waveguide surface. (**c**) Images of (i) the SiP MRR sensor chip functionalized by Peserico et al. [202] via tip-mold μCP, (ii) the optical fiber tip with an unpatterned PDMS cladding used as the μCP tool, and (iii) example of bioreceptor application on MRR using μCP. Parts (**b**,**c**) are adapted with permission from Ref. [202]. Copyright 2017 The Institution of Engineering and Technology.

Table 36. Comparison of bioreceptor patterning techniques based on SiP biofunctionalization needs.

Patterning Technique	Achievable Resolution	Ease of Multiplexing	Spot quality/ Uniformity	Reproducibility	Throughput	Reagent Consumption	Simplicity	Compatibility with System-Level Integration	Cost/ Availability
Microcontact printing (μCP)	0.1–0.5 μm [32,77]	Poor	Poor	Moderate	High throughput for patterning with 1 bioreceptor; low throughput for multiplexing	Low	Good	Poor; risk of surface and system damage due to stamp contact	Low cost, widely available (e.g., stamp preparation via standard soft lithography techniques; commercial μCP services available via companies including ThunderNIL and BALTFAB) *
Pin printing	~1–100 μm [313–315]	Moderate	Poor	Moderate	Low-high, depending on pin design	Low	Poor	Poor; risk of surface and system damage due to pin contact	Expensive for commercial pin printers (e.g., Bioforce Nano eNabler, BioOdyssey Calligrapher miniarrayer, Arrayit microarrayers) *
Microfluidic patterning in channels	~1–10 μm [316,317]	Moderate	Good	Very good	Moderate; simultaneous bioreceptor deposition possible, but limited number of uniquely addressable locations	High	Good	Moderate; μFN design and placement require careful design to avoid system damage; microfluidics often required for assays too	Low cost, widely available (e.g., μFN preparation via standard soft lithography techniques; commercial microfluidics fabrication services available via companies including ThunderNIL and MicruX Technologies) *
Inkjet printing	~30–150 μm [318,319]	Good	Poor	Moderate	Very high	Low	Poor	Good	Expensive, commercial options available (e.g., Scienion sciFLEXARRAYER, Fujifilm Dimatix DMP-2831) *
Microfluidic probe (μFP)	10 μm [83]	Good	Good	Very good	High	Low-moderate	Poor	Moderate; non-contact, but small risk of system damage due to probe motion	No commercial products available

μFN: microfluidic network. * These are not exhaustive lists of vendors. Vendors listed are based on an exploratory search and are not endorsed or suggested by the authors.

The first step of μCP is fabricating the elastomeric stamp. Polydimethylsiloxane (PDMS) is the most popular stamp material for μCP because it is easy to mold, flexible, chemically inert, and impermeable to biomolecules like proteins [32,320,328]. In μCP, the geometry of the stamp is defined by casting it in a master mold, prepared by photolithography or micromachining [320,321]. Once the stamp has been cast, it is "inked" with the bioreceptor solution to be deposited on the substrate. The ink adheres to the stamp via passive adsorption, which can be tuned by modifying the stamp's surface wettability with plasma or ozone treatment [32,328]. The inked stamp can be dried prior to stamping or used wet [329]. Next, the stamp is contacted with the substrate under a load, which can be achieved robotically or using a micropositioner to ensure precise alignment. The stamp is removed, leaving behind a 2D pattern of bioreceptors. The transfer of ink from the stamp to the substrate depends on the differential wettability between the stamp and substrate; in particular, the substrate must have greater wettability and, therefore, greater affinity toward the ink compared to the stamp [32,316].

A notable advantage of μCP is its excellent resolution. Patterns with critical dimensions down to 0.1–0.5 μm can be achieved [32,77]. This resolution is more than sufficient for patterning biomolecules on SiP surfaces, where the patterned sensing structures, like MRRs, typically have dimensions on the order of 10 μm. Some other advantages of μCP include its procedural simplicity, low cost, and good reproducibility [77,321,328,329]. PDMS stamps are robust and can be reused many times without significant loss of performance, but they are also sufficiently inexpensive and easy to fabricate that they can be treated as disposable when sample contamination is a concern [77,321]. Compared to printing techniques that address one spot on a substrate surface at a time, μCP is high-throughput, as a complex 2D pattern can be printed with only a single inking and application step [329].

While μCP is suitable for efficiently creating complex 2D patterns of a single bioreceptor, it is poorly suited to creating multiplexed arrays with many different bioreceptors [32]. Multiple cycles of inking and printing and careful stamp alignment would be required to print multiple bioreceptors, making this a time-consuming and cumbersome process. Another challenge is that bioreceptor immobilization strategies often include surface modifications, like silanization, which increase surface hydrophobicity prior to bioreceptor attachment [258]. This can reduce the differential wettability between the stamp and substrate, which may, in turn, reduce the efficiency of bioreceptor transfer to the substrate. Materials like PDMS can also transfer unwanted materials like residual uncured oligomers to the regions of the chip that they contact during stamping, potentially contaminating the surface and complicating bioreceptor patterning and subsequent assay steps [330–332]. Other limitations of μCP include a potential reduction in bioreceptor binding activity due to drying [322,329], patterning accuracy issues due to PDMS deformation under loads and swelling in the presence of some solvents [328], the requirement for cleanroom facility access to fabricate stamp master molds [77,329], and potential damage to the fragile sensor surface resulting from direct contact with the stamp.

To date, μCP has not been widely used to pattern SiP biosensors, though Peserico et al. [202] used a "tip-mold microcontact printing" technique to functionalize silicon nitride MRRs with ssDNA probes in a spatially defined manner (Table 37, Figure 17b,c). Instead of using a traditional stamp, a PDMS μCP probe was prepared by casting a thin layer of PDMS over the tip of a 125 μm-diameter optical fiber. The probe tip was treated with hydrochloric acid and hydrogen peroxide to enhance its hydrophilicity, then inked in a solution of amine-modified ssDNA. Using a micrometric positioner, the inked probe tip was contacted with the MRR of interest for 45 min in a humidified environment. This allowed sufficient time for the probes to covalently link to the amine-reactive sensor surface, which had previously been modified with APTES and a succinic anhydride/EDC linker. The authors reported that the printed ssDNA probes retained good hybridization efficiency toward their targets. Overall, a resolution of 100 μm was reported for this μCP method, which was suitable for the 200 μm-diameter MRRs used. While this variant of μCP could be used for multiplexed functionalization if parallelized with multiple tip-mold probes or multiple cycles of inking and printing, such a process would be cumbersome, time-consuming, and generally unsuitable for high-throughput biosensor preparation.

Table 37. Demonstration of bioreceptor patterning using μCP for the functionalization of SiP sensors.

Patterning Technique	Sensor Type	Printed Bioreceptors	Multiplexed Bioreceptor Patterning (i.e., 4-Plex, 8-Plex ...)	Ref.
Tip-mold reactive microcontact printing	Si_3N_4 MRR	ssDNA	-	[202]

4.2. Pin and Pipette Spotting

Nano- and micropipettes filled with a bioreceptor solution can be used in contact mode to deposit small drops of reagent on a substrate by capillarity [33]. Manual spotting of bioreceptor solutions with a micropipette, potentially accompanied by a microscope or stereoscope for improved positional accuracy, is a simple and low-cost technique for spatially controlling the deposition of different bioreceptor solutions on specified regions of a SiP chip in the research setting. However, this low-resolution technique has limited reproducibility, accuracy, and throughput. This technique could be adapted to a high throughput multiplexed dispense format using a pipetting robot [333]. Commercially available pipetting robots, however, typically have minimum dispense volumes of 200–500 nL [334–336], which is approximately three orders of magnitude greater than the dispense volumes achievable with pin- and inkjet-based dispensing. Thus, this strategy would still be limited by poor resolution.

Pin-based spotting or pin printing is a similar technique whereby a robotically controlled pin is loaded with the printing solution, then tapped on the sensor surface to deposit picoliter- to nanoliter-scale droplets (Figure 18) [77,329]. Pin printing has been widely used for the preparation of DNA microarrays, and commercial arrayed pin printers are available for this purpose [33]. This technique offers low sample consumption and good resolution, with minimum spot sizes in the range of 1–100 μm, depending on the pin geometry [33,77,313].

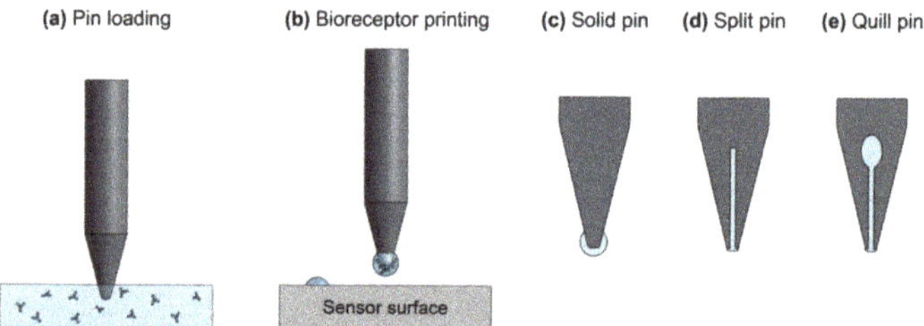

Figure 18. (**a**,**b**) Illustration of pin printing, showing (**a**) pin loading with the bioreceptor solution (antibody solution illustrated here as an example) and (**b**) printing of bioreceptors on a sensor surface using the loaded pin. (**c**–**e**) Different pin geometries, including (**c**) a solid pin, (**d**) split pin, and (**e**) quill pin.

Variations of pin printing include contact printing with solid, split and quill pins (Figure 18c–e) [77,329,337]. Solid pins are usually fabricated from micromachined stainless steel, tungsten, or titanium, and have convex, flat, or concave tips. They are loaded by dipping the pin tip in a reservoir filled with the bioreceptor solution and must be reloaded every few spots [77]. Commercially available solid microarraying pins available from Arrayit Corporation can print spots down to ~90–100 μm in diameter [315]. Solid pins are suitable for printing viscous liquids. This is valuable for protein solutions which are often prepared with viscous additives like glycerol, concentrated sugars, or high molecular weight polymers [77,329]. However, the requirement for frequent pin reloading makes solid pin printing very time-consuming. This limitation is addressed by split and quill pin designs, which permit serial printing of many spots from a single load. Split pins are fabricated with a 10–100 μm-diameter microchannel that is filled by capillary action during sample loading [77]. During printing, the split pin must be impacted on the substrate to overcome surface tension and eject picoliter- to nanoliter-scale droplets [77,337]. Quill pins have a similar design to split pins, but with a larger fluid reservoir [338]. Consequently, they can print hundreds of spots from a single load. Unlike split pins, quill pins only require a small tapping force to eject sample droplets onto the substrate [329,337]. Commercially available split and quill microarraying pins can achieve spot volumes down to ~350 pL and spot sizes down to ~37.5 μm in diameter [339–341]. Split and quill pins are best suited to low-viscosity solutions because they are susceptible to clogging with viscous liquids, which hinders spot reproducibility [77,329].

Split and quill pins can be micromachined from metal, but they have also been fabricated from silicon using standard microfabrication techniques that offer lower cost and smaller pin dimensions for improved resolution [33,77]. The BioForce Nano eNabler (Bioforce Nanosciences, Virginia Beach, VA, USA) is a commercial automated pin-based printer which uses a microfabricated silicon cantilever, called a Surface Patterning Tool (SPT), to deposit 1–60 μm droplets with 20 nm positional accuracy in the x, y, and z directions [313,314]. The SPT cantilever includes an integrated microfluidic network consisting of a reservoir to hold 0.5 μL of sample and a microchannel through which the sample flows to the tip via capillary action [313]. The droplet size is controlled by the contact time and contact force of the cantilever tip with the surface [313].

Pin-based functionalization of biosensors can be multiplexed by replacing or washing the printing needle when switching solutions [77]. In general, solid pins are easier to clean than split or quill pins, which usually require ultrasonication (for metal pins) or heating with a propane torch (for silicon pins) to thoroughly remove contamination [337]. Regarding its suitability for patterning SiP sensors, pin printing is inherently a contact technique that may damage fragile SiP structures [329]. A major challenge associated with pin printing is that optimizing spot size and reproducibility is a highly multifactorial problem [338]. Namely, the printing performance is highly dependent on the fluid properties, surface wettability, pin geometry, surface contact force, robotic controls, and environmental conditions [338]. Temperature and humidity control are typically required to slow evaporation of the sample, lower the risk of pin clogging, facilitate covalent bioreceptor immobilization on the surface, and preserve bioreceptor activity [329]. Further, spot reproducibility may deteriorate over time as a pin deforms from repeated contact with the substrate or as a split or quill pin's reservoir is depleted [77]. All of these considerations must be accounted for when designing a protocol for reliable SiP biosensor functionalization.

In the literature pipette and pin spotting have been widely used to pattern bioreceptors on SiP biosensors (Table 38). Several works have used spotting with a micropipette to pattern SiP sensors with 0.1–10 μL-scale droplets of antibodies [17,166,168,170], ssDNA probes [109,194–196,198,303], and lipid nanodiscs [148]. These strategies have been used to create 2- [198] to 9-plex [148] multiplexed biosensors. Several other works have employed the BioForce Nano eNabler to pattern SiP sensors with bovine serum albumin [313], ssDNA [163], glycans [132,133], and lipid nanodiscs [149]. These works have reported the successful preparation of 2- [133] to 8-plex [163] biosensors. Angelopoulou et al. [238] spot printed antibodies and peptides on a silicon nitride MZI sensor chip with a contact printing arrayer using solid (375 μm tip, 12 nL per spot) and quill (62.5 μm tip, 0.5 nL per spot) pins. The spotting design required multiple overlapping spots to coat the waveguides with the solid pins taking 7 times as long to print despite depositing more liquid per spot compared to the quill pins. The authors found no significant difference in the sensor response between the solid versus quill pin tips.

Table 38. Demonstrations of bioreceptor patterning using pin and pipette spotting for the functionalization of SiP sensors.

Patterning Technique	Sensor Type	Printed Bioreceptors	Multiplexed Bioreceptor Patterning (i.e., 4-Plex, 8-Plex ...)	Ref.
Hand spotting with micropipette (0.2 μL/drop)	Si MRR	Antibodies	2-plex, including control	[17]
Hand spotting with micropipette (1 μL/drop)	Si MRR	Antibodies	Up to 4-plex, including controls	[168]
Hand-spotting with micropipette	Si MRR	ssDNA probes	4-plex	[109]
Hand-spotting with micropipette	Si MRR	ssDNA probes	4-plex	[303]
Hand spotting with micropipette (1 μL/drop)	Si MRR	ssDNA probes	2-plex	[198]
Hand spotting with micropipette (0.1–0.2 μL/spot)	Si MRR	Lipid nanodiscs	9-plex	[148]
Hand spotting with micropipette	Si MRR	ssDNA probes	3-plex	[196]
Hand-spotting with micropipette	Si MRR	ssDNA probes	4-plex	[194]
BioForce Nano eNabler	Si_3N_4 MRR	Glycans	2-plex	[133]
BioForce Nano eNabler	Si_3N_4 MRR	Glycan	-	[132]
BioForce Nano eNabler	Si_3N_4 bimodal waveguide interferometric biosensor	BSA	-	[313]
BioForce Nano eNabler	Si MRR	ssDNA (subsequently used to immobilize DNA-conjugated antibodies)	8-plex	[163]
BioForce Nano eNabler	Si MRR	Lipid nanodiscs	7-plex	[149]
BioOdyssey Calligrapher miniarrayer	Si_3N_4 MZI	SARS-CoV-2 peptide	2-plex, 20–46 overlapping spots	[238]

4.3. Microfluidic Patterning in Channels

Microfluidic patterning in channels is a soft lithography technique whereby a gasket fabricated with microchannels, also called a microfluidic network (μFN), is reversibly bonded to a solid substrate and the bioreceptor solution is drawn through the microchannels (Figure 19) [33,77,321,342]. Bioreceptors are, therefore, patterned on the substrate according to the channel geometry. The μFN is usually made of molded PDMS, though laser-cut Mylar gaskets have also been used [161]. Biopatterning with μFNs was first demonstrated in 1997 by Delamarche et al. [342] for the deposition of biomolecules on solid substrates. In this work, immunoglobulins were patterned on gold, glass, and polystyrene with submicron resolution using PDMS μFNs. The channels were rendered hydrophilic with oxygen plasma and filled by capillarity to deposit the biomolecules.

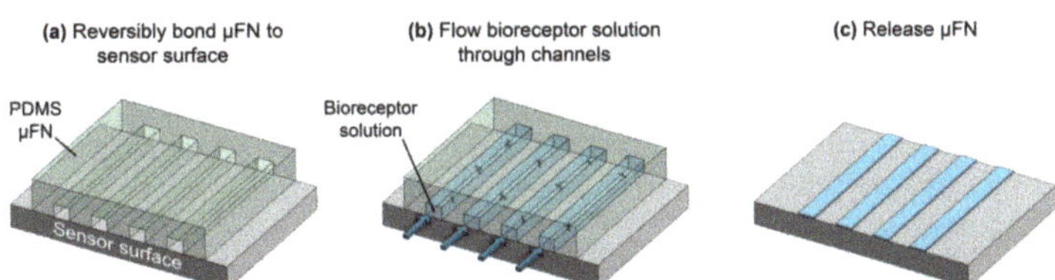

Figure 19. Illustration of the process of bioreceptor patterning on a sensor surface using a microfluidic network (μFN). (**a**) The PDMS μFN is reversibly bonded to the sensor surface, then (**b**) the bioreceptor solution is flowed through the microchannels via capillary or pressure-driven flow. This may be followed by rinsing and blocking steps. (**c**) Lastly, the μFN is released to reveal the bioreceptor-patterned surface.

μFNs using capillary flow, like those used by Delamarche et al. [342], can achieve micron-scale pattern resolution, as the microfluidic channels can be prepared with micron-scale cross section dimensions [316]. Microfluidic patterning in μFNs can also be performed using pressure-driven flow, but this requires larger channels with cross section dimensions on the order of 10 μm due to high hydraulic resistance [316,317]. Indeed, this yields poorer pattern resolution than capillary flow. However, pressure-driven flow permits the easy exchange of patterning fluids. For example, sequential surface modification steps, including crosslinker attachment, bioreceptor immobilization, rinsing, and post-processing with blocking molecules to prevent non-specific binding, can all be performed in the μFN without surface drying or removing the flow cell [32,329]. Another valuable feature of this technique is that sensing elements designed to operate in liquid media can be probed throughout the patterning process for real-time biofunctionalization monitoring [313]. Beyond biopatterning, μFNs are often used to facilitate miniaturized, simultaneous, and highly localized multi-step binding assays on functionalized sensors [321,342].

Multiplexing is typically achieved using μFNs with multiple parallel channels. Different bioreceptor solutions can be simultaneously drawn through the individually addressable channels, creating a one-dimensional array. However, this method is not well-suited to creating discrete two-dimensional patterns of bioreceptors, which would require complicated multilayer fluidics with three-dimensional flow paths [32,316]. Further, any change to the SiP sensor layout would require a redesign of the μFN [83]. Therefore, microfluidic patterning in channels has less multiplexing flexibility than pin printing, inkjet printing, and microfluidic probe-based patterning, which can localize chemical processes to arbitrary locations on a substrate [32]. Another limitation of μFNs is that reagent consumption can be high, depending on the microchannel volume, and bioreceptor molecules can be lost to microchannel walls due to nonspecific adsorption [83,316]. This is particularly undesirable when using costly bioreceptors. Similarly to μCP, materials like PDMS, which are used to fabricate the μFN, can leach uncured oligomers, which may contaminate the sensor surface and complicate functionalization and subsequent assay steps [330–332]. However, a major advantage of this technique compared to pin and inkjet printing is that bioreceptors are maintained in a controlled liquid environment throughout the patterning process. This ensures good uniformity of the biofunctionalized regions and prevents activity loss of environmentally sensitive bioreceptors due to drying [329,342]. Other advantages of this technique are that it is low cost, exceptionally simple, and unlikely to damage SiP sensing elements.

In the literature, μFNs are a popular choice for patterning SiP devices with bioreceptors (Table 39). μFNs have been used to pattern SiP sensors with antibodies [18,22,161,162,169,174,295], aptamers [174], ssDNA [174], lipid nanodiscs [150], and BSA [174,313]. This technique has been used to confine bioreceptors to select sensing structures on a single SiP device, while leaving other structures bare to control for nonspecific binding, temperature, and instrument drift [18,161]. It has also been used to compare different bioreceptor immobilization strategies. For example, Byeon et al. [295] used a 2-channel microfluidic gasket to compare bioreceptor immobilization in the presence and absence of a chemical catalyst, while González-Guerrero et al. [313] used two microfluidic channels to compare covalent and adsorption-based bioreceptor immobilization on a single sensor. Finally, μFNs have been used to create multiplexed MRR sensors with different microrings functionalized with different bioreceptors [22,150,162,169,174].

Table 39. Demonstrations of bioreceptor patterning using microfluidic patterning in channels for the functionalization of SiP sensors.

Patterning Technique	Sensor Type	Printed Bioreceptors	Multiplexed Bioreceptor Patterning (i.e., 4-Plex, 8-Plex ...)	Ref.
Microfluidic patterning in channels using 4-channel Mylar gasket	Si MRR	Antibody, DNA aptamer, ssDNA (control sequence), and BSA (control)	4-plex, including 2 controls	[174]
Microfluidic patterning in channels using 6-channel PDMS gasket	Si MRR	Antibodies	6-plex, including one control	[22]
Microfluidic patterning in channels using 2-channel Mylar gasket	Si MRR	Antibody	Patterning used to functionalize half of the rings with antibody and leave the other half bare for temperature corrections	[161]
Microfluidic patterning in channels using 2-channel Mylar gasket	Si MRR	Antibody	Patterning used to functionalize some rings with antibody and leave the rest bare to control for nonspecific binding, temperature, and instrumental drift	[18]
Microfluidic patterning in channels using 2-channel Mylar gasket	Si MRR	Antibody	μFN used to perform functionalization in the presence of catalyst on some rings and without catalyst on others	[295]
Microfluidic patterning in channels using 1-, 2-, and 4-channel Mylar gaskets	Si MRR	Lipid nanodiscs	Up to 4-plex	[150]
Microfluidic patterning in channels using 2- and 4-channel gaskets	Si MRR	Antibodies	Up to 4-plex	[169]
Microfluidic patterning in channels using Mylar gasket	Si MRR	Antibodies	4-plex	[162]
Microfluidic patterning in channels using 4-channel PDMS gasket	Si_3N_4 bimodal waveguide interferometric biosensor	BSA	2-plex to compare adsorption- and covalent-based BSA immobilization	[313]

4.4. Inkjet Printing

In contrast to the contact-based deposition systems discussed in Sections 4.1–4.3, which can expose the silicon waveguides and other structures to damage, non-contact inkjet systems use piezoelectric actuation for deposition without touching the sensor surface. Non-contact inkjet based printer systems were developed in the late 1990's where off-the-shelf desktop inkjet printers were repurposed to dispense controllable volumes of reagents in the ~80 pL range [343]. Initial development using home-built ink-jetting exposed the inkjet solution to heat resulting in a loss of functionality by denaturation or decomposition of biomolecules.

Piezoelectrically actuated non-contact inkjet devices have come to the forefront for localized reagent deposition by leveraging the control provided by a piezoelectrically actuated glass capillary that is capable of depositing droplets that are on the order of one pL to a few hundred pL in size [246,318], as illustrated in Figure 20a. These systems have x-y spatial accuracies ~15–20 µm while dispensing highly accurate volumes of assay reagents without any heat source affecting the samples [344,345]. The capillary tips hover above the etched resonator area while a voltage source piezoelectrically compresses a collar surrounding the nozzle to create pressure waves within the fluid that result in expulsion of <1 nL droplets onto the sensor surface.

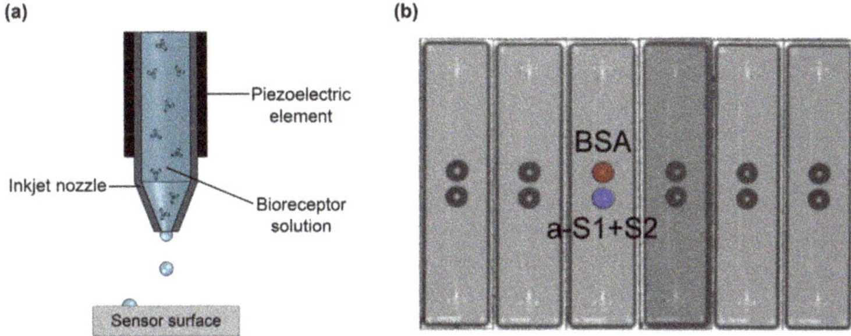

Figure 20. (**a**) Illustration of piezoelectric inkjet printing of bioreceptors on a sensor surface. (**b**) Image of antibody/antigen and BSA solutions spotted on silicon nitride photonic ring resonators using a Scienion SX microarrayer. The top (control) ring is spotted with BSA solution, and the bottom (test) ring is spotted with anti-(SARS-CoV-2 spike protein) polyclonal antibody solution (a-S1 + S2). Part (**b**) is reproduced from Ref. [167] in accordance with the Creative Commons Attribution 4.0 International (CC BY 4.0) license.

While the high spatial and volumetric controllability of piezoelectrically actuated inkjet systems is desirable, the disadvantages must be mitigated which vary for each assay solution. Nonspecific adsorption of proteins onto the borosilicate glass capillary can cause protein loss when depositing <1 nL of low concentration (<20 µg/mL) protein solutions. Delehanty and Lingler found that both the ionic strength of the printing buffer and presence of a carrier protein greatly affected the amount of biotinylated Cy5-labeled IgG that adsorbed to the capillary surface, thus influencing the amount of IgG that was dispensed from the capillary [346]. The authors' results showed an inverse relationship between the ionic strength of the buffer (PBS) and amount of IgG protein dispensed from the capillary, which was attributed to the nonspecific adsorption of proteins to borosilicate glass. Moreover, they found that with the addition of a carrier protein (BSA), the ionic strength effect could be completely mitigated while increasing the total concentration of IgG that was dispensed up to 44-fold.

On-board cameras and positioning software allow for spot printing to be carried out in a systematic fashion by fiducial mark recognition whereby ~300 pL drops of SS-A antigen at 200 µg/ml can be spot printed using a sciFLEXARRAYER S5 (Scienion, AG, Berlin, Germany) on 128 rings with PDC-70 nozzle [293]. Kirk et al. [246] illustrated the high throughput capabilities and low assay reagent consumption by printing 10 array chips with 6 microrings per chip in 9 s, consuming a total < 25 nL of reagent. In addition to functionalizing SiP chips with multiple bioreceptors for multiplexed analyte detection, inkjet printing can be leveraged to include reference sensors. As previously discussed, the functionality of a microring resonator is based on its resonance wavelength shift and is often measured with respect to a nearby reference resonator. Positioning the reference microring resonator nearby the sensing resonator helps to eliminate shift stemming from thermal gradients across the chip. Other sources contribute to anomalous background wavelength shift, such as non-specific binding, which may be useful to control for using reference rings. Therefore, while some reference rings may remain buried under an oxide cladding, it may be beneficial in some applications to also include protein coated reference resonators. Cognetti and Miller [167] fabricated a ring resonator set as shown in Figure 20b. One ring was functionalized with anti-SARS-CoV-2 RBD + SARS-CoV-2 RBD (a-S1 + S2) and another was functionalized with 0.1% BSA as a control for non-specific binding, illustrated by the blue and red dots, respectively. The piezoelectric inkjet process allowed for controlled deposition of the assay reagents as isolated elements and showed the relative (BSA ring-subtracted) wavelength shift in response to the SARS-CoV-2 spike protein [167].

Other types of detection mechanisms have been demonstrated through inkjet printing of assay reagents. For instance, Laplatine et al. [319] used a Scienion sciFELXARRAYER S12 to deposit an array of 64 different peptides in buffer on MZIs (spot size of ~150 µm). The MZI array was used to measure volatile organic chemicals (VOCs) with limits in the ppm range as the basis for a silicon olfactory sensor. Ness et al. [318] used a FUJIFILM Dimatix DMP-2831 materials piezoelectric inkjet printer (FUJIFILM Dimatix, Inc., Santa Clara, CA, USA) with 1 pL dispensing DMC-11601 cartridges to deposit ~30 µm diameter spots by optimizing the functional fluid to have a higher viscosity and lower surface tension which was achieved by the addition of glycerol and a surfactant, respectively. A Dimatix materials printer was also used to deposit a functional biotin-modified polymer and porous hydrogel on MZIs, whereby the functional polymer was able to sense the specific binding of protein streptavidin and the benzophenone dextran (benzo-dextran) porous hydrogel was shown to hinder the non-specific binding of BSA on the sensor surface [347]. Table 40 summarizes several demonstrations of inkjet-based bioreceptor deposition on SiP sensors.

Table 40. Demonstrations of bioreceptor patterning using inkjet printing for the functionalization of SiP sensors.

Patterning Technique	Sensor Type	Printed Bioreceptors	Multiplexed Bioreceptor Patterning (i.e., 4-Plex, 8-Plex ...)	Ref.
Piezoelectric non-contact printing (Scienion sciFLEXARRAYER S5)	Si MRR	SS-A antigen	-	[293]
Piezoelectric non-contact printing (Scienion sciFLEXARRAYER S3)	Si MRR	Glycoconjugates and fluorescently labeled streptavidin	Up to 4-plex	[246]
Piezoelectric non-contact printing (Scienion SciFLEXARRAYER SX)	Si_3N_4 MRR	Antibody, antigen, BSA	2-plex (performed on 2-ring chips; one ring spotted with antibody or antigen as the receptor and another with BSA as a control)	[167]
Piezoelectric non-contact printing (Scienion sciFLEXARRAYER S12)	Si_3N_4 MZI	Peptides for volatile chemicals	64 MZI sensor array	[319]
Piezoelectric non-contact printing (FUJIFILM Dimatix DMP-2831)	SiP microcantilevers	Biotin	-	[318]
Piezoelectric non-contact printing (FUJIFILM Dimatix DMP-2831)	Silicon nitride MZI	Biotin-modified polyethyleneimine functional polymer and benzophenone dextran hydrogel	-	[347]

4.5. Microfluidic Probes

Microfluidic probes (μFPs) (Figure 21), which were first demonstrated in 2005 by Juncker et al. [348], combine the features of microfluidics and scanning probes to deliver biomolecules to surfaces. μFPs are classified as "open space microfluidics", as they confine nanoliter volumes of processing liquids on substrates without solid-walled microchannels [83,317]. This is achieved through hydrodynamic flow confinement (HFC) of the processing liquid, which is made possible by the microscale dimensions of the system and the resulting laminar flow regime [32,348].

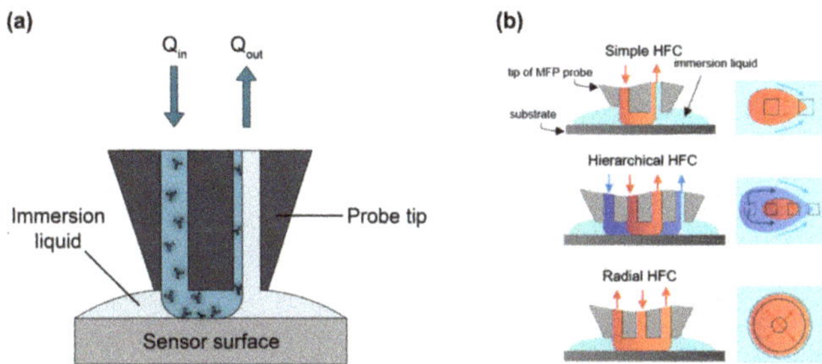

Figure 21. (**a**) Illustration of simple microfluidic printing (μFP) probe used to pattern a sensor surface with a bioreceptor solution (antibody solution illustrated here as an example). (**b**) Comparison of μFP using simple hydrodynamic flow confinement (HFC), hierarchical HFC, which permits recirculation of the patterning solution in the μFP head, and radial HFC, which produces circular, rather than teardrop-shaped spots. The processing (bioreceptor) solution is shown in red, the immersion liquid is shown in light blue, and the shaping liquid used for HFC is shown in dark blue. Insets on the right show the printing footprints for each μFP type. Part (**b**) is adapted with permission from Ref. [32]. Copyright 2021 American Chemical Society.

The tip of the μFP may be fabricated from silicon [348], silicon and glass [317], or PDMS [349]. It consists of coplanar injection and aspiration microapertures and is placed 10–200 μm above the substrate [32]. An immersion liquid fills the gap between the probe tip and substrate, while processing liquid is injected from the injection aperture and collected by the aspiration aperture. The processing liquid is confined above and below by the probe tip and substrate, while it is confined laterally by hydrodynamic boundaries formed by the immersion liquid [32]. In a simple μFP configuration, the flow rate of the injected fluid, Q_I, must be lower than the flow rate of aspirated fluid, Q_A, to maintain flow confinement [348]. The ratio Q_A/Q_I can be varied, along with the distance between the probe tip and surface, to tune the shape and size of the region where the processing fluid contacts the surface [348]. Typically, this impingement area has a teardrop shape, but an alternative radial probe tip design can be used to create a circular impingement area [32]. The μFP is mobile and can scan over a substrate to create complex patterns; depending on the direction and speed of travel relative to the microfluidic flow, continuous shapes or discrete spots can be patterned [349]. Spot sizes as small as 10×10 μm^2 are possible [83].

μFP-based bioreceptor patterning has not yet been demonstrated on SiP biosensors, but it may be a promising technique for future application. Firstly, μFPs are suitable for multiplexed patterning, as processing fluids can be rapidly switched using an external valve system [32]. Further, the probe can follow an arbitrary scan path, allowing for flexible and customized patterning of sensors with non-standard layouts [348]. Given that this is a non-contact technique, it is unlikely to damage fragile SiP surfaces. Unlike inkjet and pin printing, μFPs pattern surfaces in a liquid environment, which prevents uncontrolled wetting and drying effects, thus improving spot uniformity and homogeneity, while preventing aggregation or denaturation of printed biomolecules [317,348]. For example, Autebert et al. [83] demonstrated less than 6% variation in spot homogeneity for an array of 170 spots of IgG printed on polystyrene. While a simple μFP configuration typically requires large volumes of processing fluids, a 10-fold decrease in μFP reagent consumption has been achieved using hierarchical flow confinement and recirculation, making it comparable to pin and inkjet printing (e.g., 1.6 μL to print 170 spots of IgG, each with a 50 × 100 μm^2 footprint).

The main challenges of this patterning technique are its low throughput and limited commercial availability [32]. Using a simple μFP configuration, only one spot can be addressed at a time, each requiring a residence time defined by the kinetics of the bioreceptor's immobilization reaction. Multiple spots could be patterned simultaneously using probe tips with microfluidic channel bifurcations to increase throughput, but this would only be suitable for SiP sensors with highly standardized layouts, as aperture spacing would need to match the spacing of SiP sensing structures [83]. The accessibility of this technique is limited, as commercial μFP-based patterning systems are not yet available. Another potential challenge is that, when applied to SiP sensor surfaces, this technique may suffer from perturbations in hydrodynamic flow confinement due to the three-dimensional topography introduced by the patterned silicon structures [317]. This, in turn, may result in reduced spot homogeneity.

4.6. *Summary and Future Directions*

Here, we have discussed several strategies for preparing patterns of bioreceptors on SiP sensor surfaces for multiplexed detection. In general, non-contact patterning techniques are attractive for SiP sensor biofunctionalization, as they prevent damage to the sensor surface and integrated optical and electronic components. Of the strategies discussed here, inkjet printing is a promising strategy for biopatterning multiplexed SiP sensors. Inkjet printing is a flexible, high throughput, low-waste, and multiplexable non-contact patterning strategy that can achieve sufficient resolution for the functionalization of most SiP devices [329]. However, printing protocols (e.g., actuation waveform design, environmental controls, additives to bioreceptor "ink", etc.) must be optimized for replicable deposition of uniform spots. Future studies using inkjet-based biopatterning of SiP sensors should quantify and optimize inter- and intra-spot uniformity, along with inter-spot and run-to-run replicability to validate reliable performance of this patterning technique. μFP is another flexible non-contact patterning technique, which can achieve improved spot uniformity and replicability compared to other printing methods, and may be a promising option for SiP sensors [317,348]. Nevertheless, this technique must still be validated for bioreceptor patterning on SiP surfaces.

5. Critical Comparative Analysis of Solutions and Discussion of the Interplay between the Three Aspects of Biofunctionalization

This review has provided a detailed overview of strategies that have been or can be used to functionalize SiP biosensors in terms of bioreceptor selection, immobilization chemistry, and patterning strategy. We have benchmarked potential strategies for each of these three aspects of biofunctionalization against a set of performance criteria relevant to SiP sensing. In addition to assessing the tradeoffs of individual solutions in the context of the anticipated biosensor use case, the compatibilities and incompatibilities between solutions to each of the three aspects of biofunctionalization are an essential consideration. Moreover, the interplay between bioreceptors, immobilization chemistries, and patterning techniques can affect what is considered suitable performance for a given biofunctionalization need. For example, when using a patterning technique with very low reagent consumption, bioreceptors with a greater cost per milligram may still permit very low reagent cost per sensor. This underscores the importance of considering these three aspects of biofunctionalization in concert.

The first step in designing a biofunctionalization protocol once the application of the biosensor is defined and the target(s) known, is bioreceptor selection. As discussed in Section 2, different bioreceptors are suitable for different targets. For many targets (proteins, small molecules, viruses, bacteria, etc.) antibodies, aptamers, MIPs and PCCs may be suitable. Despite being very cost-effective and stable, currently available MIPs cannot achieve sufficient binding affinity and/or selectivity to achieve detection at clinically relevant levels for many targets. Of the other three options, antibodies are the most readily available and well-characterized, but their poor stability and high cost limit their suitability for POC use. Moreover, in our group's experience, batch-to-batch variability has been a notable roadblock in the design of replicable biosensing assays using antibodies.

Synthetic antibody analogs like aptamers and PCCs, which can achieve similar affinity and specificity to antibodies, are appealing and versatile options for POC sensors. Currently, a significant roadblock in the widespread adoption of aptamers and PCCs for biosensing applications is the relatively limited availability of pre-designed products for ready use against a diverse range of targets, though this challenge can be mitigated in coming years with further research and development [30,89,234]. Additionally, aptamers often require careful sample preparation (buffering, filtering, or tight temperature control) to avoid their folding or denaturing prematurely during use. Robust aptamer formulations need to be screened with these factors in mind, given each sensing device's use case. Regardless, their low cost, good stability, and highly reproducible and scalable production are important advantages of aptamers and PCCs for POC biosensing.

For nucleic acid targets, nucleic acid probes (hybridization-based detection), HCCD, and CRISPR-dCas9-mediated detection may be suitable bioreceptor options. In applications requiring highly multiplexed nucleic acid sensing, simple hybridization-based detection with nucleic acid probes offers the greatest flexibility and assay simplicity. When exceptionally high sensitivity and selectivity are required for very low-concentration targets, HCCD or CRISPR-dCas9-mediated detection may be preferable. It should be recognized; however, these are very early- stage approaches with limited precedent for use on SiP platforms and have yet to be validated for sensing in complex samples.

Lastly, glycans, lectins, and lipid nanodiscs are valuable for the study of carbohydrate-protein and cell membrane interactions, respectively, but their often-poor affinity and selectivity limit their applications beyond such studies.

In addition to these considerations, the immobilization chemistries that are compatible with each type of bioreceptor should be kept in mind during bioreceptor selection, with particular attention paid to compatibility of the immobilization chemistries with other steps of biosensor fabrication and integration with sample fluid delivery. Broadly, passive adsorption of bioreceptors leads to poor stability of the functionalized surface and diminishes the bioreceptor's binding activity. One exception is lipid nanodiscs, which adsorb well to silicon dioxide surfaces to yield reproducible, regenerable, and stable functional layers [148,149,275,276]. For other bioreceptors, passive adsorption is not recommended, aside from in preliminary sensor validation experiments where simplicity and rapid assay design are priorities. Nevertheless, novel polymeric coating materials (e.g., PAcrAm™ and AziGrip4™ from SuSoS AG) may permit stabler and more oriented bioreceptor immobilization with similar simplicity to passive adsorption techniques, potentially comprising a valuable future research direction [309–311].

Among the various covalent and bioaffinity-based immobilization strategies explored in this review, different immobilization methods can produce very different results depending on the bioreceptor. For example, many covalent methods can readily achieve predictable and oriented aptamer and nucleic acid probe immobilization by targeting terminal functional groups incorporated into these bioreceptors during synthesis; this ensures good binding site availability for target capture. Conversely, when used for antibody immobilization, these covalent strategies typically target native functional groups that are abundant on the antibody surface, leading to random antibody orientation and reduced target-binding capacity.

When antibody binding capacity must be optimized, bioaffinity-based strategies using antibody-binding proteins, like Protein A, may be a preferable choice, though these strategies involve tradeoffs in terms of stability, regenerability, and cost. It should also be noted that Protein A-based antibody immobilization may compromise the specificity of immunoassays using amplification with a secondary antibody [350]. In our experience, using Protein A in sandwich immunoassays was correlated with a considerable non-specific signal during the secondary antibody amplification step, which was not observed in immunoassays prepared using simple passive adsorption of the detection antibody on the SiP sensor surface [351]. This non-specific signal may be related to the unwanted capture of secondary antibodies by unoccupied Fc-binding sites on the Protein A-coated sensor surface (potentially due to incomplete functionalization with capture antibody or due to unbinding of capture antibody during the course of the experiment) [350]. One solution may be to choose secondary antibodies that do not bind well to Protein A, though this may be challenging, as Protein A and Protein G bind well with antibodies from many common host species (cow, goat, mouse, rabbit, and sheep) that are used in immunoassays [350]. Protein L, which does not bind with cow, goat and sheep antibodies and binds weakly to rabbit antibodies, may offer greater flexibility in the choice of secondary antibody, potentially making it a preferable antibody-binding protein for sandwich assays [350]. Silane-based covalent strategies have also been successfully used to immobilize capture antibodies on SiP sensors for assays using amplification with a secondary antibody [17,18,162,163]. This highlights that the assay format (label-free/labeled) and its synergy with the biofunctionalization strategy should be carefully considered.

Next, the selection of a patterning technique should take into consideration factors such as the bioreceptor cost, fluid properties of the bioreceptor solution, and changes in sensor surface hydrophilicity caused by the immobilization chemistry. For instance, patterning in microfluidic channels is a simple and popular choice in SiP biosensor functionalization protocols, but it typically has high reagent consumption. As an example, in our group's previous work, we deposited 20 μg/mL solutions of capture antibodies on SiP sensors via microfluidic channels using pressure-driven flow at 30 μL/min for 45 min [351]. This consumed a total of 27 μg of antibody, which costs roughly CAD $135, assuming an antibody cost of ~CAD 500/100 μg. Further, bioreceptors may be lost to adsorption on the channel walls during patterning, and this inefficient reagent use is particularly undesirable for costly bioreceptors, such as antibodies. In assays using in-flow patterning followed by sample introduction using the same μFN, targets in the sample may bind to bioreceptors coating the channel walls and non-sensing regions of the SiP chip. This can deplete target molecules from the sample more rapidly than if the sensing regions, alone, were functionalized. Consequently, this may worsen the limit of detection [352]. Offline patterning of bioreceptors to ensure that they are only localized to the sensing regions is, therefore, particularly beneficial for detecting precious targets at very low concentrations.

The fluid properties of the bioreceptor solution can also dictate the success of a patterning technique. In particular, μCP, pin printing, and inkjet printing strategies are sensitive to the viscosity and surface tension of the bioreceptor solution. Required additions to bioreceptor solutions, such as glycerol to slow evaporation, must be accounted for when optimizing the patterning protocol. Surface modifications used for different bioreceptor immobilization chemistries affect the hydrophilicity of the sensor surface, which, in turn, influences the efficacy and resolution of the patterning strategy [338]. For example, silanization decreases the hydrophilicity of the SiP sensor surface [267]. In the context of μCP, this may inhibit the transfer of bioreceptor "ink" from the stamp to the sensor surface. On the other hand, this decreased surface hydrophilicity will decrease the spreading of droplets of aqueous bioreceptor solutions. This may improve the resolution of patterning techniques such as pin and inkjet printing.

While not a focus of this review, antifouling strategies must typically be integrated with biofunctionalization protocols in order to prevent non-specific adsorption of sample matrix components to the sensor surface [29]. Antifouling strategies can be included in covalent bioreceptor immobilization protocols through the use of linkers that include polyethylene glycol (PEG) chains (e.g., SM(PEG)12 [133], BS(PEG)9 [132]), which increase the hydrophilicity of the surface coating to reduce non-specific protein adsorption [29]. Other approaches include coating the surface via passive adsorption with bovine serum albumin [264,353] or commercial blockers, such as StartingBlock [109,163,196], BlockAid, and StabilCoat [167], after bioreceptor immobilization. It is important to consider how to best fit antifouling strategies into biofunctionalization workflows. For further details about antifouling strategies for SiP biosensors, readers are directed to ref. [29].

Lastly, the entire biofunctionalization procedure must be considered in the context of the overall sensor system design. While some functionalization strategies may be suitable for the SiP sensor chip itself, they may not be suitable for systems including chip-mounted electronic/photonic inputs and outputs, which can be used to translate this technology to a commercial POC platform (Figure 22) [12,13,66,354]. For example, immobilization chemistries requiring solution-phase reactions may be unsuitable for sensor designs including photonic wire bonds that connect optical inputs and outputs to the on-chip waveguides. Solvents or other chemicals used in functionalization may damage or swell the photonic wire bond or low-index photonic wire bond cladding materials, resulting in damage to the fine optical connection [355,356]. In this case, immobilization chemistries employing vapor-phase surface modifications or direct crosslinking of bioreceptors (e.g., UV crosslinking of nucleic acids or aptamers) to the unmodified surface may be preferable. Similarly, plasma or UV/ozone treatment are likely more suitable surface pre-treatment techniques than immersion in piranha solution for integrated SiP systems. For these systems, the surface topography and locations of chip-mounted components should inform the selection and design of the patterning strategy. In general, non-contact patterning techniques (e.g., inkjet printing) can permit flexible bioreceptor pattern design, while preventing damage to the system, making them preferable to techniques that require contact between the patterning tool and surface.

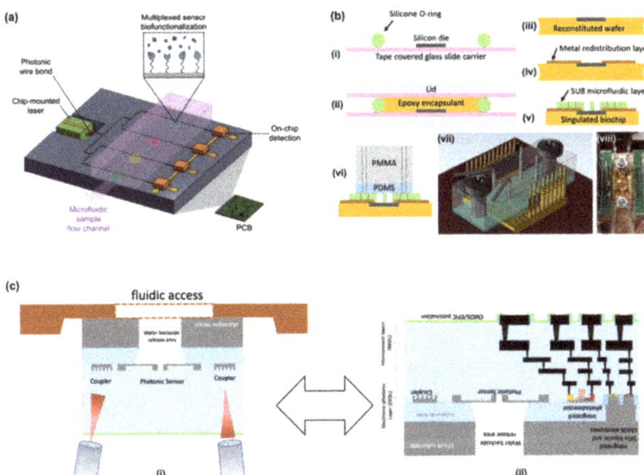

Figure 22. Integration approaches for SiP biosensors. (**a**) Multiplexed biofunctionalization of integrated SiP sensor system for POC use, which includes on-chip photonic inputs (chip-mounted fixed wavelength laser), outputs (on-chip detectors), photonic wire bonds, and microfluidics. See Ref. [13] for further information about this integration approach. (**b**) System-level integration of active SiP sensor by Laplatine et al. [66] using fan-out wafer-level packaging, showing (i–vi) schematics of the packaging process, (vii) 3D illustration of the packaged chip, and (viii) photograph of the experimental biochip setup. Part (**b**) is adapted with permission from Ref. [66]. Copyright 2018 Elsevier. (**c**) Photonic integrated circuit sensor chip presented by Mai et al. [354] using local backside release to enable integration with fluidics on one side of the chip and (i) optical coupling or (ii) optical coupling and electrical interconnects on the other. This allows for a more compact form factor than chips using front side integration only. Part (**c**) is adapted with permission from Ref. [354]. Copyright 2022 Elsevier.

In summary, it is important to consider the interplay between the three constituents of biofunctionalization as well as the silicon photonic device, fluidics, and detection assay when designing a biofunctionalization strategy. The examples above highlight the importance of considering and addressing the relationships between different bioreceptors, immobilization strategies, and patterning techniques and their suitability for different assay formats and integrated sensing architectures. This discussion aims to bring attention to the importance of considering and addressing these relationships in order to design successful biofunctionalization protocols for SiP biosensors.

6. Conclusions

When combined with carefully designed biofunctionalization strategies, SiP sensors have the potential to permit accurate and information-rich decentralized diagnostic testing for a diverse range of clinical applications. We have identified and evaluated different strategies for SiP sensor biofunctionalization in terms of bioreceptor selection, immobilization strategy, and patterning technique. Different solutions for each aspect of biofunctionalization have been benchmarked against a set of critical performance criteria relevant to multiplexed SiP biosensing and examples from the literature have been discussed and categorized. In addition to providing critical discussion about solutions for each aspect of biofunctionalization, we have also identified the interplay between these three aspects to help inform the design of SiP functionalization protocols and have highlighted additional functionalization process constraints relevant to SiP system integration for POC biosensing.

Broadly, several classes of synthetic bioreceptors (e.g., aptamers, PCCs, nucleic acids) offer excellent potential for multiplexed POC biosensing, as they can achieve high affinity and specificity, and offer scalable and cost-effective production, good stability, and regenerability. However, the availability of ready-to-use reagents remains a roadblock for the use of synthetic antibody analogs. In terms of immobilization strategies, covalent methods offer stable, scalable, and highly tailorable bioreceptor immobilization, but their success often depends highly on the reaction conditions and bioreceptor type, underscoring the potential value of developing standardized and reliable reaction protocols that are optimized for SiP surfaces. Regarding patterning, pin and inkjet-based printing are popular techniques that offer good flexibility and resolution, while inkjet printing has the additional advantages of exceptionally high throughput and being a non-contact method that will not damage the SiP surface or integrated electronic/photonic structures. μFP-based patterning is another attractive potential solution for flexible bioreceptor patterning that may achieve improved spot uniformity, though this technique has yet to be tested on SiP platforms. Overall, this review serves as a detailed overview of the biofunctionalization options available and previously tested on SiP platforms. This can help guide the design of new functionalization protocols, which must also be individually tailored for the specific target analyte(s), assay format, system architecture, and intended operating environment.

Author Contributions: Conceptualization, L.S.P., S.M.G., J.M.M., J.R.B., L.C., S.S. and K.C.C.; Investigation, L.S.P., S.M.G. and J.M.M.; Resources, L.C., S.S. and K.C.C.; Data Curation, L.S.P., S.M.G. and J.M.M.; Writing—Original Draft Preparation, L.S.P., S.M.G. and J.M.M.; Writing—Review and Editing, L.S.P., S.M.G., J.M.M., J.R.B., L.C., S.S. and K.C.C.; Visualization, L.S.P., S.M.G. and J.M.M.; Supervision, L.C., S.S. and K.C.C.; Project Administration, L.S.P., S.M.G., L.C., S.S. and K.C.C.; Funding Acquisition, L.C., S.S. and K.C.C. All authors have read and agreed to the published version of the manuscript.

Funding: This project was supported by Canada's Digital Technology Supercluster, a Digital Technology Supercluster Expansion project "Preventing COVID19 spread through workplace health and safety—Canada's Digital Technology Supercluster (CDTS)", the generosity of Eric and Wendy Schmidt by recommendation of the Schmidt Futures program, and the MITACS, Inc./Elevate Postdoctoral Fellowship Program, IT25501, In collaboration with Dream Photonics Inc. and Innovation, Science and Economic Development Canada.

Institutional Review Board Statement: Not applicable.

Informed Consent Statement: Not applicable.

Data Availability Statement: Not applicable.

Acknowledgments: The authors are grateful to work and live on land that is the traditional, ancestral, and unceded territory of the Coast Salish Peoples, including the territories of the Musqueam, Squamish, and Tsleil-Waututh First Nations. The authors would also like to acknowledge support from the Natural Sciences and Engineering Research Council of Canada (NSERC) (graduate scholarship supported by the NSERC Canada Graduate Scholarships—Master's Program) and the Silicon Electronics-Photonics Integrated Circuits Fabrication (SiEPICfab) consortium. The authors sincerely appreciate helpful discussions and feedback from Matthew B. Coppock, Avineet Randhawa, Matthew Mitchell, and Heather M. Robison.

Conflicts of Interest: The authors declare no conflict of interest.

References

1. Luan, E.; Shoman, H.; Ratner, D.M.; Cheung, K.C.; Chrostowski, L. Silicon Photonic Biosensors Using Label-Free Detection. *Sensors* **2018**, *18*, 3519. [CrossRef] [PubMed]
2. Stratis-Cullum, D.N.; Finch, A.S. Current trends in ubiquitous biosensing. *J. Anal. Bioanal. Tech.* **2013**, *S7*, 9. [CrossRef]
3. Blevins, M.G.; Fernandez-Galiana, A.; Hooper, M.J.; Boriskina, S.V. Roadmap on Universal Photonic Biosensors for Real-Time Detection of Emerging Pathogens. *Photonics* **2021**, *8*, 342. [CrossRef]
4. Chrostowski, L.; Grist, S.; Flueckiger, J.; Shi, W.; Wang, X.; Ouellet, E.; Yun, H.; Webb, M.; Nie, B.; Liang, Z.; et al. Silicon photonic resonator sensors and devices. In *Laser Resonators, Microresonators, and Beam Control XIV*; Kudryashov, A.V., Paxton, A.H., Ilchenko, V.S., Eds.; SPIE Proceedings; SPIE: Bellingham, WA, USA, 2012; Volume 8236, p. 823620.
5. Pohanka, M.; Skládal, P. Electrochemical biosensors—Principles and applications. *J. Appl. Biomed.* **2008**, *6*, 57–64. [CrossRef]
6. Pohanka, M. Overview of piezoelectric biosensors, immunosensors and DNA sensors and their applications. *Materials* **2018**, *11*, 448. [CrossRef] [PubMed]
7. Zhang, J.X.J.; Hoshino, K. Mechanical transducers: Cantilevers, acoustic wave sensors, and thermal sensors. In *Molecular Sensors and Nanodevices*; Elsevier: Amsterdam, The Netherlands, 2019; pp. 311–412. ISBN 9780128148624.
8. Ciminelli, C.; Dell'Olio, F.; Conteduca, D.; Armenise, M.N. Silicon photonic biosensors. *IET Optoelectronics* **2018**, *13*, 48–54. [CrossRef]
9. Soref, R. The past, present, and future of silicon photonics. *IEEE J. Select. Topics Quantum Electron.* **2006**, *12*, 1678–1687. [CrossRef]
10. Bailey, R.C.; Washburn, A.L.; Qavi, A.J.; Iqbal, M.; Gleeson, M.; Tybor, F.; Gunn, L.C. A robust silicon photonic platform for multiparameter biological analysis. In *Silicon Photonics IV*; Kubby, J.A., Reed, G.T., Eds.; SPIE Proceedings; SPIE: Bellingham, WA, USA, 2009; Volume 7220, p. 72200N.
11. Chrostowski, L.; Hochberg, M. *Silicon Photonics Design*; Cambridge University Press: Cambridge, UK, 2015; ISBN 9781316084168.
12. Steglich, P.; Lecci, G.; Mai, A. Surface plasmon resonance (SPR) spectroscopy and photonic integrated circuit (PIC) biosensors: A comparative review. *Sensors* **2022**, *22*, 2901. [CrossRef] [PubMed]
13. Chrostowski, L.; Leanne, D.; Matthew, M.; Connor, M.; Luan, E.; Al-Qadasi, M.; Avineet, R.; Mojaver, H.R.; Lyall, E.; Gervais, A.; et al. A silicon photonic evanescent-field sensor architecture using a fixed-wavelength laser. In *Optical Interconnects XXI*; Schröder, H., Chen, R.T., Eds.; SPIE: Bellingham, WA, USA, 2021; Volume 11692.
14. Steglich, P.; Hülsemann, M.; Dietzel, B.; Mai, A. Optical Biosensors Based on Silicon-On-Insulator Ring Resonators: A Review. *Molecules* **2019**, *24*, 519. [CrossRef] [PubMed]
15. Iqbal, M.; Gleeson, M.A.; Spaugh, B.; Tybor, F.; Gunn, W.G.; Hochberg, M.; Baehr-Jones, T.; Bailey, R.C.; Gunn, L.C. Label-Free Biosensor Arrays Based on Silicon Ring Resonators and High-Speed Optical Scanning Instrumentation. *IEEE J. Select. Topics Quantum Electron.* **2010**, *16*, 654–661. [CrossRef]
16. Puumala, L.S.; Grist, S.M.; Wickremasinghe, K.; Al-Qadasi, M.A.; Chowdhury, S.J.; Liu, Y.; Mitchell, M.; Chrostowski, L.; Shekhar, S.; Cheung, K.C. An Optimization Framework for Silicon Photonic Evanescent-Field Biosensors Using Sub-Wavelength Gratings. *Biosensors* **2022**, *12*, 840. [CrossRef] [PubMed]
17. Valera, E.; Shia, W.W.; Bailey, R.C. Development and validation of an immunosensor for monocyte chemotactic protein 1 using a silicon photonic microring resonator biosensing platform. *Clin. Biochem.* **2016**, *49*, 121–126. [CrossRef] [PubMed]
18. Luchansky, M.S.; Bailey, R.C. Silicon photonic microring resonators for quantitative cytokine detection and T-cell secretion analysis. *Anal. Chem.* **2010**, *82*, 1975–1981. [CrossRef] [PubMed]
19. Arshavsky-Graham, S.; Massad-Ivanir, N.; Segal, E.; Weiss, S. Porous Silicon-Based Photonic Biosensors: Current Status and Emerging Applications. *Anal. Chem.* **2019**, *91*, 441–467. [CrossRef] [PubMed]
20. Dincer, C.; Bruch, R.; Kling, A.; Dittrich, P.S.; Urban, G.A. Multiplexed Point-of-Care Testing—xPOCT. *Trends Biotechnol.* **2017**, *35*, 728–742. [CrossRef] [PubMed]
21. Jarockyte, G.; Karabanovas, V.; Rotomskis, R.; Mobasheri, A. Multiplexed nanobiosensors: Current trends in early diagnostics. *Sensors* **2020**, *20*, 6890. [CrossRef] [PubMed]
22. Washburn, A.L.; Luchansky, M.S.; Bowman, A.L.; Bailey, R.C. Quantitative, label-free detection of five protein biomarkers using multiplexed arrays of silicon photonic microring resonators. *Anal. Chem.* **2010**, *82*, 69–72. [CrossRef]

23. Shin, Y.; Perera, A.P.; Kee, J.S.; Song, J.; Fang, Q.; Lo, G.-Q.; Park, M.K. Label-free methylation specific sensor based on silicon microring resonators for detection and quantification of DNA methylation biomarkers in bladder cancer. *Sens. Actuators B Chem.* **2013**, *177*, 404–411. [CrossRef]
24. Park, M.K.; Kee, J.S.; Quah, J.Y.; Netto, V.; Song, J.; Fang, Q.; La Fosse, E.M.; Lo, G.-Q. Label-free aptamer sensor based on silicon microring resonators. *Sens. Actuators B Chem.* **2013**, *176*, 552–559. [CrossRef]
25. Chalyan, T.; Pasquardini, L.; Gandolfi, D.; Guider, R.; Samusenko, A.; Zanetti, M.; Pucker, G.; Pederzolli, C.; Pavesi, L. Aptamer- and Fab′- Functionalized Microring Resonators for Aflatoxin M1 Detection. *IEEE J. Select. Topics Quantum Electron.* **2017**, *23*, 350–357. [CrossRef]
26. Adamopoulos, C.; Buchbinder, S.; Zarkos, P.; Bhargava, P.; Gharia, A.; Niknejad, A.; Anwar, M.; Stojanovic, V. Fully Integrated Electronic–Photonic Biosensor for Label-Free Real-Time Molecular Sensing in Advanced Zero-Change CMOS-SOI Process. *IEEE Solid-State Circuits Lett.* **2021**, *4*, 198–201. [CrossRef]
27. Robison, H.M.; Bailey, R.C. A Guide to Quantitative Biomarker Assay Development using Whispering Gallery Mode Biosensors. *Curr. Protoc. Chem. Biol.* **2017**, *9*, 158–173. [CrossRef] [PubMed]
28. Bhalla, N.; Jolly, P.; Formisano, N.; Estrela, P. Introduction to biosensors. *Essays Biochem.* **2016**, *60*, 1–8. [CrossRef] [PubMed]
29. Soler, M.; Lechuga, L.M. Biochemistry strategies for label-free optical sensor biofunctionalization: Advances towards real applicability. *Anal. Bioanal. Chem.* **2021**, *414*, 5071–5085. [CrossRef] [PubMed]
30. Morales, M.A.; Halpern, J.M. Guide to selecting a biorecognition element for biosensors. *Bioconjug. Chem.* **2018**, *29*, 3231–3239. [CrossRef] [PubMed]
31. Bañuls, M.-J.; Puchades, R.; Maquieira, Á. Chemical surface modifications for the development of silicon-based label-free integrated optical (IO) biosensors: A review. *Anal. Chim. Acta* **2013**, *777*, 1–16. [CrossRef]
32. Delamarche, E.; Pereiro, I.; Kashyap, A.; Kaigala, G.V. Biopatterning: The art of patterning biomolecules on surfaces. *Langmuir* **2021**, *37*, 9637–9651. [CrossRef]
33. Nicu, L.; Leïchlé, T. *Micro- and Nanoelectromechanical Biosensors*; John Wiley & Sons Inc.: Hoboken, NJ, USA, 2013; pp. 65–91. ISBN 9781118760857.
34. Baker, J.E.; Sriram, R.; Miller, B.L. Two-dimensional photonic crystals for sensitive microscale chemical and biochemical sensing. *Lab Chip* **2015**, *15*, 971–990. [CrossRef]
35. Byrne, B.; Stack, E.; Gilmartin, N.; O'Kennedy, R. Antibody-based sensors: Principles, problems and potential for detection of pathogens and associated toxins. *Sensors* **2009**, *9*, 4407–4445. [CrossRef]
36. Sajid, M.; Kawde, A.-N.; Daud, M. Designs, formats and applications of lateral flow assay: A literature review. *Journal of Saudi Chemical Society* **2015**, *19*, 689–705. [CrossRef]
37. Arya, S.K.; Estrela, P. Recent Advances in Enhancement Strategies for Electrochemical ELISA-Based Immunoassays for Cancer Biomarker Detection. *Sensors* **2018**, *18*, 2010. [CrossRef]
38. Narita, F.; Wang, Z.; Kurita, H.; Li, Z.; Shi, Y.; Jia, Y.; Soutis, C. A Review of Piezoelectric and Magnetostrictive Biosensor Materials for Detection of COVID-19 and Other Viruses. *Adv. Mater.* **2021**, *33*, e2005448. [CrossRef]
39. Huertas, C.S.; Calvo-Lozano, O.; Mitchell, A.; Lechuga, L.M. Advanced Evanescent-Wave Optical Biosensors for the Detection of Nucleic Acids: An Analytic Perspective. *Front. Chem.* **2019**, *7*, 724. [CrossRef] [PubMed]
40. Fard, S.T.; Donzella, V.; Schmidt, S.A.; Flueckiger, J.; Grist, S.M.; Talebi Fard, P.; Wu, Y.; Bojko, R.J.; Kwok, E.; Jaeger, N.A.F.; et al. Performance of ultra-thin SOI-based resonators for sensing applications. *Opt. Express* **2014**, *22*, 14166–14179. [CrossRef] [PubMed]
41. Oliverio, M.; Perotto, S.; Messina, G.C.; Lovato, L.; De Angelis, F. Chemical functionalization of plasmonic surface biosensors: A tutorial review on issues, strategies, and costs. *ACS Appl. Mater. Interfaces* **2017**, *9*, 29394–29411. [CrossRef] [PubMed]
42. Chen, Y.; Liu, J.; Yang, Z.; Wilkinson, J.S.; Zhou, X. Optical biosensors based on refractometric sensing schemes: A review. *Biosens. Bioelectron.* **2019**, *144*, 111693. [CrossRef] [PubMed]
43. Wijaya, E.; Lenaerts, C.; Maricot, S.; Hastanin, J.; Habraken, S.; Vilcot, J.-P.; Boukherroub, R.; Szunerits, S. Surface plasmon resonance-based biosensors: From the development of different SPR structures to novel surface functionalization strategies. *Curr. Opin. Solid State Mater. Sci.* **2011**, *15*, 208–224. [CrossRef]
44. Luan, E. Improving the Performance of Silicon Photonic Optical Resonator-based Sensors for Biomedical Applications. Doctoral Dissertation, University of British Columbia, Vancouver, BC, Canada, 2020.
45. Donzella, V.; Sherwali, A.; Flueckiger, J.; Talebi Fard, S.; Grist, S.M.; Chrostowski, L. Sub-wavelength grating components for integrated optics applications on SOI chips. *Opt. Express* **2014**, *22*, 21037–21050. [CrossRef] [PubMed]
46. Donzella, V.; Sherwali, A.; Flueckiger, J.; Grist, S.M.; Fard, S.T.; Chrostowski, L. Design and fabrication of SOI micro-ring resonators based on sub-wavelength grating waveguides. *Opt. Express* **2015**, *23*, 4791–4803. [CrossRef]
47. Huang, Y.; Zhang, L.; Zhang, H.; Li, Y.; Liu, L.; Chen, Y.; Qiu, X.; Yu, D. Development of a portable SPR sensor for nucleic acid detection. *Micromachines* **2020**, *11*, 526. [CrossRef]
48. Harpaz, D.; Koh, B.; Marks, R.S.; Seet, R.C.S.; Abdulhalim, I.; Tok, A.I.Y. Point-of-Care Surface Plasmon Resonance Biosensor for Stroke Biomarkers NT-proBNP and S100β Using a Functionalized Gold Chip with Specific Antibody. *Sensors* **2019**, *19*, 2533. [CrossRef] [PubMed]
49. Masson, J.-F. Portable and field-deployed surface plasmon resonance and plasmonic sensors. *Analyst* **2020**, *145*, 3776–3800. [CrossRef]

50. Ouellet, E.; Lausted, C.; Lin, T.; Yang, C.W.T.; Hood, L.; Lagally, E.T. Parallel microfluidic surface plasmon resonance imaging arrays. *Lab Chip* **2010**, *10*, 581–588. [CrossRef] [PubMed]
51. Luo, Y.; Yu, F.; Zare, R.N. Microfluidic device for immunoassays based on surface plasmon resonance imaging. *Lab Chip* **2008**, *8*, 694–700. [CrossRef]
52. Endo, T.; Kerman, K.; Nagatani, N.; Hiepa, H.M.; Kim, D.-K.; Yonezawa, Y.; Nakano, K.; Tamiya, E. Multiple label-free detection of antigen-antibody reaction using localized surface plasmon resonance-based core-shell structured nanoparticle layer nanochip. *Anal. Chem.* **2006**, *78*, 6465–6475. [CrossRef] [PubMed]
53. Tokel, O.; Inci, F.; Demirci, U. Advances in plasmonic technologies for point of care applications. *Chem. Rev.* **2014**, *114*, 5728–5752. [CrossRef] [PubMed]
54. Breault-Turcot, J.; Masson, J.-F. Nanostructured substrates for portable and miniature SPR biosensors. *Anal. Bioanal. Chem.* **2012**, *403*, 1477–1484. [CrossRef] [PubMed]
55. Song, C.; Zhang, J.; Jiang, X.; Gan, H.; Zhu, Y.; Peng, Q.; Fang, X.; Guo, Y.; Wang, L. SPR/SERS dual-mode plasmonic biosensor via catalytic hairpin assembly-induced AuNP network. *Biosens. Bioelectron.* **2021**, *190*, 113376. [CrossRef] [PubMed]
56. Barchiesi, D.; Grosges, T.; Colas, F.; de la Chapelle, M.L. Combined SPR and SERS: Otto and Kretschmann configurations. *J. Opt.* **2015**, *17*, 114009. [CrossRef]
57. Meyer, S.A.; Le Ru, E.C.; Etchegoin, P.G. Combining surface plasmon resonance (SPR) spectroscopy with surface-enhanced Raman scattering (SERS). *Anal. Chem.* **2011**, *83*, 2337–2344. [CrossRef]
58. Bontempi, N.; Vassalini, I.; Danesi, S.; Ferroni, M.; Donarelli, M.; Colombi, P.; Alessandri, I. Non-Plasmonic SERS with Silicon: Is It Really Safe? New Insights into the Optothermal Properties of Core/Shell Microbeads. *J. Phys. Chem. Lett.* **2018**, *9*, 2127–2132. [CrossRef] [PubMed]
59. Juan-Colás, J.; Parkin, A.; Dunn, K.E.; Scullion, M.G.; Krauss, T.F.; Johnson, S.D. The electrophotonic silicon biosensor. *Nat. Commun.* **2016**, *7*, 12769. [CrossRef] [PubMed]
60. Kaushik, A.; Mujawar, M.A. Point of care sensing devices: Better care for everyone. *Sensors* **2018**, *18*, 4303. [CrossRef] [PubMed]
61. Morales, J.M.; Cho, P.; Bickford, J.R.; Pellegrino, P.M.; Leake, G.; Fanto, M.L. Development of army relevant wearable Photonic Integrated Circuit (PIC) biosensors. In *Chemical, Biological, Radiological, Nuclear, and Explosives (CBRNE) Sensing XXII*; Guicheteau, J.A., Howle, C.R., Eds.; SPIE: Bellingham, WA, USA, 2021; p. 15.
62. Zan, G.; Wu, T.; Zhu, F.; He, P.; Cheng, Y.; Chai, S.; Wang, Y.; Huang, X.; Zhang, W.; Wan, Y.; et al. A biomimetic conductive super-foldable material. *Matter* **2021**, *4*, 3232–3247. [CrossRef]
63. Steglich, P.; Bondarenko, S.; Mai, C.; Paul, M.; Weller, M.G.; Mai, A. CMOS-Compatible Silicon Photonic Sensor for Refractive Index Sensing Using Local Back-Side Release. *IEEE Photon. Technol. Lett.* **2020**, *32*, 1241–1244. [CrossRef]
64. Steglich, P.; Paul, M.; Mai, C.; Böhme, A.; Bondarenko, S.; Weller, M.G.; Mai, A. A monolithically integrated micro fluidic channel in a silicon-based photonic-integrated-circuit technology for biochemical sensing. In *Optical Sensors 2021*; Lieberman, R.A., Baldini, F., Homola, J., Eds.; SPIE: Bellingham, WA, USA, 2021; p. 3.
65. Zhou, Z.; Yin, B.; Michel, J. On-chip light sources for silicon photonics. *Light Sci. Appl.* **2015**, *4*, e358. [CrossRef]
66. Laplatine, L.; Luan, E.; Cheung, K.; Ratner, D.M.; Dattner, Y.; Chrostowski, L. System-level integration of active silicon photonic biosensors using Fan-Out Wafer-Level-Packaging for low cost and multiplexed point-of-care diagnostic testing. *Sens. Actuators B Chem.* **2018**, *273*, 1610–1617. [CrossRef]
67. Steglich, P.; Rabus, D.G.; Sada, C.; Paul, M.; Weller, M.G.; Mai, C.; Mai, A. Silicon Photonic Micro-Ring Resonators for Chemical and Biological Sensing: A Tutorial. *IEEE Sens. J.* **2021**, *22*, 10089–10105. [CrossRef]
68. Gao, S.; Guisán, J.M.; Rocha-Martin, J. Oriented immobilization of antibodies onto sensing platforms—A critical review. *Anal. Chim. Acta* **2022**, *1189*, 338907. [CrossRef]
69. Welch, N.G.; Scoble, J.A.; Muir, B.W.; Pigram, P.J. Orientation and characterization of immobilized antibodies for improved immunoassays (Review). *Biointerphases* **2017**, *12*, 02D301. [CrossRef]
70. Teles, F.; Fonseca, L. Trends in DNA biosensors. *Talanta* **2008**, *77*, 606–623. [CrossRef]
71. Nimse, S.B.; Song, K.; Sonawane, M.D.; Sayyed, D.R.; Kim, T. Immobilization techniques for microarray: Challenges and applications. *Sensors* **2014**, *14*, 22208–22229. [CrossRef] [PubMed]
72. Chiappini, A.; Pasquardini, L.; Bossi, A.M. Molecular imprinted polymers coupled to photonic structures in biosensors: The state of art. *Sensors* **2020**, *20*, 5069. [CrossRef] [PubMed]
73. Chambers, J.P.; Arulanandam, B.P.; Matta, L.L.; Weis, A.; Valdes, J.J. Biosensor recognition elements. *Curr. Issues Mol. Biol.* **2008**, *10*, 1–12. [CrossRef]
74. Vashist, S.K.; Lam, E.; Hrapovic, S.; Male, K.B.; Luong, J.H.T. Immobilization of antibodies and enzymes on 3-aminopropyltriethoxysilane-functionalized bioanalytical platforms for biosensors and diagnostics. *Chem. Rev.* **2014**, *114*, 11083–11130. [CrossRef] [PubMed]
75. Shen, M.; Rusling, J.; Dixit, C.K. Site-selective orientated immobilization of antibodies and conjugates for immunodiagnostics development. *Methods* **2017**, *116*, 95–111. [CrossRef]
76. Mauriz, E.; García-Fernández, M.C.; Lechuga, L.M. Towards the design of universal immunosurfaces for SPR-based assays: A review. *TrAC Trends in Analytical Chemistry* **2016**, *79*, 191–198. [CrossRef]
77. Barbulovic-Nad, I.; Lucente, M.; Sun, Y.; Zhang, M.; Wheeler, A.R.; Bussmann, M. Bio-microarray fabrication techniques–a review. *Crit. Rev. Biotechnol.* **2006**, *26*, 237–259. [CrossRef]

78. Nicu, L.; Leïchlé, T. Biosensors and tools for surface functionalization from the macro- to the nanoscale: The way forward. *J. Appl. Phys.* **2008**, *104*, 111101. [CrossRef]
79. Amirjani, A.; Rahbarimehr, E. Recent advances in functionalization of plasmonic nanostructures for optical sensing. *Mikrochim. Acta* **2021**, *188*, 57. [CrossRef]
80. Sassolas, A.; Leca-Bouvier, B.D.; Blum, L.J. DNA biosensors and microarrays. *Chem. Rev.* **2008**, *108*, 109–139. [CrossRef] [PubMed]
81. Ravina; Kumar, D.; Prasad, M.; Mohan, H. Biological recognition elements. In *Electrochemical Sensors*; Elsevier: Amsterdam, The Netherlands, 2022; pp. 213–239. ISBN 9780128231487.
82. Goode, J.A.; Rushworth, J.V.H.; Millner, P.A. Biosensor regeneration: A review of common techniques and outcomes. *Langmuir* **2015**, *31*, 6267–6276. [CrossRef] [PubMed]
83. Autebert, J.; Cors, J.F.; Taylor, D.P.; Kaigala, G.V. Convection-Enhanced Biopatterning with Recirculation of Hydrodynamically Confined Nanoliter Volumes of Reagents. *Anal. Chem.* **2016**, *88*, 3235–3242. [CrossRef]
84. Abbas MBBS, A.K.; Lichtman MD PhD, A.H.; Pillai MBBS PhD, S. *Cellular and Molecular Immunology*, 10th ed.; Elsevier: Amsterdam, The Netherlands, 2021; p. 618. ISBN 978-0-323-75748-5.
85. Nimjee, S.M.; Rusconi, C.P.; Sullenger, B.A. Aptamers: An emerging class of therapeutics. *Annu. Rev. Med.* **2005**, *56*, 555–583. [CrossRef] [PubMed]
86. Landry, J.P.; Ke, Y.; Yu, G.-L.; Zhu, X.D. Measuring affinity constants of 1450 monoclonal antibodies to peptide targets with a microarray-based label-free assay platform. *J. Immunol. Methods* **2015**, *417*, 86–96. [CrossRef]
87. Kaur, H. Overview of monoclonal antibodies. In *Monoclonal Antibodies*; Elsevier: Amsterdam, The Netherlands, 2021; pp. 1–29. ISBN 9780128223185.
88. Singh, A.; Mishra, A.; Verma, A. Antibodies: Monoclonal and polyclonal. In *Animal Biotechnology*; Elsevier: Amsterdam, The Netherlands, 2020; pp. 327–352. ISBN 9780128117101.
89. Zhou, J.; Rossi, J. Aptamers as targeted therapeutics: Current potential and challenges. *Nat. Rev. Drug Discov.* **2017**, *16*, 181–202. [CrossRef]
90. Lipman, N.S.; Jackson, L.R.; Trudel, L.J.; Weis-Garcia, F. Monoclonal versus polyclonal antibodies: Distinguishing characteristics, applications, and information resources. *ILAR J.* **2005**, *46*, 258–268. [CrossRef]
91. Layouni, R.; Cao, T.; Coppock, M.B.; Laibinis, P.E.; Weiss, S.M. Peptide-Based Capture of Chikungunya Virus E2 Protein Using Porous Silicon Biosensor. *Sensors* **2021**, *21*, 8248. [CrossRef]
92. Edfors, F.; Hober, A.; Linderbäck, K.; Maddalo, G.; Azimi, A.; Sivertsson, Å.; Tegel, H.; Hober, S.; Szigyarto, C.A.-K.; Fagerberg, L.; et al. Enhanced validation of antibodies for research applications. *Nat. Commun.* **2018**, *9*, 4130. [CrossRef]
93. Antibodypedia. Available online: https://www.antibodypedia.com/ (accessed on 29 March 2022).
94. Global Industry Analysts Research Antibodies. Available online: https://www.marketresearch.com/Global-Industry-Analysts-v1039/Research-Antibodies-32538641/ (accessed on 2 December 2022).
95. Grand View Research Research Antibodies Market Size, Share & Trends Analysis Report By Product (Primary, Secondary), By Type, By Technology, By Source, By Application, By End-use, By Region, And Segment Forecasts, 2022–2030. Available online: https://www.grandviewresearch.com/industry-analysis/research-antibodies-market (accessed on 2 December 2022).
96. Balamurugan, S.; Obubuafo, A.; Soper, S.A.; Spivak, D.A. Surface immobilization methods for aptamer diagnostic applications. *Anal. Bioanal. Chem.* **2008**, *390*, 1009–1021. [CrossRef]
97. Available online: https://www.sigmaaldrich.com/CA/en/search/antibodies?facet=facet_clonality%3Amonoclonal&focus=products&page=1&perpage=30&sort=relevance&term=antibodies&type=product_name; www.sigmaaldrich.com/CA/en/search/antibodies (accessed on 19 July 2022).
98. Primary Antibodies | Abcam. Available online: https://www.abcam.com/nav/primary-antibodies (accessed on 19 July 2022).
99. Primary Antibodies | Thermo Fisher Scientific—CA. Available online: https://www.thermofisher.com/ca/en/home/life-science/antibodies/primary-antibodies.html?icid=ab-search-primary-icons (accessed on 19 July 2022).
100. Tombelli, S.; Minunni, M.; Mascini, M. Analytical applications of aptamers. *Biosens. Bioelectron.* **2005**, *20*, 2424–2434. [CrossRef]
101. Sun, H.; Zu, Y. A highlight of recent advances in aptamer technology and its application. *Molecules* **2015**, *20*, 11959–11980. [CrossRef]
102. Lakhin, A.V.; Tarantul, V.Z.; Gening, L.V. Aptamers: Problems, solutions and prospects. *Acta Naturae* **2013**, *5*, 34–43. [CrossRef] [PubMed]
103. Grand View Research DNA Synthesis Market Size, Share & Trends Analysis Report By Service Type (Gene Synthesis, Oligonucleotide Synthesis), By Application (Research And Development, Therapeutics), By End-use, By Region, And Segment Forecasts, 2023–2030. Available online: https://www.grandviewresearch.com/industry-analysis/dna-synthesis-market-report (accessed on 2 December 2022).
104. LP Information Inc. Global DNA Synthesis Service Market Growth (Status and Outlook) 2022–2028. Available online: https://www.marketresearch.com/LP-Information-Inc-v4134/Global-DNA-Synthesis-Service-Growth-32230582/ (accessed on 2 December 2022).
105. Yoo, H.; Jo, H.; Oh, S.S. Detection and beyond: Challenges and advances in aptamer-based biosensors. *Mater. Adv.* **2020**, *1*, 2663–2687. [CrossRef]
106. Large Scale DNA Synthesis Services–Bio-Synthesis. Available online: https://www.biosyn.com/large-scale-dna-synthesis.aspx (accessed on 19 October 2022).

107. Bielec, K.; Sozanski, K.; Seynen, M.; Dziekan, Z.; Ten Wolde, P.R.; Holyst, R. Kinetics and equilibrium constants of oligonucleotides at low concentrations. Hybridization and melting study. *Phys. Chem. Chem. Phys.* **2019**, *21*, 10798–10807. [CrossRef] [PubMed]
108. Hu, W.; Fu, G.; Kong, J.; Zhou, S.; Scafa, N.; Zhang, X. Advancement of nucleic acid biosensors based on morpholino. *Am. J. Biomed. Sci.* **2015**, *7*, 40–51. [CrossRef]
109. Qavi, A.J.; Kindt, J.T.; Gleeson, M.A.; Bailey, R.C. Anti-DNA:RNA antibodies and silicon photonic microring resonators: Increased sensitivity for multiplexed microRNA detection. *Anal. Chem.* **2011**, *83*, 5949–5956. [CrossRef] [PubMed]
110. Yousuf, S.; Kim, J.; Orozaliev, A.; Dahlem, M.S.; Song, Y.-A.; Viegas, J. Label-Free Detection of Morpholino-DNA Hybridization Using a Silicon Photonics Suspended Slab Micro-Ring Resonator. *IEEE Photonics J.* **2021**, *13*, 1–9. [CrossRef]
111. Wages, J.M. NUCLEIC ACIDS | immunoassays. In *Encyclopedia of Analytical Science*; Elsevier: Amsterdam, The Netherlands, 2005; pp. 408–417. ISBN 9780123693976.
112. Kosuri, S.; Church, G.M. Large-scale de novo DNA synthesis: Technologies and applications. *Nat. Methods* **2014**, *11*, 499–507. [CrossRef]
113. DNA and RNA Molecular Weights and Conversions | Thermo Fisher Scientific—CA. Available online: https://www.thermofisher.com/ca/en/home/references/ambion-tech-support/rna-tools-and-calculators/dna-and-rna-molecular-weights-and-conversions.html (accessed on 22 October 2022).
114. Mandelkern, M.; Elias, J.G.; Eden, D.; Crothers, D.M. The dimensions of DNA in solution. *J. Mol. Biol.* **1981**, *152*, 153–161. [CrossRef]
115. PNA OLIGONUCLEOTIDES. Available online: https://www.biomers.net/Media/Preislisten/biomers_net_pna_pricelist_euro.pdf (accessed on 20 July 2022).
116. Affinity Plus DNA & RNA Oligonucleotides. Available online: https://www.idtdna.com/pages/products/custom-dna-rna/dna-oligos/affinity-plus-dna-rna-oligonucleotides (accessed on 20 July 2022).
117. GENE TOOLS PRICE LIST. Available online: https://www.gene-tools.com/sites/default/files/Price_list_01_Jul_2022.pdf (accessed on 19 July 2022).
118. Zamora-Gálvez, A.; Morales-Narváez, E.; Mayorga-Martinez, C.C.; Merkoçi, A. Nanomaterials connected to antibodies and molecularly imprinted polymers as bio/receptors for bio/sensor applications. *Appl. Mater. Today* **2017**, *9*, 387–401. [CrossRef]
119. BelBruno, J.J. Molecularly Imprinted Polymers. *Chem. Rev.* **2019**, *119*, 94–119. [CrossRef] [PubMed]
120. Pavan, S.; Berti, F. Short peptides as biosensor transducers. *Anal. Bioanal. Chem.* **2012**, *402*, 3055–3070. [CrossRef] [PubMed]
121. Silva, M.L.S. Lectin biosensors in cancer glycan biomarker detection. *Adv. Clin. Chem.* **2019**, *93*, 1–61. [CrossRef]
122. Van Breedam, W.; Pöhlmann, S.; Favoreel, H.W.; de Groot, R.J.; Nauwynck, H.J. Bitter-sweet symphony: Glycan-lectin interactions in virus biology. *FEMS Microbiol. Rev.* **2014**, *38*, 598–632. [CrossRef]
123. Ward, E.M.; Kizer, M.E.; Imperiali, B. Strategies and Tactics for the Development of Selective Glycan-Binding Proteins. *ACS Chem. Biol.* **2021**, *16*, 1795–1813. [CrossRef]
124. Hideshima, S.; Hayashi, H.; Hinou, H.; Nambuya, S.; Kuroiwa, S.; Nakanishi, T.; Momma, T.; Nishimura, S.-I.; Sakoda, Y.; Osaka, T. Glycan-immobilized dual-channel field effect transistor biosensor for the rapid identification of pandemic influenza viral particles. *Sci. Rep.* **2019**, *9*, 11616. [CrossRef]
125. Lim, S.Y.; Ng, B.H.; Li, S.F.Y. Glycans in blood as biomarkers for forensic applications. *TrAC Trends in Analytical Chemistry* **2020**, *133*, 116084. [CrossRef]
126. Shang, J.; Cheng, F.; Dubey, M.; Kaplan, J.M.; Rawal, M.; Jiang, X.; Newburg, D.S.; Sullivan, P.A.; Andrade, R.B.; Ratner, D.M. An organophosphonate strategy for functionalizing silicon photonic biosensors. *Langmuir* **2012**, *28*, 3338–3344. [CrossRef]
127. Seeberger, P.H.; Overkleeft, H.S. Chapter 53: Chemical synthesis of glycans and glycoconjugates. In *Essentials of Glycobiology*; Varki, A., Cummings, R.D., Esko, J.D., Stanley, P., Hart, G.W., Aebi, M., Darvill, A.G., Kinoshita, T., Packer, N.H., Prestegard, J.H., et al., Eds.; Cold Spring Harbor Laboratory Press: Cold Spring Harbor, NY, USA, 2015.
128. Mulloy, B.; Dell, A.; Stanley, P.; Prestegard, J.H. Structural analysis of glycans. In *Essentials of Glycobiology*; Varki, A., Cummings, R.D., Esko, J.D., Stanley, P., Hart, G.W., Aebi, M., Darvill, A.G., Kinoshita, T., Packer, N.H., Prestegard, J.H., et al., Eds.; Cold Spring Harbor Laboratory Press: Cold Spring Harbor, NY, USA, 2015.
129. Zhang, Q.; Li, Z.; Song, X. Preparation of complex glycans from natural sources for functional study. *Front. Chem.* **2020**, *8*, 508. [CrossRef]
130. Lam, S.K.; Ng, T.B. Lectins: Production and practical applications. *Appl. Microbiol. Biotechnol.* **2011**, *89*, 45–55. [CrossRef] [PubMed]
131. Yaghoubi, M.; Rahimi, F.; Negahdari, B.; Rezayan, A.H.; Shafiekhani, A. A lectin-coupled porous silicon-based biosensor: Label-free optical detection of bacteria in a real-time mode. *Sci. Rep.* **2020**, *10*, 16017. [CrossRef] [PubMed]
132. Ghasemi, F.; Eftekhar, A.A.; Gottfried, D.S.; Song, X.; Cummings, R.D.; Adibi, A. Self-referenced silicon nitride array microring biosensor for toxin detection using glycans at visible wavelength. In *Nanoscale Imaging, Sensing, and Actuation for Biomedical Applications X*; Cartwright, A.N., Nicolau, D.V., Eds.; SPIE Proceedings; SPIE: Bellingham, WA, USA, 2013; Volume 8594.
133. Ghasemi, F.; Hosseini, E.S.; Song, X.; Gottfried, D.S.; Chamanzar, M.; Raeiszadeh, M.; Cummings, R.D.; Eftekhar, A.A.; Adibi, A. Multiplexed detection of lectins using integrated glycan-coated microring resonators. *Biosens. Bioelectron.* **2016**, *80*, 682–690. [CrossRef] [PubMed]

134. Lectins and Other Carbohydrate-Binding Proteins—Section 7.7 | Thermo Fisher Scientific—CA. Available online: https://www.thermofisher.com/ca/en/home/references/molecular-probes-the-handbook/antibodies-avidins-lectins-and-related-products/lectins-and-other-carbohydrate-binding-proteins.html (accessed on 21 July 2022).
135. Glycan | Sigma-Aldrich. Available online: https://www.sigmaaldrich.com/CA/en/search/glycan?focus=products&page=1&perpage=30&sort=relevance&term=glycan&type=product_name (accessed on 24 May 2022).
136. Available online: https://www.sigmaaldrich.com/CA/en/search/lectin?focus=products&page=1&perpage=30&sort=relevance&term=lectin&type=product_name; www.sigmaaldrich.com/CA/en/search/lectin (accessed on 22 October 2022).
137. Layouni, R.; Dubrovsky, M.; Bao, M.; Chung, H.; Du, K.; Boriskina, S.V.; Weiss, S.M.; Vermeulen, D. High contrast cleavage detection for enhancing porous silicon sensor sensitivity. *Opt. Express* **2021**, *29*, 1–11. [CrossRef] [PubMed]
138. Dubrovsky, M.; Blevins, M.; Boriskina, S.V.; Vermeulen, D. High contrast cleavage detection. *Opt. Lett.* **2021**, *46*, 2593–2596. [CrossRef]
139. Liu, L.; Dubrovsky, M.; Gundavarapu, S.; Vermeulen, D.; Du, K. Viral nucleic acid detection with CRISPR-Cas12a using high contrast cleavage detection on micro-ring resonator biosensors. In *Frontiers in Biological Detection: From Nanosensors to Systems XIII*; Miller, B.L., Weiss, S.M., Danielli, A., Eds.; SPIE: Bellingham, WA, USA, 2021; p. 8.
140. Alt-R CRISPR-Cas12a (Cpf1) Genome Editing. Available online: https://www.idtdna.com/pages/products/crispr-genome-editing/alt-r-crispr-cpf1-genome-editing (accessed on 19 October 2022).
141. Gold Nanoparticles: Properties and Applications. Available online: https://www.sigmaaldrich.com/CA/en/technical-documents/technical-article/materials-science-and-engineering/biosensors-and-imaging/gold-nanoparticles (accessed on 7 December 2022).
142. Quantum Dots. Available online: https://www.sigmaaldrich.com/CA/en/technical-documents/technical-article/materials-science-and-engineering/biosensors-and-imaging/quantum-dots (accessed on 7 December 2022).
143. Koo, B.; Kim, D.-E.; Kweon, J.; Jin, C.E.; Kim, S.-H.; Kim, Y.; Shin, Y. CRISPR/dCas9-mediated biosensor for detection of tick-borne diseases. *Sens. Actuators B Chem.* **2018**, *273*, 316–321. [CrossRef]
144. Start Genome Editing with CRISPR-Cas9 | IDT. Available online: https://www.idtdna.com/pages/products/crispr-genome-editing/alt-r-crispr-cas9-system (accessed on 19 October 2022).
145. Kaminski, M.M.; Abudayyeh, O.O.; Gootenberg, J.S.; Zhang, F.; Collins, J.J. CRISPR-based diagnostics. *Nat. Biomed. Eng.* **2021**, *5*, 643–656. [CrossRef]
146. Rosser, A.; Rollinson, D.; Forrest, M.; Webster, B.L. Isothermal Recombinase Polymerase amplification (RPA) of Schistosoma haematobium DNA and oligochromatographic lateral flow detection. *Parasit. Vectors* **2015**, *8*, 446. [CrossRef]
147. Daher, R.K.; Stewart, G.; Boissinot, M.; Bergeron, M.G. Recombinase polymerase amplification for diagnostic applications. *Clin. Chem.* **2016**, *62*, 947–958. [CrossRef]
148. Medfisch, S.M.; Muehl, E.M.; Morrissey, J.H.; Bailey, R.C. Phosphatidylethanolamine-phosphatidylserine binding synergy of seven coagulation factors revealed using Nanodisc arrays on silicon photonic sensors. *Sci. Rep.* **2020**, *10*, 17407. [CrossRef]
149. Muehl, E.M.; Gajsiewicz, J.M.; Medfisch, S.M.; Wiersma, Z.S.B.; Morrissey, J.H.; Bailey, R.C. Multiplexed silicon photonic sensor arrays enable facile characterization of coagulation protein binding to nanodiscs with variable lipid content. *J. Biol. Chem.* **2017**, *292*, 16249–16256. [CrossRef] [PubMed]
150. Sloan, C.D.K.; Marty, M.T.; Sligar, S.G.; Bailey, R.C. Interfacing lipid bilayer nanodiscs and silicon photonic sensor arrays for multiplexed protein-lipid and protein-membrane protein interaction screening. *Anal. Chem.* **2013**, *85*, 2970–2976. [CrossRef] [PubMed]
151. Denisov, I.G.; Sligar, S.G. Nanodiscs for structural and functional studies of membrane proteins. *Nat. Struct. Mol. Biol.* **2016**, *23*, 481–486. [CrossRef]
152. Nanodisc Products Protein Research Cube Biotech. Available online: https://cube-biotech.com/us/products/nanodisc-products/ (accessed on 19 October 2022).
153. Borrebaeck, C.A. Antibodies in diagnostics—From immunoassays to protein chips. *Immunol. Today* **2000**, *21*, 379–382. [CrossRef] [PubMed]
154. Janeway, C.A., Jr.; Travers, P.; Walport, M.; Shlomchik, M.J. The structure of a typical antibody molecule. In *Immunobiology: The Immune System in Health and Disease*; Garland Science: New York, NY, USA, 2001.
155. Antibody production | Abcam. Available online: https://www.abcam.com/protocols/antibody-production (accessed on 7 November 2022).
156. Marx, V. Finding the right antibody for the job. *Nat. Methods* **2013**, *10*, 703–707. [CrossRef] [PubMed]
157. Frenzel, A.; Hust, M.; Schirrmann, T. Expression of recombinant antibodies. *Front. Immunol.* **2013**, *4*, 217. [CrossRef]
158. Grand View Research Antibody Fragments Market Size, Share & Trends Analysis Report by Specificity (Monoclonal Antibodies, Polyclonal Antibodies), by Type, by Therapy, by Application, by Region, and Segment Forecasts, 2022–2030. Available online: https://www.grandviewresearch.com/industry-analysis/antibody-fragments-market-report (accessed on 6 December 2022).
159. Transparency Market Research Antibody Fragments Market Trend Shows a Rapid Growth by 2024 according to New Research Report | BioSpace. Available online: https://www.biospace.com/article/antibody-fragments-market-trend-shows-a-rapid-growth-by-2024-according-to-new-research-report/ (accessed on 5 December 2022).
160. Cox, K.L.; Devanarayan, V.; Kriauciunas, A.; Manetta, J.; Montrose, C.; Sittampalam, S. Immunoassay Methods. In *Assay Guidance Manual*; Sittampalam, G.S., Coussens, N.P., Nelson, H., Arkin, M., Auld, D., Austin, C., Bejcek, B., Glicksman, M., Inglese, J.,

Iversen, P.W., et al., Eds.; Eli Lilly & Company and the National Center for Advancing Translational Sciences: Bethesda, MD, USA, 2004.

161. Washburn, A.L.; Gunn, L.C.; Bailey, R.C. Label-free quantitation of a cancer biomarker in complex media using silicon photonic microring resonators. *Anal. Chem.* **2009**, *81*, 9499–9506. [CrossRef]
162. Kindt, J.T.; Luchansky, M.S.; Qavi, A.J.; Lee, S.-H.; Bailey, R.C. Subpicogram per milliliter detection of interleukins using silicon photonic microring resonators and an enzymatic signal enhancement strategy. *Anal. Chem.* **2013**, *85*, 10653–10657. [CrossRef]
163. Washburn, A.L.; Shia, W.W.; Lenkeit, K.A.; Lee, S.-H.; Bailey, R.C. Multiplexed cancer biomarker detection using chip-integrated silicon photonic sensor arrays. *Analyst* **2016**, *141*, 5358–5365. [CrossRef]
164. Christenson, C.; Baryeh, K.; Ahadian, S.; Nasiri, R.; Dokmeci, M.R.; Goudie, M.; Khademhosseini, A.; Ye, J.Y. Enhancement of label-free biosensing of cardiac troponin I. In *Label-Free Biomedical Imaging and Sensing (LBIS)*; SPIE: Bellingham, WA, USA, 2020; Volume 11251. [CrossRef]
165. Zhang, B.; Tamez-Vela, J.M.; Solis, S.; Bustamante, G.; Peterson, R.; Rahman, S.; Morales, A.; Tang, L.; Ye, J.Y. Detection of Myoglobin with an Open-Cavity-Based Label-Free Photonic Crystal Biosensor. *J. Med. Eng.* **2013**, *2013*, 808056. [CrossRef]
166. Arnfinnsdottir, N.B.; Chapman, C.A.; Bailey, R.C.; Aksnes, A.; Stokke, B.T. Impact of silanization parameters and antibody immobilization strategy on binding capacity of photonic ring resonators. *Sensors* **2020**, *20*, 3163. [CrossRef] [PubMed]
167. Cognetti, J.S.; Miller, B.L. Monitoring Serum Spike Protein with Disposable Photonic Biosensors Following SARS-CoV-2 Vaccination. *Sensors* **2021**, *21*, 5857. [CrossRef] [PubMed]
168. Shia, W.W.; Bailey, R.C. Single domain antibodies for the detection of ricin using silicon photonic microring resonator arrays. *Anal. Chem.* **2013**, *85*, 805–810. [CrossRef] [PubMed]
169. McClellan, M.S.; Domier, L.L.; Bailey, R.C. Label-free virus detection using silicon photonic microring resonators. *Biosens. Bioelectron.* **2012**, *31*, 388–392. [CrossRef] [PubMed]
170. Pal, S.; Yadav, A.R.; Lifson, M.A.; Baker, J.E.; Fauchet, P.M.; Miller, B.L. Selective virus detection in complex sample matrices with photonic crystal optical cavities. *Biosens. Bioelectron.* **2013**, *44*, 229–234. [CrossRef]
171. Griol, A.; Peransi, S.; Rodrigo, M.; Hurtado, J.; Bellieres, L.; Ivanova, T.; Zurita, D.; Sánchez, C.; Recuero, S.; Hernández, A.; et al. Design and development of photonic biosensors for swine viral diseases detection. *Sensors* **2019**, *19*, 3985. [CrossRef]
172. Ramachandran, A.; Wang, S.; Clarke, J.; Ja, S.J.; Goad, D.; Wald, L.; Flood, E.M.; Knobbe, E.; Hryniewicz, J.V.; Chu, S.T.; et al. A universal biosensing platform based on optical micro-ring resonators. *Biosens. Bioelectron.* **2008**, *23*, 939–944. [CrossRef]
173. Yoshimoto, K.; Nishio, M.; Sugasawa, H.; Nagasaki, Y. Direct observation of adsorption-induced inactivation of antibody fragments surrounded by mixed-PEG layer on a gold surface. *J. Am. Chem. Soc.* **2010**, *132*, 7982–7989. [CrossRef]
174. Byeon, J.-Y.; Bailey, R.C. Multiplexed evaluation of capture agent binding kinetics using arrays of silicon photonic microring resonators. *Analyst* **2011**, *136*, 3430–3433. [CrossRef]
175. Xu, J.; Suarez, D.; Gottfried, D.S. Detection of avian influenza virus using an interferometric biosensor. *Anal. Bioanal. Chem.* **2007**, *389*, 1193–1199. [CrossRef]
176. Chhasatia, R.; Sweetman, M.J.; Harding, F.J.; Waibel, M.; Kay, T.; Thomas, H.; Loudovaris, T.; Voelcker, N.H. Non-invasive, in vitro analysis of islet insulin production enabled by an optical porous silicon biosensor. *Biosens. Bioelectron.* **2017**, *91*, 515–522. [CrossRef] [PubMed]
177. Barrios, C.A.; Bañuls, M.J.; González-Pedro, V.; Gylfason, K.B.; Sánchez, B.; Griol, A.; Maquieira, A.; Sohlström, H.; Holgado, M.; Casquel, R. Label-free optical biosensing with slot-waveguides. *Opt. Lett.* **2008**, *33*, 708–710. [CrossRef] [PubMed]
178. Uhlen, M.; Bandrowski, A.; Carr, S.; Edwards, A.; Ellenberg, J.; Lundberg, E.; Rimm, D.L.; Rodriguez, H.; Hiltke, T.; Snyder, M.; et al. A proposal for validation of antibodies. *Nat. Methods* **2016**, *13*, 823–827. [CrossRef] [PubMed]
179. Bunka, D.H.J.; Stockley, P.G. Aptamers come of age—At last. *Nat. Rev. Microbiol.* **2006**, *4*, 588–596. [CrossRef] [PubMed]
180. Kohlberger, M.; Gadermaier, G. SELEX: Critical factors and optimization strategies for successful aptamer selection. *Biotechnol. Appl. Biochem.* **2022**, *69*, 1771–1792. [CrossRef] [PubMed]
181. Cambio—Excellence in Molecular Biology—Aptamers. Available online: https://www.cambio.co.uk/products/apps/?id=28 (accessed on 27 May 2022).
182. Custom Oligos. Available online: https://www.idtdna.com/pages/products/custom-dna-rna (accessed on 27 May 2022).
183. Bernhardt, H.S.; Tate, W.P. Primordial soup or vinaigrette: Did the RNA world evolve at acidic pH? *Biol. Direct* **2012**, *7*, 4. [CrossRef]
184. Schasfoort, R.B.M. Chapter 1. introduction to surface plasmon resonance. In *Handbook of Surface Plasmon Resonance*; Schasfoort, R.B.M., Ed.; Royal Society of Chemistry: Cambridge, UK, 2017; pp. 1–26. ISBN 978-1-78262-730-2.
185. Guider, R.; Gandolfi, D.; Chalyan, T.; Pasquardini, L.; Samusenko, A.; Pederzolli, C.; Pucker, G.; Pavesi, L. Sensitivity and Limit of Detection of biosensors based on ring resonators. *Sens. Bio-Sens. Res.* **2015**, *6*, 99–102. [CrossRef]
186. Kusser, W. Chemically modified nucleic acid aptamers for in vitro selections: Evolving evolution. *Rev. Mol. Biotechnol.* **2000**, *74*, 27–38. [CrossRef]
187. Wang, T.; Chen, C.; Larcher, L.M.; Barrero, R.A.; Veedu, R.N. Three decades of nucleic acid aptamer technologies: Lessons learned, progress and opportunities on aptamer development. *Biotechnol. Adv.* **2019**, *37*, 28–50. [CrossRef]
188. Paniel, N.; Baudart, J.; Hayat, A.; Barthelmebs, L. Aptasensor and genosensor methods for detection of microbes in real world samples. *Methods* **2013**, *64*, 229–240. [CrossRef]

189. Wang, S.; Kool, E.T. Origins of the large differences in stability of DNA and RNA helices: C-5 methyl and 2′-hydroxyl effects. *Biochemistry* **1995**, *34*, 4125–4132. [CrossRef] [PubMed]
190. Hughes, R.A.; Ellington, A.D. Synthetic DNA synthesis and assembly: Putting the synthetic in synthetic biology. *Cold Spring Harb. Perspect. Biol.* **2017**, *9*, a023812. [CrossRef] [PubMed]
191. Braasch, D.A.; Corey, D.R. Locked nucleic acid (LNA): Fine-tuning the recognition of DNA and RNA. *Chem. Biol.* **2001**, *8*, 1–7. [CrossRef] [PubMed]
192. Liu, Q.; Tu, X.; Kim, K.W.; Kee, J.S.; Shin, Y.; Han, K.; Yoon, Y.-J.; Lo, G.-Q.; Park, M.K. Highly sensitive Mach–Zehnder interferometer biosensor based on silicon nitride slot waveguide. *Sens. Actuators B Chem.* **2013**, *188*, 681–688. [CrossRef]
193. Lai, Q.; Chen, W.; Zhang, Y.; Liu, Z. Application strategies of peptide nucleic acids toward electrochemical nucleic acid sensors. *Analyst* **2021**, *146*, 5822–5835. [CrossRef]
194. Qavi, A.J.; Bailey, R.C. Multiplexed detection and label-free quantitation of microRNAs using arrays of silicon photonic microring resonators. *Angew. Chem. Int. Ed.* **2010**, *49*, 4608–4611. [CrossRef]
195. Graybill, R.M.; Cardenosa-Rubio, M.C.; Yang, H.; Johnson, M.D.; Bailey, R.C. Multiplexed microRNA Expression Profiling by Combined Asymmetric PCR and Label-Free Detection using Silicon Photonic Sensor Arrays. *Anal. Methods* **2018**, *10*, 1618–1623. [CrossRef]
196. Kindt, J.T.; Bailey, R.C. Chaperone probes and bead-based enhancement to improve the direct detection of mRNA using silicon photonic sensor arrays. *Anal. Chem.* **2012**, *84*, 8067–8074. [CrossRef]
197. Shin, Y.; Perera, A.P.; Park, M.K. Label-free DNA sensor for detection of bladder cancer biomarkers in urine. *Sens. Actuators B Chem.* **2013**, *178*, 200–206. [CrossRef]
198. Scheler, O.; Kindt, J.T.; Qavi, A.J.; Kaplinski, L.; Glynn, B.; Barry, T.; Kurg, A.; Bailey, R.C. Label-free, multiplexed detection of bacterial tmRNA using silicon photonic microring resonators. *Biosens. Bioelectron.* **2012**, *36*, 56–61. [CrossRef]
199. Liu, Q.; Lim, B.K.L.; Lim, S.Y.; Tang, W.Y.; Gu, Z.; Chung, J.; Park, M.K.; Barkham, T. Label-free, real-time and multiplex detection of Mycobacterium tuberculosis based on silicon photonic microring sensors and asymmetric isothermal amplification technique (SPMS-AIA). *Sens. Actuators B Chem.* **2017**, *255*, 1595–1603. [CrossRef]
200. Sepúlveda, B.; del Río, J.S.; Moreno, M.; Blanco, F.J.; Mayora, K.; Domínguez, C.; Lechuga, L.M. Optical biosensor microsystems based on the integration of highly sensitive Mach–Zehnder interferometer devices. *J. Opt. A Pure Appl. Opt.* **2006**, *8*, S561–S566. [CrossRef]
201. Hu, S.; Zhao, Y.; Qin, K.; Retterer, S.T.; Kravchenko, I.I.; Weiss, S.M. Enhancing the Sensitivity of Label-Free Silicon Photonic Biosensors through Increased Probe Molecule Density. *ACS Photonics* **2014**, *1*, 590–597. [CrossRef]
202. Peserico, N.; Castagna, R.; Bellieres, L.; Rodrigo, M.; Melloni, A. Tip-mould microcontact printing for functionalisation of optical microring resonator. *IET Nanobiotechnol.* **2018**, *12*, 87–91. [CrossRef]
203. Toccafondo, V.; García-Rupérez, J.; Bañuls, M.J.; Griol, A.; Castelló, J.G.; Peransi-Llopis, S.; Maquieira, A. Single-strand DNA detection using a planar photonic-crystal-waveguide-based sensor. *Opt. Lett.* **2010**, *35*, 3673–3675. [CrossRef] [PubMed]
204. Rashid, J.I.A.; Yusof, N.A. The strategies of DNA immobilization and hybridization detection mechanism in the construction of electrochemical DNA sensor: A review. *Sens. Bio-Sens. Res.* **2017**, *16*, 19–31. [CrossRef]
205. Zhu, B.; Travas-Sejdic, J. PNA versus DNA in electrochemical gene sensing based on conducting polymers: Study of charge and surface blocking effects on the sensor signal. *Analyst* **2018**, *143*, 687–694. [CrossRef] [PubMed]
206. Kaisti, M.; Kerko, A.; Aarikka, E.; Saviranta, P.; Boeva, Z.; Soukka, T.; Lehmusvuori, A. Real-time wash-free detection of unlabeled PNA-DNA hybridization using discrete FET sensor. *Sci. Rep.* **2017**, *7*, 15734. [CrossRef] [PubMed]
207. El-Schich, Z.; Zhang, Y.; Feith, M.; Beyer, S.; Sternbæk, L.; Ohlsson, L.; Stollenwerk, M.; Wingren, A.G. Molecularly imprinted polymers in biological applications. *BioTechniques* **2020**, *69*, 406–419. [CrossRef] [PubMed]
208. Ertürk, G.; Mattiasson, B. Molecular imprinting techniques used for the preparation of biosensors. *Sensors* **2017**, *17*, 288. [CrossRef]
209. Ahmad, O.S.; Bedwell, T.S.; Esen, C.; Garcia-Cruz, A.; Piletsky, S.A. Molecularly imprinted polymers in electrochemical and optical sensors. *Trends Biotechnol.* **2019**, *37*, 294–309. [CrossRef] [PubMed]
210. Wackerlig, J.; Schirhagl, R. Applications of Molecularly Imprinted Polymer Nanoparticles and Their Advances toward Industrial Use: A Review. *Anal. Chem.* **2016**, *88*, 250–261. [CrossRef] [PubMed]
211. Hammond, G.D.; Vojta, A.L.; Grant, S.A.; Hunt, H.K. Integrating Nanostructured Artificial Receptors with Whispering Gallery Mode Optical Microresonators via Inorganic Molecular Imprinting Techniques. *Biosensors* **2016**, *6*, 26. [CrossRef] [PubMed]
212. Chen, Y.; Liu, Y.; Shen, X.; Chang, Z.; Tang, L.; Dong, W.-F.; Li, M.; He, J.-J. Ultrasensitive Detection of Testosterone Using Microring Resonator with Molecularly Imprinted Polymers. *Sensors* **2015**, *15*, 31558–31565. [CrossRef]
213. Eisner, L.; Wilhelm, I.; Flachenecker, G.; Hürttlen, J.; Schade, W. Molecularly Imprinted Sol-Gel for TNT Detection with Optical Micro-Ring Resonator Sensor Chips. *Sensors* **2019**, *19*, 3909. [CrossRef] [PubMed]
214. Farrell, M.E.; Coppock, M.B.; Holthoff, E.L.; Pellegrino, P.M.; Bickford, J.R.; Cho, P.S. Towards Army Relevant Sensing with Integrated Molecularly Imprinted Polymer Photonic (IMIPP) Devices. In Proceedings of the 2018 IEEE Research and Applications of Photonics In Defense Conference (RAPID), Miramar Beach, FL, USA, 22–24 August 2018; pp. 1–4.
215. Farrell, M.E.; Holthoff, E.L.; Bickford, J.; Cho, P.S.; Pellegrino, P.M. Development of army relevant integrated photonics MIP platform. In *Smart Biomedical and Physiological Sensor Technology XVI*; Cullum, B.M., McLamore, E.S., Kiehl, D., Eds.; SPIE: Bellingham, WA, USA, 2019; p. 13.

216. Lorenzo, R.A.; Carro, A.M.; Alvarez-Lorenzo, C.; Concheiro, A. To remove or not to remove? The challenge of extracting the template to make the cavities available in Molecularly Imprinted Polymers (MIPs). *Int. J. Mol. Sci.* **2011**, *12*, 4327–4347. [CrossRef] [PubMed]
217. Han, X.-Y.; Wu, Z.-L.; Yang, S.-C.; Shen, F.-F.; Liang, Y.-X.; Wang, L.-H.; Wang, J.-Y.; Ren, J.; Jia, L.-Y.; Zhang, H.; et al. Recent progress of imprinted polymer photonic waveguide devices and applications. *Polymers* **2018**, *10*, 603. [CrossRef]
218. Refaat, D.; Aggour, M.G.; Farghali, A.A.; Mahajan, R.; Wiklander, J.G.; Nicholls, I.A.; Piletsky, S.A. Strategies for Molecular Imprinting and the Evolution of MIP Nanoparticles as Plastic Antibodies-Synthesis and Applications. *Int. J. Mol. Sci.* **2019**, *20*, 6304. [CrossRef]
219. Xie, Z.; Cao, Z.; Liu, Y.; Zhang, Q.; Zou, J.; Shao, L.; Wang, Y.; He, J.; Li, M. Highly-sensitive optical biosensor based on equal FSR cascaded microring resonator with intensity interrogation for detection of progesterone molecules. *Opt. Express* **2017**, *25*, 33193. [CrossRef]
220. Marfà, J.; Pupin, R.R.; Sotomayor, M.P.T.; Pividori, M.I. Magnetic-molecularly imprinted polymers in electrochemical sensors and biosensors. *Anal. Bioanal. Chem.* **2021**, *413*, 6141–6157. [CrossRef]
221. Kupai, J.; Razali, M.; Buyuktiryaki, S.; Kecili, R.; Szekely, G. Long-term stability and reusability of molecularly imprinted polymers. *Polym. Chem.* **2017**, *8*, 666–673. [CrossRef] [PubMed]
222. Nicholls, I.A.; Andersson, H.S.; Charlton, C.; Henschel, H.; Karlsson, B.C.G.; Karlsson, J.G.; O'Mahony, J.; Rosengren, A.M.; Rosengren, K.J.; Wikman, S. Theoretical and computational strategies for rational molecularly imprinted polymer design. *Biosens. Bioelectron.* **2009**, *25*, 543–552. [CrossRef] [PubMed]
223. Umpleby, R.J.; Baxter, S.C.; Chen, Y.; Shah, R.N.; Shimizu, K.D. Characterization of Molecularly Imprinted Polymers with the Langmuir−Freundlich Isotherm. *Anal. Chem.* **2001**, *73*, 4584–4591. [CrossRef] [PubMed]
224. Ndunda, E.N. Molecularly imprinted polymers-A closer look at the control polymer used in determining the imprinting effect: A mini review. *J. Mol. Recognit.* **2020**, *33*, e2855. [CrossRef] [PubMed]
225. Reddy, S.M.; Hawkins, D.M.; Phan, Q.T.; Stevenson, D.; Warriner, K. Protein detection using hydrogel-based molecularly imprinted polymers integrated with dual polarisation interferometry. *Sens. Actuators B Chem.* **2013**, *176*, 190–197. [CrossRef]
226. Harris, L.J.; Skaletsky, E.; McPherson, A. Crystallographic structure of an intact IgG1 monoclonal antibody. *J. Mol. Biol.* **1998**, *275*, 861–872. [CrossRef]
227. Carothers, J.M.; Davis, J.H.; Chou, J.J.; Szostak, J.W. Solution structure of an informationally complex high-affinity RNA aptamer to GTP. *RNA* **2006**, *12*, 567–579. [CrossRef]
228. Nishimasu, H.; Ran, F.A.; Hsu, P.D.; Konermann, S.; Shehata, S.I.; Dohmae, N.; Ishitani, R.; Zhang, F.; Nureki, O. Crystal structure of Cas9 in complex with guide RNA and target DNA. *Cell* **2014**, *156*, 935–949. [CrossRef] [PubMed]
229. McKenzie, E.A.; Abbott, W.M. Expression of recombinant proteins in insect and mammalian cells. *Methods* **2018**, *147*, 40–49. [CrossRef]
230. Chandrudu, S.; Simerska, P.; Toth, I. Chemical methods for peptide and protein production. *Molecules* **2013**, *18*, 4373–4388. [CrossRef]
231. Merrifield, R.B. Solid phase synthesis (nobel lecture). *Angew. Chem. Int. Ed. Engl.* **1985**, *24*, 799–810. [CrossRef]
232. Tian, J.; Li, Y.; Ma, B.; Tan, Z.; Shang, S. Automated peptide synthesizers and glycoprotein synthesis. *Front. Chem.* **2022**, *10*, 896098. [CrossRef]
233. Monty, O.B.C.; Simmons, N.; Chamakuri, S.; Matzuk, M.M.; Young, D.W. Solution-Phase Fmoc-Based Peptide Synthesis for DNA-Encoded Chemical Libraries: Reaction Conditions, Protecting Group Strategies, and Pitfalls. *ACS Comb. Sci.* **2020**, *22*, 833–843. [CrossRef] [PubMed]
234. Agnew, H.D.; Coppock, M.B.; Idso, M.N.; Lai, B.T.; Liang, J.; McCarthy-Torrens, A.M.; Warren, C.M.; Heath, J.R. Protein-Catalyzed Capture Agents. *Chem. Rev.* **2019**, *119*, 9950–9970. [CrossRef] [PubMed]
235. Ma, H.; Ó'Fágáin, C.; O'Kennedy, R. Antibody stability: A key to performance—Analysis, influences and improvement. *Biochimie* **2020**, *177*, 213–225. [CrossRef]
236. Martínez-Ceron, M.C.; Giudicessi, S.L.; Saavedra, S.L.; Gurevich-Messina, J.M.; Erra-Balsells, R.; Albericio, F.; Cascone, O.; Camperi, S.A. Latest advances in OBOC peptide libraries. improvements in screening strategies and enlarging the family from linear to cyclic libraries. *Curr. Pharm. Biotechnol.* **2016**, *17*, 449–457. [CrossRef]
237. Bozovičar, K.; Bratkovič, T. Evolving a peptide: Library platforms and diversification strategies. *Int. J. Mol. Sci.* **2019**, *21*, 215. [CrossRef]
238. Angelopoulou, M.; Makarona, E.; Salapatas, A.; Misiakos, K.; Synolaki, E.; Ioannidis, A.; Chatzipanagiotou, S.; Ritvos, M.A.; Pasternack, A.; Ritvos, O.; et al. Directly immersible silicon photonic probes: Application to rapid SARS-CoV-2 serological testing. *Biosens. Bioelectron.* **2022**, *215*, 114570. [CrossRef]
239. Martucci, N.M.; Rea, I.; Ruggiero, I.; Terracciano, M.; De Stefano, L.; Migliaccio, N.; Palmieri, C.; Scala, G.; Arcari, P.; Rendina, I.; et al. A new strategy for label-free detection of lymphoma cancer cells. *Biomed. Opt. Express* **2015**, *6*, 1353–1362. [CrossRef]
240. Cao, T.; Layouni, R.; Coppock, M.B.; Laibinis, P.E.; Weiss, S.M. Use of peptide capture agents in porous silicon biosensors. In *Frontiers in Biological Detection: From Nanosensors to Systems XII*; Miller, B.L., Weiss, S.M., Danielli, A., Eds.; SPIE: Bellingham, WA, USA, 2020; p. 21.
241. Lamberti, A.; Sanges, C.; Migliaccio, N.; De Stefano, L.; Rea, I.; Orabona, E.; Scala, G.; Rendina, I.; Arcari, P. Silicon-Based Technology for Ligand-Receptor Molecular Identification. *J. At. Mol. Opt. Phys.* **2012**, *2012*, 948390. [CrossRef]

242. Chen, S.; Liu, L.; Zhou, J.; Jiang, S. Controlling Antibody Orientation on Charged Self-Assembled Monolayers. *Langmuir* **2003**, *19*, 2859–2864. [CrossRef]
243. Coppock, M.B.; Warner, C.R.; Dorsey, B.; Orlicki, J.A.; Sarkes, D.A.; Lai, B.T.; Pitram, S.M.; Rohde, R.D.; Malette, J.; Wilson, J.A.; et al. Protein catalyzed capture agents with tailored performance for in vitro and in vivo applications. *Biopolymers* **2017**, *108*, e22934. [CrossRef] [PubMed]
244. Belický, Š.; Katrlík, J.; Tkáč, J. Glycan and lectin biosensors. *Essays Biochem.* **2016**, *60*, 37–47. [CrossRef]
245. Overkleeft, H.S.; Seeberger, P.H. Chapter 54: Chemoenzymatic synthesis of glycans and glycoconjugates. In *Essentials of Glycobiology*; Varki, A., Cummings, R.D., Esko, J.D., Stanley, P., Hart, G.W., Aebi, M., Darvill, A.G., Kinoshita, T., Packer, N.H., Prestegard, J.H., et al., Eds.; Cold Spring Harbor Laboratory Press: Cold Spring Harbor, NY, USA, 2015.
246. Kirk, J.T.; Fridley, G.E.; Chamberlain, J.W.; Christensen, E.D.; Hochberg, M.; Ratner, D.M. Multiplexed inkjet functionalization of silicon photonic biosensors. *Lab Chip* **2011**, *11*, 1372–1377. [CrossRef] [PubMed]
247. Van Landuyt, L.; Lonigro, C.; Meuris, L.; Callewaert, N. Customized protein glycosylation to improve biopharmaceutical function and targeting. *Curr. Opin. Biotechnol.* **2019**, *60*, 17–28. [CrossRef] [PubMed]
248. Chen, B.; Li, Y.; Xu, F.; Yang, X. Powerful CRISPR-Based Biosensing Techniques and Their Integration With Microfluidic Platforms. *Front. Bioeng. Biotechnol.* **2022**, *10*, 851712. [CrossRef]
249. Li, Y.; Li, S.; Wang, J.; Liu, G. CRISPR/Cas Systems towards Next-Generation Biosensing. *Trends Biotechnol.* **2019**, *37*, 730–743. [CrossRef] [PubMed]
250. Leung, R.K.-K.; Cheng, Q.-X.; Wu, Z.-L.; Khan, G.; Liu, Y.; Xia, H.-Y.; Wang, J. CRISPR-Cas12-based nucleic acids detection systems. *Methods* **2021**, *203*, 276–281. [CrossRef] [PubMed]
251. Chung, H.; Park, J.; Boriskina, S.V. Inverse-designed waveguide-based biosensor for high-sensitivity, single-frequency detection of biomolecules. *Nanophotonics* **2022**, *11*, 1427–1442. [CrossRef]
252. Lobato, I.M.; O'Sullivan, C.K. Recombinase polymerase amplification: Basics, applications and recent advances. *Trends Analyt. Chem.* **2018**, *98*, 19–35. [CrossRef]
253. Bruch, R.; Urban, G.A.; Dincer, C. Crispr/cas powered multiplexed biosensing. *Trends Biotechnol.* **2019**, *37*, 791–792. [CrossRef] [PubMed]
254. McCarty, N.S.; Graham, A.E.; Studená, L.; Ledesma-Amaro, R. Multiplexed CRISPR technologies for gene editing and transcriptional regulation. *Nat. Commun.* **2020**, *11*, 1281. [CrossRef] [PubMed]
255. Vatankhah, M.; Azizi, A.; Sanajouyan Langeroudi, A.; Ataei Azimi, S.; Khorsand, I.; Kerachian, M.A.; Motaei, J. CRISPR-based biosensing systems: A way to rapidly diagnose COVID-19. *Crit. Rev. Clin. Lab. Sci.* **2021**, *58*, 225–241. [CrossRef] [PubMed]
256. Schuler, M.A.; Denisov, I.G.; Sligar, S.G. Nanodiscs as a new tool to examine lipid-protein interactions. *Methods Mol. Biol.* **2013**, *974*, 415–433. [CrossRef]
257. Morita, M.; Ohmi, T.; Hasegawa, E.; Kawakami, M.; Ohwada, M. Growth of native oxide on a silicon surface. *J. Appl. Phys.* **1990**, *68*, 1272–1281. [CrossRef]
258. Aissaoui, N.; Bergaoui, L.; Landoulsi, J.; Lambert, J.-F.; Boujday, S. Silane layers on silicon surfaces: Mechanism of interaction, stability, and influence on protein adsorption. *Langmuir* **2012**, *28*, 656–665. [CrossRef]
259. Coen, M.C.; Lehmann, R.; Gröning, P.; Bielmann, M.; Galli, C.; Schlapbach, L. Adsorption and bioactivity of protein A on silicon surfaces studied by AFM and XPS. *J. Colloid Interface Sci.* **2001**, *233*, 180–189. [CrossRef]
260. Cuddy, M.F.; Poda, A.R.; Brantley, L.N. Determination of isoelectric points and the role of pH for common quartz crystal microbalance sensors. *ACS Appl. Mater. Interfaces* **2013**, *5*, 3514–3518. [CrossRef]
261. Jönsson, U.; Malmqvist, M.; Rönnberg, I. Immobilization of immunoglobulins on silica surfaces. Stability. *Biochem. J.* **1985**, *227*, 363–371. [CrossRef]
262. Lin, J.N.; Andrade, J.D.; Chang, I.N. The influence of adsorption of native and modified antibodies on their activity. *J. Immunol. Methods* **1989**, *125*, 67–77. [CrossRef] [PubMed]
263. Vashist, S.K.; Dixit, C.K.; MacCraith, B.D.; O'Kennedy, R. Effect of antibody immobilization strategies on the analytical performance of a surface plasmon resonance-based immunoassay. *Analyst* **2011**, *136*, 4431–4436. [CrossRef] [PubMed]
264. Flueckiger, J.; Schmidt, S.; Donzella, V.; Sherwali, A.; Ratner, D.M.; Chrostowski, L.; Cheung, K.C. Sub-wavelength grating for enhanced ring resonator biosensor. *Opt. Express* **2016**, *24*, 15672–15686. [CrossRef] [PubMed]
265. Hermanson, G.T. Avidin–Biotin Systems. In *Bioconjugate Techniques*; Elsevier: Amsterdam, The Netherlands, 2008; pp. 900–923. ISBN 9780123705013.
266. Yalcin, A.; Popat, K.C.; Aldridge, J.C.; Desai, T.A.; Hryniewicz, J.; Chbouki, N.; Little, B.E.; King, O.; Van, V.; Chu, S.; et al. Optical sensing of biomolecules using microring resonators. *IEEE J. Select. Topics Quantum Electron.* **2006**, *12*, 148–155. [CrossRef]
267. Yadav, A.R.; Sriram, R.; Carter, J.A.; Miller, B.L. Comparative study of solution-phase and vapor-phase deposition of aminosilanes on silicon dioxide surfaces. *Mater. Sci. Eng. C Mater. Biol. Appl.* **2014**, *35*, 283–290. [CrossRef]
268. Zhu, M.; Lerum, M.Z.; Chen, W. How to prepare reproducible, homogeneous, and hydrolytically stable aminosilane-derived layers on silica. *Langmuir* **2012**, *28*, 416–423. [CrossRef] [PubMed]
269. Hunt, H.K.; Soteropulos, C.; Armani, A.M. Bioconjugation strategies for microtoroidal optical resonators. *Sensors* **2010**, *10*, 9317–9336. [CrossRef]

270. Cattani-Scholz, A.; Pedone, D.; Dubey, M.; Neppl, S.; Nickel, B.; Feulner, P.; Schwartz, J.; Abstreiter, G.; Tornow, M. Organophosphonate-based PNA-functionalization of silicon nanowires for label-free DNA detection. *ACS Nano* **2008**, *2*, 1653–1660. [CrossRef] [PubMed]
271. Kolb, H.C.; Finn, M.G.; Sharpless, K.B. Click Chemistry: Diverse Chemical Function from a Few Good Reactions. *Angew. Chem. Int. Ed.* **2001**, *40*, 2004–2021. [CrossRef]
272. Chen, R.; Surman, C.; Potyrailo, R.; Pris, A.; Holwitt, E.A.; Sorola, V.K.; Kiel, J.L. Immobilization of aptamers onto unmodified glass surfaces for affordable biosensors. In *Photonic Microdevices/Microstructures for Sensing III*; Xiao, H., Fan, X., Wang, A., Eds.; SPIE Proceedings; SPIE: Bellingham, WA, USA, 2011; Volume 8034, p. 803405.
273. Gudnason, H.; Dufva, M.; Duong Bang, D.; Wolff, A. An inexpensive and simple method for thermally stable immobilization of DNA on an unmodified glass surface: UV linking of poly(T)10-poly(C)10-tagged DNA probes. *BioTechniques* **2008**, *45*, 261–271. [CrossRef]
274. Antoniou, M.; Tsounidi, D.; Petrou, P.S.; Beltsios, K.G.; Kakabakos, S.E. Functionalization of silicon dioxide and silicon nitride surfaces with aminosilanes for optical biosensing applications. *Med. Devices Sens.* **2020**, *3*, e10072. [CrossRef]
275. Anderson, T.H.; Min, Y.; Weirich, K.L.; Zeng, H.; Fygenson, D.; Israelachvili, J.N. Formation of supported bilayers on silica substrates. *Langmuir* **2009**, *25*, 6997–7005. [CrossRef]
276. Goluch, E.D.; Shaw, A.W.; Sligar, S.G.; Liu, C. Microfluidic patterning of nanodisc lipid bilayers and multiplexed analysis of protein interaction. *Lab Chip* **2008**, *8*, 1723–1728. [CrossRef]
277. Seo, J.; Lee, S.; Poulter, C.D. Regioselective covalent immobilization of recombinant antibody-binding proteins A, G, and L for construction of antibody arrays. *J. Am. Chem. Soc.* **2013**, *135*, 8973–8980. [CrossRef]
278. Ikeda, T.; Hata, Y.; Ninomiya, K.-I.; Ikura, Y.; Takeguchi, K.; Aoyagi, S.; Hirota, R.; Kuroda, A. Oriented immobilization of antibodies on a silicon wafer using Si-tagged protein A. *Anal. Biochem.* **2009**, *385*, 132–137. [CrossRef]
279. Choe, W.; Durgannavar, T.A.; Chung, S.J. Fc-Binding Ligands of Immunoglobulin G: An Overview of High Affinity Proteins and Peptides. *Materials* **2016**, *9*, 994. [CrossRef]
280. Åkerström, B.; Björck, L. Protein L: An Immunoglobulin Light Chain-binding Bacterial Protein. *J. Biol. Chem.* **1989**, *264*, 19740–19746. [CrossRef]
281. Taniguchi, K.; Nomura, K.; Hata, Y.; Nishimura, T.; Asami, Y.; Kuroda, A. The Si-tag for immobilizing proteins on a silica surface. *Biotechnol. Bioeng.* **2007**, *96*, 1023–1029. [CrossRef]
282. Fukuyama, M.; Nishida, M.; Abe, Y.; Amemiya, Y.; Ikeda, T.; Kuroda, A.; Yokoyama, S. Detection of Antigen–Antibody Reaction Using Si Ring Optical Resonators Functionalized with an Immobilized Antibody-Binding Protein. *Jpn. J. Appl. Phys.* **2011**, *50*, 04DL07. [CrossRef]
283. Anderson, G.P.; Jacoby, M.A.; Ligler, F.S.; King, K.D. Effectiveness of protein A for antibody immobilization for a fiber optic biosensor. *Biosens. Bioelectron.* **1997**, *12*, 329–336. [CrossRef]
284. Knoglinger, C.; Zich, A.; Traxler, L.; Poslední, K.; Friedl, G.; Ruttmann, B.; Schorpp, A.; Müller, K.; Zimmermann, M.; Gruber, H.J. Regenerative biosensor for use with biotinylated bait molecules. *Biosens. Bioelectron.* **2018**, *99*, 684–690. [CrossRef]
285. Choi, H.W.; Takahashi, H.; Ooya, T.; Takeuchi, T. Label-free detection of glycoproteins using reflectometric interference spectroscopy-based sensing system with upright episcopic illumination. *Anal. Methods* **2011**, *3*, 1366. [CrossRef]
286. Lü, H.; Zhao, Y.; Ma, J.; Li, W.; Lu, Z. Characterization of DNA hybridization on the optical fiber surface. *Colloids Surf. A Physicochem. Eng. Asp.* **2000**, *175*, 147–152. [CrossRef]
287. Holmberg, A.; Blomstergren, A.; Nord, O.; Lukacs, M.; Lundeberg, J.; Uhlén, M. The biotin-streptavidin interaction can be reversibly broken using water at elevated temperatures. *Electrophoresis* **2005**, *26*, 501–510. [CrossRef] [PubMed]
288. Bonanno, L.M.; DeLouise, L.A. Whole blood optical biosensor. *Biosens. Bioelectron.* **2007**, *23*, 444–448. [CrossRef]
289. Bonanno, L.M.; Delouise, L.A. Steric crowding effects on target detection in an affinity biosensor. *Langmuir* **2007**, *23*, 5817–5823. [CrossRef]
290. Hermanson, G.T. Silane Coupling Agents. In *Bioconjugate Techniques*; Elsevier: Amsterdam, The Netherlands, 2008; pp. 562–581. ISBN 9780123705013.
291. Escorihuela, J.; Bañuls, M.J.; García Castelló, J.; Toccafondo, V.; García-Rupérez, J.; Puchades, R.; Maquieira, Á. Chemical silicon surface modification and bioreceptor attachment to develop competitive integrated photonic biosensors. *Anal. Bioanal. Chem.* **2012**, *404*, 2831–2840. [CrossRef]
292. De Vos, K.; Bartolozzi, I.; Schacht, E.; Bienstman, P.; Baets, R. Silicon-on-Insulator microring resonator for sensitive and label-free biosensing. *Opt. Express* **2007**, *15*, 7610–7615. [CrossRef]
293. Mudumba, S.; de Alba, S.; Romero, R.; Cherwien, C.; Wu, A.; Wang, J.; Gleeson, M.A.; Iqbal, M.; Burlingame, R.W. Photonic ring resonance is a versatile platform for performing multiplex immunoassays in real time. *J. Immunol. Methods* **2017**, *448*, 34–43. [CrossRef]
294. Ksendzov, A.; Lin, Y. Integrated optics ring-resonator sensors for protein detection. *Opt. Lett.* **2005**, *30*, 3344–3346. [CrossRef]
295. Byeon, J.-Y.; Limpoco, F.T.; Bailey, R.C. Efficient bioconjugation of protein capture agents to biosensor surfaces using aniline-catalyzed hydrazone ligation. *Langmuir* **2010**, *26*, 15430–15435. [CrossRef]
296. Arkles, B. *Silane Coupling Agents: Connecting Across Boundaries*, 3rd ed.; Gelest Inc.: Morrisville, PA, USA, 2014.
297. Zhang, F.; Srinivasan, M.P. Self-assembled molecular films of aminosilanes and their immobilization capacities. *Langmuir* **2004**, *20*, 2309–2314. [CrossRef] [PubMed]

298. Smith, E.A.; Chen, W. How to prevent the loss of surface functionality derived from aminosilanes. *Langmuir* **2008**, *24*, 12405–12409. [CrossRef] [PubMed]
299. Bañuls, M.-J.; González-Pedro, V.; Barrios, C.A.; Puchades, R.; Maquieira, A. Selective chemical modification of silicon nitride/silicon oxide nanostructures to develop label-free biosensors. *Biosens. Bioelectron.* **2010**, *25*, 1460–1466. [CrossRef] [PubMed]
300. Hermanson, G.T. Homobifunctional Crosslinkers. In *Bioconjugate Techniques*; Elsevier: Amsterdam, The Netherlands, 2013; pp. 275–298. ISBN 9780123822390.
301. Sheng, J.C.-C.; De La Franier, B.; Thompson, M. Assembling Surface Linker Chemistry with Minimization of Non-Specific Adsorption on Biosensor Materials. *Materials* **2021**, *14*, 472. [CrossRef]
302. Xia, B.; Xiao, S.-J.; Guo, D.-J.; Wang, J.; Chao, J.; Liu, H.-B.; Pei, J.; Chen, Y.-Q.; Tang, Y.-C.; Liu, J.-N. Biofunctionalisation of porous silicon (PS) surfaces by using homobifunctional cross-linkers. *J. Mater. Chem* **2006**, *16*, 570–578. [CrossRef]
303. Qavi, A.J.; Mysz, T.M.; Bailey, R.C. Isothermal discrimination of single-nucleotide polymorphisms via real-time kinetic desorption and label-free detection of DNA using silicon photonic microring resonator arrays. *Anal. Chem.* **2011**, *83*, 6827–6833. [CrossRef]
304. Qavi, A.J.; Meserve, K.; Aman, M.J.; Vu, H.; Zeitlin, L.; Dye, J.M.; Froude, J.W.; Leung, D.W.; Yang, L.; Holtsberg, F.W.; et al. Rapid detection of an Ebola biomarker with optical microring resonators. *Cell Rep. Methods* **2022**, *2*, 100234. [CrossRef]
305. Cardenosa-Rubio, M.C.; Graybill, R.M.; Bailey, R.C. Combining asymmetric PCR-based enzymatic amplification with silicon photonic microring resonators for the detection of lncRNAs from low input human RNA samples. *Analyst* **2018**, *143*, 1210–1216. [CrossRef]
306. Karyakin, A.A.; Presnova, G.V.; Rubtsova, M.Y.; Egorov, A.M. Oriented immobilization of antibodies onto the gold surfaces via their native thiol groups. *Anal. Chem.* **2000**, *72*, 3805–3811. [CrossRef]
307. Hanson, E.L.; Schwartz, J.; Nickel, B.; Koch, N.; Danisman, M.F. Bonding self-assembled, compact organophosphonate monolayers to the native oxide surface of silicon. *J. Am. Chem. Soc.* **2003**, *125*, 16074–16080. [CrossRef]
308. Hermanson, G.T. Chemoselective ligation: Bioorthogonal reagents. In *Bioconjugate Techniques*; Elsevier: Amsterdam, The Netherlands, 2013; pp. 757–785. ISBN 9780123822390.
309. SuSoS Functional Coatings for Glass, Polymeric, Metallic Surfaces. Available online: https://susos.com/products/ (accessed on 27 October 2022).
310. Andreatta, G.A.L.; Hendricks, N.R.; Grivel, A.; Billod, M. Grafted-to Polymeric Layers Enabling Highly Adhesive Copper Films Deposited by Electroless Plating on Ultra-Smooth Three-Dimensional-Printed Surfaces. *ACS Appl. Electron. Mater.* **2022**, *4*, 1864–1874. [CrossRef]
311. Weydert, S.; Zürcher, S.; Tanner, S.; Zhang, N.; Ritter, R.; Peter, T.; Aebersold, M.J.; Thompson-Steckel, G.; Forró, C.; Rottmar, M.; et al. Easy to Apply Polyoxazoline-Based Coating for Precise and Long-Term Control of Neural Patterns. *Langmuir* **2017**, *33*, 8594–8605. [CrossRef] [PubMed]
312. Garg, H.; Nair, P.R. Stochastic modeling of steric hindrance effects in biosensors. In Proceedings of the 2018 IEEE SENSORS, New Delhi, India, 28–31 October 2018; pp. 1–4.
313. González-Guerrero, A.B.; Alvarez, M.; García Castaño, A.; Domínguez, C.; Lechuga, L.M. A comparative study of in-flow and micro-patterning biofunctionalization protocols for nanophotonic silicon-based biosensors. *J. Colloid Interface Sci.* **2013**, *393*, 402–410. [CrossRef] [PubMed]
314. Arshak, K.; Korostynska, O.; Cunniffe, C. Nanopatterning using the bioforce nanoenabler. In *Functionalized Nanoscale Materials, Devices and Systems*; Vaseashta, A., Mihailescu, I.N., Eds.; NATO Science for Peace and Security Series B: Physics and Biophysics; Springer: Dordrecht, The Netherlands, 2008; pp. 299–304. ISBN 978-1-4020-8902-2.
315. ChipMaker 3 Microarray Printing Pins (90–100 Micron Features). Available online: https://shop.arrayit.com/chipmaker3 microspottingpin90-100micronfeatures.aspx (accessed on 6 October 2022).
316. Folch, A. *Introduction to BioMEMS*; CRC Press: Boca Raton, FL, USA, 2016; ISBN 9780429194047.
317. Kaigala, G.V.; Lovchik, R.D.; Drechsler, U.; Delamarche, E. A vertical microfluidic probe. *Langmuir* **2011**, *27*, 5686–5693. [CrossRef] [PubMed]
318. Ness, S.J.; Kim, S.; Woolley, A.T.; Nordin, G.P. Single-sided inkjet functionalization of silicon photonic microcantilevers. *Sensors and Actuators B: Chemical* **2012**, *161*, 80–87. [CrossRef]
319. Laplatine, L.; Fournier, M.; Gaignebet, N.; Hou, Y.; Mathey, R.; Herrier, C.; Liu, J.; Descloux, D.; Gautheron, B.; Livache, T. Silicon photonic olfactory sensor based on an array of 64 biofunctionalized Mach-Zehnder interferometers. *Opt. Express* **2022**, *30*, 33955–33968. [CrossRef] [PubMed]
320. Xia, Y.; Whitesides, G.M. Soft Lithography. *Angew. Chem. Int. Ed.* **1998**, *37*, 550–575. [CrossRef]
321. Kane, R.S.; Takayama, S.; Ostuni, E.; Ingber, D.E.; Whitesides, G.M. Patterning proteins and cells using soft lithography. *Biomaterials* **1999**, *20*, 2363–2376. [CrossRef]
322. Graber, D.J.; Zieziulewicz, T.J.; Lawrence, D.A.; Shain, W.; Turner, J.N. Antigen binding specificity of antibodies patterned by microcontact printing. *Langmuir* **2003**, *19*, 5431–5434. [CrossRef]
323. Lange, S.A.; Benes, V.; Kern, D.P.; Hörber, J.K.H.; Bernard, A. Microcontact printing of DNA molecules. *Anal. Chem.* **2004**, *76*, 1641–1647. [CrossRef] [PubMed]
324. Thibault, C.; Le Berre, V.; Casimirius, S.; Trévisiol, E.; François, J.; Vieu, C. Direct microcontact printing of oligonucleotides for biochip applications. *J. Nanobiotechnol.* **2005**, *3*, 7. [CrossRef] [PubMed]

325. Wang, Y.; Goh, S.H.; Bi, X.; Yang, K.-L. Replication of DNA submicron patterns by combining nanoimprint lithography and contact printing. *J. Colloid Interface Sci.* **2009**, *333*, 188–194. [CrossRef] [PubMed]
326. Idil, N.; Hedström, M.; Denizli, A.; Mattiasson, B. Whole cell based microcontact imprinted capacitive biosensor for the detection of Escherichia coli. *Biosens. Bioelectron.* **2017**, *87*, 807–815. [CrossRef] [PubMed]
327. Buhl, M.; Traboni, S.; Körsgen, M.; Lamping, S.; Arlinghaus, H.F.; Ravoo, B.J. On surface O-glycosylation by catalytic microcontact printing. *Chem. Commun.* **2017**, *53*, 6203–6206. [CrossRef]
328. Qiu, S.; Ji, J.; Sun, W.; Pei, J.; He, J.; Li, Y.; Li, J.J.; Wang, G. Recent advances in surface manipulation using micro-contact printing for biomedical applications. *Smart Mater. Med.* **2021**, *2*, 65–73. [CrossRef]
329. Romanov, V.; Davidoff, S.N.; Miles, A.R.; Grainger, D.W.; Gale, B.K.; Brooks, B.D. A critical comparison of protein microarray fabrication technologies. *Analyst* **2014**, *139*, 1303–1326. [CrossRef]
330. Ong, J.K.Y.; Moore, D.; Kane, J.; Saraf, R.F. Negative printing by soft lithography. *ACS Appl. Mater. Interfaces* **2014**, *6*, 14278–14285. [CrossRef]
331. Yunus, S.; de Looringhe, C.D.C.; Poleunis, C.; Delcorte, A. Diffusion of oligomers from polydimethylsiloxane stamps in microcontact printing: Surface analysis and possible application. *Surf. Interface Anal.* **2007**, *39*, 922–925. [CrossRef]
332. Yang, L.; Shirahata, N.; Saini, G.; Zhang, F.; Pei, L.; Asplund, M.C.; Kurth, D.G.; Ariga, K.; Sautter, K.; Nakanishi, T.; et al. Effect of surface free energy on PDMS transfer in microcontact printing and its application to ToF-SIMS to probe surface energies. *Langmuir* **2009**, *25*, 5674–5683. [CrossRef]
333. Fluent®Workstation-Tecan. Available online: https://www.tecan.com/fluent-automated-workstation?utm_term=robotic%20pipette&utm_campaign=SO-Liquid+Handling&utm_source=adwords&utm_medium=ppc&hsa_net=adwords&hsa_tgt=kwd-1223309053168&hsa_ad=605529119671&hsa_acc=9279258943&hsa_grp=121576188140&hsa_mt=p&hsa_cam=12736641036&hsa_kw=robotic%20pipette&hsa_ver=3&hsa_src=g&gclid=Cj0KCQjw1vSZBhDuARIsAKZlijS14MpBw3DcUBTV9B0GbtZIDM8IKOPaTo58M70vwGtNy3tyXpkfbdsaAp9FEALw_wcB (accessed on 5 October 2022).
334. Pipetting Robot Using Electronic Pipettes for Liquid Handling-Andrew+ -. Available online: https://www.andrewalliance.com/pipetting-robot/ (accessed on 27 October 2022).
335. Pipetting Robots | INTEGRA. Available online: https://www.integra-biosciences.com/canada/en/pipetting-robots (accessed on 27 October 2022).
336. Pipetting Robot—Automated Pipettor Robots for Liquid Handling. Available online: https://hudsonrobotics.com/products/liquid-handling/solo-liquid-handling/ (accessed on 27 October 2022).
337. Austin, J.; Holway, A.H. Contact printing of protein microarrays. *Methods Mol. Biol.* **2011**, *785*, 379–394. [CrossRef]
338. Wu, D.; Song, L.; Chen, K.; Liu, F. Modelling and hydrostatic analysis of contact printing microarrays by quill pins. *International Journal of Mechanical Sciences* **2012**, *54*, 206–212. [CrossRef]
339. Arrayit 946 Microarray Pins—Printing Spotting Robot Automation Microfluidic Chip RT-PCR PCR DNA Sequencing. Available online: https://shop.arrayit.com/microarray_printing_spotting_pins.aspx (accessed on 6 October 2022).
340. Arrayit Stealth Microarray Pins and Printheads—Printing Spotting Robot Automation Microfluidic Chip RT-PCR PCR DNA Sequencing. Available online: https://shop.arrayit.com/microarray_pins_stealth.aspx (accessed on 6 October 2022).
341. Arrayit Green Microarray Pins and Printheads—Printing Spotting Robot Automation Microfluidic Chip RT-PCR PCR DNA Sequencing. Available online: https://shop.arrayit.com/greenmicroarraypins.aspx (accessed on 6 October 2022).
342. Delamarche, E.; Bernard, A.; Schmid, H.; Michel, B.; Biebuyck, H. Patterned delivery of immunoglobulins to surfaces using microfluidic networks. *Science* **1997**, *276*, 779–781. [CrossRef] [PubMed]
343. Silzel, J.W.; Dodson, C.; Obremski, R.J.; Tsay, T. Mass-Sensing Multianalyte Micro-Array Based Immunoassay by Direct Nir Fluorescence and Quantitative Image Analysis. In *Micro Total Analysis Systems '98*; Harrison, D.J., van den Berg, A., Eds.; Springer: Dordrecht, The Netherlands, 1998; pp. 65–68. ISBN 978-94-010-6225-1.
344. SCIENION | Global Leader in Precision Dispensing | SCIENION. Available online: https://www.scienion.com/ (accessed on 31 October 2022).
345. Micro-Dispensing and Inkjet Printing Applications—Microdrop Technologies GmbH. Available online: https://www.microdrop.de/ (accessed on 31 October 2022).
346. Delehanty, J.B.; Ligler, F.S. Method for printing functional protein microarrays. *BioTechniques* **2003**, *34*, 380–385. [CrossRef] [PubMed]
347. Melnik, E.; Strasser, F.; Muellner, P.; Heer, R.; Mutinati, G.C.; Koppitsch, G.; Lieberzeit, P.; Laemmerhofer, M.; Hainberger, R. Surface modification of integrated optical MZI sensor arrays using inkjet printing technology. *Procedia Eng.* **2016**, *168*, 337–340. [CrossRef]
348. Juncker, D.; Schmid, H.; Delamarche, E. Multipurpose microfluidic probe. *Nat. Mater.* **2005**, *4*, 622–628. [CrossRef]
349. Queval, A.; Ghattamaneni, N.R.; Perrault, C.M.; Gill, R.; Mirzaei, M.; McKinney, R.A.; Juncker, D. Chamber and microfluidic probe for microperfusion of organotypic brain slices. *Lab Chip* **2010**, *10*, 326–334. [CrossRef]
350. Seo, J.; Poulter, C.D. Sandwich antibody arrays using recombinant antibody-binding protein L. *Langmuir* **2014**, *30*, 6629–6635. [CrossRef]
351. Randhawa, A. Exploring the Microfluidic Organ-On-Chip Platform for Aerosol Exposure Study. Master's Thesis, University of British Columbia, Vancouver, BC, Canada, 2022.

352. Squires, T.M.; Messinger, R.J.; Manalis, S.R. Making it stick: Convection, reaction and diffusion in surface-based biosensors. *Nat. Biotechnol.* **2008**, *26*, 417–426. [CrossRef]
353. Luan, E.; Yun, H.; Laplatine, L.; Dattner, Y.; Ratner, D.M.; Cheung, K.C.; Chrostowski, L. Enhanced Sensitivity of Subwavelength Multibox Waveguide Microring Resonator Label-Free Biosensors. *IEEE J. Select. Topics Quantum Electron.* **2019**, *25*, 7300211. [CrossRef]
354. Mai, A.; Mai, C.; Steglich, P. From Lab-on-chip to Lab-in-App: Challenges towards silicon photonic biosensors product developments. *Results Opt.* **2022**, *9*, 100317. [CrossRef]
355. Koos, C.; Freude, W.; Lindenmann, N.; Leuthold, J. *Photonic wire bonds*, 2013.
356. Kappert, E.J.; Raaijmakers, M.J.T.; Tempelman, K.; Cuperus, F.P.; Ogieglo, W.; Benes, N.E. Swelling of 9 polymers commonly employed for solvent-resistant nanofiltration membranes: A comprehensive dataset. *J. Memb. Sci.* **2019**, *569*, 177–199. [CrossRef]

Review

Optical Methods for Label-Free Detection of Bacteria

Pengcheng Wang [1], Hao Sun [2], Wei Yang [2] and Yimin Fang [2,*]

[1] Jiangsu Province Key Laboratory of Anesthesiology, Xuzhou Medical University, Xuzhou 221004, China
[2] Key Laboratory of Cardiovascular & Cerebrovascular Medicine, School of Pharmacy, Nanjing Medical University, Nanjing 211166, China
* Correspondence: yfang@njmu.edu.cn

Abstract: Pathogenic bacteria are the leading causes of food-borne and water-borne infections, and one of the most serious public threats. Traditional bacterial detection techniques, including plate culture, polymerase chain reaction, and enzyme-linked immunosorbent assay are time-consuming, while hindering precise therapy initiation. Thus, rapid detection of bacteria is of vital clinical importance in reducing the misuse of antibiotics. Among the most recently developed methods, the label-free optical approach is one of the most promising methods that is able to address this challenge due to its rapidity, simplicity, and relatively low-cost. This paper reviews optical methods such as surface-enhanced Raman scattering spectroscopy, surface plasmon resonance, and dark-field microscopic imaging techniques for the rapid detection of pathogenic bacteria in a label-free manner. The advantages and disadvantages of these label-free technologies for bacterial detection are summarized in order to promote their application for rapid bacterial detection in source-limited environments and for drug resistance assessments.

Keywords: bacteria detection; dark-field microscopy; Raman spectroscopy; surface plasmon resonance; label-free; rapid detection

Citation: Wang, P.; Sun, H.; Yang, W.; Fang, Y. Optical Methods for Label-Free Detection of Bacteria. *Biosensors* **2022**, *12*, 1171. https://doi.org/10.3390/bios12121171

Received: 13 November 2022
Accepted: 13 December 2022
Published: 15 December 2022

1. Introduction

Bacteria are the most abundant, widely distributed, diverse microorganisms in nature and of a special type. After a long period of natural evolution, bacteria have established complex antagonistic or symbiotic relationships with various species [1]. Although most of the bacteria are harmless, bacterial and viral infections account for approximately 70% of all human pathogenic diseases [2]. Bacterial pathogens can be obtained from food, water, animals, and even clinical settings including hospitals and other healthcare facilities. Pathogenic bacteria such as Salmonella, *Escherichia coli* (*E. coli*), Staphylococcus, etc. are the main causes of foodborne illness, which poses a constant threat to food safety. Bacterial infection is considered to be a common and costly global public health problem [3,4]. Bacteria not only cause some specific diseases in the host, but also act as opportunistic pathogens. When the host's immunity is low, the immune barrier is destroyed, flora imbalance or bacterial translocation occurs, which releases many virulent factors causing the host infection [5,6]. Treatment with antibiotics is the most effective and frequently used solution to this problem. Nevertheless, with the increasing use of antibiotics, the emergence of bacterial resistance to antibiotics is rising, which reduces the effectiveness of antibiotics for bacterial infection treatment, leading to increasing morbidity, mortality, and medical costs. According to the World Health Organization, antibiotic resistance kills 700,000 people every year, and if this problem is not addressed, the number of deaths resulting from antibiotic resistance will increase to 10 million by 2050 [7]. At present, bacterial resistance has become an increasingly serious global challenge, as well as a worldwide concern to governments and society [8]. According to the U.S. Centers for Disease Control and Prevention, about 2.8 million infections in the U.S. each year are related to antimicrobial resistance, implying significantly increasing treatment times and costs as well as mortality from bacterial infections [9].

The effectiveness of antibiotic treatment can be largely retained with the rational use of antibiotics. Rapid identification of pathogens is particularly important in clinical diagnosis, not only to minimize risks to patients, but also to provide a basis for physicians to prescribe pathogen-specific antibiotics rather than broad-spectrum antibiotics to reduce irrational use of antibiotics. However, rapid bacterial detection is quite a challenging task due to the large variety of bacteria and severe interference from the complex matrix in the growth environment [10]. Traditional methods, such as bacterial culture, PCR, and enzyme-linked immunosorbent assay (ELISA) are frequently used, but these methods have their own disadvantages. The bacterial culture method is the golden standard method for bacterial detection, but it is quite time-consuming, and easily contaminated by non-target bacteria. Detection of some clinically relevant pathogens by this method can take up to five days to develop an adequate culture [11]. PCR is a molecular biology technique used to amplify specific nucleic acid fragments. It replicates nucleic acid exponentially at a very low concentration [12] to a detectable amount within hours. Therefore, it has been widely used in bacteria detection. However, contamination of the test sample and erroneous DNA amplification can lead to false positive or negative results. PCR is relatively expensive and takes hours which is not rapid enough for regular use in antibiotic prescription. Immunoassays rely on the specific reaction of antigens and antibodies and are also used for the detection of bacteria [13,14] but are less sensitive and require a large amount of clinical samples.

To overcome these difficulties, more sensitive and rapid methods for bacterial detection have been extensively studied. In recent years, applications based on biosensors, which are analytical devices that convert biological responses into measurable signals, have become increasingly widespread [15]. Such an application usually consists of three parts: (1) ligands attached to the surface of the biosensor to recognize the target through specific interactions; (2) a sensor that converts biometric identification generated on the sensor surface into quantifiable physical signals such as light, electricity, heat, and voltage, etc.; (3) a signal detector. Biosensors have become an important tool for the rapid, sensitive, and selective detection of microorganisms. These methods include biosensor-based electrochemical methods [16–25], fluorescence detection methods [20–26], and spectroscopy methods [27–39]. However, most of the biosensing methods require labeling of target objects for signal reading, which significantly increases the measurement time and cost. Moreover, the presence of dyes and labels tends to interfere with the normal physiological function of bacteria, which does not reflect the true state of the bacteria, especially in the evaluation of antibiotic resistance. Therefore, label-free methods are advantageous in rapid pathogen detection and drug resistance evaluation.

Compared with the labeling methods, which generally require a long incubation time, label-free approaches are much simpler, faster, and cost-effective, making them good candidates for rapid bacterial detection in clinical application. Efforts have been made in this direction, among which the optical methods, such as Raman spectroscopy and single-particle imaging approaches, are the most promising approaches due to their high sensitivity, simplicity, and low-cost for label-free detection of bacteria [40–42]. In this review, we describe the advantages and disadvantages of optical methods such as Raman spectroscopy, SPR, and dark-field microscopy for label-free detection of bacteria and their applications in clinical detection and drug resistance evaluation.

2. Surface Plasmon Resonance for Bacteria Detection

2.1. Principle of Surface Plasmon Resonance

A typical optical system of planar SPR is mainly composed of a polarized excitation light source, a prism and a glass sensor chip coated with a thin gold film (~50 nm). The incident light passes through the prism in total internal reflection mode. The reflected light significantly decreases at a specific angle (defined as the resonance angle), while the wave vector matches the surface plasma frequency of the gold film in the propagation direction as shown in Figure 1a. The shift of the SPR angle is very sensitive to the refractive index change at the metal–liquid interface, making it a powerful tool for real-time monitoring of molecular and particle binding at the interface in a label-free manner [43]. It has been used to analyze binding specificity between molecules [44–46], the concentration of target molecules [47,48], kinetic parameters of association and dissociation [49,50], etc. More recently, with the development of SPR microscopy as shown in Figure 1b, which can directly monitor the nanoscale motion of single bacteria at the interface, SPR microscopy has become a powerful tool for rapid drug resistance evaluation [51]. In contrast to conventional SPR biosensors such as BIAcore, which provide an average signal of the designed area on the surface of the sensor chip, SPR microscopy enables the detection of areas or particles of interest on the chip surface, facilitating the detection of bacteria at the single cell level This process can be accomplished by recording an SPR image of the chip surface with a charge-coupled device (CCD) or a complementary metal oxide semiconductor camera. In addition, high spatial resolution of the perceived surface can be obtained by introducing a lens or a high numerical aperture (NA) objective into the SPR image system to replace the prism [52,53]. In addition to SPR microscopy, the use of an SPR image to detect bacteria has also been widely reported. For example, Tripathi et al. [54] proposed coating the gold surface of traditional SPR biosensors with graphene to improve the adhesion of bacteria on the surface of the sensor and applied it to the detection of *Pseudomonas* and *Pseudomason*-like bacteria. Park et al. [55,56] immobilized antibodies onto the sensor chip via EDC mediated coupling and realized the label-free and highly sensitive detection of foodborne *Salmonella* at low PH (4.6) and high antibody concentrations (up to 1000 μg/mL).

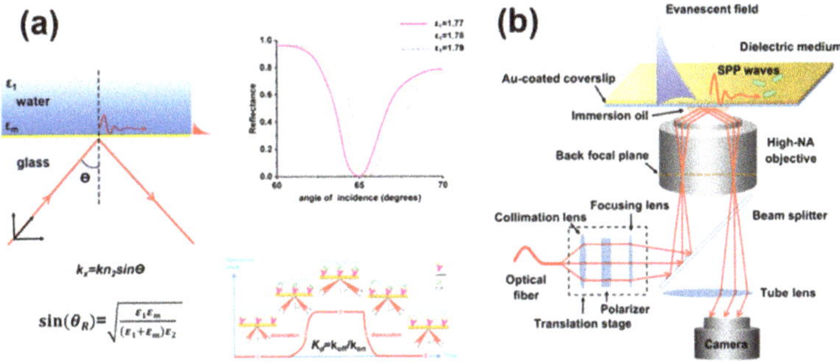

Figure 1. Schematic diagrams of (**a**) SPR optical system and (**b**) SPR microscopy.

2.2. Method and Application of SPR Technology for Label-Free Detection of Bacteria

The direct detection of bacteria by SPR requires specific antibodies against the target bacteria, which are immobilized on the surface of the gold film and specifically bind to the target bacteria to generate SPR signals. When the bacteria-containing solution flows to the sensor surface with specific antibody immobilization, the target bacteria bind to the gold film, which is then flushed to remove nonspecific interaction. As the SPR signal is positively correlated with the concentration of target bacteria, the number of target bacteria can be determined by setting up a calibration curve of bacterial concentration versus SPR signal intensity. The immobilization of antibodies on the sensor surface is a critical step for the detection of bacteria, which can improve the sensitivity and selectivity of bacterial SPR detection [57]. Physical adsorption and covalent binding are the main methods to fix the antibody on the sensor surface.

(i) Physical adsorption. Physical adsorption is a simple method of coating a surface that utilizes non-covalent bond interactions such as van der Waals forces, hydrogen bonds, electrostatic forces, and hydrophobic interactions to adsorb the target to the detection chip.

Capturing bacteria on the surface creates a refractive index (RI) change, and RI is used to quantify the presence and quantity of the bacteria. Jarvis et al. [58] used SPR technology to track in real time the attachment of *Pseudom onas aeruginosa* bacteria to bare gold film. This study showed that the adsorption of wild-type and mutant bacteria and the concentration of bacteria in bacterial suspension could be distinguished by physical adsorption. The results of this method were compared with those of crystal violet assay for different mutant bacteria, and it was found that there was qualitative correlation between them. Another method of physical adsorption of bacteria is to first modify hydrophobic or hydrophilic compounds or biologically active molecules on the surface of the gold chip, and then incubate the bacteria with the modified surface of the gold sheet, so that it can be adsorbed to the surface of the gold chip in a non-covalent interaction. Livache et al. [59] used pyrrole co-electropolymerization to attach different types of carbohydrates to the surface of gold film. Because different carbohydrate types have different physical adsorption capacities compared to the five closely related *E. coli* strains, different types of *E. coli* were incubated and grown on the substrates modified with different carbohydrate strains. SPR imaging was used to detect their interactions with bacteria during culture. This method can detect and identify tested bacteria from an initial bacterial concentration of 10^2 CFU/mL.

(II) Covalent immobilization. The measurement of SPR is based on the change of refractive index. However, because the gold film itself is not selective, it is not possible to distinguish the target in the complex mixture directly on the gold chip. SPR sensors specific to an analyte can be obtained by grafting an antibody that is specifically recognized by the analyte onto the surface of the gold chip. A reasonable method of immobilization of antibodies is to chemically conjugate antibodies to the surface of the sensor; immobilization of antibodies based on self-assembled monomolecular membrane (SAM) is the most studied method at present. SAM is an ordered single molecular structure formed by the adsorption of mercapto, amine, silane, or carboxylic acid components onto the solid surface in solution [43]. SAM can help control antibody binding direction, reduce nonspecific adsorption, and provide stable and directed analyte curing [60]. Thiolate compounds with different properties can easily be prepared with monolayers of different surface properties (such as wettability). SAM can be covalently bound to the primary amine of the ligand when it contains a carboxyl group at its end. This coupling is widely used for protein fixation. During the covalent binding of ligands, the non-specific binding of ligands on gold chips hinders the active functional groups in SAMs, which reduces the specificity. Therefore, a blocking agent, such as ethanolamine, is used to block the carboxyl groups remaining on the surface. In addition, bovine serum albumin is commonly used to block the gold surface to reduce the nonspecific interaction. Srikhirin et al. [61] developed an immunosensor based on SPR imaging using specific monoclonal antibody 11E5 (MAb 11E5) for the detection of seed-borne bacterium Acidovorax avenae subsp. citrulli (Aac). Aac was detected by self-assembly of MAb 11E5 mixed with monolayers (SAM). This method can be applied to multiplex detection, and it shows good selectivity for Aac with a limit of detection (LOD) of 10^6 CFU/mL. Evoyet et al. [62] used cysteine labeling and mercaptan chemistry to modify a specific caudate protein (tsp) on the surface of gold film for specific capture of *Salmonella typhi* with a detection limit of 10^3 CFU/mL. Chen et al. applied polyclonal anti-*E. coli* O157:H7 antibody to an NHS/EDC-activated surface by activating a SAM-coated chip with a mixture of NHS and EDC to generate an NHS ester receptor capable of binding to the amino group of the antibody via an amide bond [63]. Roupioz et al. used an antigen–antibody fixation method to modify the antibodies of a series of different bacteria in different regions of the gold sheet, and then cultured the advantages of this microarray on the chip with contaminated food. The culture of the bacteria results in an increase in the concentration of the target bacteria around the specific antibody, and then surface plasma resonance imaging is used to detect the growth of the bacteria. This single-step assay method enabled multiplex testing of *Cronobacterium* and *Salmonella* in less than a day and demonstrated that both bacteria were detected in 25 g of milk powder with as few as 30 CFU cells [64].

Tao et al. modified the gold chip with a layer of PEG/PEG-COOH self-assembled monomolecular layer, and then activated PEG-COOH by NHS and EDC to generate NHS ester receptors that react with the primary amine group on the antibody by amide bonds. The polyclonal anti-*E. coli* O157: H7 IgG antibodies have been applied to NHS/EDC-activated surfaces so that bacteria can be specifically attached to the surface. By using SPR microscopy, the nanoscale-motion of bacteria can be sensitively monitored at the gold chip surface as show in Figure 2. As the nanoscale-motion of bacteria is related to their activities, Tao's group developed a culture-free antimicrobial susceptibility test (AST) by tracking the motion using SPR microscopy, facilitating rapid antimicrobial resistance testing [51].

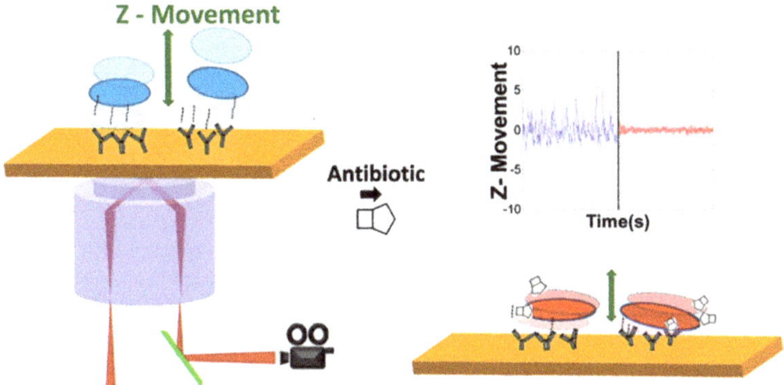

Figure 2. Schematic diagram of the rapid antimicrobial susceptibility test at single bacteria level using SPR microscopy [51]. Adapted with permission from Ref. [51]. Copyright © 2015 American Chemical Society.

In addition to the above commonly used bacterial label-free detection, new methods have also been developed in recent years for SPR methods for bacterial label-free detection. Culture–Capture–Measure (CCM): The protein is covalently bound to the pyrrole monomer on the chip, and then different types of antibodies are modified on the chip in the form of microarrays. Bacteria are cultured on the surface of the chip and then combined with sensitive SPR assays, which enables rapid and specific detection of bacteria on the protein microarrays. This culture–capture–measurement method can significantly reduce the processing steps of bacterial detection and the overall analysis time of bacterial detection [64–66]. For example, Thierry et al. combined microbial incubation on chips with SPR detection to achieve rapid specific detection of *Salmonella enterica* serovar Enteritidis, *Streptococcus pneumoniae* and *E. coli* O157:H7 cultured on protein microarrays [65]. Several methods have also been proposed to further improve the sensitivity: A highly sensitive sensor based on surface material modification was constructed by modifying nanomaterials [67–69] (graphene, molybdenum disulfide, barium titanate) or organic compounds [59] (carbohydrate) on the surface of a gold chip, which can significantly improve the sensitivity of bacterial detection. Livache et al. detected their interactions with bacteria by efficiently grafting simple carbohydrates onto the surface of a gold sheet and then using surface plasma resonance imaging during the process of culturing the bacteria on the surface. It was found that each type of bacteria interacts with carbohydrate chips in different ways. Compared with the detection limit of 1.0×10^4 CFU/mL for other electrochemical methods, the detection limit of this method can reach 1.2×10^2 CFU/mL [59].

Besides the antibodies, the surface of the gold chip is modified with small molecules such as bacteriophages, polymyxin B, aptamers, etc., as a bacterial identification element [69–71]. For example, Michel Meunier et al. used l-cysteine SAM to coat a gold sheet, and then linked the T4 bacteriophage and BP14 bacteriophage to the self-assembled membrane respectively to specifically detect *E. coli* and Methicillin-resistant *S. aureus*. This method does not require the prior step of labeling or enriching bacteria and can detect concentrations of 10^3 CFU/mL in less than 20 min [69]. In addition, the target bacteria were isolated and purified from complex samples by magnetic separation technology before SPR detection. Veli et al. developed a rapid and efficient magnetic separation step followed by the rapid detection of *B. melitensis* contamination in milk samples by SPR. Two aptamers with high affinity and specificity for *B. malitensis* were selected by a complete bacteria-SELEX procedure. The high-affinity aptamer (B70 aptamer) was immobilized on the surface of magnetic silica core-shell nanoparticles for the initial purification of target bacterial cells from the milk matrix. Another aptamer with high specificity for *B. melitensis* cells (B46 aptamer) was used to prepare SPR sensor chips for the sensitive determination of Brucella in magnetic purification eluted samples. This method can rapidly detect *B. melitensis* contamination in 1 mL milk samples by SPR, with LOD values as low as 27 ± 11 cells [72].

3. Raman Spectroscopy for Pathogen Bacteria Detection

3.1. The Principle of Raman Spectroscopy

Raman scattering can be defined as the inelastic scattering of photons from molecules. For every 10^6 photons scattered from the molecules, approximately one photon is inelastically scattered (Raman scattering). The detection of inelastic scattering photons from a molecule produces a spectrum of Raman shifts by the acquisition of energy differences from incident light. Each Raman shift corresponds to a specific vibration mode of molecular bonds, thus allowing molecular identification based on a specific vibrational fingerprint. Compared to fluorescence spectroscopy, Raman spectroscopy has higher resolution and narrower bandwidth, making it easy for the multiplex detection of different analytes. An advantage of Raman spectroscopy for bacterial detection is that Raman scattering can occur at any wavelength. This allows free choice of the excitation wavelength to meet the needs of biological Raman spectroscopy acquisition, especially in reducing the significant background from fluorescence. Raman excitation using visible wavelengths can be integrated into standard light microscopes. This shorter wavelength excitation allows higher spatial resolution compared with infrared microscopy, allowing smaller sample volumes or even the detection of individual bacteria.

3.2. Label-Free Detection of Bacteria by Raman Spectroscopy

Raman spectroscopy has been used for many years to probe the biochemistry of various biomolecules, and more recently for disease detection. Specifically, Raman spectroscopy has been used to characterize bacteria in microbial colonies to detect their presence in smaller sample sizes with rapidity. However, most bacterial detection using RS relies on microspectral identification of reference strains or clinical isolates [73–75]. Raman mic rospectroscopy can detect bacterial cells in liquid suspensions, and it can identify bacteria directly from patient body fluids without culture. Sandra et al. conducted two studies in which isolation protocols from filtration [76] and centrifugation [77] were both developed to extract bacteria from patient sputum and urine, respectively. The type of causative strain was determined by Raman spectroscopy. By combining Raman spectroscopy with hierarchical cluster analysis (HCA), Jiirgen et al. directly detected individual bacterial cells from cerebrospinal fluid samples of meningococcal patients without any sample preparation steps [78].

The major limitation is that Raman scattering is extremely weak, resulting in relatively poor sensitivity compared with other optical methods such as autofluorescence and absorption [78]. This means that collecting vibrational spectra via spontaneously generated Raman photons requires extremely sensitive detection hardware, long exposure times, and relatively high excitation power compared to other optical techniques. In recent years, surface-enhanced Raman spectroscopy (SERS) has been extensively studied in the detection of chemical and biological agents with its rapid and ultra-sensitive characteristics [79,80].

3.3. Label-Free Detection of Bacteria by SERS

SERS is a combination of Raman spectroscopy and nanotechnology. It retains the advantages of fast acquisition of RS, less sample consumption, and fingerprint spectra for specific analytes. In addition, SERS significantly enhances the sensitivity of Raman spectroscopy over several orders, thus reducing the interference from self-fluorescence. The weak Raman scattering intensity of the sample is greatly enhanced by placing the sample on the nanoscale rough noble metal surface or mixing the sample with the noble metal colloidal suspension. In SERS, the average enhancement coefficient was between 10^4 and 10^8, and it could reach 10^{11} in some cases [81–84].

The SERS effect can be explained by two enhancement mechanisms: electromagnetic and chemical. The former is the enhancement of electromagnetic field due to local surface plasmon resonance (LSPR) [85,86], while the latter is chemical enhancement due to the charge transfer process between metal nanoparticles and analytes [86], although the contribution of this mechanism has been shown to be much lower than that of electromagnetic enhancement. Two SERS methods have been developed, the label-based method and the label-free method. However, despite the high sensitivity, label-based methods only provide information about reporter molecules and lose the intrinsic information of bacterial cells. The accuracy of the label-based method is entirely dependent on the specificity of recognition molecules. In addition, the labeling will significantly increase the sample analysis time. Compared with the label-based SERS method, the label-free method is rapid and easy to operate without any external labeling [87]. Label-free methods can detect bacteria by measuring the SERS pattern inherent in the cell wall, allowing for direct bacteria identification. However, the sensitivity of the label-free SERS method largely depends on the SERS substrate, the bacterial species, and the sample preparation methods.

Noble metal nanoparticles such as gold and silver are the usually preferred light intensifiers in SERS. The plasmonic characteristics of these noble metal nanoparticles, namely LSPR and the electromagnetic field generated on the surface, are mainly determined by the size, shape, and mutual assembly of the metal nanoparticles and the dielectric properties of the surrounding medium [88]. In general, silver-based SERS substrates have higher SERS enhancement effects than gold. However, silver is less stable, and has a biotoxic effect on living organisms, which limits its application in living organisms. Gold is much more stable, strongly chemical inert, and less biotoxic than silver. The nanostructure of gold is stable, facilitating better control of the size and shape of particles with higher biocompatibility. In order to achieve highly sensitive and repeatable SERS detection, the size, shape, and stability of metal particles should be reasonably controlled. The aggregate of nanoparticles was found to exhibit a larger Raman-enhanced signal than individual nanoparticles due to the generation of hot spots in the gaps between nanoparticles. Additionally, the nanostructure with sharp tips can also significantly enhance the SERS intensity. The generation of hot spots is highly sensitive to the size, shape, and gap-distance of nanoparticles [87]. Therefore, top-down lithography methods and bottom-up self-assembly methods have been developed to control the shapes, arrangements, and assemblies of nanoparticles [89–91].

In general, there are several strategies that have been developed for the direct label-free SERS detection of bacteria, which are summarized as follows.

3.3.1. In Situ Formation of Colloidal Silver/Gold on the Surface of Bacteria

The common methods for forming colloidal silver/gold on the surface or inside of bacteria are achieved by soaking the bacteria in sodium borohydride solution, then resuspending in silver nitrate or chloroauric acid ($HAuCl_4$). The metal ions outside the cell wall react with reducing agents released from the cell, resulting in the colloids formation on the cell wall. Tamitake et al. employed a focused near-infrared laser beam to capture individual bacteria in aqueous Ag nitrate; Ag nanoaggregates were generated on *E. coli* by an additional green laser beam stimulation. In this way, the Raman scattering signal of *E. coli* was obtained by the Raman tweezer technique at single cell level [92].

3.3.2. Direct Bacteria Detection on a Planar SERS Surface

The planar SERS substrate can be gold-plated glass slides with high roughness or self-assembled SERS active substrate through rational design. The bacterial suspension is dropped on the substrate and allowed to dry for bacterial detection [93–95]. Wang et al. [96] prepared Ag/AAO SERS substrates embedding Ag nanoparticles in anodic aluminum oxide (AAO) nanochannels. This substrate possesses high reproducibility, therefore can be further analyzed by principal component analysis (PCA), linear discriminant analysis (LDA), and support vector machine (SVM) to detect Staphylococcus Aureus (Gram-positive bacterium), Klebsiella Pneumoniae (Gram-negative bacterium), and Mycobacterium Smegmatis (Mycobacterium) and other bacteria, providing a good strategy for clinical microbial detection.

Andrei et al. reported that with the modification of anti-fimbrial antibodies onto the polyethylene glycol (OEG12) molecular layer on the amorphous hydrogenated silicon (a-Si:H) film. The fimbriated *E. coli* was specifically captured onto the surface as shown in Figure 3a. The positively charged gold nanorods (Au NRs) were attracted to the negatively charged *E. coli* on the film, facilitating the reading of the SERS signals. This method has high repeatability for the detection of bacteria, due to the uniform coverage of Au NRs on the bacterial membrane [97].

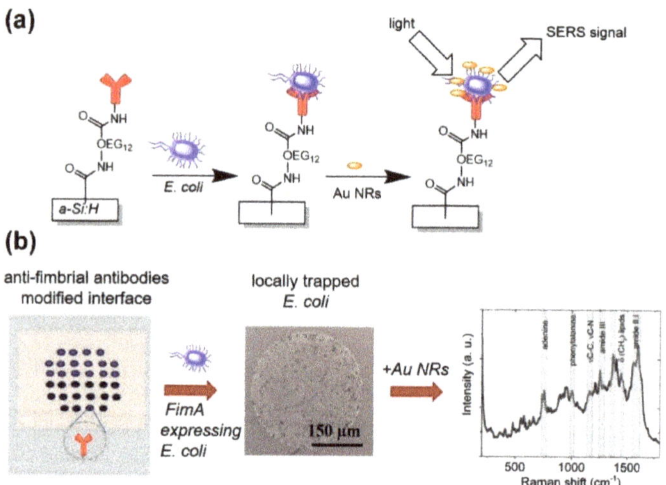

Figure 3. Schematic detection principle of *E. coli* hydrogenated amorphous silicon a-Si:H surface modified with anti-fimbrial antibodies against the major pilin protein fimA. (**a**) Surface structures of *E. coli* expressing fimA selectively captured and positively charged Au-NRs incubated with *E. coli* for SERS sensing. (**b**) Anti-fimbriae modified array, optical imaging of spots after interaction with *E. coli* and SERS spectra after capturing bacteria [97]. Adapted with permission from Ref. [97]. Copyright © 2020 Elsevier B.V.

Lv et al. used glycidyl methacrylate and ethylene dimethacrylate to prepare a convex substrate using a concave glass mold. The surface was treated with mercaptan to capture the Au nanoparticles on the surface as shown in Figure 4. The bacterial suspension is dropped on the SERS substrate, and the SERS spectrum of *E. coli* can be obtained after the sample dries naturally, as shown in Figure 4d. This simple SERS substrate preparation method proposed in this study was able to generate homogeneous and reproducible SERS active substrates over a large area, which has significantly improved the sensitivity of Raman spectroscopy. In this experiment, propanethiol, 3-mercaptopropionic acid, and cysteamine were modified on the surface of gold nanoparticles to improve the preferential adsorption ability of bacteria in very diluted thallus solution, while the SERS spectrum was used for the direct detection of the captured microorganisms as shown in Figure 4d [98].

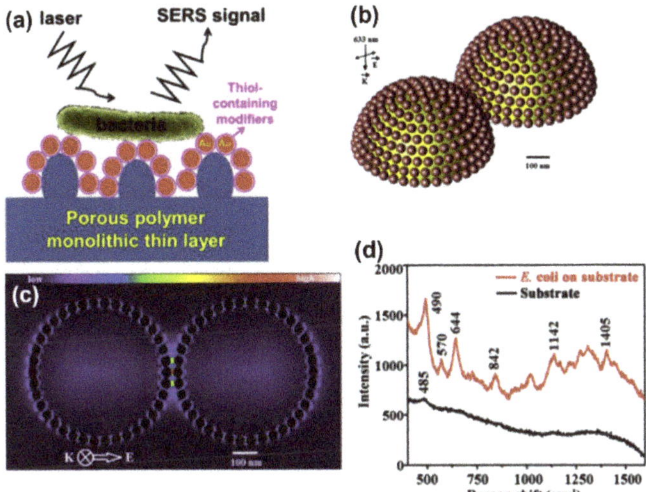

Figure 4. Schematic and detection principle of GNP/monolith modified substrate for the capture of *E. coli*. (**a**) Cross-sectional view of *E. coli* captured on gold nanoparticles modified substrates. (**b**) SERS enhancement factor of porous substrate functionalized with 40 nm gold nanoparticles simulated by FDTD. (**c**) In the simulation, the geometry of the model is reduced to two hemispheres coated with 40 nm spherical gold nanoparticles, separated by 10 nm; the electric field intensity distributions in x-y plane and y-z plane of gold on porous monolithic substrate excited by 633 nm laser are calculated. (**d**) SERS spectra of 40 nm gold nanoparticles/substrate functionalized with cysteamine [98]. Adapted with permission from Ref. [98]. Copyright © 2015 Elsevier B.V.

3.3.3. Direct Bacteria Detection in SERS Suspension

Bacteria detection can be achieved in the suspension by directly mixing the bacteria with colloid. By optimizing the volume ratio of bacterial suspension to colloidal silver. Davis et al. were able to detect *E. coli* as low as 10^3 CFU/mL by correcting the Raman spectrum of the wide vibrational OH band in water [99]. Jennifer developed a bacterial SERS detection platform that can detect bacteria in a controlled liquid environment that maintains the viability of bacteria in a liquid environment. Plasmon resonance nanorods with different longitudinal lengths were used to detect Gram-negative *E. coli*, *Staphylococcus epidermidis*, *Serratia marcescens*, and Gram-positive *S. aureus*. The SERS signal was much higher with the higher surface charge density of the bacteria, indicating that the higher SERS-enhanced signal comes from the electrostatic attraction between the positively charged nanorods and the negatively charged bacteria. This label-free liquid-SERS assay provides a promising strategy for bacterial identification and AST testing in living organisms [100].

4. Label-Free Detection of Bacteria by Dark-Field Microscopy

4.1. Dark-Field Microscopy Imaging Principle

Dark-field microscopy is a microscopy technique that obliquely illuminates a sample by attaching a circular opaque baffle to a condenser to prevent the incident light from directing into the camera [101]. When the incident light enters the condenser, the center part is blocked by the baffle, leaving the edge light to pass through. The annular beam formed by the incident light turns into a hollow conical beam after the light is concentrated through the condenser, and illuminates the sample, thus stimulating the scattering of sample particles. In this setting, only scattering light from objects in the medium enters the objective lens, creating a bright scattering pattern in a dark background [102]. Due to the

Tyndall effect, particles far below the resolution limit of typical light microscopes can be observed using dark-field microscopes [103].

4.2. Label and Label-Free Detection of Bacteria by Dark-Field Microscopy

Dark-field microscopy is an interesting optical technique that has been successfully used to image bacteria [40,104–114] and protozoa [102,115] due to its very low background, simple construction, portability, and low cost. Since plasma nanoparticles exhibit strong scattering to visible light, dark-field microscopy is a powerful tool for imaging and localization of noble metal nanoparticles in single cell analysis [101,109,116,117]. For example, hollow gold-silver nanoparticles are used as an alternative, less invasive contrast agent to assess the uptake process of malignant lymphocytes [118]. When the nanoparticles were modified by ligand and specifically bound to the cell membrane or internalized into the organelles, bright spots of different sizes and strengths could be observed on the surface of the target bacteria or around the organelles. Bacteria can be identified or counted based on the location and intensity of the bright spots. For example, Li et al. [104] developed a simple and fast bacterial count method based on dark-field light scattering imaging of a bacteria using gold nanoparticles as reporters. Zhou et al. [119] functionalized magnetic nanoparticles (MNP) using specific antibodies, which then formed a ring structure around *E. coli*, facilitating the counting of MNP conjugated *E. coli* under a dark-field microscope, as shown in Figure 5. In a similar way, Watanabe et al. [112] used phages as biometric elements, and aggregation-induced light scattering signals from silica nanospheres assembled by gold nanoparticles as signal transducers. After mixing the samples with the phage scattering probe of *S. aureus*, the detection limit of *S. aureus* was 8×10^4 CFU/mL within 15–20 min.

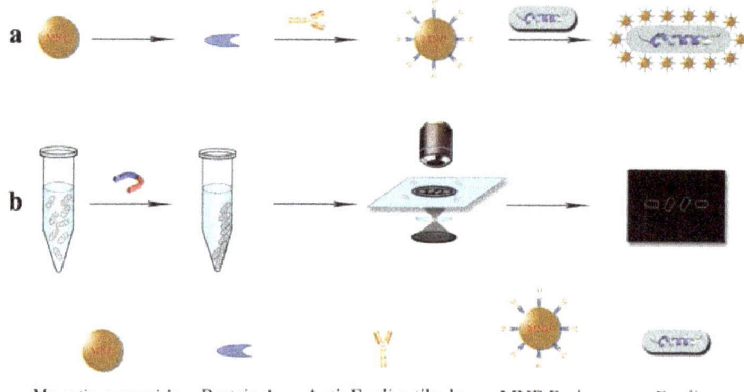

Figure 5. Schematic diagram of counting *E. coli* under dark-field, using antibody functionalization of MNP to form a gold ring structure around *E. coli*. (**a**) MNP probe was obtained by culture of *E. coli* antibody onto MNP. *E. coli* samples are first mixed with MNP probes to form probe-*E. coli* complexes. (**b**)The complex of *E. coli* and MNP probes was separated by a magnet and then counted under a dark-field microscope. [119]. Adapted with permission from Ref. [119]. Copyright © 2018 The Author(s).

Shiigi et al. [117] developed a novel molecular imprinting polymer (MIP) particle coated with gold nanoparticles (AuNPs) that can act as an acceptor and an optical signal transmitter in biological systems after modifying specific antibodies on its surface. Due to the coating of AuNPs, MIP particles produce a strong scattered light signal, and the binding of MIP particles increases the light intensity of the target bacteria. This allows bacteria to be clearly visible under darkfield microscopy, allowing them to be quantified using scattered light intensity. Using this technique, they successfully quantified *E. coli* O157 cells in meat samples.

Although powerful, the above-mentioned methods require the use of nanoparticles for signal reading of bacteria via dark-field microscopy, which affects the original physiological activity state of the bacteria detected and cannot reflect the real physiological activity and quantity of the bacteria [40,117,120–122]. Therefore, it is more desirable to detect bacteria in a label-free, rapid manner as the scattering intensity of bacteria is strong enough for direct dark-field imaging. In recent years, several methods have been developed to detect bacteria label-free using dark-field microscopy. For example, Colpo et al. [40] established a sensing platform for the rapid detection of bacteria in field samples using specific antibodies as recognition elements and dark-field microscopy as detection technology. By covering a gold layer on the polished silicon wafer and covalently modifying polyclonal anti-*E. Coli* antibodies to the surface, the sensing chip can be used for the specific capture of *E. coli* on the surface. As shown in Figure 6, the circularity and size of the object were used to identify the captured bacteria by dark-field microscopy. The performance was tested and compared to the Colilert-18 test and the quantitative polymerase chain reaction (qPCR), which showed comparable results.

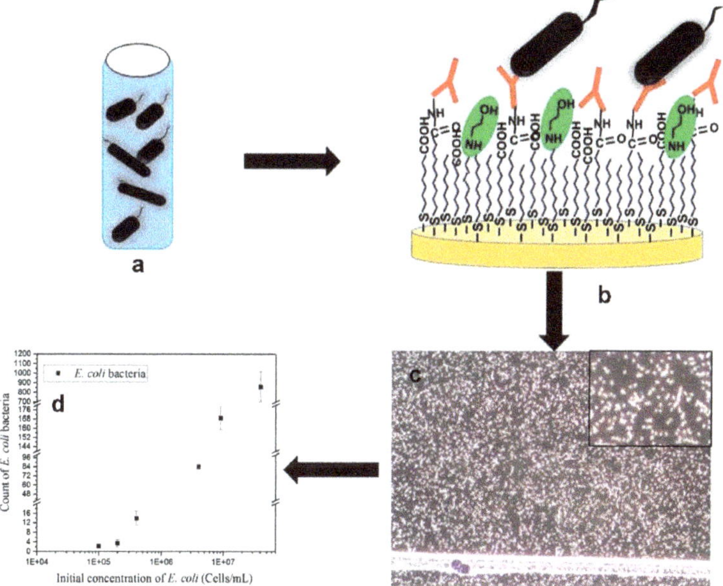

Figure 6. Schematic of detection of *E. coli* with dark-field microscopy. (**a**) Samples containing *E. coli*. (b) an anti-*E. coli* antibody functionalized gold surface. (**c**) Dark-field microscopy is used to inspect the surface of the gold sheet after 75 min incubation with the field sample and rinse with phosphate buffer solution, enlarging the image. (**d**) Statistical image analysis was used to count the bacteria captured by the antibodies [40]. Adapted with permission from Ref. [40]. Copyright © 2019 MDPI.

Creighton et al. identified Treponema Pallidum under optical microscopy with double-reflection and single-reflection dark-field condensers based on spirochetes of bacterial characteristic morphology and locomotion criteria. Ideally, this method can identify *Treponema Pallidum* using dark-field microscopy within 20 min [120].

Rapid diagnosis of bacterial infectious diseases has important clinical significance for rapid and rational use of antibiotics, so as to avoid the misuse of antibiotics. However, the detection of pathogenic bacteria generally requires molecular identification using antibodies or aptamers, which requires long incubation time, as well as complex sample pretreatment and signal amplification. To address this challenge, Fang [121] and Wang [122] used light scattering imaging methods to detect individual bacteria without labeling by the scattering intensity trajectory of particles in free solution. The scattering strength variation provides particle shape information because it is relevance to the morphological heterogeneity of the particle. The fluctuating pattern of the scattering intensity also depends on the shape and orientation of the particles in free solution, such as rod-shaped bacteria, whose scattering intensity fluctuates significantly higher than that of the spherical shape in free solution, which can be used to characterize the shape of the bacteria. Fang's group used label-free single-particle dark-field imaging for rapid and sensitive identification of bacteria in free solution by modulating the convection [121] as shown in Figure 7. Using this method, they were able to distinguish positive samples of streptococcus agalactiae from vaginal swabs within 10 min without the use of any biological reagents. In addition to the spherical shape bacteria, the optical characteristics of single bacteria with different shapes such as *E-coli* are also significantly different from the matrix, implying that the rapid detection of different types of bacteria in one clinical sample is plausible, facilitating the precise prescription of antibiotics.

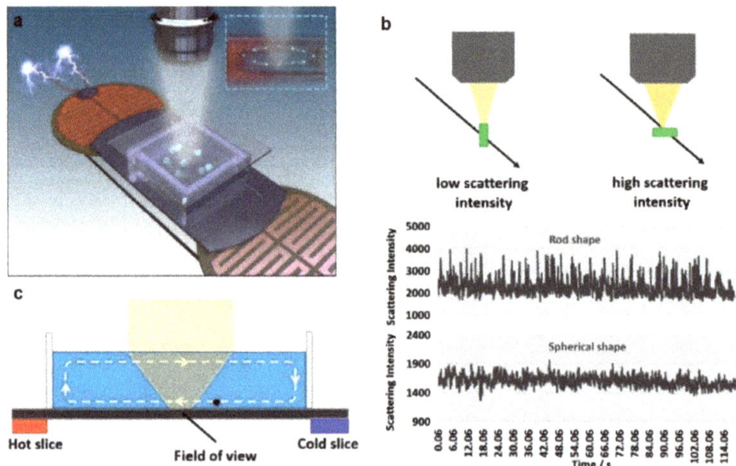

Figure 7. Bacteria detection principle by a single-particle imaging approach. (**a**) Schematic diagram of bacteria detection by single-particle imaging. (**b**) The inhomogeneity of particle morphology is identified by tracking the fluctuations of scattering intensity in free solution. (**c**) Convection induced by an electric heater was used to screen individual bacteria in a small field of view [121]. Adapted with permission from Ref. [121]. Copyright © 2022 The Author(s).

Similarly, Wang et al. used a large-volume solution scattering imaging (LVSi) system to track the scattering intensity and movement track of individual bacteria in short videos. The machine learning algorithm was used to perform aggregation analysis on their scattering intensity and movement trajectory. The presence of *E. coli* or similar bacteria in urine could be accurately determined, and bacteria could be distinguished from other common particles in urine, as shown in Figure 8. The method can detect patients with urinary tract infection within 10 min with an accuracy of 92.3%.

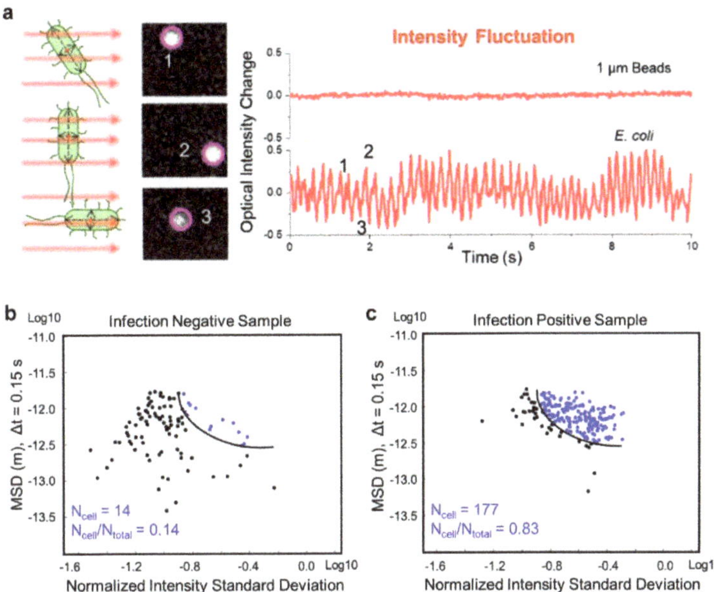

Figure 8. The principle of tracking the rapid identification of 1 um polystyrene spheres and single cell phenotypic characteristics of *E. coli*. (**a**) *E. coli* rotation-induced scattering intensity fluctuation tracking compared with 1 μm microbeads. (**b**) SVM classification result of one representative infection negative sample. (**c**) SVM classification result of one representative infection positive sample. [122]. Adapted with permission from Ref. [122]. Copyright © 2022 American Chemical Society.

5. Other Methods for Label-Free Detection of Bacteria

Other progress in the field of label-free optical biosensors is the advent of optical fiber gratings. Smietana [123] et al. first proposed a low-cost LGPs sensor that detects specific *E. coli* without labeling by physical adsorption. To further improve the sensitivity, Saurabh [124] proposed a compact ultra-sensitive long-period fiber grating (LPFGs) detection method for high-sensitivity label-free detection of specific *E. coli*. with modification of bacteriophage as shown in Figure 9. Simona [125] developed a reflective long-period fiber grating (RT-LPG) biosensor that can rapidly detect Class C β-lactamases in simple and complex biological samples. Additionally, fiber Bragg gratings (FBGs) can be used for bacterial detection [126,127].

Alternatively, the bacteria can be detected with the preparation of SERS hot spots on a fiber tip using optical fiber technology. The fiber-optic SERS probe (SERS on-a-tip) is highly controllable and reproducible [128–130].

Similarly, Biolayer interferometry (BLI) technology has been reported for bacterial detection in recent years. BLI is a label-free optical detection technique for real-time monitoring of biomolecular interactions. When an analyte binds to a ligand immobilized on the tip surface of a glass fiber-optic biosensor, its spectrum shifts with the change in the thickness of the related molecular layer. For example, Zhang et al. [131] reported a new method for the rapid, label-free real-time detection of *Salmonella enterica* using NLI incorporating antibodies as receptors, with a detection limit of 1.6×10^5 CFU/mL. Gu et al. [127] used C54A mutant LysGH15 as a receptor and combined it with BLI to establish a rapid, highly specific and label-free method for real-time detection of *Staphylococcus aureus* (*S. aureus*). This method can directly detect *S. aureus*, and its detection limit is 13·CFU/mL.

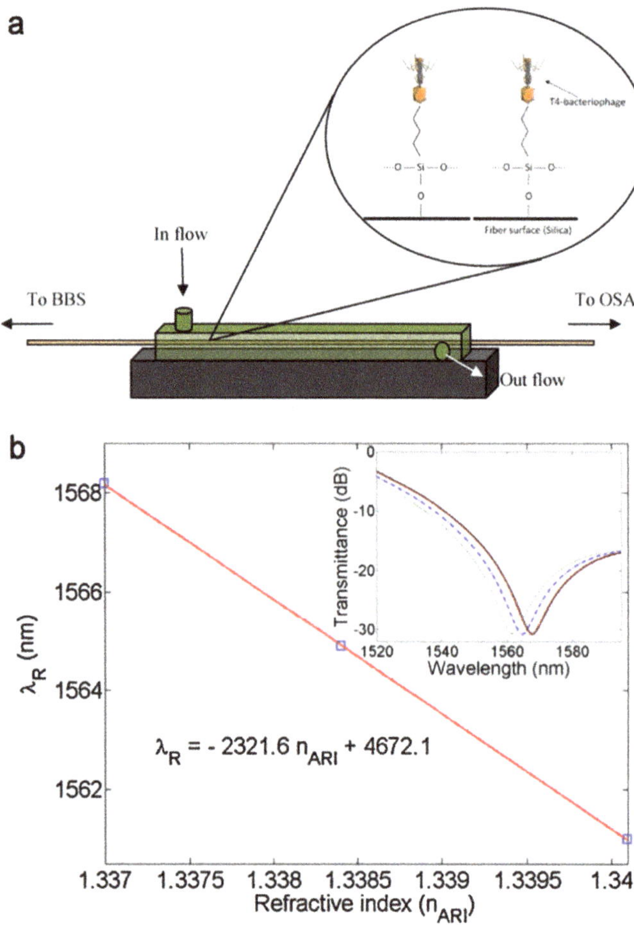

Figure 9. Schematic illustration of the experimental arrangement. (**a**) Covalent binding of phage to SiO_2 on fiber surface. (**b**) Resonance wavelength change with analyte refractive index transmission spectrum [124]. Adapted with permission from Ref. [124]. Copyright © 2012 Elsevier B.V.

6. Conclusions

In this paper, the applications of SPR, Raman spectroscopy and dark-field microscopy for the label-free detection of pathogenic bacteria are reviewed. The principle of SPR, Raman spectroscopy, dark-field microscopy as well as fiber-based methods for the label-free detection of pathogenic bacteria are considered. These label-free optical methods possess advantages of rapidity and low-cost, and are promising candidates for the clinical use for infectious disease diagnosis, facilitating the precise prescription of antibiotics to avoid the misuse of antibiotics, which is becoming a global problem.

The SPR imaging platform has been applied for high-throughput analysis, including the simultaneous detection of different bacterial species, antibiotic and bacterial interactions, etc. However, SPR generally suffers from the problem of non-specific adsorption and the direct detection of bacteria without sample preprocessing remains a challenge. Due to the high spatial resolution, SPR microscopy is able to image the bacteria at single cell level and possibly distinguish particles by their mass, and is potentially able to differentiate nonspecific adsorption. However, SPR microscopy is not commercially available, the total internal reflection fluorescence objective used for SPR microscopy is quite expensive. SERS is another label-free method for the rapid detection of bacteria with low cost based on the fingerprint vibration spectra. However, it is still a challenging task to detect bacteria in a label-free manner in complex biological environments. An obstacle lies in the large SERS background contributed from the complex matrix. Fortunately, recently developed machine learning methods are possibly to address this challenge. Direct detection of bacteria by dark-field microscopy on a substrate can be significant interfered with by nonspecific adsorption of other substances such as cell fragments and exosomes in the matrix, therefore, relatively few studies on the label-free detection of bacteria by this technique have been reported. However, the direct imaging of bacteria in free solution by dark-field microscopy is a unique approach reported recently which is quite promising in addressing this challenge due to it rapidity and low-cost, as it does not need any biological reagents or an incubation process. Despite the difficulties in differentiating bacteria with similar sizes and shapes, the recently developed image recognition and machine learning technologies are likely to address this challenge. Therefore, we believe that this dark-field imaging method for label-free bacterial detection in free solution will be widely used in bacterial detection, clinical diagnosis, and infectious disease control due to its high sensitivity, rapidity, simplicity, and low-cost.

Compared with the label-free optical methods, the paper based colorimetric methods have attracted increasing attention due to their simplicity and cost-effectiveness, as well as the rapid signal readout with the naked eye, making them a promising candidate for the development of point of care devices [132]. However, the colorimetric methods are largely compromised by relatively poor sensitivity. Signal amplification methods can be applied to further improve the sensitivity, but they require additional processes, which significantly increase the detection time. Therefore, we believe that the reagent-free dark-field imaging method for label-free bacterial detection in free solution is more advantageous and will be widely used in bacterial detection, clinical diagnosis, and infectious disease control due to its high sensitivity, rapidity, simplicity, and low-cost.

Author Contributions: Writing—original draft preparation, P.W.; writing—review and editing, H.S., W.Y. and Y.F.; funding acquisition, Y.F. All authors have read and agreed to the published version of the manuscript.

Funding: This research was funded by National Natural Science Foundation of China (NSFC, Grant No: 21874072, 22174069); and the Key R&D Program of Jiangsu Province (Grant No. BE2021373).

Acknowledgments: The authors thank the Key R&D Program of Jiangsu Province (Grant No. BE2021373) and National Natural Science Foundation of China (NSFC, Grant No: 21874072, 22174069) for funding supports.

Conflicts of Interest: The authors declare no conflict of interest.

References

1. Drlica, K.; Zhao, X. Bacterial death from treatment with fluoroquinolones and other lethal stressors. *Expert Rev. Anti-Infect. Ther.* **2021**, *19*, 601–618. [CrossRef] [PubMed]
2. Kaufmann, S.H.E.; Dorhoi, A.; Hotchkiss, R.S.; Bartenschlager, R. Host-directed therapies for bacterial and viral infections. *Nat. Rev. Drug Discov.* **2018**, *17*, 35–56. [CrossRef] [PubMed]
3. Havelaar, A.H.; Kirk, M.D.; Torgerson, P.R.; Gibb, H.J.; Hald, T.; Lake, R.J.; Praet, N.; Bellinger, D.C.; de Silva, N.R.; Gargouri, N.; et al. World Health Organization Global Estimates and Regional Comparisons of the Burden of Foodborne Disease in 2010. *PLoS Med.* **2015**, *12*, e1001923. [CrossRef]
4. Wu, X.; Han, C.; Chen, J.; Huang, Y.-W.; Zhao, Y. Rapid Detection of Pathogenic Bacteria from Fresh Produce by Filtration and Surface-Enhanced Raman Spectroscopy. *J. Miner.* **2016**, *68*, 1156–1162. [CrossRef]
5. Tsalik, E.L.; Bonomo, R.A.; Fowler, V.G., Jr. New Molecular Diagnostic Approaches to Bacterial Infections and Antibacterial Resistance. *Annu. Rev. Med.* **2018**, *69*, 379–394. [CrossRef] [PubMed]
6. Piano, S.; Bartoletti, M.; Tonon, M.; Baldassarre, M.; Chies, G.; Romano, A.; Viale, P.; Vettore, E.; Domenicali, M.; Stanco, M.; et al. Assessment of Sepsis-3 criteria and quick SOFA in patients with cirrhosis and bacterial infections. *Gut* **2018**, *67*, 1892–1899. [CrossRef]
7. Seale, A.C.; Gordon, N.C.; Islam, J.; Peacock, S.J.; Scott, J.A.G. AMR Surveillance in low and middle-income settings—A roadmap for participation in the Global Antimicrobial Surveillance System (GLASS). *Wellcome Open Res.* **2017**, *2*, 92. [CrossRef]

8. Pang, X.; Xiao, Q.; Cheng, Y.; Ren, E.; Lian, L.; Zhang, Y.; Gao, H.; Wang, X.; Leung, W.; Chen, X.; et al. Bacteria-Responsive Nanoliposomes as Smart Sonotheranostics for Multidrug Resistant Bacterial Infections. *ACS Nano* **2019**, *13*, 2427–2438. [CrossRef]
9. Maragakis, L.L.; Perencevich, E.N.; Cosgrove, S.E. Clinical and economic burden of antimicrobial resistance. *Expert Rev. Anti-Infect. Ther.* **2008**, *6*, 751–763. [CrossRef]
10. Grant, S.S.; Hung, D.T. Persistent bacterial infections, antibiotic tolerance, and the oxidative stress response. *Virulence* **2013**, *4*, 273–283. [CrossRef]
11. Sheikhzadeh, E.; Chamsaz, M.; Turner, A.P.F.; Jager, E.W.H.; Beni, V. Label-free impedimetric biosensor for Salmonella Typhimurium detection based on poly [pyrrole-co-3-carboxyl-pyrrole] copolymer supported aptamer. *Biosens. Bioelectron.* **2016**, *80*, 194–200. [CrossRef] [PubMed]
12. Chakvetadze, C.; Purcarea, A.; Pitsch, A.; Chelly, J.; Diamantis, S. Detection of Fusobacterium nucleatum in culture-negative brain abscess by broad-spectrum bacterial 16S rRNA Gene PCR. *IDCases* **2017**, *8*, 94–95. [CrossRef] [PubMed]
13. Guo, H.; Yuan, Y.H.; Niu, C.; Qiu, Y.; Wei, J.P.; Yue, T.L. Development of an indirect enzyme-linked immunosorbent assay for the detection of osmotolerant yeast Zygosaccharomyces rouxii in different food. *Food Agric. Immunol.* **2018**, *29*, 976–988. [CrossRef]
14. Jackson, N.; Wu, T.Z.; Adams-Sapper, S.; Satoorian, T.; Geisberg, M.; Murthy, N.; Lee, L.; Riley, L.W. A multiplexed, indirect enzyme-linked immunoassay for the detection and differentiation of *E. coli* from other Enterobacteriaceae and P. aeruginosa from other glucose non-fermenters. *J. Microbiol. Methods* **2019**, *158*, 52–58. [CrossRef]
15. Szunerits, S. Editorial Overview: Sensors and Biosensors (2021) Opening up the field to medical biosensors. *Curr. Opin. Electrochem.* **2021**, *30*, 100864. [CrossRef]
16. Das, R.; Chaterjee, B.; Kapil, A.; Sharma, T.K. Aptamer-NanoZyme mediated sensing platform for the rapid detection of Escherichia coli in fruit juice. *Sens. Bio-Sens. Res.* **2020**, *27*, 100313. [CrossRef]
17. Dong, X.; Shi, Z.; Xu, C.; Yang, C.; Chen, F.; Lei, M.; Wang, J.; Cui, Q. CdS quantum dots/Au nanoparticles/ZnO nanowire array for self-powered photoelectrochemical detection of *Escherichia coli* O157:H7. *Biosens. Bioelectron.* **2020**, *149*, 111843. [CrossRef]
18. Bai, H.; Bu, S.; Liu, W.; Wang, C.; Li, Z.; Hao, Z.; Wan, J.; Han, Y. An electrochemical aptasensor based on cocoon-like DNA nanostructure signal amplification for the detection of *Escherichia coli* O157:H7. *Analyst* **2020**, *145*, 7340–7348. [CrossRef]
19. Zhang, J.; Oueslati, R.; Cheng, C.; Zhao, L.; Chen, J.; Almeida, R.; Wu, J. Rapid, highly sensitive detection of Gram-negative bacteria with lipopolysaccharide based disposable aptasensor. *Biosens. Bioelectron.* **2018**, *112*, 48–53. [CrossRef]
20. Kaur, H.; Shorie, M.; Sharma, M.; Ganguli, A.K.; Sabherwal, P. Bridged Rebar Graphene functionalized aptasensor for pathogenic E. coil O78:K80:H11 detection. *Biosens. Bioelectron.* **2017**, *98*, 486–493. [CrossRef]
21. Cai, R.; Zhang, Z.; Chen, H.; Tian, Y.; Zhou, N. A versatile signal-on electrochemical biosensor for Staphylococcus aureus based on triple-helix molecular switch. *Sens. Actuators B-Chem.* **2021**, *326*, 128842. [CrossRef]
22. Ranjbar, S.; Shahrokhian, S. Design and fabrication of an electrochemical aptasensor using Au nanoparticles/carbon nanoparticles/cellulose nanofibers nanocomposite for rapid and sensitive detection of Staphylococcus aureus. *Bioelectrochemistry* **2018**, *123*, 70–76. [CrossRef] [PubMed]
23. Jiang, H.; Sun, Z.; Guo, Q.; Weng, X. Microfluidic thread-based electrochemical aptasensor for rapid detection of Vibrio parahaemolyticus. *Biosens. Bioelectron.* **2021**, *182*, 113191. [CrossRef] [PubMed]
24. Chen, Q.; Yao, C.; Yang, C.; Liu, Z.; Wan, S. Development of an in-situ signal amplified electrochemical assay for detection of Listeria monocytogenes with label-free strategy. *Food Chem.* **2021**, *358*, 129894. [CrossRef]
25. Pham, T.-T.D.; Phan, L.M.T.; Park, J.; Cho, S. Review-Electrochemical Aptasensor for Pathogenic Bacteria Detection. *J. Electrochem. Soc.* **2022**, *169*, 087501. [CrossRef]
26. Jones, L.M.; Dunham, D.; Rennie, M.Y.; Kirman, J.; Lopez, A.J.; Keim, K.C.; Little, W.; Gomez, A.; Bourke, J.; Ng, H.; et al. In vitro detection of porphyrin-producing wound bacteria with real-time fluorescence imaging. *Future Microbiol.* **2020**, *15*, 319–332. [CrossRef]
27. Sparks, W.B.; Hough, J.; Germer, T.A.; Chen, F.; DasSarma, S.; DasSarma, P.; Robb, F.T.; Manset, N.; Kolokolova, L.; Reid, N.; et al. Detection of circular polarization in light scattered from photosynthetic microbes. *Proc. Natl. Acad. Sci. USA* **2009**, *106*, 7816–7821. [CrossRef]
28. Martak, D.; Valot, B.; Sauget, M.; Cholley, P.; Thouverez, M.; Bertrand, X.; Hocquet, D. Fourier-Transform InfraRed Spectroscopy Can Quickly Type Gram-Negative Bacilli Responsible for Hospital Outbreaks. *Front. Microbiol.* **2019**, *10*, 1440. [CrossRef]
29. Bagcioglu, M.; Fricker, M.; Johler, S.; Ehling-Schulz, M. Detection and Identification of Bacillus cereus, Bacillus cytotoxicus, Bacillus thuringiensis, Bacillus mycoides and Bacillus weihenstephanensis via Machine Learning Based FTIR Spectroscopy. *Front. Microbiol.* **2019**, *10*, 902. [CrossRef]
30. Vogt, S.; Loeffler, K.; Dinkelacker, A.G.; Bader, B.; Autenrieth, I.B.; Peter, S.; Liese, J. Fourier-Transform Infrared (FTIR) Spectroscopy for Typing of Clinical Enterobacter cloacae Complex Isolates. *Front. Microbiol.* **2019**, *10*, 2582. [CrossRef]
31. Lechowicz, L.; Urbaniak, M.; Adamus-Bialek, W.; Kaca, W. The use of infrared spectroscopy and artificial neural networks for detection of uropathogenic *Escherichia coli* strains' susceptibility to cephalothin. *Acta Biochim. Pol.* **2013**, *60*, 713–718. [CrossRef] [PubMed]
32. Ayala, O.D.; Doster, R.S.; Manning, S.D.; O'Brien, C.M.; Aronoff, D.M.; Gaddy, J.A.; Mahadevan-Jansen, A. Raman microspectroscopy differentiates perinatal pathogens on ex vivo infected human fetal membrane tissues. *J. Biophotonics* **2019**, *12*, e201800449. [CrossRef] [PubMed]

33. Schroeder, U.-C.; Bokeloh, F.; O'Sullivan, M.; Glaser, U.; Wolf, K.; Pfister, W.; Popp, J.; Ducree, J.; Neugebauer, U. Rapid, culture-independent, optical diagnostics of centrifugally captured bacteria from urine samples. *Biomicrofluidics* **2015**, *9*, 044118. [CrossRef] [PubMed]
34. Ayala, O.D.; Wakeman, C.A.; Pence, I.J.; Gaddy, J.A.; Slaughter, J.C.; Skaar, E.P.; Mahadevan-Jansen, A. Drug-Resistant Staphylococcus aureus Strains Reveal Distinct Biochemical Features with Raman Microspectroscopy. *ACS Infect. Dis.* **2018**, *4*, 1197–1210. [CrossRef] [PubMed]
35. Neugebauer, U.; Schmid, U.; Baumann, K.; Holzgrabe, U.; Ziebuhr, W.; Kozitskaya, S.; Kiefer, W.; Schmitt, M.; Popp, J. Characterization of bacterial growth and the influence of antibiotics by means of UV resonance Raman spectroscopy. *Biopolymers* **2006**, *82*, 306–311. [CrossRef]
36. Wang, Y.; Lee, K.; Irudayaraj, J. Silver Nanosphere SERS Probes for Sensitive Identification of Pathogens. *J. Phys. Chem. C* **2010**, *114*, 16122–16128. [CrossRef]
37. Premasiri, W.R.; Chen, Y.; Williamson, P.M.; Bandarage, D.C.; Pyles, C.; Ziegler, L.D. Rapid urinary tract infection diagnostics by surface-enhanced Raman spectroscopy (SERS): Identification and antibiotic susceptibilities. *Anal. Bioanal. Chem.* **2017**, *409*, 3043–3054. [CrossRef]
38. Boardman, A.K.; Wong, W.S.; Premasiri, W.R.; Ziegler, L.D.; Lee, J.C.; Miljkovic, M.; Klapperich, C.M.; Sharon, A.; Sauer-Budge, A.F. Rapid Detection of Bacteria from Blood with Surface-Enhanced Raman Spectroscopy. *Anal. Chem.* **2016**, *88*, 8026–8035. [CrossRef]
39. Wang, J.; Wu, X.; Wang, C.; Rong, Z.; Ding, H.; Li, H.; Li, S.; Shao, N.; Dong, P.; Xiao, R.; et al. Facile Synthesis of Au-Coated Magnetic Nanoparticles and Their Application in Bacteria Detection via a SERS Method. *ACS Appl. Mater. Interfaces* **2016**, *8*, 19958–19967. [CrossRef]
40. La Spina, R.; António, D.C.; Desmet, C.; Valsesia, A.; Bombera, R.; Norlén, H.; Lettieri, T.; Colpo, P. Dark Field Microscopy-Based Biosensors for the Detection of *E. coli* in Environmental Water Samples. *Sensors* **2019**, *19*, 4652. [CrossRef]
41. Wang, B.; Park, B. Immunoassay Biosensing of Foodborne Pathogens with Surface Plasmon Resonance Imaging: A Review. *J. Agric. Food Chem.* **2020**, *68*, 12927–12939. [CrossRef] [PubMed]
42. Bodelón, G.; Montes-García, V.; Pérez-Juste, J.; Pastoriza-Santos, I. Surface-Enhanced Raman Scattering Spectroscopy for Label-Free Analysis of P. aeruginosa Quorum Sensing. *Front. Cell. Infect. Microbiol.* **2018**, *8*, 143. [CrossRef] [PubMed]
43. Zhang, P.; Chen, Y.P.; Wang, W.; Shen, Y.; Guo, J.S. Surface plasmon resonance for water pollutant detection and water process analysis. *Trac-Trends Anal. Chem.* **2016**, *85*, 153–165. [CrossRef]
44. Taghipour, P.; Zakariazadeh, M.; Sharifi, M.; Ezzati Nazhad Dolatabadi, J.; Barzegar, A. Bovine serum albumin binding study to erlotinib using surface plasmon resonance and molecular docking methods. *J. Photochem. Photobiol. B Biol.* **2018**, *183*, 11–15. [CrossRef]
45. Lewis, T.; Giroux, E.; Jovic, M.; Martic-Milne, S. Localized surface plasmon resonance aptasensor for selective detection of SARS-CoV-2 S1 protein. *Analyst* **2021**, *146*, 7207–7217. [CrossRef]
46. Xue, J.; Bai, Y.; Liu, H. Hybrid methods of surface plasmon resonance coupled to mass spectrometry for biomolecular interaction analysis. *Anal. Bioanal. Chem.* **2019**, *411*, 3721–3729. [CrossRef]
47. Stojanović, I.; Ruivo, C.F.; van der Velden, T.J.G.; Schasfoort, R.B.M.; Terstappen, L. Multiplex Label Free Characterization of Cancer Cell Lines Using Surface Plasmon Resonance Imaging. *Biosensors* **2019**, *9*, 70. [CrossRef]
48. Zhou, J.; Qi, Q.; Wang, C.; Qian, Y.; Liu, G.; Wang, Y.; Fu, L. Surface plasmon resonance (SPR) biosensors for food allergen detection in food matrices. *Biosens. Bioelectron.* **2019**, *142*, 111449. [CrossRef]
49. Choi, H.J.; Chung, B.H.; Kim, Y. Analysis of Protein-Protein Interactions by Surface Plasmon Resonance Imaging-based Microwell and Microfluidic Chip. *Bull. Korean Chem. Soc.* **2016**, *37*, 752–755. [CrossRef]
50. Day, C.J.; Poole, J.; Pluschke, G.; Jennings, M.P. Investigation of Mycobacterium ulcerans Glycan Interactions Using Glycan Array and Surface Plasmon Resonance. *Methods Mol. Biol.* **2022**, *2387*, 29–40.
51. Syal, K.; Iriya, R.; Yang, Y.; Yu, H.; Wang, S.; Haydel, S.E.; Chen, H.Y.; Tao, N. Antimicrobial Susceptibility Test with Plasmonic Imaging and Tracking of Single Bacterial Motions on Nanometer Scale. *ACS Nano* **2016**, *10*, 845–852. [CrossRef] [PubMed]
52. Shinohara, H.; Sakai, Y.; Mir, T.A. Real-time monitoring of intracellular signal transduction in PC12 cells by two-dimensional surface plasmon resonance imager. *Anal. Biochem.* **2013**, *441*, 185–189. [CrossRef] [PubMed]
53. Yanase, Y.; Hiragun, T.; Yanase, T.; Kawaguchi, T.; Ishii, K.; Hide, M. Evaluation of peripheral blood basophil activation by means of surface plasmon resonance imaging. *Biosens. Bioelectron.* **2012**, *32*, 62–68. [CrossRef] [PubMed]
54. Verma, A.; Prakash, A.; Tripathi, R. Performance analysis of graphene based surface plasmon resonance biosensors for detection of pseudomonas-like bacteria. *Opt. Quantum Electron.* **2015**, *47*, 1197–1205. [CrossRef]
55. Chen, J.; Park, B. Label-free screening of foodborne Salmonella using surface plasmon resonance imaging. *Anal. Bioanal. Chem.* **2018**, *410*, 5455–5464. [CrossRef]
56. Park, B.; Wang, B.; Chen, J. Label-Free Immunoassay for Multiplex Detections of Foodborne Bacteria in Chicken Carcass Rinse with Surface Plasmon Resonance Imaging. *Foodborne Pathog. Dis.* **2021**, *18*, 202–209. [CrossRef]
57. Skottrup, P.D.; Nicolaisen, M.; Justesen, A.F. Towards on-site pathogen detection using antibody-based sensors. *Biosens. Bioelectron.* **2008**, *24*, 339–348. [CrossRef]
58. Jenkins, A.T.A.; Ffrench-Constant, R.; Buckling, A.; Clarke, D.J.; Jarvis, K. Study of the attachment of Pseudomonas aeruginosa on gold and modified gold surfaces using surface plasmon resonance. *Biotechnol. Prog.* **2004**, *20*, 1233–1236. [CrossRef]

59. Bulard, E.; Bouchet-Spinelli, A.; Chaud, P.; Roget, A.; Calemczuk, R.; Fort, S.; Livache, T. Carbohydrates as New Probes for the Identification of Closely Related *Escherichia coli* Strains Using Surface Plasmon Resonance Imaging. *Anal. Chem.* **2015**, *87*, 1804–1811. [CrossRef]
60. Shankaran, D.R.; Gobi, K.V.A.; Miura, N. Recent advancements in surface plasmon resonance immunosensors for detection of small molecules of biomedical, food and environmental interest. *Sens. Actuators B-Chem.* **2007**, *121*, 158–177. [CrossRef]
61. Puttharugsa, C.; Wangkam, T.; Huangkamhang, N.; Gajanandana, O.; Himananto, O.; Sutapun, B.; Amarit, R.; Somboonkaew, A.; Srikhirin, T. Development of surface plasmon resonance imaging for detection of Acidovorax avenae subsp. citrulli (Aac) using specific monoclonal antibody. *Biosens. Bioelectron.* **2011**, *26*, 2341–2346. [CrossRef] [PubMed]
62. Singh, A.; Arya, S.K.; Glass, N.; Hanifi-Moghaddam, P.; Naidoo, R.; Szymanski, C.M.; Tanha, J.; Evoy, S. Bacteriophage tailspike proteins as molecular probes for sensitive and selective bacterial detection. *Biosens. Bioelectron.* **2010**, *26*, 131–138. [CrossRef] [PubMed]
63. Syal, K.; Wang, W.; Shan, X.; Wang, S.; Chen, H.Y.; Tao, N. Plasmonic imaging of protein interactions with single bacterial cells. *Biosens. Bioelectron.* **2015**, *63*, 131–137. [CrossRef] [PubMed]
64. Morlay, A.; Piat, F.; Mercey, T.; Roupioz, Y. Immunological detection of Cronobacter and Salmonella in powdered infant formula by plasmonic label-free assay. *Lett. Appl. Microbiol.* **2016**, *62*, 459–465. [CrossRef] [PubMed]
65. Bouguelia, S.; Roupioz, Y.; Slimani, S.; Mondani, L.; Casabona, M.G.; Durmort, C.; Vernet, T.; Calemczuk, R.; Livache, T. On-chip microbial culture for the specific detection of very low levels of bacteria. *Lab. Chip* **2013**, *13*, 4024–4032. [CrossRef]
66. Mondani, L.; Roupioz, Y.; Delannoy, S.; Fach, P.; Livache, T. Simultaneous enrichment and optical detection of low levels of stressed *Escherichia coli* O157:H7 in food matrices. *J. Appl. Microbiol.* **2014**, *117*, 537–546. [CrossRef]
67. Raikwar, S.; Prajapati, Y.K.; Srivastava, D.K.; Maurya, J.B.; Saini, J.P. Detection of Leptospirosis Bacteria in Rodent Urine by Surface Plasmon Resonance Sensor Using Graphene. *Photonic Sens.* **2021**, *11*, 305–313. [CrossRef]
68. Maurya, J.B.; Prajapati, Y.K.; Tripathi, R. Effect of Molybdenum Disulfide Layer on Surface Plasmon Resonance Biosensor for the Detection of Bacteria. *Silicon* **2018**, *10*, 245–256. [CrossRef]
69. Arya, S.K.; Singh, A.; Naidoo, R.; Wu, P.; McDermott, M.T.; Evoy, S. Chemically immobilized T4-bacteriophage for specific *Escherichia coli* detection using surface plasmon resonance. *Analyst* **2011**, *136*, 486–492. [CrossRef]
70. Tawil, N.; Sacher, E.; Mandeville, R.; Meunier, M. Surface plasmon resonance detection of *E. coli* and methicillin-resistant S. aureus using bacteriophages. *Biosens. Bioelectron.* **2012**, *37*, 24–29. [CrossRef]
71. Li, Y.; Zhu, J.; Zhang, H.; Liu, W.; Ge, J.; Wu, J.; Wang, P. High sensitivity gram-negative bacteria biosensor based on a small-molecule modified surface plasmon resonance chip studied using a laser scanning confocal imaging-surface plasmon resonance system. *Sens. Actuators B-Chem.* **2018**, *259*, 492–497. [CrossRef]
72. Dursun, A.D.; Borsa, B.A.; Bayramoglu, G.; Arica, M.Y.; Ozalp, V.C. Surface plasmon resonance aptasensor for Brucella detection in milk. *Talanta* **2022**, *239*, 123074. [CrossRef] [PubMed]
73. Rebrosova, K.; Siler, M.; Samek, O.; Ruzicka, F.; Bernatova, S.; Hola, V.; Jezek, J.; Zemanek, P.; Sokolova, J.; Petras, P. Rapid identification of staphylococci by Raman spectroscopy. *Sci. Rep.* **2017**, *7*, 14846. [CrossRef] [PubMed]
74. Willemse-Erix, D.F.M.; Scholtes-Timmerman, M.J.; Jachtenberg, J.-W.; van Leeuwen, W.B.; Horst-Kreft, D.; Schut, T.C.B.; Deurenberg, R.H.; Puppels, G.J.; van Belkum, A.; Vos, M.C.; et al. Optical Fingerprinting in Bacterial Epidemiology: Raman Spectroscopy as a Real-Time Typing Method. *J. Clin. Microbiol.* **2009**, *47*, 652–659. [CrossRef] [PubMed]
75. Ayala, O.D.; Wakeman, C.A.; Pence, I.J.; O'Brien, C.M.; Werkhaven, J.A.; Skaar, E.P.; Mahadevan-Jansen, A. Characterization of bacteria causing acute otitis media using Raman microspectroscopy. *Anal. Methods* **2017**, *9*, 1864–1871. [CrossRef] [PubMed]
76. Kloss, S.; Lorenz, B.; Dees, S.; Labugger, I.; Roesch, P.; Popp, J. Destruction-free procedure for the isolation of bacteria from sputum samples for Raman spectroscopic analysis. *Anal. Bioanal. Chem.* **2015**, *407*, 8333–8341. [CrossRef]
77. Kloss, S.; Kampe, B.; Sachse, S.; Roesch, P.; Straube, E.; Pfister, W.; Kiehntopf, M.; Popp, J. Culture Independent Raman Spectroscopic Identification of Urinary Tract Infection Pathogens: A Proof of Principle Study. *Anal. Chem.* **2013**, *85*, 9610–9616. [CrossRef]
78. Harz, M.; Kiehntopf, M.; Stoeckel, S.; Roesch, P.; Straube, E.; Deufel, T.; Popp, J. Direct analysis of clinical relevant single bacterial cells from cerebrospinal fluid during bacterial meningitis by means of micro-Raman spectroscopy. *J. Biophotonics* **2009**, *2*, 70–80. [CrossRef]
79. Hakonen, A.; Andersson, P.O.; Schmidt, M.S.; Rindzevicius, T.; Kall, M. Explosive and chemical threat detection by surface-enhanced Raman scattering: A review. *Anal. Chim. Acta* **2015**, *893*, 1–13. [CrossRef]
80. Popp, J.; Mayerhöfer, T. Surface-enhanced Raman spectroscopy. *Anal. Bioanal. Chem.* **2009**, *394*, 1717–1718. [CrossRef]
81. Nikoobakht, B.; Wang, J.P.; El-Sayed, M.A. Surface-enhanced Raman scattering of molecules adsorbed on gold nanorods: Off-surface plasmon resonance condition. *Chem. Phys. Lett.* **2002**, *366*, 17–23. [CrossRef]
82. Fromm, D.P.; Sundaramurthy, A.; Kinkhabwala, A.; Schuck, P.J.; Kino, G.S.; Moerner, W.E. Exploring the chemical enhancement for surface-enhanced Raman scattering with Au bowtie nanoantennas. *J. Chem. Phys.* **2006**, *124*, 61101. [CrossRef] [PubMed]
83. Zhao, L.L.; Jensen, L.; Schatz, G.C. Surface-enhanced Raman scattering of pyrazine at the junction between two Ag-20 nanoclusters. *Nano Lett.* **2006**, *6*, 1229–1234. [CrossRef] [PubMed]
84. Ko, H.; Singamaneni, S.; Tsukruk, V.V. Nanostructured Surfaces and Assemblies as SERS Media. *Small* **2008**, *4*, 1576–1599. [CrossRef]

85. Halas, N.J.; Lal, S.; Chang, W.-S.; Link, S.; Nordlander, P. Plasmons in Strongly Coupled Metallic Nanostructures. *Chem. Rev.* **2011**, *111*, 3913–3961. [CrossRef]
86. Schlücker, S. Surface-Enhanced Raman spectroscopy: Concepts and chemical applications. *Angew. Chem. Int. Ed.* **2014**, *53*, 4756–4795. [CrossRef]
87. Prucek, R.; Ranc, V.; Kvítek, L.; Panáček, A.; Zbořil, R.; Kolář, M. Reproducible discrimination between gram-positive and gram-negative bacteria using surface enhanced Raman spectroscopy with infrared excitation. *Analyst* **2012**, *137*, 2866–2870. [CrossRef]
88. Yu, R.; Liz-Marzán, L.M.; García de Abajo, F.J. Universal analytical modeling of plasmonic nanoparticles. *Chem. Soc. Rev.* **2017**, *46*, 6710–6724. [CrossRef]
89. Gwo, S.; Chen, H.Y.; Lin, M.H.; Sun, L.; Li, X. Nanomanipulation and controlled self-assembly of metal nanoparticles and nanocrystals for plasmonics. *Chem. Soc. Rev.* **2016**, *45*, 5672–5716. [CrossRef]
90. Hamon, C.; Liz-Marzán, L.M. Colloidal design of plasmonic sensors based on surface enhanced Raman scattering. *J. Colloid Interface Sci.* **2018**, *512*, 834–843. [CrossRef]
91. Mosier-Boss, P.A. Review of SERS Substrates for Chemical Sensing. *Nanomaterials* **2017**, *7*, 142. [CrossRef] [PubMed]
92. Kitahama, Y.; Itoh, T.; Ishido, T.; Hirano, K.; Ishikawa, M. Surface-Enhanced Raman Scattering from Photoreduced Ag Nanoaggregates on an Optically Trapped Single Bacterium. *Bull. Chem. Soc. Jpn.* **2011**, *84*, 976–978. [CrossRef]
93. Kogler, M.; Ryabchikov, Y.V.; Uusitalo, S.; Popov, A.; Popov, A.; Tselikov, G.; Valimaa, A.-L.; Al-Kattan, A.; Hiltunen, J.; Laitinen, R.; et al. Bare laser-synthesized Au-based nanoparticles as nondisturbing surface-enhanced Raman scattering probes for bacteria identification. *J. Biophotonics* **2018**, *11*, e201700225. [CrossRef] [PubMed]
94. Kahraman, M.; Yazici, M.M.; Sahin, F.; Bayrak, O.F.; Culha, M. Reproducible surface-enhanced Raman scattering spectra of bacteria on aggregated silver nanoparticles. *Appl. Spectrosc.* **2007**, *61*, 479–485. [CrossRef]
95. Zhu, A.; Ali, S.; Xu, Y.; Qin, O.; Wang, Z.; Chen, Q. SERS-based Au@Ag NPs Solid-phase substrate combined with chemometrics for rapid discrimination of multiple foodborne pathogens. *Spectrochim. Acta Part A-Mol. Biomol. Spectrosc.* **2022**, *270*, 120814. [CrossRef]
96. Liu, T.Y.; Chen, Y.; Wang, H.H.; Huang, Y.L.; Chao, Y.C.; Tsai, K.T.; Cheng, W.C.; Chuang, C.Y.; Tsai, Y.H.; Huang, C.Y.; et al. Differentiation of bacteria cell wall using Raman scattering enhanced by nanoparticle array. *J. Nanosci. Nanotechnol.* **2012**, *12*, 5004–5008. [CrossRef]
97. Andrei, C.-C.; Moraillon, A.; Lau, S.; Felidj, N.; Yamakawa, N.; Bouckaert, J.; Larquet, E.; Boukherroub, R.; Ozanam, F.; Szunerits, S.; et al. Rapid and sensitive identification of uropathogenic *Escherichia coli* using a surface-enhanced-Raman-scattering-based biochip. *Talanta* **2020**, *219*, 121174. [CrossRef]
98. Cao, Y.; Lv, M.; Xu, H.; Svec, F.; Tan, T.; Lv, Y. Planar monolithic porous polymer layers functionalized with gold nanoparticles as large-area substrates for sensitive surface-enhanced Raman scattering sensing of bacteria. *Anal. Chim. Acta* **2015**, *896*, 111–119. [CrossRef]
99. Sengupta, A.; Mujacic, M.; Davis, E.J. Detection of bacteria by surface-enhanced Raman spectroscopy. *Anal. Bioanal. Chem.* **2006**, *386*, 1379–1386. [CrossRef]
100. Tadesse, L.F.; Ho, C.S.; Chen, D.H.; Arami, H.; Banaei, N.; Gambhir, S.S.; Jeffrey, S.S.; Saleh, A.A.E.; Dionne, J. Plasmonic and Electrostatic Interactions Enable Uniformly Enhanced Liquid Bacterial Surface-Enhanced Raman Scattering (SERS). *Nano Lett.* **2020**, *20*, 7655–7661. [CrossRef]
101. Fakhrullin, R.; Nigamatzyanova, L.; Fakhrullina, G. Dark-field/hyperspectral microscopy for detecting nanoscale particles in environmental nanotoxicology research. *Sci. Total Environ.* **2021**, *772*, 145478. [CrossRef] [PubMed]
102. Belini, V.L.; Souza Freitas, B.L.; Sabogal-Paz, L.P.; Branco, N.; Bueno Franco, R.M. Label-Free Darkfield-Based Technique to Assist in the Detection of Giardia Cysts. *Water Air Soil Pollut.* **2018**, *229*, 195. [CrossRef]
103. Mock, J.J.; Barbic, M.; Smith, D.R.; Schultz, D.A.; Schultz, S. Shape effects in plasmon resonance of individual colloidal silver nanoparticles. *J. Chem. Phys.* **2002**, *116*, 6755–6759. [CrossRef]
104. Xu, X.; Chen, Y.; Wei, H.; Xia, B.; Liu, F.; Li, N. Counting Bacteria Using Functionalized Gold Nanoparticles as the Light-Scattering Reporter. *Anal. Chem.* **2012**, *84*, 9721–9728. [CrossRef]
105. Liao, X.-W.; Xu, Q.-Y.; Tan, Z.; Liu, Y.; Wang, C. Recent Advances in Plasmonic Nanostructures Applied for Label-free Single-cell Analysis. *Electroanalysis* **2022**, *34*, 923–936. [CrossRef]
106. Hernandez-Rodriguez, P.; Diaz, C.A.; Dalmau, E.A.; Quintero, G.M. A comparison between polymerase chain reaction (PCR) and traditional techniques for the diagnosis of leptospirosis in bovines. *J. Microbiol. Methods* **2011**, *84*, 1–7. [CrossRef]
107. Cameron, C.E. Leptospiral Structure, Physiology, and Metabolism. In *Leptospira and Leptospirosis*; Springer: Berlin/Heidelberg, Germany, 2015; Volume 387, pp. 21–41.
108. Gupta, S.; Jain, P.K.; Kumra, M.; Rehani, S.; Mathias, Y.; Gupta, R.; Mehendiratta, M.; Chander, A. Bacterial Viability within Dental Calculus: An Untrodden, Inquisitive Clinico-Patho- Microbiological Research. *J. Clin. Diagn. Res. JCDR* **2016**, *10*, ZC71-5. [CrossRef]
109. Rodriguez-Fajardo, V.; Sanz, V.; de Miguel, I.; Berthelot, J.; Acimovic, S.S.; Porcar-Guezenec, R.; Quidant, R. Two-color dark-field (TCDF) microscopy for metal nanoparticle imaging inside cells. *Nanoscale* **2018**, *10*, 4019–4027. [CrossRef]
110. Zheng, L.; Wen, Y.; Ren, W.; Duan, H.; Lin, J.; Irudayaraj, J. Hyperspectral dark-field microscopy for pathogen detection based on spectral angle mapping. *Sens. Actuators B-Chem.* **2022**, *367*, 132042. [CrossRef]

111. Benacer, D.; Woh, P.Y.; Zain, S.N.M.; Amran, F.; Thong, K.L. Pathogenic and Saprophytic Leptospira Species in Water and Soils from Selected Urban Sites in Peninsular Malaysia. *Microbes Environ.* **2013**, *28*, 135–140. [CrossRef]
112. Imai, M.; Mine, K.; Tomonari, H.; Uchiyama, J.; Matuzaki, S.; Niko, Y.; Hadano, S.; Watanabe, S. Dark-Field Microscopic Detection of Bacteria using Bacteriophage-Immobilized SiO2@AuNP Core-Shell Nanoparticles. *Anal. Chem.* **2019**, *91*, 12352–12357. [CrossRef] [PubMed]
113. Nakao, H.; Saito, K.; Tomita, S.; Magariyama, Y.; Kaizuka, Y.; Takeda, Y. Direct Imaging of Carboxymethyl Cellulose-mediated Aggregation of Lactic Acid Bacteria Using Dark-field Microscopy. *Anal. Sci.* **2016**, *32*, 1047–1051. [CrossRef] [PubMed]
114. Zaki, A.M.; Hod, R.; Shamsusah, N.A.; Isa, Z.M.; Bejo, S.K.; Agustar, H.K. Detection ofLeptospira kmetyiat recreational areas in Peninsular Malaysia. *Environ. Monit. Assess.* **2020**, *192*, 703. [CrossRef] [PubMed]
115. Fotso Fotso, A.; Drancourt, M. Laboratory Diagnosis of Tick-Borne African Relapsing Fevers: Latest Developments. *Front. Public Health* **2015**, *3*, 254. [CrossRef] [PubMed]
116. Wang, X.; Cui, Y.; Irudayaraj, J. Single-Cell Quantification of Cytosine Modifications by Hyperspectral Dark-Field Imaging. *ACS Nano* **2015**, *9*, 11924–11932. [CrossRef] [PubMed]
117. Tanabe, S.; Itagaki, S.; Sun, S.; Matsui, K.; Kinoshita, T.; Nishii, S.; Yamamoto, Y.; Sadanaga, Y.; Shiigi, H. Quantification of Enterohemorrhagic *Escherichia coli* via Optical Nanoantenna and Temperature-responsive Artificial Antibodies. *Anal. Sci.* **2021**, *37*, 1597–1601. [CrossRef] [PubMed]
118. Nagy-Simon, T.; Tatar, A.-S.; Craciun, A.-M.; Vulpoi, A.; Jurj, M.-A.; Florea, A.; Tomuleasa, C.; Berindan-Neagoe, I.; Astilean, S.; Boca, S. Antibody Conjugated, Raman Tagged Hollow Gold-Silver Nanospheres for Specific Targeting and Multimodal Dark-Field/SERS/Two Photon-FLIM Imaging of CD19(+) B Lymphoblasts. *ACS Appl. Mater. Interfaces* **2017**, *9*, 21155–21168. [CrossRef]
119. Xu, H.; Tang, F.; Dai, J.; Wang, C.; Zhou, X. Ultrasensitive and rapid count of *Escherichia coli* using magnetic nanoparticle probe under dark-field microscope. *BMC Microbiol.* **2018**, *18*, 100. [CrossRef]
120. Theel, E.S.; Katz, S.S.; Pillay, A. Molecular and Direct Detection Tests for Treponema pallidum Subspecies pallidum: A Review of the Literature, 1964–2017. *Clin. Infect. Dis.* **2020**, *71*, S4–S12. [CrossRef]
121. Chen, S.; Su, Y.-W.; Sun, J.; Chen, T.; Zheng, Y.; Sui, L.-J.; Yang, S.; Liu, C.; Wang, P.; Li, T.; et al. Label-free single-particle imaging approach for ultra-rapid detection of pathogenic bacteria in clinical samples. *Proc. Natl. Acad. Sci. USA* **2022**, *119*, e2206990119. [CrossRef]
122. Zhang, F.; Mo, M.; Jiang, J.; Zhou, X.; McBride, M.; Yang, Y.; Reilly, K.S.; Grys, T.E.; Haydel, S.E.; Tao, N.; et al. Rapid Detection of Urinary Tract Infection in 10 min by Tracking Multiple Phenotypic Features in a 30 s Large-Volume Scattering Video of Urine Microscopy. *ACS Sens.* **2022**, *7*, 2262–2272. [CrossRef] [PubMed]
123. Smietana, M.; Bock, W.J.; Mikulic, P.; Ng, A.; Chinnappan, R.; Zourob, M. Detection of bacteria using bacteriophages as recognition elements immobilized on long-period fiber gratings. *Opt. Express* **2011**, *19*, 7971–7978. [CrossRef] [PubMed]
124. Tripathi, S.M.; Bock, W.J.; Mikulic, P.; Chinnappan, R.; Ng, A.; Tolba, M.; Zourob, M. Long period grating based biosensor for the detection of *Escherichia coli* bacteria. *Biosens. Bioelectron.* **2012**, *35*, 308–312. [CrossRef] [PubMed]
125. Zuppolini, S.; Quero, G.; Consales, M.; Diodato, L.; Vaiano, P.; Venturelli, A.; Santucci, M.; Spyrakis, F.; Costi, M.P.; Giordano, M.; et al. Label-free fiber optic optrode for the detection of class C β-lactamases expressed by drug resistant bacteria. *Biomed. Opt. Express* **2017**, *8*, 5191–5205. [CrossRef]
126. Lo Presti, D.; Massaroni, C.; Jorge Leitao, C.S.; De Fatima Domingues, M.; Sypabekova, M.; Barrera, D.; Floris, I.; Massari, L.; Oddo, C.M.; Sales, S.; et al. Fiber Bragg Gratings for Medical Applications and Future Challenges: A Review. *IEEE Access* **2020**, *8*, 156863–156888. [CrossRef]
127. Srinivasan, R.; Umesh, S.; Murali, S.; Asokan, S.; Gorthi, S.S. Bare fiber Bragg grating immunosensor for real-time detection of *Escherichia coli* bacteria. *J. Biophotonics* **2017**, *10*, 224–230. [CrossRef]
128. Manago, S.; Quero, G.; Zito, G.; Tullii, G.; Galeotti, F.; Pisco, M.; De Luca, A.C.; Cusano, A. Tailoring lab-on-fiber SERS optrodes towards biological targets of different sizes. *Sens. Actuators B-Chem.* **2021**, *339*, 129321. [CrossRef]
129. Kim, J.A.; Wales, D.J.; Thompson, A.J.; Yang, G.-Z. Fiber-Optic SERS Probes Fabricated Using Two-Photon Polymerization For Rapid Detection of Bacteria. *Adv. Opt. Mater.* **2020**, *8*, 1901934. [CrossRef]
130. Hunter, R.; Sohi, A.N.; Khatoon, Z.; Berthiaume, V.R.; Alarcon, E.I.; Godin, M.; Anis, H. Optofluidic label-free SERS platform for rapid bacteria detection in serum. *Sens. Actuators B-Chem.* **2019**, *300*, 126907. [CrossRef]
131. Zang, C.-L.; Zhang, M.-D.; Zhang, Y.; Li, Y.-S.; Liu, K.; Xie, N.-N.; Sun, C.-Y.; Zhang, X.-G. Rapid label-free detection of Salmonella enterica with biolayer interferometry. *J. Food Saf.* **2021**, *41*, e12896. [CrossRef]
132. Kim, D.M.; Yoo, S.M. Colorimetric Systems for the Detection of Bacterial Contamination: Strategy and Applications. *Biosensors* **2022**, *12*, 532. [CrossRef] [PubMed]

 biosensors

Review

New Horizons for MXenes in Biosensing Applications

Decheng Lu [1], Huijuan Zhao [1,2], Xinying Zhang [1], Yingying Chen [1] and Lingyan Feng [1,3,*]

1 Department of Materials Genome Institute, Shanghai University, Shanghai 200444, China
2 Qing Wei Chang College, Shanghai University, Shanghai 200444, China
3 Shanghai Engineering Research Center of Organ Repair, Shanghai 200444, China
* Correspondence: lingyanfeng@t.shu.edu.cn

Abstract: Over the last few decades, biosensors have made significant advances in detecting non-invasive biomarkers of disease-related body fluid substances with high sensitivity, high accuracy, low cost and ease in operation. Among various two-dimensional (2D) materials, MXenes have attracted widespread interest due to their unique surface properties, as well as mechanical, optical, electrical and biocompatible properties, and have been applied in various fields, particularly in the preparation of biosensors, which play a critical role. Here, we systematically introduce the application of MXenes in electrochemical, optical and other bioanalytical methods in recent years. Finally, we summarise and discuss problems in the field of biosensing and possible future directions of MXenes. We hope to provide an outlook on MXenes applications in biosensing and to stimulate broader interests and research in MXenes across different disciplines.

Keywords: two-dimensional material; MXenes; biosensor; electrochemistry; optics

Citation: Lu, D.; Zhao, H.; Zhang, X.; Chen, Y.; Feng, L. New Horizons for MXenes in Biosensing Applications. *Biosensors* **2022**, *12*, 820. https://doi.org/10.3390/bios12100820

Received: 7 September 2022
Accepted: 28 September 2022
Published: 2 October 2022

1. Introduction

The main two-dimensional (2D) material is a solid crystal consisting of a single or several atomic layers, a sheet thickness of 1–10 Å, and a lateral size ranging from 100 nm to several µm [1]. Two-dimensional materials with properties such as large specific surface area and unique electronics are focuses of research in many research fields [2]. Since 2004, Novoselov et al. performed exfoliation to obtain graphene nanostructures; since then, the two-dimensional material has attracted much attention [3]. In 2011, Gogotsi et al. prepared a two-dimensional Ti_3C_2 nanosheet named MXenes [4]. MXenes are typically a few µm laterally and 1 nm thick or less [5]. It shows superior physicochemical properties compared to other two-dimensional nanomaterials [6].

The precursor of MXenes is the MAX phase. MAX consists of $M_{n+1}X_n$ units and an alternately stacked "A" element single atomic plane, expressed as $M_{n+1}AX_n$. The unique crystal structure of the MAX phase combines the excellent properties of ceramics and metals [7]. Etching the "A" element of the MAX phase yields two-dimensional nanomaterial MXenes with a structural formula of $M_{n+1}X_nT_x$ [8]. MXenes can be expressed as M_2XF_2, $M_2X(OH)_2$, M_2XO_2, etc. M is a transition metal; "A" is an element of Groups 13 and 14 of the periodic table; X is boron, carbon, or nitrogen; n includes integers from 1 to 3; T_x denotes surface groups [9] (Figure 1A,B). A list of the significant syntheses and processes in the field of MXenes research over the last decade, as well as the development of new MXenes core components and surface group control techniques, is illustrated in Figure 1C. Compared to the precursor MAX phase, derivative MXenes retain metallic and electrical conductivity benefits of MAX but also offer smaller lateral dimensions and thicknesses, as well as unique physical and chemical properties [10,11].

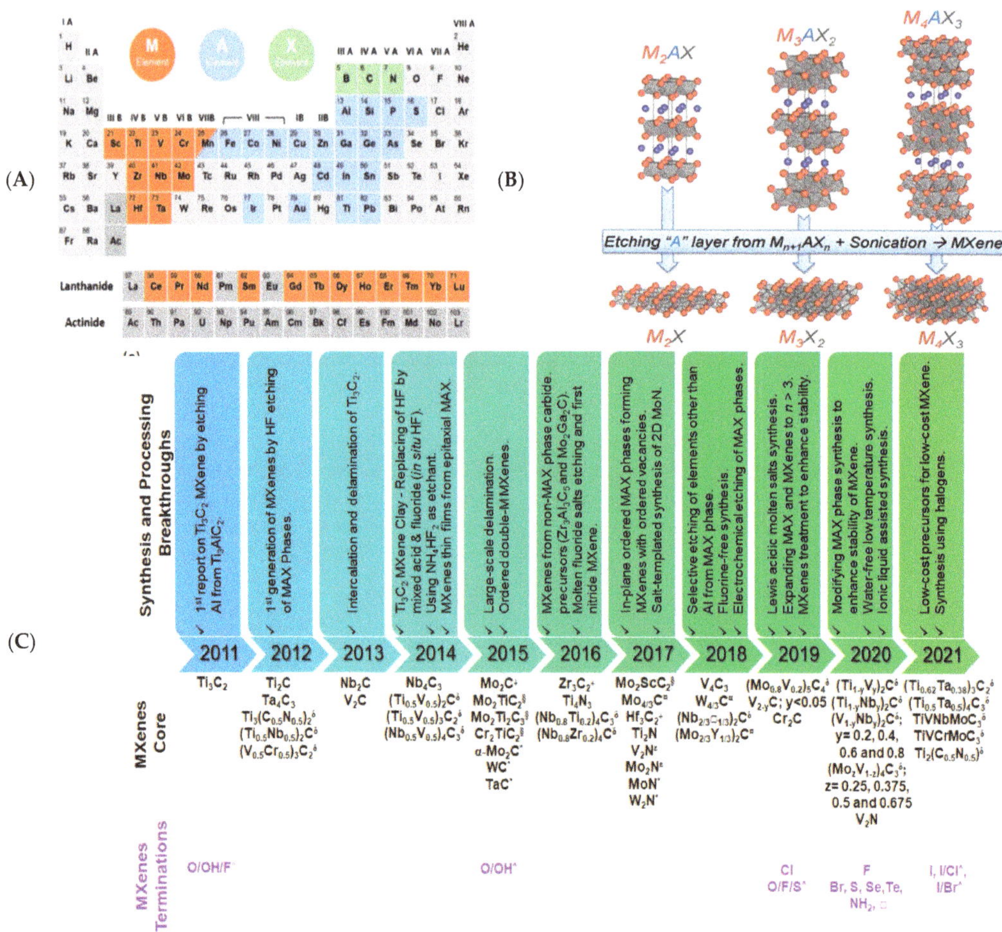

Figure 1. (**A**) Elements represented by M, A, and X in the MAX phase. (**B**) The structure of the MAX phase and its corresponding MXenes. Reprinted with permission from Ref. [9]. Copyright 2013, Wiley-VCH. (**C**) Chronological presentations of progress in the field of MXenes. List of the main synthesis and processing breakthroughs over the first 10 years of MXenes' research and new MXenescore compositions discovered in that decade and progress in surface terminations control. $^{\delta}$ Solid solution MXenes; $^{\square}$ MXenesfrom non-MAX phase precursors; § out-of-plane ordered double transition metal MXene; $^{\square}$ MXenes from in-plane ordered double transition metal MAX phase analogues; * 2D carbides and nitrides produced by bottom-up approaches; $^{\varepsilon}$ nitride MXenes produced by the post-synthesis treatment of carbide MXene; $^{\square}$ vacancy; $^{\wedge}$ mixed terminations. Reprinted with permission from Ref. [12]. Copyright 2021, Wiley-VCH.

The central area of current advanced biosensing research studies is developing biosensors for detecting biological and chemical molecules that affect disease or are damaging to the human body. The most advanced biosensors can accurately and rapidly detect the target, predict the onset of the disease in time, and receive immediate medical attention [13]. Hence, high sensitivity and selectivity are significant for the design of biosensors. Due to its unique mechanical, hydrophilic, biocompatibility, and other excellent properties, MXenes are frequently used as a new biosensing platform. Electrochemical biosensors are essential for biological, environmental, and pharmaceutical fields. It offers high sensitivity, long-term reliability and high accuracy, rapidity, low cost, and easy miniaturisation [14]. In addition, electrochemical biosensors offer a further path for creating next-generation point-of-care testing devices [15]. With advancing nanotechnology with respect to MXene-based optical biosensors, unprecedented progress has been made in optical analysis. Optical analysis has advantages of high sensitivity, high selectivity, fast analysis, and good reproducibility. It has been widely used in biochemistry and biomedical and environmental analysis and has received increasing attention [16]. The synthesis of MXenes and their application in biosensing are reflected in Scheme 1. We will review and summarize published studies on biosensing since the development of MXenes, including those mainly classifying biosensors into electrochemical, optical biosensors and some derivative biosensors. In addition, we will also discuss the challenges of MXenes in preparing biosensors and future perspectives on applying MXenes in biosensing.

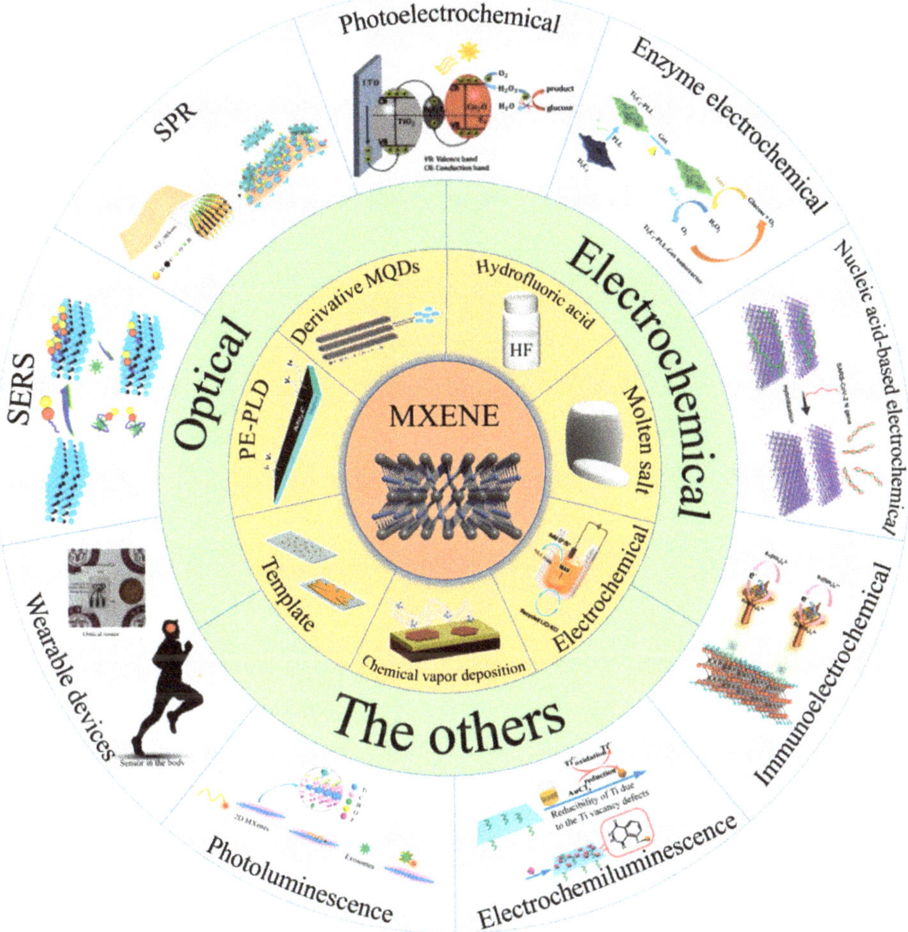

Scheme 1. MXenes cover both top-down and bottom-up methods of synthesis. They play an irreplaceable position in enzyme-, nucleic-acid-, immune-based electrochemical biosensing; photoluminescence; electrochemiluminescence; photoelectric effect-based optical biosensors; and other biosensors such as wearable biosensors, surface plasmon resonance, and surface-enhanced Raman spectroscopy.

2. Synthesis and Structures of MXenes

2.1. Synthesis of MXenes

There are two methods for the synthesis of MXenes. The top-down method is the most commonly used, which can be used to exfoliate multilayer materials into a few-layer or single-layer MXenes sheet. The second method is a bottom-up approach, which focuses on the growth of Mxenes from atoms or molecules [17,18].

2.1.1. Top-Down Method

Selective etching disintegrates the strong covalent bonds between the MX and the A layers in the MAX phase. The primary method is etching with hydrofluoric acid (HF), molten salts, etc. In this process, oxygen (O), hydroxyl (OH), and fluorine (F) replace the M-A strong metal bond [17]. There are two main steps to gain 2D MXenes by HF: etching and exfoliation. Although the direct use of HF is straightforward and practical, it causes environmental pollution and damages to the human body [4]. In situ HF can be obtained by reacting a fluorinated salt with mild acid, which is less toxic to MXenes surfaces [19]. Researchers explored new synthetic methods (Figure 2). The typical chemical reaction equation for the synthesis of MXenes in the MAX phase is as follows [9].

$$M_{n+1}AX_n + 3HF \rightarrow AF_3 + M_{n+1}X_n + \frac{3}{2}H_2 \quad (1)$$

$$M_{n+1}X_n + 2H_2O \rightarrow M_{n+1}X_n(OH)_2 + H_2 \quad (2)$$

$$M_{n+1}X_n + 2HF \rightarrow M_{n+1}X_nF_2 + H_2 \quad (3)$$

MXenes must undergo an exfoliation process to obtain nanosheet structures: The surface groups of MXenes result in the layers being linked by hydrogen and Van der Waals forces [3]. Exfoliation enhances the interlayer spacing by weakening interactions between layers using various molecular and ionic processes [20].

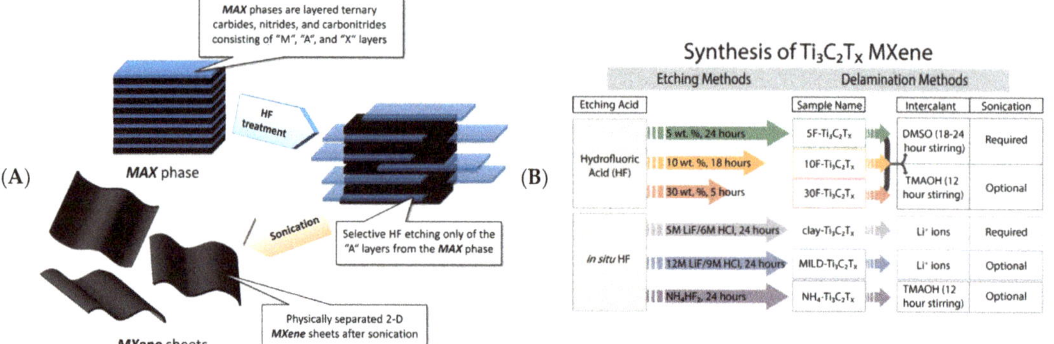

Figure 2. (**A**) Schematic diagram of the process of preparing MXenes by HF. Reprinted with permission from Ref. [21]. Copyright 2012, American Chemical Society. (**B**) A guide to Ti_3C_2 MXenes synthesis using HF. Reprinted with permission from Ref. [22]. Copyright 2017, American Chemical Society.

The molten salt method uses fluorinated molten salts, Lewis salts [23]. The synthesis does not involve fluoride, reducing the risk of synthesis [8,24]. The mechanism of MXenes formation in molten salts is similar to that of conventional HF methods: $ZnCl_2$ and $CuCl_2$ high-temperature molten salts strip a more comprehensive range of MAX phase materials [8] (Figure 3A). In the molten salt of Lewis acids, Zn^{2+}, Cu^{2+}, and Cl^- are consistent with acting H^+ and F^- in HF. Minimally intensive layer delamination (MILD) and electrochemical etching can also be used for MAX etching, producing high-quality, non-toxic MXenes [25,26].

2.1.2. Bottom-Up Method

Bottom-up synthesis methods have been reported, such as chemical vapor deposition (CVD) [27], template [28], and plasma-enhanced pulsed laser deposition (PE-PLD) [29] (Figure 3B). MXenes produced by this method possess good crystalline quality and controllable structure and size [18].

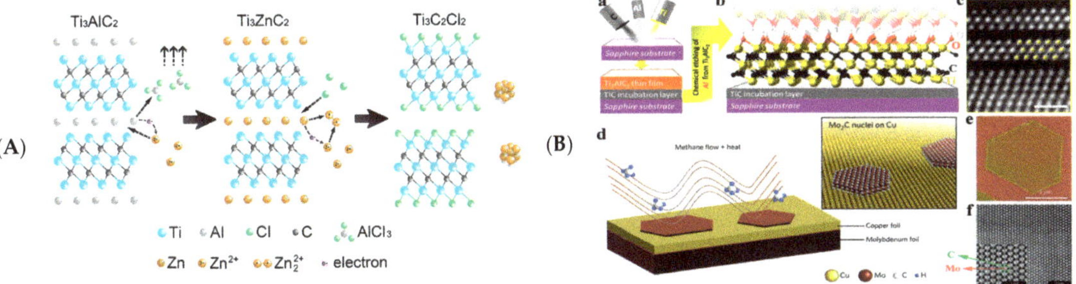

Figure 3. (**A**) Preparation mechanism of $Ti_3C_2Cl_2$ etched by $ZnCl_2$. Reprinted with permission from Ref. [30]. Copyright 2019, American Chemical Society. (**B**) Bottom-up approach to obtain MXenes. Atomic layer deposition method: steps to prepare Ti_3AlC_2 MAX films by sputtering Ti, Al and C on a sapphire substrate (**a**), schematic diagram of $Ti_3C_2T_x$ (**b**) and STEM images (**c**). CVD method: schematic diagram of the Mo_2C synthesis process (**d**), AFM images of hexagonal ultra-thin Mo_2C crystals (**e**) and STEM images (**f**). Reprinted with permission from Ref. [27]. Copyright 2020, American Chemical Society.

Xu et al. used CVD to synthesize high-quality Mo_2C crystals [27]. The synthesis of Mo_2C MXene/graphene heterostructures and Mo_2C MXene-graphene hybrid films by this method has been reported [29,31]. Compared to CVD, the template method has a relatively high yield of MXenes. Two-dimensional MXenes are mainly obtained by carbonizing or nitriding two-dimensional transition metal oxide (TMO) nanosheet templates. Xia et al. prepared hexagonal-structured 2D h-MoN nanosheets using precursor MoO_2 nanosheets [28]. PE-PLD is a successful method for preparing large-area ultra-thin face-centered cubic (FCC) Mo_2C MXene [29].

The stability of MXenes is an important property and limits its application to a certain extent. Researchers have tried to improve its stability. High concentrations of HF accelerate the degradation of MXenes and affect its structure, so relatively mild reaction conditions are necessary [32]. Organic solvents mitigate the oxidation of MXenes. Contact with water should be avoided as much as possible to prevent oxidation [33]. The oxidation of MXenes is quicker in liquid media than in solid media, and this degradation process is exacerbated by photocatalysis and thermocatalysis [34]. The storage of MXenes in Ar-sealed vials at 4 °C exhibits high stability at room temperatures [35].

2.2. Strustures of Mxenes

The crystal structure within a 2D material can affect its properties [18]. There are six types of MXenes structures (Figure 4A): (1) single transition metal MXenes (Ti_3C_2 and Nb_4C_3); (2) solid solution MXenes ($(Ti, V)_3C_2$ and $(Cr, V)_3C_2$); (3) sequential planar internal and external bimetal MXenes with one transition metal occupying the outer layer (Cr and Mo); the central metal is another metal (Nb and Ta) [36,37]; (4) ordered double-transition metals MXenes ($(Cr_2V) C_2$); (5) orrderly double vacancy MXenes ($Mo_{1.33}CTx$) [38]; (6) random empty space MXenes ($Nb_{1.33}CTx$) [39].

Computational simulation studies have been reported to identify novel stable MXenes structures, contributing to exploratory studies [40]. The properties and applications of these materials can be adapted by various parameters for composition, surface modification by heat treatment or chemical pathways, and structural adjustments [41]. MXenes have two—dimensional structures (a), one—dimensional structures (b) and (c), three—dimensional structures (d), and zero—dimensional structures (e) (Figure 4B).

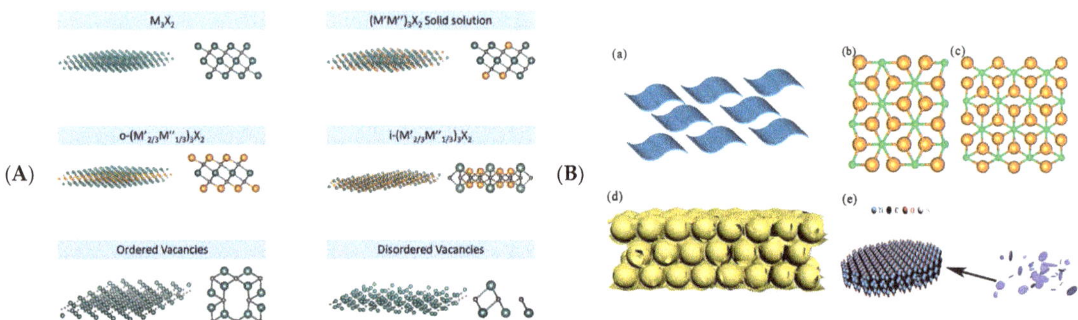

Figure 4. (**A**) Different types of MXenes structures. Reprinted with permission from Ref. [42]. Copyright 2019, Elsevier. (**B**) 2D, 1D, 3D, and 0D structures of MXenes. Adapted with permission from Ref. [18]. Copyright 2021, Wiley-VCH.

3. MXenes in Biosensing

Several strategies involving MXenes in analytical nanoscience, biosensing, and other areas have been reported. MXenes exhibit hydrophilicity due to surface groups such as OH, O, and F. Its surface can interact with most biomolecules through hydrogen bonding, Van der Waals forces, electrostatic interactions, and ligand binding, rendering it an excellent carrier for biosensors applications [43–45]. Several different MXenes compositions have been proved to be biocompatible and non-cytotoxic [46,47].

We summarized the composition and analytical performance of some MXene-based electrochemical biosensors, optical biosensors, and other biosensors and attached them to the subsections. These cases demonstrate the broad applicability of MXenes in the fabrication of biosensors. Readers can easily extract MXene-based biosensing research and measurement data from these tables.

3.1. Electrochemical Biosensing

Due to their high electronic conductivity, MXenes can drive most electrochemical reactions, which is of great significance for the application in electrochemical biosensing [48]. The electrical properties of MXenes can be improved by changing elemental compositions and surface groups [18]. In particular, the external transition metal layer of MXenes plays a more critical role in the electronic properties than the internal layer [49]. The number and thickness of the layers of MXenes also affect electrical properties [3,50].

Biosensors based on electrical signals change the electrochemical properties of the sensor surface by binding to essential substances in the organism, such as proteins, amino acids, nucleic acids, antibodies, etc. (Figure 5). The development of MXenes for electrochemical biosensors has been intensively investigated because of their excellent properties, such as high conductivity, electrochemical activity, and large surface area. The classification of electrochemical biosensors is as follows: enzyme electrochemical, nucleic acid electrochemical, and immunoelectrochemical biosensing.

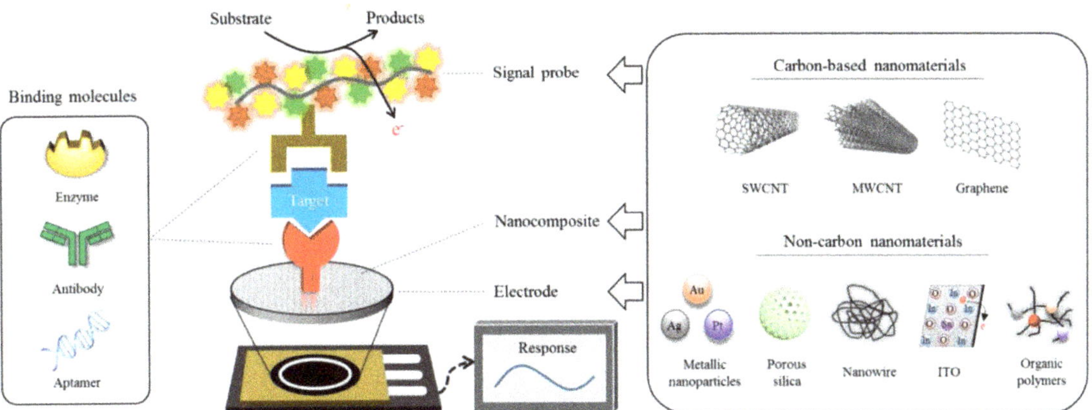

Figure 5. Schematic diagram of the analytical principle of an electrochemical biosensor. Reprinted with permission from Ref. [51]. Copyright 2020, Springer Nature.

3.1.1. Enzyme-Based Electrochemical Biosensing

Enzyme electrochemical biosensors with higher efficiencies and substrate specificities in mild conditions have been extensively explored over the last few years (Table 1). The basic principle is the direct electron transfer (DET) process between the enzyme and the electrode. The immobilisation of enzymes on the bare electrode surface can render the enzymes biologically inactive, making it extremely difficult to perform DET on the electrode's surface [52]. MXenes can be used as a strategy for enhancing DET because of their large specific surface area, excellent electrical conductivity, and good biocompatibility.

Much of the literature has shown that MXenes or MXenes composite materials can maintain the activity of enzymes after complexing enzymes due to the various properties and unique structures of MXenes. This demonstrates that MXenes can be magnificent structures for enzyme-based biosensors. Xu et al. mixed Ti_3C_2 MXene and HRP enzyme directly to fabricate a biosensor for the detection of H_2O_2 to analyse the levels of serum samples from AMI patients before and after surgery [53]. Ma et al. fabricated a low detection limit enzyme biosensor for the detection of H_2O_2 using a chitosan complex of Ti_3C_2 MXene-loaded HRP enzyme and successfully used it to detect trace amounts of H_2O_2 in foods [54].

Table 1. Enzyme-based electrochemical biosensors for identifying units, target, and analytical parameters.

MXenes Composite	Identify Units	Target	LOD	Range	Ref.
Au/Ti_3C_2	glucose oxidase	glucose	5.9 μM	0.1–18 mM	[55]
PLL/Ti_3C_2	glucose oxidase	glucose	2.6 μM	4.0–20 μM	[56]
PEDOT: SCX/$Ti_3C_2T_x$	glucose oxidase	glucose	22.5 μM	0.5–8 mM	[57]
Ti_3C_2/Nafions	Horse Radish Peroxidase	H_2O_2	1 μM	5–8000 μM	[53]
MXene/chitosan	Horse Radish Peroxidase	H_2O_2	0.74 μM	5–1650 μM	[54]
Chit/ChOx/$Ti_3C_2T_x$	cholesterol oxidase	cholesterol	0.11 nM	0.3–4.5 nM	[58]
Ti_3C_2	tyrosinase	phenol	12 nmol L^{-1}	50 nM–15.5 μM	[59]
CS-$Ti_3C_2T_x$	acetylcholinesterase	acetylthiocholine chloride	3 fM	10 nM–10 fM	[60]
GA/Nb_2CT_x	acetylcholinesterase	phosmet	144 pM	200 pM–1 μM	[26]

In addition, there are several reports on using MXenes in different compound types of enzymes, such as glucose oxidase [57], cholesterol oxidase [58], acetylcholinesterase [60], tyrosinase [59], etc. Wu et al. proposed a hybrid PLL/Ti_3C_2/glucose oxidase glucose biosensor that accelerates the breakdown of H_2O_2 generated during glucose oxidation by catalysing a cascade reaction [56] (Figure 6A). Xia et al. developed a Chit/cholesterol oxidase/$Ti_3C_2T_x$ composite cholesterol oxidase biosensor [55]. Chit/$Ti_3C_2T_x$ served as a support matrix for immobilising the enzyme. Gold nanoparticles anchored on $Ti_3C_2T_x$ MXene nanosheets enhanced the electron transfer between the enzyme and the electrode. The relative current sensitivity and LOD were 0.3–4.5 nM and 0.11 nM, respectively. Song et al. derived electrochemical etching to derive fluorine-free Nb_2CT_x with low cytotoxicity and constructed a Nb_2CT_x/acetylcholinesterase biosensor to detect sulfoxide [26] (Figure 6B). Moreover, the sensor's enzymatic activity and electron transfer are superior to the corresponding V_2C and Ti_3C_2 MXenes biosensors. Wu et al. used Ti_3C_2 MXene as a new substrate to immobilise tyrosinase and facilitated the direct electron transfer process for the sensitive and rapid detection of phenol [59]. Therefore, Ti_3C_2 MXene can be a phenolic biosensor with high recovery and long-term stability. The biosensor exhibits good analytical performance over a wide linear range of 0.05–15.5 μM, with detection limits as low as 12 nM.

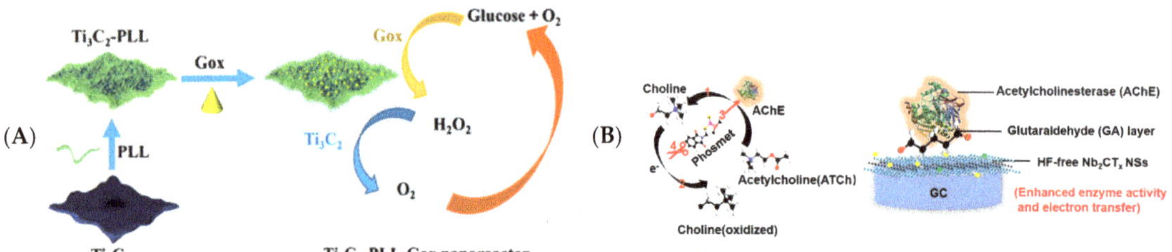

Figure 6. (**A**) Schematic diagram of the detection of the Ti_3C_2-PLL-Gox nanoreactor Reprinted with permission from Ref. [56]. Copyright 2020, Elsevier. (**B**) Schematic illustration of the enzymatic inhibition of sulfoxide detection by the HF-free Nb_2CT_x/AChE biosensor Reprinted with permission from Ref. [26]. Copyright 2020, Wiley-VCH.

The above examples and the contents show that it is feasible to combine enzymes directly on MXenes or with other materials to improve the performance of enzyme electrochemical biosensors.

3.1.2. Nucleic Acid-Based Electrochemical Biosensing

Using nucleic acids as recognition elements allows the specific recognition of the target and the generation of some signal changes [61]. Nucleic acid is a stable and easy-to-handle biomolecule, so it has excellent detection performances [62]. Nucleic-acid-based electrochemical biosensors offer advantages of both nucleic acid probes and electrochemical detection, enabling the sensitive detection of analytes such as nucleic acid, ref. [63] proteins [64], biological molecules [65], inorganic ions [66], and cells [67] (Table 2). Nucleic acid electrochemical biosensors are based on five conformations: double-stranded, triple-stranded, quadruple-stranded, DNA nanostructures, and single-stranded DNA functionalisation (hairpin structure, aptamers, and DNAzyme) [68]. Unlike enzymes, nucleic acids possess little redox capacity. The development of nucleic acid electrochemical biosensors generally relies on molecules with redox properties, such as methylene blue (MB) and ferrocene (Fc), or through charge changes that occur during nucleic acid hybridisation [69]. The nucleic acid electrochemical biosensor has various applications in genetics, clinical medicine, and biosensing due to its rapid detection, simple experimental procedures, high sensitivity, and low cost [70]. There are two types of nucleic acid biosensors.

The first type of nucleic acid electrochemical biosensor follows the Watson–Crick pairing principle, which hybridizes a nucleic acid sequence with a complementary nucleic acid sequence through base pairing [61]. The detection principle works by immobilising nucleic acids on the electrode's surface to capture complementary nucleic acid sequences, thus obtaining an altered electrical signal for specific detection [71]. There are many reports using specific nucleic acid sequences to create biosensors for the detection of disease-predicting miRNAs and DNA, and some electrochemical biosensors have been validated for point-of-care detection. Duan et al. developed a Ti_3C_2/FePc QDs MXene nanocomposite nucleic acid biosensor with good biocompatibility [72]. The Ti_3C_2/FePc QDs composite material was used as a carrier to detect miRNA-155 by using a change in electrochemical impedance caused by DNA modifications. Mohammadniaei et al. used double screen-printed gold electrodes modified with MXenes and AuNPs and single-stranded DNA-functionalised magnetic particles to detect miRNA-21 and miRNA-141 by using duplex-specific nuclease (DSN) amplification assay strategy [73] (Figure 7A). This biosensor can continue to be upgraded to quantify more analytes, forming a device for point-of-care testing (POC) cancer screening. Chen et al. fabricated a DNA electrochemical biosensor using MXene-based [74]. The surface groups were covered using ssDNA adsorbed on Ti_3C_2 MXene to attenuate conductivity. When target DNA and ssDNA are hybridized and desorbed from Ti_3C_2 MXene, the fast, simple, and sensitive detection of N-gene sequences in SARS-Cov-2 was possible (Figure 7B). The feasibility of DNA-functionalised MXenes in developing real-time monitoring diagnostic devices for clinical testing can be demonstrated.

Table 2. Nucleic acid-based electrochemical biosensors identify units, target, and analytical parameters.

MXenes Composite	Identify Units	Target	LOD	Range	Ref.
MoS_2 /Au NPs/Ti_3C_2	DNA probe	miRNA-182	0.43 fM	1 fM–0.1 nM	[63]
Au/Ti_3C_2	DNA probe	miRNA-21, 141	204 aM 138 aM	500 aM–50 nM	[73]
$Ti_3C_2T_x$ @FePcQDs	DNA probe	miRNA-155	4.3 aM	0.01 fM–10 pM	[72]
MCH/CP/AuNPs/$Ti_3C_2T_x$	DNA probe	BCR/ABL fusion gene	0.05 fM	0.2 fM–20 nM	[75]
$Ti_3C_2T_x$	DNA probe	SARS-Cov-2 N gene	10^5 copies mL^{-1}	10^5–10^9 copies mL^{-1}	[74]
PMo_{12}/PPy@$Ti_3C_2T_x$	Aptamer	Osteopontin	0.98 fg mL^{-1}	0.05–10,000 pg mL^{-1}	[64]
AuNPs/Ti_3C_2	Aptamer	Mucin 1	0.72 pg mL^{-1}	5 pg mL^{-1}–50 ng mL^{-1}	[76]
Ti_3C_2	Aptamer	gliotoxin	5 pM	5 pM–10 nM	[65]
Ti_3C_2	Aptamer	HER2-positive CTCs	47 cell mL^{-1}	20–200 cells mL^{-1}	[77]
CoCu-ZIF@ Ti_3C_2 CDs	Aptamer	B16-F10 cell	33 cells·mL^{-1}	1×10^2–1×10^5 cells·mL^{-1}	[67]
Au@$Nb_4C_3T_x$	Aptamer	Pb^{2+}	4 nM	10 nM–5 μM	[66]

The second nucleic acid electrochemical biosensor uses single-stranded DNA (ss-DNA) or RNA to bind to various biomolecules for analyte detection, including proteins, small biomolecules, cells, etc. [78,79]. Electrochemical biosensors made up of aptamers are easy, reliable, quick in responding, low in price, and possess acceptable repeatability [80]. Geng Xue of our research group cleverly used conformational changes of aptamers before and after capturing serotonin to construct an aptamer biosensor [81]. The interaction between aptamer and serotonin was destroyed by guanidine hydrochloride, and 98.2% of the signal was recovered, showing acceptable repeatability. Zhou et al. synthesized intercalating polypyrrole (PPy) $Ti_3C_2T_x$ MXene and phosphomolybdic acid (PMo_{12}) composites with a strong synergistic effect, promoting the anchoring of RNA aptamers on the composites [64] (Figure 7C). The G-quadruplex formed by osteopontin (OPN) and aptamer exhibits stable and high sensitivity, which proves the excellent performance of this MXene composite aptamer biosensor. Li et al. created a nuclease-driven DNA walker cascade signal amplification strategy to construct electrochemical aptamer biosensors on Au nanoparticles/MXene-modified electrodes for mucin 1 [76]. A DNA nanostructure-modified Ti_3C_2 MXene nanosheet biosensor was developed by Wang et al. for the detection of gliotoxins [65]. Tetrahedral DNA nanostructures were quickly immobilised on the surface of MXenes nanosheets, thus avoiding the tedious and expensive modification of DNA probes. HB5 aptamer immobilised on the MXenes layer via electrostatic interactions was highly selective for HER-2-positive cells, as reported by Vajhadin et al. Sandwich-like structures formed between magnetically captured cells, and functionalised MXenes electrodes effectively shield electron transfers, allowing quantitative cell detection with changes in the current [77] (Figure 7D).

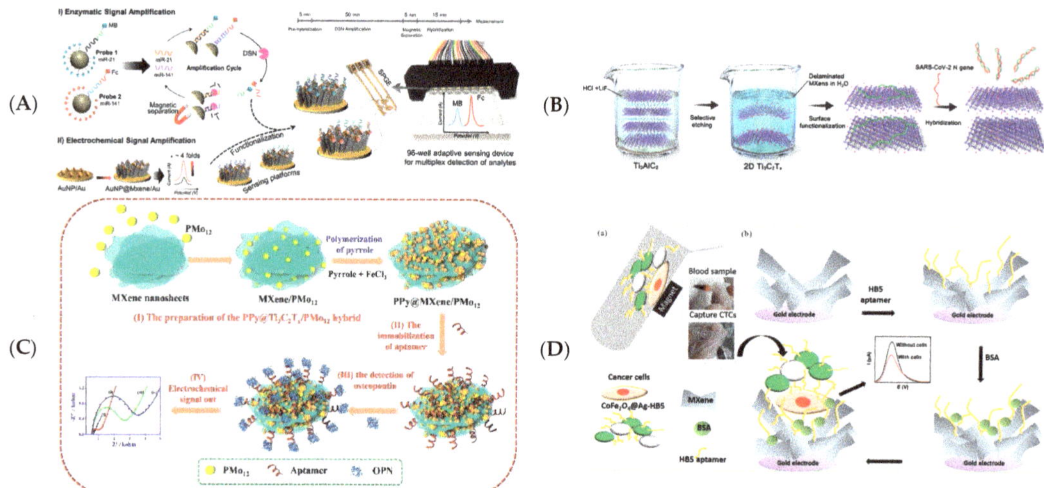

Figure 7. (**A**) Schematic diagram representing the entire assay procedure for multiplex and concurrent detections of miRNA. Reprinted with permission from Ref. [73]. Copyright 2020, Elsevier. (**B**) Schematic of the ssDNA/$Ti_3C_2T_x$ for the detection of the SARS-Cov-2 nucleocapsid gene. Reprinted with permission from Ref. [74]. Copyright 2022, American Chemical Society. (**C**) Schematic diagram of PPy@Ti_3C_2/PMo_{12} aptamer biosensor for OPN detection. Reprinted with permission from Ref. [64]. Copyright 2019, Elsevier. (**D**) Schematic diagram of the MXenes based cell sensor for the detection of SK-BR-3 cells: magnetic cell separation using $CoFe_2O_4$@Ag-HB5 (**a**) and electrochemical cell detection on a functionalised MXenes surface (**b**). Adapted with permission from Ref. [77]. Copyright 2021, Elsevier.

3.1.3. Immunoelectrochemical Biosensing

Electrochemical immunosensors are coupled to the sensor via antigen–antibody interactions. The accessibility of antibody to a wide range of molecules and the high selectivity and sensitivity renders immunochemical methods valuable for clinical diagnosis. These electrochemical biosensors for bioanalysis have advantages of small reagent volumes, high sensitivity and specificity, and portability [82]. As observed from the contents, immunoelectrochemical biosensors offer tremendous advantages in the specific detection of biomolecules (Table 3).

Table 3. Immunoelectrochemical biosensors identify units, target, and analytical parameters.

MXenes Composite	Identify Units	Target	LOD	Range	Ref.
Ti_3C_2@CuAu-LDH	Antibody	Carcinoembryonic antigen	0.033 pg mL^{-1}	0.0001–80 ng mL^{-1}	[83]
Ti_3C_2	Antibody	CEA	18 fg mL^{-1}	0.0001–2000 ngmL^{-1}	[84]
Au/$Ti_3C_2T_x$	Antibody	PSA	3.0 fg mL^{-1}	0.01–1.0 pg mL^{-1}	[85]
M/NTO/PEDOT/AuNPs	Antibody	PSA	0.03 pg L^{-1}	0.0001–20 ng mL^{-1}	[86]
CuPtRh/NH_2-Ti_3C_2	Antibody	cardiac troponin I	8.3 fg mL^{-1}	25 fg mL^{-1}–100 ng mL^{-1}	[87]

In 2018, Kumar et al. fabricated the first MXene-based immunoelectrochemical sensor to detect carcinoembryonic antigens (CEAs) [84]. Aminosilane-functionalised MXenes offered more binding sites for bioreceptors than GCE, and the CEA antigen is better immobilised on Ti_3C_2 MXene (Figure 8A). Xu et al. synthesized a composite of 3D sodium titanate nanoribbons, anchored poly(3,4-ethylene dioxythiophene), and gold nanoparticles by oxidizing and alkalizing Ti_3C_2 Mxene [86]. The composites described above were used to immobilise prostate-specific substance (PSA) antibodies to create a facile electrochemical label-free immunosensor for the sensitive detection of PSA (Figure 8B).

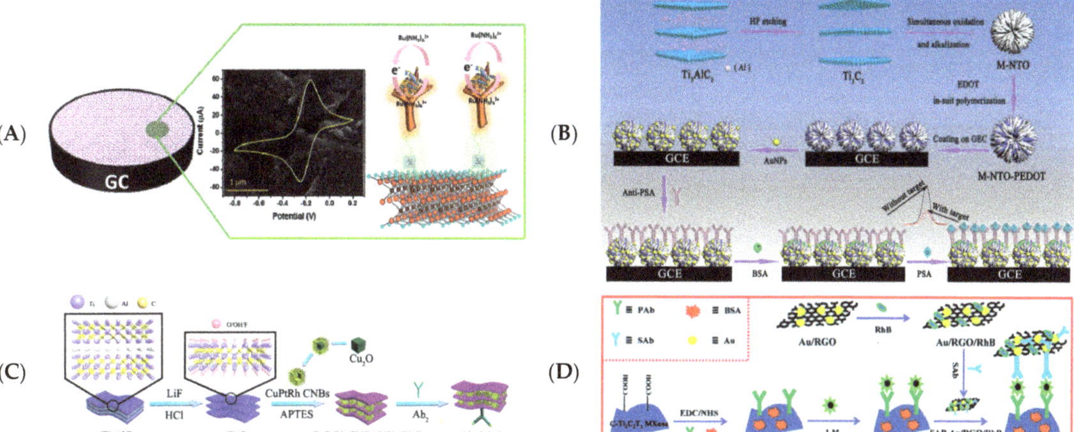

Figure 8. (A) Schematic diagram of the detection mechanism of electrochemical CEA. Reprinted with permission from Ref. [84]. Copyright 2018, Elsevier. (B) The fabrication and detection steps of the immunosensor. Reprinted with permission from Ref. [86]. Copyright 2020, Elsevier. (C) Schematic diagram of the fabrication of the working electrode of the immunosensor. Reprinted with permission from Ref. [87]. Copyright 2020, American Chemical Society. (D) Illustrations for constructing the sandwich-like immunosensor of LM based on $Ti_3C_2T_x$ MXene. Reprinted with permission from Ref. [88]. Copyright 2021, Royal Society of Chemistry.

Dong et al. used a CuPtRh/NH_2-Ti_3C_2 nanocomposite composed of trimetallic hollow CuPtRh cubic nanoboxes (CNBs) and laminated ammoniated Ti_3C_2 flakes to fabricate a sandwich-type immunosensor to detect cardiac troponin I (CTnI) [87]. Aminated Ti_3C_2 provides abundant binding sites for both CuPtRh CNBs and antibodies, while CuPtRh CNBs can prevent Ti_3C_2 from stacking again (Figure 8C). In addition, MXenes can serve to detect bacteria. Niu et al. constructed a sensing platform with carboxylated $Ti_3C_2T_x$ MXene and rhodamine B/gold/reduced graphene oxide as the signal [88] (Figure 8D). A sandwich electrochemical immunosensing platform for detecting Listeria monocytogenes was also developed by them.

3.2. Optical Biosensing

Optical properties, including absorption, transmission, photoluminescence, scattering, and emission, are essential for applying MXenes. The surface groups, doping, and defects affect the energy band's structure [89]. A thin layer of $Ti_3C_2T_x$ has been reported to absorb photons in the UV-visible region between 300 and 500 nm with a transmission of 91%. O-functionalised Ti_3C_2 MXene has a higher light absorption efficiency [90]. The optical properties are also affected by the thickness of the film and the distance between MXenes layers. Intercalation with hydrazine, urea, methyl ammonium hydroxide, and DMSO changes the interlayer distance of $Ti_3C_2T_x$, decreasing light transmittance [50].

MXenes have excellent hydrophilicity, biocompatibility, and optical characteristics, making them appropriate for all sorts of biosensing applications. It was discovered as a fluorescence quenching agent and a carrier for biomedical and imaging applications, contributing to high-performance optical biosensors. The interaction of light and materials is central to the optical inspection principle. It identifies samples by non-destructively monitoring changes in the intensity or spectral shift of light [91]. This section will summarize MXenes biosensing applications in photoluminescence, electrochemiluminescence, and photoelectrochemical applications.

3.2.1. Photoluminescence (PL)

MXenes possess features that make MXenes excellent for fluorescence biosensors, such as larger absorption bands, higher energy levels, etc., which causes fluorescence quenches in fluorescent substances [91]. Hence, changes in fluorescence intensity can be employed as indications for biological analytes detection (Table 4). MXene quantum dots (MQDs) are luminous, extremely water-soluble, dispersible, and biocompatible [92]. As a consequence, searching for photoluminescent biosensors based on MXenes and MQDs has emerged as a widespread research issue [93,94].

Table 4. MXene-based photoluminescence biosensors identify units, target, and analytical parameters.

MXenes Composite	Identify Units	Target	LOD	Range	Ref.
DSPE-PEG/Ti_3C_2	Gox/RCDs	Glucose	50 μM	0.1–20 mM	[94]
Ti_3C_2	DNA probe	HPV-18 DNA	100 pM	0.5 nM–50 nM	[95]
Ti_3C_2	Aptamer, DNA probe	MUC1, miRNA-21	6 nM, 0.8 nM	0–60 nM, 0–25 nM	[96]
Ag@Ti3C_2/GQDs	Antibody NSE	neuron-specific enolase (NSE)	0.05 pg mL^{-1}	0.0001–1500 ng mL^{-1}	[97]
PL-Ti_3C_2	rhodamine B	Phospholipase D	0.10 U L^{-1}	0.5–50 U L^{-1}	[98]
$Ti_3C_2T_x$	quantum dots	glutathione	3.0 μM	5.0–100 μM	[99]
Ti3C_2 QD	quantum dots	RAW264.7 cells, Zn^{2+}	-	-	[100]
N,P-Ti3C_2 QDs	quantum dots	Cu^{2+}	2 μM	2–100 μM, 250–5000 μM	[101]
Ti_3C_2 MQDs	quantum dots	alkaline phosphatase	0.02 U L^{-1}	0.1–2.0 U L^{-1}	[102]
epsilon-poly-L-lysine (PLL)/Ti3C_2 QDs	quantum dots	cytochrome c and trypsin	20.5 nM and 0.1 μg mL^{-1}	0.2–40 μM and 0.5–80.0 μg mL^{-1}	[103]
Ti_3C_2 QDs	quantum dots	Intracellular pH	-	pH 6.0–8.0	[104]

Because of MXenes' strong and broad absorption in the visible and near-infrared regions, MXenes generally act as an acceptor designed to quench the fluorescence signal emitted by sensing probes, such as metal nanoclusters, quantum dots, fluorescent dyes, etc. [94]. Due to the two-dimensional planer structure and hydrophilic surface groups of MXenes, the abundant binding sites and hydrophilic groups on MXenes provide more possibilities for biomolecular interactions [44].

Shi et al. detected glutathione by combining copper nanoclusters (Cu NCs)-functionalised MXenes [99]. MXenes quenches the fluorescence of Cu NCs through the internal filtering effect (IFE), and glutathione can analyze MXenes and Cu NCs, resulting in fluorescence recovery. Ti_3C_2 MXene nanosheets combined with red-emitting carbon dots (RCDs) area unit effective and selective fluorescence stimulant sensors were used for glucose detection by Zhu et al. Ti_3C_2 nanosheets impassively quenched the fluorescence intensity of RCDs (>96%) through IFE [94] (Figure 9A). Kalkal et al. constructed a fluorescent biosensing system based on Ag/Ti_3C_2 to quench the fluorescence signal on antibody/amino-graphene quantum dots [97]. The fluorescence recovered when antigen was added. It can be used to detect neuron-specific enolase with good reproducibility.

MXenes are used as fluorescence quenchers to construct optical sensors for monitoring enzyme activity and biomolecules. Similarly to the previous section, the Fc and MB of the nucleic acid biosensor can be replaced with some fluorescent materials that can be used, which are more practical for this type of biosensor. Zhu et al. reported a Ti_3C_2 MXene-based fluorescent biosensor to detect phospholipase D by FRET quenching of rhodamine B (RhB)-labeled phospholipids [98]. Phospholipase D cleaves phospholipids, causing RhB-labeled phospholipids to detach from Ti_3C_2 MXenes and re-reflorescence. Peng et al. used the affinity difference between single-stranded and double-stranded DNA on MXenes to construct fluorescent signal detection for human papillomavirus HPV-18 DNA on ultra-thin Ti_3C_2 MXene [95] (Figure 9B). Wang et al. presented dual-signal-labelled DNA-functionalised Ti_3C_2 MXene nanoprobes to achieve a dual analysis of MUC 1 and miRNA-21 at low concentrations in vitro, and the in situ imaging of MCF-7 breast cancer cells [96]. Furthermore, cell imaging can provide multiple layers of information, such as biomarkers' expression levels and spatial distribution.

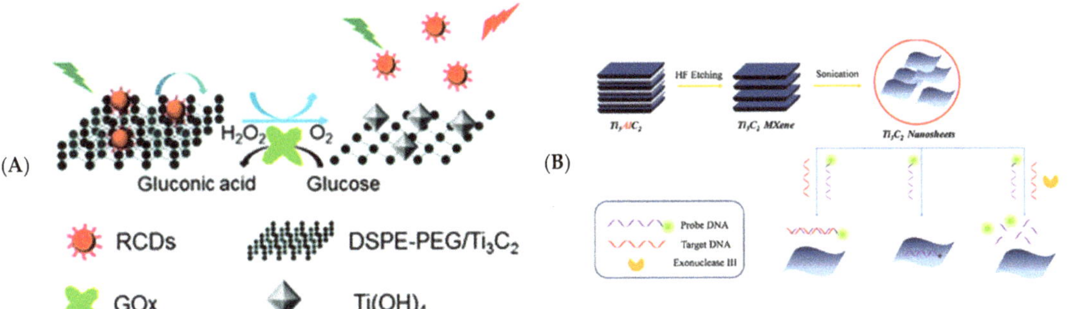

Figure 9. (**A**) Schematic diagram of MXene-based glucose oxidase fluorescent biosensor. Reprinted with permission from Ref. [94]. Copyright 2019, Royal Society of Chemistry. (**B**) Illustration of analysis of HPV-18 type DNA using Ti_3C_2 MXene. Reprinted with permission from Ref. [95]. Copyright 2019, Elsevier.

When the thickness dimensions of 2D nanomaterials are less than 100 nm, MXenes can be converted into quantum dots with quantum confinement and optical properties [105]. MQDs, with an average lateral size ranging from 1.8 to 16 nm, can be obtained by hydrothermal processes [100], acidic oxidation, and chemical stripping [106]. Charge transfer is enhanced, and fluorescence is enhanced by utilizing heteroatom doping [101]. MQDs have similar properties to MXenes, such as high dispersion and good biocompatibility. The small band gaps of MXenes can expand their band gap through quantum effects, contributing to their strong fluorescence effect [107]. Some researchers synthesized MQDs that exhibited different fluorescence effects in different solvents under 365 nm UV light irradiation [93,104].

On account of their tunable size, photoluminescence, and photostability, MQDs can be applied as fluorescent probes and can also be functionalised with natural biomolecules [107]. The performance of MQDs as fluorescent agents or signals can be improved, and the application of MXenes in biosensing can be widely expanded [91,108]. MQDs have contributed enormously to detecting metal ions, biomolecules, and cellular imaging. The first MQD-based fluorescence sensor is based on the coordination of Zn^{2+} through hydroxyl groups on the surface of MQDs with selective quenching [100] (Figure 10B). Heteroatom-doped MQDs can be the detector for the fluorescence detection of different metal particles, such as Cu^{2+} [101], Ag^{+}, and Mn^{2+} [109].

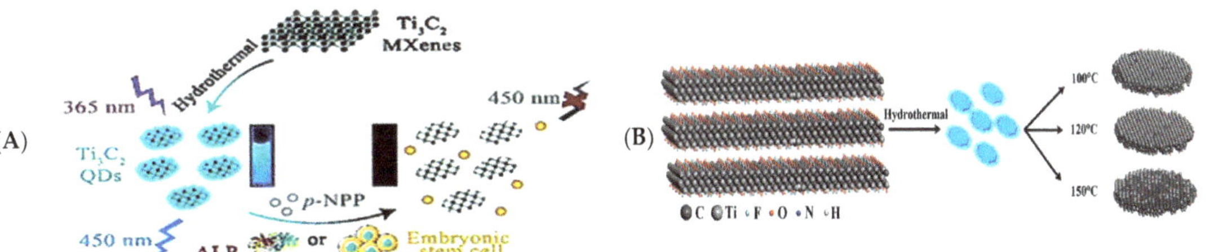

Figure 10. (**A**) Schematic diagram of fluorescence assay of alkaline phosphatase activity of MQDs. Reprinted with permission from Ref. [102]. Copyright 2018, Royal Society of Chemistry. (**B**) Schematic diagram of hydrothermal preparation of MQDs. Reprinted with permission from Ref. [100]. Copyright 2017, Wiley-VCH.

MQDs can be implemented to detect some biomolecules because they have absorption bands that overlap with the excitation and/or emission spectra of MQDs. Guo et al. designed an MQD-based fluorometric strategy for alkaline phosphatase activities and embryonic stem cell identification [102] (Figure 10A). The effective quenching of MQD fluorescence was obtained by p-nitrophenol produced by the alkaline phosphatase-catalysed dephosphorylation of p-nitrophenyl phosphate. It can also be used as an IFE-based method to analyse ESC biomarker ALP in ESC lysates accurately. Liu et al. described a fluorescent platform for detecting cytochrome c and trypsin [103]. The fluorescence of MQDs was burst by cytochrome c through the IFE. Meanwhile, cytochrome could be degraded by trypsin, and MQDs' fluorescence could be restored. Chen et al. constructed a fluorescent sensor with the pH-dependent emission of blue fluorescence from MQDs for ratiometric MQDs probes to detect cellular pH [104].

3.2.2. Electrochemiluminescence (ECL)

As a mixture of electrochemistry and optics, electrochemiluminescence is a new method for evaluations and detections. Because of its low background signal, excellent sensitivity, controllability, speed, and low cost, it is frequently employed in biochemistry for proteins, nucleic acids, enzymes, and clinical diagnostics [91]. MXenes have been proven viable as working electrodes for ECL biosensors, with improved ECL characteristics compared to glassy carbon electrodes [110]. The ECL biosensor is well suited for the analysis of nucleic acids or gene fragments, biomolecules, biomarkers, and even cells (Table 5).

In 2018, Fang et al. fabricated an ECL biosensor of $Ru(bpy)_3^{2+}$ functionalised $Ti_3C_2T_x$ MXene to detect unlabelled single nucleotide mismatches in human urine, using tripropylamine as a co-reactant [111]. Exposed bases in mismatched DNA bind to $Ru(bpy)_3^{2+}$ on the $Ti_3C_2T_x$ MXene and are more prone to electrochemical oxidation in enhancing ECL intensities. Zhuang et al. constructed ECL nanoprobes via $Ti_3C_2T_x$-mediated in situ formations of Au NPs and the anchoring of luminol and utilised the catalytic hairpin assembly (CHA) amplification of signalling to fabricate ECL biosensors for miRNA-155 detection [112] (Figure 11A). Yao et al. detected the SARS-Cov-2 gene by MXenes/PEI adsorbed Au and $Ru(bpy)_3^{2+}$ DNA walkers [113]. After the DNA walker excised hairpin DNA under the action of Nb.BbvCl endonuclease, template DNA-Ag hybridized with hairpin DNA and decreased the signal of ECL (Figure 11B). Zhang et al. modified DNA probes on MXenes/PEI composites by $Ru(bpy)_3^{2+}$ and AuNP and used the CRISPR-Cas12a strategy to construct an ECL signal on/off biosensor for the detection of SARS-Cov-2 (RdRp) gene [114].

Table 5. MXene-based electrochemiluminescence biosensors identify units, target, and analytical parameters.

MXenes Composite	Identify Units	Target	LOD	Range	Ref.
$g\text{-}C_3N_4/Ti_3C_2$	Ti_3C_2	Protein Kinase	1.0 mU mL^{-1}	0.015–40 U mL^{-1}	[115]
$Ti_3C_2T_x$	$Ru(bpy)_3^{2+}$	nucleotide mismatch	5 nM	-	[111]
$Au@Ti_3C_2@PEI\text{-}Ru(dcbpy)_3^{2+}$	Model DNA-AgNCs	SARS-Cov-2 Gene	0.21 fM	1 fM–100 pM	[113]
$AuNPs/Ti_3C_2$/Luminol	sDNA	miRNA-155	0.15 fM	0.3 fM–1 nM	[112]
$Ru@Ti_3C_2@AuNPs$	Fc-DNA	SARS-Cov-2 gene	12.8 aM	-	[114]
Ti_3C_2/PEI	aptamer	MCF-7	125 particles μL^{-1}	5×10^2–5×10^6 particles μL^{-1}	[116]
Ti_3C_2/Au	aptamer	CD63	30 particles μL^{-1}	10^2–10^5 particles μL^{-1}	[117]
$AuNPs/Ti_3C_2$	aptamer	cardiac troponin I	0.04 fM	0.1 fM–1 pM	[118]
$AuNPs\text{-}Ru\text{-}Arg@Ti_3C_2$	antibody	CEA	1.5 pg mL^{-1}	0.01–150 ng mL^{-1}	[119]
$R6G\text{-}Ti_3C_2T_x@AuNRs/ABEI$	antibody	Vibrio vulnificus	1 CFU mL^{-1}	1–10^8 CFU mL^{-1}	[120]

Strategies for detecting biomolecules can be implemented with aptamers, resulting in higher ECL signal intensities. Sun et al. proposed PEI-functionalised MXenes and $g\text{-}C_3N_4$ composites as detection probes, and kemptide chelated with Ti in the composites after protein kinase A (PKA) phosphorylation to promote electron transfers at the electrode's interface, enhancing strategies for ECL signalling [115]. Moreover, this biosensor enables inhibitor screening and PKA activity monitoring in MCF-7 cell lysates. Mi et al. reported a method for the quantitative detection of cardiac troponin (CTnI) by electrochemical and ECL dual signals using tetrahedral DNAs (TDs) and in situ hybrid chain reaction (HPR) on Au/Ti_3C_2 MXene [118] (Figure 11C). Both ECL Dox-Luminol/Current Dox and Current MB/Current Dox dual signals can be used for the quantitative detection of CTnI, which is expected to be used in screen critically ill patients with COVID-19. Zhang et al. used MXenes to generate AuNPs in situ and modified aptamers and constructed an ECL biosensor to detect exosomes CD63 [117]. Zhang and colleagues developed an exosome-selective ECL biosensor using aptamer-modified Ti_3C_2 MXenes as probes with an LOD of 125 μL particles^{-1} [116].

Immunochemical methods are also highly selective and sensitive in the field of ECL. Luo et al. constructed an MXene-based substrate using $[Ru(bpy)_2(mcpbpy)]Cl_2$ and L-arginine as co-reactants to detect carcinoembryonic proteins (CEA) by antigen [119]. Upon antigen binding to the antibody, spatial site resistance leads to a decline in the rate of electron transfer and electrolyte diffusion at the electrode's surface, resulting in a decrease in ECL signal intensities (Figure 11D). Wei et al. constructed an ECL/SERS dual-signal biosensor to detect the causative agent of Vibrio vulnificus (VV) [120]. The pathogenic bacteria. VV is captured by Fe_3O_4@Ab_1 as the capture unit. Ab_2, R6G, and ABEI bind to AuNR as the signal unit to capture VV through Au-S and Au-N, forming a Faraday cage structure.

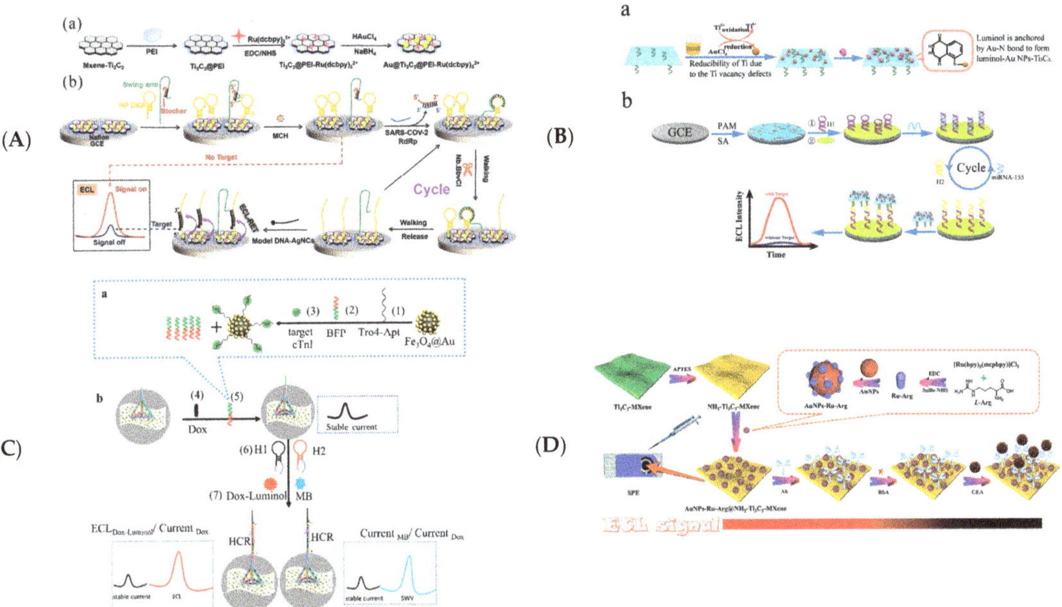

Figure 11. (**A**) Schematic diagram of the preparation of Au@Ti_3C_2@PEI-$Ru(dcbpy)_3^{2+}$ nanocomposites (**a**); Combined unilateral DNA walker amplification strategy based on nanocomposites for ECL biosensor detection of SARS-Cov-2 RdRp gene (**b**). Reprinted with permission from Ref. [113]. Copyright 2021, American Chemistry Society. (**B**) Strategy of stable luminol-Au NPs-Ti_3C_2 (**a**) and construction of the proposed ECL biosensor (**b**). Reprinted with permission from Ref. [112]. Copyright 2021, Springer Nature. (**C**) Schematic representation of specific target recognition and BFP release (**a**) and the ratiometric biosensing mechanism of cTnI (**b**). Reprinted with permission from Ref. [118]. Copyright 2021, Elsevier. (**D**) Schematic representation showing the detection principle of the prepared ECL biosensor. Reprinted with permission from Ref. [119]. Copyright 2022, Royal Society of Chemistry.

3.2.3. Photoelectrochemical (PEC)

Similarly to electrochemiluminescence sensing, photoelectrochemical sensing is a practical analytical method that integrates optical and electrochemical analyses. MXenes also promise photoelectrochemical sensors with their excellent optical and electronic properties [121] (Table 6).

Table 6. MXene-based photoelectrochemical biosensors identify units, target, and analytical parameters.

MXenes Composite	Identify Units	Target	LOD	Range	Ref.
Ti_3C_2/Cu_2O	non-enzymatic	glucose	0.17 nM	0.5 nM–0.5 mM	[122]
$TiO_2/Ti_3C_2T_x/Cu_2O$	non-enzymatic	glucose	33.75 nM	100 nM–10 μM	[123]
Bi_2S_3/Ti_3C_2	DNA probe	methyltransferase	0.003 U mL^{-1}	0.01–30 U mL^{-1}	[124]
Ti_3C_2/CdS	Aptamer	Exosomes	7.875×10^4 particles mL^{-1}	7.3×10^5–3.285×10^8 particles mL^{-1}	[125]
$Ti_3C_2@ReS_2$	DNA probe	miRNA-141	2.4 aM	0.1 fM–1 nM	[126]
APTES/ Ti_3C_2	Antibody	5hmCTP	4.21 pM	0.008–100 nM	[127]
(IOPCs)/Ti_3C_2 QDs	quantum dots	glutathione	9.0 nM	0.1–1000 μM	[128]

Li et al. took advantage of Ti_3C_2 MXene, readily forming PN junctions with photosensitive semiconductors and, therefore, used Ti_3C_2/Cu_2O heterostructures for the high-sensitivity detection of glucose [122]. The in situ growth of Cu_2O on MXenes improves photoelectrochemical performances compared to pure Cu_2O. In order to improve the photocurrent conversion efficiency and detection sensitivity of glucose, a Z-type heterostructure based on $TiO_2/Ti_3C_2T_x/Cu_2O$ was proposed in addition to the construction of a Schottky-junction-based PEC sensor [123]. A DNA probe can also achieve the label-free PEC determination of methyltransferase (MTase) for label-free Bi_2S_3/Ti_3C_2 PN junctions [124]. For exosomes, the enzyme-induced deposition of CdS on Ti_3C_2 MXene forms a Ti_3C_2 MXene/CdS composite, creating a built-in electric field in the tight interface between CdS and Ti_3C_2 MXene, enabling highly accurate detection [125]. In addition to this, the use of $Ti_3C_2@ReS_2$ to immobilise DNA probes and perform specific PEC detections of miRNA-141 has excellent performance [126]. For the detection of 5hmCTP on APTES/Ti_3C_2, the use of antibodies for PEC is also feasible [127]. Chen et al. developed a photoelectrochemical biosensor for the sensitive and selective detection of glutathione based on MQDs [128].

3.3. Other Biosensing

3.3.1. Wearable Biosensing

The covalent between the M transition metal and the X element in MXenes, the terminal surface groups, and the thickness of the atomic layers resulted in excellent mechanical properties [3,18]. Numerous theoretical findings on the mechanical properties of MXenes have been reported. Kurtoglu et al. predicted a higher elasticity coefficients for various pristine and functionalised MXenes than their precursors due to the greater density of charge density in the $M_{n+1}X_n$ layer [129]. The excellent mechanical properties provide favourable conditions for fabricating wearable biosensors.

Some studies have used wearable nanoelectronics to detect health-related physiological activities, such as physical or chemical stimulation, micropressure, and changes in physiological signals. Stretchable mechanical properties, high gauge factor, flexible materials integrating flexible bio-electronic interfaces, and miniaturized signalling systems need to be investigated to meet the required sensitivity of sensor devices and to improve the usability of wearable devices [130]. Recently, ultrathin MXenes comprised high-performance materials for stretchable and bendable conductive coatings [131] (Table 7). Conductive and conformal MXenes multilayers can withstand up to 2.5 mm bending and 40% tensile stretching, with recoverable electrical resistances, while maintaining a conductivity of 2000 S m^{-1} [10].

Table 7. MXene-based wearable biosensors identify units, target, and analytical parameters.

MXenes Composite	Detection Technology	Target	LOD	Range	Ref.
$Ti_3C_2T_x$/LBG/PDMS	Vitro Perspiration Analysis	cortisol	88 pM	0.01–100 nM	[132]
$Ti_3C_2T_x$/PB	Vitro Perspiration Analysis	glucose and lactate	35.3 and 11.4 $\mu A\ mm^{-1}\ cm^{-2}$	-	[133]
$Ti_3C_2T_x$/MB	Vitro Perspiration Analysis	glucose and lactate	2.4 nA μM^{-1} and 0.49 $\mu A\ mM^{-1}$	-	[134]
$Ti_3C_2T_x$/MWNTS	Vitro Perspiration Analysis	K^+	-	1–32 mM	[135]
$Ti_3C_2T_x$	Vitro Perspiration Analysis	Na^+, protein	-	-	[136]
F-$Ti_3C_2T_x$/PANI	Vitro Perspiration Analysis	human sweat pH	-	-	[137]

Piezoresistive wearable biosensors are designed to detect weak movements of the human body using stretch changes in materials. The development of MXene-based piezoresistive biosensors has been reported to change the resistance of the biosensor by varying the MXene's layer spacing under external pressure [138]. The biosensor can monitor physical stimulation processes, such as blinking, throat swallowing, and knee bending release through electrical current. Strain sensors were also fabricated by using Ti_3C_2 MXenes nanocomposites with single-walled carbon nanotubes (SWCNT) obtained by layer-by-layer (LBL) spraying [139]. The multifunctional force-sensing sensor for acoustic monitoring consists of two Au electrodes on a polyethylene terephthalate (PET) substrate at the top and bottom, an intermediate MXenes layer, and a fingerprint structure on the substrate in a combined arrangement [140]. The manufactured sensor is versatile and capable of sensing sound, micro-motion, and acceleration in a single device. This biosensor can be flexibly attached to a person's throat and wrist and is used to detect a person's vocalisation and pulse. The sensor can record relevant peaks when saying "hello" and "sensor" or when detecting a steady heartbeat pulse signal. The biosensor has shown excellent sensitivity in detecting subtle human activity and other weak stresses (Figure 12). It offers a new research direction for portable and wearable sensing devices in biosensing and human behaviour analyses.

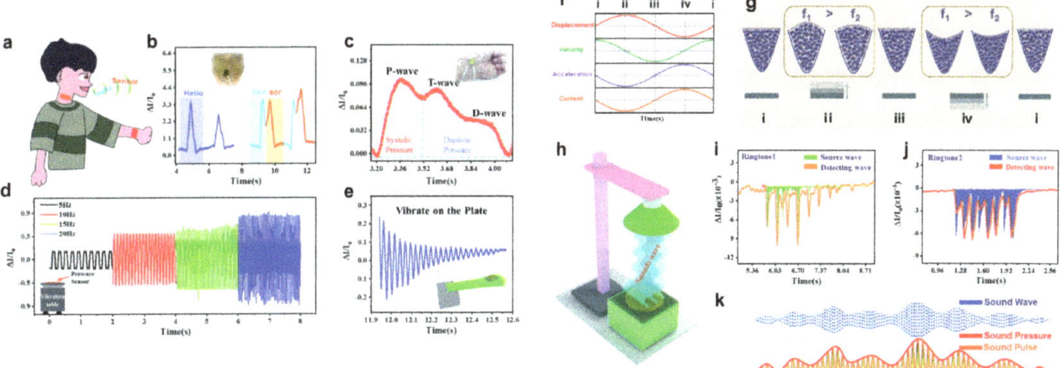

Figure 12. Schematic diagram of a pressure sensor as a wearable device for physiological signal detection (**a**). Corresponding signals for saying "hello" and "sensor" (**b**). Corresponding signal for the pulse on the medial wrist (**c**). Corresponding signal of the pressure sensor vibrating at different frequencies on the shaker (**d**). Corresponding signals of the pressure sensor vibrating on plate (**e**). Corresponding relationship between current values, displacement, velocity and acceleration in the vibration mode (**f**). Schematic diagram of the density of MXenes at different vibration stages (**g**). Diagram of a sound wave hitting a pressure transducer (**h**). Schematic diagram of the acoustic pressure on the sensor surface associated with sound waves and sound pulses (**i**). Corresponding detection signals and source waves for the two ringtones (**j**–**k**). Reprinted with permission from Ref. [140]. Copyright 2020, Wiley-VCH.

Wearable microfluidic biosensors were originally designed to integrate biological identifiers (enzymes, nucleic acids, enzymes, or cellular receptors) into the sensor operation. Non-invasive biomarker detection platforms via biofluids such as sweat, saliva, tears, or interstitial fluid are more practical [141]. Such wearable sensors provide real-time biochemical information about the wearer's health and offer effective disease detection and body function management [142]. A 3D electrode network electrochemical impedance immunosensor based on loaded laser-burned graphene (LBG) loaded with $Ti_3C_2T_x$ was fabricated for the non-invasive monitoring of cortisol biomarkers in human sweat [132]. The sensor has a detection limit and linearity of 88 pM and 0.01–100 nM, respectively (Figure 13). In addition, microfluidic wearable biosensors can be used to detect K^+, Na^+ ions [135,136], glucose, lactate [134,143], pH, and other human biochemical information in biological fluids [137].

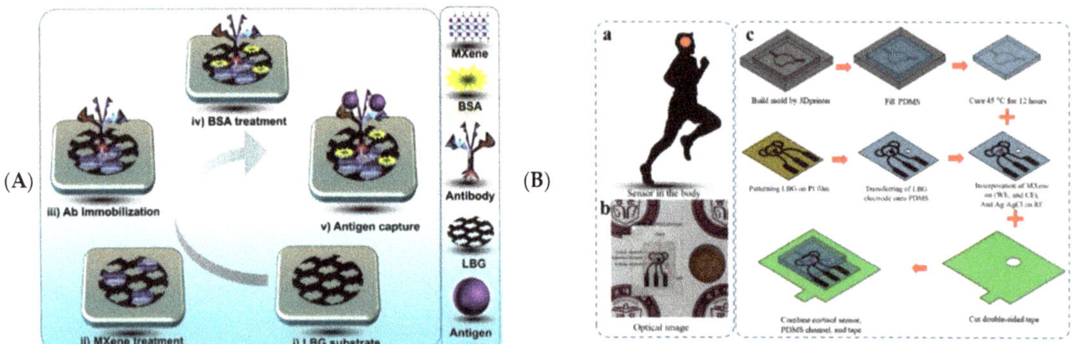

Figure 13. (**A**) Schematic representation of the Ti_3C_2Tx MXene-loaded/LBG-based cortisol biomarker assay. (**B**) Patch sensor attached to the body position (**a**), optical image of the wearable patch (**b**) and fabrication sequence of a wearable patch cortisol sensor integrated with a microfluidic system (**c**). Reprinted with permission from Ref. [132]. Copyright 2020, Elsevier.

In various wearable biosensors, sensing electrodes play an essential role in the design of wearable biosensors. MXenes offer the ability to immobilise biomolecules as a sensitive detection platform. Nevertheless, the mechanical friction and deformation of wearable devices against human skin over time leads to mechanical failure and requires a re-structuring of the device. Moreover, the attainment of signal-to-noise ratios and the stability required to achieve this device are highly challenging [133].

3.3.2. Surface-Enhanced Raman Spectroscopy (SERS)

The hydrophilic nature of the MXenes surface provides a good site for Raman labelling. It serves as a potential material for SERS and provides an effective method for the ultra-sensitive determination of targets (Table 8). Sarycheva et al. showed that composites of metals and MXenes could be used as SERS subestrates for rhodamine 6G [144]. Integrating noble metal nanoparticles with MXenes exhibits an empathetic, sensitive SERS response in detecting several common dye molecules. This extends MXenes composites for visible light SERS in the sensor field. A reliable substrate for MXenes/AuNR composites was prepared by Xie et al. with high sensitivities for determining common organic dyes, such as Rh6G, crystalline violet, and peacock green [145]. It can detect organic contaminants and shows high sensitivities for more complex organic pesticides and contaminants. A ratiometric SERS aptamer sensor for ochratoxin A was developed by Zhao et al. 2-Mercaptobenzimidazole-5-carboxylic acid ligands and Au-Ag Janus nanoparticles were used as Raman signal molecules to amplify the SERS signal efficiently [146]. Liu et al. used the $Ti_3C_2T_x$-PDDA-Ag NPs hybrid platform as a sensitive and homogeneous biosensor for the label-free quantification of the biomolecule adenine based on the SERS method [147]. These studies demonstrate that MXenes can be well-suited for SERS.

Table 8. MXene-based other biosensors identify units, target, and analytical parameters.

MXenes Composite	Detection Technology	Target	LOD	Range	Ref.
$Ti_3C_2T_x$/AuNRs	-	R6G, crystal violet, malachite green	1 pM, 1 pM, 10 nM	-	[145]
$Ti_3C_2O_2$	Aptamer	Ochratoxin A	1.28 pM	-	[146]
$Ti_3C_2T_x$-PDDA-AgNPs	Electrostatic	Dopamine	10 nM	5 µM–50 nM	[147]
R6G-$Ti_3C_2T_x$@AuNRs/ABEI	Antibody Vibrio vulnificus	Vibrio vulnificus	10^2 CFU mL^{-1}	10^2–10^8 CFU mL^{-1}	[120]

3.3.3. Surface Plasmon Resonance (SPR)

SPR sensing is a non-destructive, label-free, real-time detection method. Nanomaterials modify the sensor's surface to enhance the signal and demonstrate high sensitivities to low concentration targets [148] (Table 9). MXenes-based SPR biosensors for the ultrasensitive detection of carcinoembryonic antigens (CEAs) were fabricated by Wu et al. These SPR biosensors demonstrate good reproducibility and high selectivity in human serum samples, providing a potential method for early clinical diagnoses and cancer surveillance [149]. Their group also constructed SPR biosensors for CEA using amino-functionalised MXenes [150]. MXenes were assembled on Au films to immobilise monoclonal anti-CEA antibody sensing materials covalently. MXenes were used as a substrate for binding hollow gold nanoparticles (HGNPs), and they were modified with SPA. MXenes/HGNPs/SPA/Ab_2 nanocomplexes act as signal enhancers for SPR-sensing components. The sensor offers a wide linear detection range, ultra-low detection limits, and good selectivity for CEA in human serums. Chen et al. anchored targeting aptamers with thiol-modified niobium carbide MXene quantum dots and could specifically bind the SARS-Cov-2 N gene, resulting in a change in the SPR signal for laser irradiation at a wavelength of 633 nm [151]. The above studies indicate the development of MXene-based biomarker sensing chips or devices in the field of SPR to broaden the area of MXenes biosensing applications.

Table 9. MXene-based SPR biosensors identify units, target, and analytical parameters.

MXenes Composite	Detection Technology	Target	LOD	Range	Ref.
Ti_3C_2/AuNPs	Antibody	carcinoembryonic antigen	0.2 fM–20 nM	0.07 fM	[149]
APTES/Ti_3C_2/HGNPs	Antibody	carcinoembryonic antigen	0.001–1000 pM	0.15 fM	[150]
Nb_2C-SH QDs	Aptamer	N-gene of SARS CoV-2	4.9 pg mL^{-1}	0.05–100 ng mL^{-1}	[151]

4. Conclusions and Outlook

From its discovery to the present moment, MXenes, an emerging two-dimensional biosensing material, has achieved unprecedented rapid development. Although the development time is relatively short, synthesis methods of MXenes are constantly innovating and developing towards the direction of green environmental protection, simplicity and convenience, and controllable surface groups. MXenes are known for their tunable surface groups; good hydrophilicity and biocompatibility; excellent mechanical, electrical, and optical properties; and specific morphological structures. They have become a focus of research in electrochemical and optical biosensing.

HF synthesis is undoubtedly convenient and quick, giving access to abundant hydrophilic surface groups. However, using F^- can cause a certain amount of pollution to humans, the environment, and even the ecosystem. For the synthesis of MXenes, synthetic research has always progressed in the direction of less or no fluorine. Different synthetic methods and raw materials can adjust different surface groups. The structure determines the properties, and mechanical, hydrophilic, biocompatible, electrical, and optical properties can be adjusted according to different research needs using different synthetic routes.

From its large specific surface area, high electrical conductivity, large surface groups, and fast electron transfer properties, MXenes are often used as modification and immobilisation materials in electrochemical biosensors to improve electrocatalytic performances and detection sensitivities and to reduce redox potentials in order to obtain high-performance composites. However, most of the current MXene-based electrochemical biosensors use composite materials adsorbed on MXene's surface. There are relatively few reports on surface groups on MXenes as the adsorption media for the adsorption of biomolecules to prepare electrochemical biosensors. With the development of the tunable functionalisation of MXenes surface groups, we believe there is more significant potential for the direct binding of MXenes surfaces to biomolecules, such as proteins and nucleic acids. Although optical biosensors have been less studied than electrochemical biosensors, optical sensing is also a vital detection strategy in biosensing. By benefitting from the fluorescence quenching effect of MXenes and MQD's fluorescence enhancement effect, photoluminescence has been reported relatively more often in optical biosensing work. From the above studies, it can be concluded that nucleic acid-based biosensors are less selective than enzyme and immune-based biosensors. However, nucleic acids are smaller than antibodies and enzymes, so nucleic-acid-based biosensors have a higher density of surface modifications and have a higher sensitivity and reproducibility.

In contrast, SPR, SERS, and other optical sensing have been reported less often. Even so, the sparseness and singularity of functional groups on MXene's surface led to the reduced binding of biomolecules or other organic molecules. Meanwhile, the study of MQDs with high fluorescence efficiency, quantum yield, and different luminescence wavelengths has become an urgent problem.

About the main challenges mentioned above, we believe that several aspects can be studied for future developments and trends in MXenes biosensing.

(1) MXenes and MQDs synthesis methods are supposed to develop with adjustable size and surface groups. It is necessary to study further study the interaction between the composition, structure, properties, and biomolecules of MXenes in order to understand the appropriate MXene-adaptation-related biosensors that can be developed according to the actual situation. Furthermore, it can provide theoretical support for the development and development of MXenes morphological structure, size, and surface groups, and the study of interaction with biomolecules with the help of machine learning methods.

(2) The direct construction of biosensors on MXenes surfaces: The abundant surface groups of MXenes can be complexed or combined with many biomolecules, and enzyme-based electrochemical sensors have been partially verified. Nevertheless, there are relatively few studies on nucleic acid, antibodies, and other biomolecules combined with MXenes and constructed in sensors. Moreover, the natural binding process can shorten the experimental time and reduce errors caused by too many experimental procedures. At the same time, it is also very convenient for developed point-of-care tests. During the COVID-19 epidemic, point-of-care was necessary for rapid diagnoses and the timely treatment of patients.

(3) Utilise the catalytic properties and the reduction of MXenes: For electrochemical biosensing, MXenes can reduce precious metal nanoparticles in situ. They can also be used to catalyse redox reactions of O_2 in systems that promote free radical reactions in ECL, and improving the ECL signal is well worth exploring. MXenes are also used as semiconductor materials during electrochemical reactions and can be used to enhance the electrochemical signal. Therefore, reducing the detection limit, extending the detection range, and improving the sensitivity during electrochemical biosensing are possible.

(4) Develop MQDs with high fluorescence efficiency, quantum yield, and different fluorescence emission wavelengths: As a result, the sensitivity and detection limit of fluorescence detections can be improved, and they can be used in optical biosensors

with different wavelengths. MQDs can be extended to cell imaging, photothermal therapy, and other biomedical tissue applications.

Author Contributions: Conceptualization, D.L. and H.Z.; writing—original draft preparation, D.L.; writing—review and editing, D.L., X.Z. and H.Z.; project administration, Y.C. and L.F.; supervision, L.F. All authors have read and agreed to the published version of the manuscript.

Funding: This work was supported by the National Natural Science Foundation of China (No. 21705106) and the National Science Foundation of China (No. 22177067); the Program for Professor of Special Appointment (Eastern Scholar) at Shanghai Institutions of Higher Learning (No. TP2016023); and the Shanghai Sailing Program (No. 20YF1413000).

Institutional Review Board Statement: Not applicable.

Informed Consent Statement: Not applicable.

Data Availability Statement: Not applicable.

Conflicts of Interest: The authors declare no conflict of interest.

References

1. Yang, F.; Song, P.; Ruan, M.; Xu, W. Recent progress in two-dimensional nanomaterials: Synthesis, engineering, and applications. *FlatChem* **2019**, *18*, 100133. [CrossRef]
2. Choi, S.-J.; Kim, I.-D. Recent Developments in 2D Nanomaterials for Chemiresistive-Type Gas Sensors. *Electron. Mater. Lett.* **2018**, *14*, 221–260. [CrossRef]
3. Deshmukh, K.; Kovářík, T.; Khadheer Pasha, S.K. State of the art recent progress in two dimensional MXenes based gas sensors and biosensors: A comprehensive review. *Coord. Chem. Rev.* **2020**, *424*, 213514. [CrossRef]
4. Naguib, M.; Kurtoglu, M.; Presser, V.; Lu, J.; Niu, J.; Heon, M.; Hultman, L.; Gogotsi, Y.; Barsoum, M.W. Two-dimensional nanocrystals produced by exfoliation of Ti3AlC2. *Adv. Mater.* **2011**, *23*, 4248–4253. [CrossRef] [PubMed]
5. Huang, X.; Wu, P. A Facile, High-Yield, and Freeze-and-Thaw-Assisted Approach to Fabricate MXene with Plentiful Wrinkles and Its Application in On-Chip Micro-Supercapacitors. *Adv. Funct. Mater.* **2020**, *30*, 1910048. [CrossRef]
6. Ma, Y.; Li, B.; Yang, S. Ultrathin two-dimensional metallic nanomaterials. *Mater. Chem. Front.* **2018**, *2*, 456–467. [CrossRef]
7. Ghazaly, A.E.; Ahmed, H.; Rezk, A.R.; Halim, J.; Persson, P.O.Å.; Yeo, L.Y.; Rosen, J. Ultrafast, One-Step, Salt-Solution-Based Acoustic Synthesis of Ti_3C_2 MXene. *ACS Nano* **2021**, *15*, 4287–4293. [CrossRef]
8. Qing, H.; Mian, L.I. Recent Progress and Prospects of Ternary Layered Carbides/Nitrides MAX Phases and Their Derived Two-Dimensional Nanolaminates MXenes. *J. Inorg. Mater.* **2019**, *35*, 1–7. [CrossRef]
9. Naguib, M.; Mochalin, V.N.; Barsoum, M.W.; Gogotsi, Y. 25th anniversary article: MXenes: A new family of two-dimensional materials. *Adv. Mater.* **2014**, *26*, 992–1005. [CrossRef]
10. Li, X.; Lu, Y.; Liu, Q. Electrochemical and optical biosensors based on multifunctional MXene nanoplatforms: Progress and prospects. *Talanta* **2021**, *235*, 122726. [CrossRef]
11. Wang, L.; Zhang, H.; Zhuang, T.; Liu, J.; Sojic, N.; Wang, Z. Sensitive electrochemiluminescence biosensing of polynucleotide kinase using the versatility of two-dimensional Ti_3C_2TX MXene nanomaterials. *Anal. Chim. Acta* **2022**, *1191*, 339346. [CrossRef] [PubMed]
12. Naguib, M.; Barsoum, M.W.; Gogotsi, Y. Ten Years of Progress in the Synthesis and Development of MXenes. *Adv. Mater.* **2021**, *33*, 2103393. [CrossRef] [PubMed]
13. Roointan, A.; Ahmad Mir, T.; Ibrahim Wani, S.; Mati Ur, R.; Hussain, K.K.; Ahmed, B.; Abrahim, S.; Savardashtaki, A.; Gandomani, G.; Gandomani, M.; et al. Early detection of lung cancer biomarkers through biosensor technology: A review. *J. Pharm. Biomed. Anal.* **2019**, *164*, 93–103. [CrossRef] [PubMed]
14. Zhou, Y.; Kubota, L.T. Trends in Electrochemical Sensing. *ChemElectroChem* **2020**, *7*, 3684–3685. [CrossRef]
15. Mohammadniaei, M.; Nguyen, H.V.; Tieu, M.V.; Lee, M.-H. 2D Materials in Development of Electrochemical Point-of-Care Cancer Screening Devices. *Micromachines* **2019**, *10*, 662. [CrossRef]
16. Wrobel, T.P.; Bhargava, R. Infrared Spectroscopic Imaging Advances as an Analytical Technology for Biomedical Sciences. *Anal. Chem.* **2018**, *90*, 1444–1463. [CrossRef]
17. Kumar, J.A.; Prakash, P.; Krithiga, T.; Amarnath, D.J.; Premkumar, J.; Rajamohan, N.; Vasseghian, Y.; Saravanan, P.; Rajasimman, M. Methods of synthesis, characteristics, and environmental applications of MXene: A comprehensive review. *Chemosphere* **2022**, *286 Pt 1*, 131607. [CrossRef]
18. Huang, J.; Li, Z.; Mao, Y.; Li, Z. Progress and biomedical applications of MXenes. *Nano Sel.* **2021**, *2*, 1480–1508. [CrossRef]
19. Zhang, J.; Li, Y.; Duan, S.; He, F. Highly electrically conductive two-dimensional Ti_3C_2 Mxenes-based 16S rDNA electrochemical sensor for detecting Mycobacterium tuberculosis. *Anal. Chim. Acta* **2020**, *1123*, 9–17. [CrossRef]
20. Lu, L.; Han, X.; Lin, J.; Zhang, Y.; Qiu, M.; Chen, Y.; Li, M.; Tang, D. Ultrasensitive fluorometric biosensor based on Ti_3C_2 MXenes with Hg(2+)-triggered exonuclease III-assisted recycling amplification. *Analyst* **2021**, *146*, 2664–2669. [CrossRef]

21. Naguib, M.; Mashtalir, O.; Carle, J.; Presser, V.; Lu, J.; Hultman, L.; Gogotsi, Y.; Barsoum, M.W. Two-Dimensional Transition Metal Carbides. *ACS Nano* **2012**, *6*, 1322–1331. [CrossRef] [PubMed]
22. Alhabeb, M.; Maleski, K.; Anasori, B.; Lelyukh, P.; Clark, L.; Sin, S.; Gogotsi, Y. Guidelines for Synthesis and Processing of Two-Dimensional Titanium Carbide (Ti_3C_2Tx MXene). *Chem. Mater.* **2017**, *29*, 7633–7644. [CrossRef]
23. Ghidiu, M.; Lukatskaya, M.R.; Zhao, M.Q.; Gogotsi, Y.; Barsoum, M.W. Conductive two-dimensional titanium carbide 'clay' with high volumetric capacitance. *Nature* **2014**, *516*, 78–81. [CrossRef]
24. Sang, X.; Xie, Y.; Lin, M.W.; Alhabeb, M.; Van Aken, K.L.; Gogotsi, Y.; Kent, P.R.C.; Xiao, K.; Unocic, R.R. Atomic Defects in Monolayer Titanium Carbide (Ti_3C_2Tx) MXene. *ACS Nano* **2016**, *10*, 9193–9200. [CrossRef] [PubMed]
25. Shahzad, F.; Alhabeb, M.; Hatter Christine, B.; Anasori, B.; Man Hong, S.; Koo Chong, M.; Gogotsi, Y. Electromagnetic interference shielding with 2D transition metal carbides (MXenes). *Science* **2016**, *353*, 1137–1140. [CrossRef] [PubMed]
26. Song, M.; Pang, S.Y.; Guo, F.; Wong, M.C.; Hao, J. Fluoride-Free 2D Niobium Carbide MXenes as Stable and Biocompatible Nanoplatforms for Electrochemical Biosensors with Ultrahigh Sensitivity. *Adv. Sci* **2020**, *7*, 2001546. [CrossRef] [PubMed]
27. Gao, L.; Li, C.; Huang, W.; Mei, S.; Lin, H.; Ou, Q.; Zhang, Y.; Guo, J.; Zhang, F.; Xu, S.; et al. MXene/Polymer Membranes: Synthesis, Properties, and Emerging Applications. *Chem. Mater.* **2020**, *32*, 1703–1747. [CrossRef]
28. Tian, Z.; Wei, C.; Sun, J. Recent advances in the template-confined synthesis of two-dimensional materials for aqueous energy storage devices. *Nanoscale Adv.* **2020**, *2*, 2220–2233. [CrossRef]
29. Zhang, F.; Zhang, Z.; Wang, H.; Chan, C.H.; Chan, N.Y.; Chen, X.X.; Dai, J.-Y. Plasma-enhanced pulsed-laser deposition of single-crystalline Mo2C ultrathin superconducting films. *Phys. Rev. Mater.* **2017**, *1*, 034002. [CrossRef]
30. Li, M.; Lu, J.; Luo, K.; Li, Y.; Chang, K.; Chen, K.; Zhou, J.; Rosen, J.; Hultman, L.; Eklund, P.; et al. Element Replacement Approach by Reaction with Lewis Acidic Molten Salts to Synthesize Nanolaminated MAX Phases and MXenes. *J. Am. Chem. Soc.* **2019**, *141*, 4730–4737. [CrossRef]
31. Sun, W.; Wang, X.; Feng, J.; Li, T.; Huan, Y.; Qiao, J.; He, L.; Ma, D. Controlled synthesis of 2D Mo2C/graphene heterostructure on liquid Au substrates as enhanced electrocatalytic electrodes. *Nanotechnology* **2019**, *30*, 385601. [CrossRef] [PubMed]
32. Srivastava, P.; Mishra, A.; Mizuseki, H.; Lee, K.R.; Singh, A.K. Mechanistic Insight into the Chemical Exfoliation and Functionalization of Ti_3C_2 MXene. *ACS Appl Mater. Interfaces* **2016**, *8*, 24256–24264. [CrossRef] [PubMed]
33. Li, X.; Huang, Z.; Zhi, C. Environmental Stability of MXenes as Energy Storage Materials. *Front. Mater.* **2019**, *6*, 312. [CrossRef]
34. Huang, S.; Mochalin, V.N. Hydrolysis of 2D Transition-Metal Carbides (MXenes) in Colloidal Solutions. *Inorg. Chem.* **2019**, *58*, 1958–1966. [CrossRef]
35. Zhang, C.J.; Pinilla, S.; McEvoy, N.; Cullen, C.P.; Anasori, B.; Long, E.; Park, S.-H.; Seral-Ascaso, A.; Shmeliov, A.; Krishnan, D.; et al. Oxidation Stability of Colloidal Two-Dimensional Titanium Carbides (MXenes). *Chem. Mater.* **2017**, *29*, 4848–4856. [CrossRef]
36. Tan, T.L.; Jin, H.M.; Sullivan, M.B.; Anasori, B.; Gogotsi, Y. High-Throughput Survey of Ordering Configurations in MXene Alloys Across Compositions and Temperatures. *ACS Nano* **2017**, *11*, 4407–4418. [CrossRef] [PubMed]
37. Cheng, Y.-W.; Dai, J.-H.; Zhang, Y.-M.; Song, Y. Two-Dimensional, Ordered, Double Transition Metal Carbides (MXenes): A New Family of Promising Catalysts for the Hydrogen Evolution Reaction. *J. Phys. Chem. C* **2018**, *122*, 28113–28122. [CrossRef]
38. Tao, Q.; Dahlqvist, M.; Lu, J.; Kota, S.; Meshkian, R.; Halim, J.; Palisaitis, J.; Hultman, L.; Barsoum, M.W.; Persson, P.O.A.; et al. Two-dimensional Mo1.33C MXene with divacancy ordering prepared from parent 3D laminate with in-plane chemical ordering. *Nat. Commun.* **2017**, *8*, 14949. [CrossRef]
39. Halim, J.; Palisaitis, J.; Lu, J.; Thörnberg, J.; Moon, E.J.; Precner, M.; Eklund, P.; Persson, P.O.Å.; Barsoum, M.W.; Rosen, J. Synthesis of Two-Dimensional Nb1.33C (MXene) with Randomly Distributed Vacancies by Etching of the Quaternary Solid Solution (Nb2/3Sc1/3)2AlC MAX Phase. *ACS Appl. Nano Mater.* **2018**, *1*, 2455–2460. [CrossRef]
40. Zhang, P.; Yang, X.J.; Li, P.; Zhao, Y.; Niu, Q.J. Fabrication of novel MXene (Ti_3C_2)/polyacrylamide nanocomposite hydrogels with enhanced mechanical and drug release properties. *Soft Matter* **2020**, *16*, 162–169. [CrossRef]
41. Khaledialidusti, R.; Anasori, B.; Barnoush, A. Temperature-dependent mechanical properties of Tin+1CnO2 (n = 1, 2) MXene monolayers: A first-principles study. *Phys. Chem Chem Phys.* **2020**, *22*, 3414–3424. [CrossRef]
42. Ronchi, R.M.; Arantes, J.T.; Santos, S.F. Synthesis, structure, properties and applications of MXenes: Current status and perspectives. *Ceram. Int.* **2019**, *45*, 18167–18188. [CrossRef]
43. Wu, X.; Ma, P.; Sun, Y.; Du, F.; Song, D.; Xu, G. Application of MXene in Electrochemical Sensors: A Review. *Electroanalysis* **2021**, *33*, 1827–1851. [CrossRef]
44. Manzanares-Palenzuela, C.L.; Pourrahimi, A.M.; Gonzalez-Julian, J.; Sofer, Z.; Pykal, M.; Otyepka, M.; Pumera, M. Interaction of single- and double-stranded DNA with multilayer MXene by fluorescence spectroscopy and molecular dynamics simulations. *Chem. Sci.* **2019**, *10*, 10010–10017. [CrossRef]
45. Huang, Z.; Liu, B.; Liu, J. Mn(2+)-Assisted DNA Oligonucleotide Adsorption on Ti2C MXene Nanosheets. *Langmuir* **2019**, *35*, 9858–9866. [CrossRef]
46. Vural, M.; Zhu, H.; Pena-Francesch, A.; Jung, H.; Allen, B.D.; Demirel, M.C. Self-Assembly of Topologically Networked Protein–Ti_3C_2Tx MXene Composites. *ACS Nano* **2020**, *14*, 6956–6967. [CrossRef]
47. Huang, R.; Chen, X.; Dong, Y.; Zhang, X.; Wei, Y.; Yang, Z.; Li, W.; Guo, Y.; Liu, J.; Yang, Z.; et al. MXene Composite Nanofibers for Cell Culture and Tissue Engineering. *ACS Appl. Bio Mater.* **2020**, *3*, 2125–2131. [CrossRef]

48. Shahzad, F.; Zaidi, S.A.; Naqvi, R.A. 2D Transition Metal Carbides (MXene) for Electrochemical Sensing: A Review. *Crit Rev. Anal. Chem.* **2020**, *52*, 848–864. [CrossRef] [PubMed]
49. Anasori, B.; Xie, Y.; Beidaghi, M.; Lu, J.; Hosler, B.C.; Hultman, L.; Kent, P.R.C.; Gogotsi, Y.; Barsoum, M.W. Two-Dimensional, Ordered, Double Transition Metals Carbides (MXenes). *ACS Nano* **2015**, *9*, 9507–9516. [CrossRef]
50. Verger, L.; Natu, V.; Carey, M.; Barsoum, M.W. MXenes: An Introduction of Their Synthesis, Select Properties, and Applications. *Trends Chem.* **2019**, *1*, 656–669. [CrossRef]
51. Cho, I.H.; Kim, D.H.; Park, S. Electrochemical biosensors: Perspective on functional nanomaterials for on-site analysis. *Biomater. Res.* **2020**, *24*, 6. [CrossRef] [PubMed]
52. Chia, H.L.; Mayorga-Martinez, C.C.; Antonatos, N.; Sofer, Z.; Gonzalez-Julian, J.J.; Webster, R.D.; Pumera, M. MXene Titanium Carbide-based Biosensor: Strong Dependence of Exfoliation Method on Performance. *Anal. Chem.* **2020**, *92*, 2452–2459. [CrossRef]
53. Xu, W.; Sakran, M.; Fei, J.; Li, X.; Weng, C.; Yang, W.; Zhu, G.; Zhu, W.; Zhou, X. Electrochemical Biosensor Based on HRP/Ti_3C_2/Nafion Film for Determination of Hydrogen Peroxide in Serum Samples of Patients with Acute Myocardial Infarction. *ACS Biomater. Sci. Eng.* **2021**, *7*, 2767–2773. [CrossRef]
54. Ma, B.K.; Li, M.; Cheong, L.Z.; Weng, X.C.; Shen, C.; Huang, Q. Enzyme-MXene Nanosheets: Fabrication and Application in Electrochemical Detection of H_2O_2. *J. Inorg. Mater.* **2020**, *35*, 131. [CrossRef]
55. Rakhi, R.B.; Nayak, P.; Xia, C.; Alshareef, H.N. Novel amperometric glucose biosensor based on MXene nanocomposite. *Sci. Rep.* **2016**, *6*, 36422. [CrossRef] [PubMed]
56. Wu, M.; Zhang, Q.; Fang, Y.; Deng, C.; Zhou, F.; Zhang, Y.; Wang, X.; Tang, Y.; Wang, Y. Polylysine-modified MXene nanosheets with highly loaded glucose oxidase as cascade nanoreactor for glucose decomposition and electrochemical sensing. *J. Colloid Interface Sci.* **2021**, *586*, 20–29. [CrossRef] [PubMed]
57. Murugan, P.; Annamalai, J.; Atchudan, R.; Govindasamy, M.; Nallaswamy, D.; Ganapathy, D.; Reshetilov, A.; Sundramoorthy, A.K. Electrochemical Sensing of Glucose Using Glucose Oxidase/PEDOT:4-Sulfocalix [4]arene/MXene Composite Modified Electrode. *Micromachines* **2022**, *13*, 304. [CrossRef] [PubMed]
58. Xia, T.; Liu, G.; Wang, J.; Hou, S.; Hou, S. MXene-based enzymatic sensor for highly sensitive and selective detection of cholesterol. *Biosens. Bioelectron.* **2021**, *183*, 113243. [CrossRef] [PubMed]
59. Wu, L.; Lu, X.; Dhanjai; Wu, Z.-S.; Dong, Y.; Wang, X.; Zheng, S.; Chen, J. 2D transition metal carbide MXene as a robust biosensing platform for enzyme immobilization and ultrasensitive detection of phenol. *Biosens. Bioelectron.* **2018**, *107*, 69–75. [CrossRef] [PubMed]
60. Zhou, L.Y.; Zhang, X.; Ma, L.; Gao, J.; Jiang, Y.J. Acetylcholinesterase/chitosan-transition metal carbides nanocomposites-based biosensor for the organophosphate pesticides detection. *Biochem. Eng. J.* **2017**, *128*, 243–249. [CrossRef]
61. Du, Y.; Dong, S. Nucleic Acid Biosensors: Recent Advances and Perspectives. *Anal. Chem.* **2017**, *89*, 189–215. [CrossRef]
62. Wu, Y.; Arroyo-Currás, N. Advances in nucleic acid architectures for electrochemical sensing. *Curr. Opin. Electrochem.* **2021**, *27*, 100695. [CrossRef]
63. Liu, L.; Wei, Y.; Jiao, S.; Zhu, S.; Liu, X. A novel label-free strategy for the ultrasensitive miRNA-182 detection based on MoS2/Ti_3C_2 nanohybrids. *Biosens. Bioelectron.* **2019**, *137*, 45–51. [CrossRef] [PubMed]
64. Zhou, S.; Gu, C.; Li, Z.; Yang, L.; He, L.; Wang, M.; Huang, X.; Zhou, N.; Zhang, Z. Ti_3C_2Tx MXene and polyoxometalate nanohybrid embedded with polypyrrole: Ultra-sensitive platform for the detection of osteopontin. *Appl. Surf. Sci.* **2019**, *498*, 143889. [CrossRef]
65. Wang, H.; Li, H.; Huang, Y.; Xiong, M.; Wang, F.; Li, C. A label-free electrochemical biosensor for highly sensitive detection of gliotoxin based on DNA nanostructure/MXene nanocomplexes. *Biosens. Bioelectron.* **2019**, *142*, 111531. [CrossRef]
66. Rasheed, P.A.; Pandey, R.P.; Jabbar, K.A.; Mahmoud, K.A. Nb4C3Tx(MXene)/Au/DNA Aptasensor for the Ultraselective Electrochemical Detection of Lead in Water Samples. *Electroanalysis* **2022**, *34*, 1–8. [CrossRef]
67. Liu, Y.; Huang, S.J.; Li, J.N.; Wang, M.H.; Wang, C.B.; Hu, B.; Zhou, N.; Zhang, Z.H. 0D/2D heteronanostructure-integrated bimetallic CoCu-ZIF nanosheets and MXene-derived carbon dots for impedimetric cytosensing of melanoma B16-F10 cells. *Microchim. Acta* **2021**, *188*, 69. [CrossRef]
68. Hu, Z.; Suo, Z.; Liu, W.; Zhao, B.; Xing, F.; Zhang, Y.; Feng, L. DNA conformational polymorphism for biosensing applications. *Biosens. Bioelectron.* **2019**, *131*, 237–249. [CrossRef]
69. Yoon, J.; Shin, M.; Lim, J.; Lee, J.Y.; Choi, J.W. Recent Advances in MXene Nanocomposite-Based Biosensors. *Biosensors* **2020**, *10*, 185. [CrossRef]
70. Khan, R.; Ben Aissa, S.; Sherazi, T.A.; Catanante, G.; Hayat, A.; Marty, J.L. Development of an Impedimetric Aptasensor for Label Free Detection of Patulin in Apple Juice. *Molecules* **2019**, *24*, 1017. [CrossRef]
71. Zhu, C.; Yang, G.; Li, H.; Du, D.; Lin, Y. Electrochemical sensors and biosensors based on nanomaterials and nanostructures. *Anal. Chem.* **2015**, *87*, 230–249. [CrossRef]
72. Duan, F.; Guo, C.; Hu, M.; Song, Y.; Wang, M.; He, L.; Zhang, Z.; Pettinari, R.; Zhou, L. Construction of the 0D/2D heterojunction of Ti_3C_2Tx MXene nanosheets and iron phthalocyanine quantum dots for the impedimetric aptasensing of microRNA-155. *Sens. Actuators B Chem.* **2020**, *310*, 127844. [CrossRef]
73. Mohammadniaei, M.; Koyappayil, A.; Sun, Y.; Min, J.; Lee, M.H. Gold nanoparticle/MXene for multiple and sensitive detection of oncomiRs based on synergetic signal amplification. *Biosens. Bioelectron.* **2020**, *159*, 112208. [CrossRef]

74. Chen, W.Y.; Lin, H.; Barui, A.K.; Gomez, A.M.U.; Wendt, M.K.; Stanciu, L.A. DNA-Functionalized Ti_3C_2Tx MXenes for Selective and Rapid Detection of SARS-CoV-2 Nucleocapsid Gene. *ACS Appl. Nano Mater.* **2021**, *5*, 1902–1910. [CrossRef]
75. Yu, R.J.; Xue, J.; Wang, Y.; Qiu, J.F.; Huang, X.Y.; Chen, A.Y.; Xue, J.J. Novel Ti_3C_2Tx MXene nanozyme with manageable catalytic activity and application to electrochemical biosensor. *J. Nanobiotechnol.* **2022**, *20*, 119. [CrossRef] [PubMed]
76. Li, Z.Y.; Li, D.Y.; Huang, L.; Hu, R.; Yang, T.; Yang, Y.H. An electrochemical aptasensor based on intelligent walking DNA nanomachine with cascade signal amplification powered by nuclease for Mucin 1 assay. *Anal. Chim. Acta* **2022**, *1214*, 339964. [CrossRef]
77. Vajhadin, F.; Mazloum-Ardakani, M.; Shahidi, M.; Moshtaghioun, S.M.; Haghiralsadat, F.; Ebadi, A.; Amini, A. MXene-based cytosensor for the detection of HER2-positive cancer cells using CoFe2O4@Ag magnetic nanohybrids conjugated to the HB5 aptamer. *Biosens. Bioelectron.* **2022**, *195*, 113626. [CrossRef] [PubMed]
78. Fu, Z.; Xiang, J. Aptamers, the Nucleic Acid Antibodies, in Cancer Therapy. *Int. J. Mol. Sci.* **2020**, *21*, 2793. [CrossRef]
79. Ranallo, S.; Porchetta, A.; Ricci, F. DNA-Based Scaffolds for Sensing Applications. *Anal. Chem.* **2019**, *91*, 44–59. [CrossRef]
80. Wei, X.; Wang, S.; Zhan, Y.; Kai, T.; Ding, P. Sensitive Identification of Microcystin-LR via a Reagent-Free and Reusable Electrochemical Biosensor Using a Methylene Blue-Labeled Aptamer. *Biosensors* **2022**, *12*, 556. [CrossRef] [PubMed]
81. Geng, X.; Zhang, M.; Long, H.; Hu, Z.; Zhao, B.; Feng, L.; Du, J. A reusable neurotransmitter aptasensor for the sensitive detection of serotonin. *Anal. Chim. Acta* **2021**, *1145*, 124–131. [CrossRef]
82. Fan, Y.; Shi, S.; Ma, J.; Guo, Y. A paper-based electrochemical immunosensor with reduced graphene oxide/thionine/gold nanoparticles nanocomposites modification for the detection of cancer antigen 125. *Biosens. Bioelectron.* **2019**, *135*, 1–7. [CrossRef] [PubMed]
83. Zhang, M.M.; Mei, L.S.; Zhang, L.; Wang, X.; Liao, X.C.; Qiao, X.W.; Hong, C.L. Ti_3C_2 MXene anchors CuAu-LDH multifunctional two-dimensional nanomaterials for dual-mode detection of CEA in electrochemical immunosensors. *Bioelectrochemistry* **2021**, *142*, 107943. [CrossRef] [PubMed]
84. Kumar, S.; Lei, Y.; Alshareef, N.H.; Quevedo-Lopez, M.A.; Salama, K.N. Biofunctionalized two-dimensional Ti_3C_2 MXenes for ultrasensitive detection of cancer biomarker. *Biosens. Bioelectron.* **2018**, *121*, 243–249. [CrossRef] [PubMed]
85. Medetalibeyoglu, H.; Kotan, G.; Atar, N.; Yola, M.L. A novel and ultrasensitive sandwich-type electrochemical immunosensor based on delaminated MXene@AuNPs as signal amplification for prostate specific antigen (PSA) detection and immunosensor validation. *Talanta* **2020**, *220*, 121403. [CrossRef]
86. Xu, Q.; Xu, J.K.; Jia, H.Y.; Tian, Q.Y.; Liu, P.; Chen, S.X.; Cai, Y.; Lu, X.Y.; Duan, X.M.; Lu, L.M. Hierarchical Ti_3C_2 MXene-derived sodium titanate nanoribbons/PEDOT for signal amplified electrochemical immunoassay of prostate specific antigen. *J. Electroanal. Chem.* **2022**, *860*, 113869. [CrossRef]
87. Dong, H.; Cao, L.; Tan, Z.; Liu, Q.; Zhou, J.; Zhao, P.; Wang, P.; Li, Y.; Ma, W.; Dong, Y. A Signal Amplification Strategy of CuPtRh CNB-Embedded Ammoniated Ti3C2 MXene for Detecting Cardiac Troponin I by a Sandwich-Type Electrochemical Immunosensor. *ACS Appl. Bio Mater.* **2020**, *3*, 377–384. [CrossRef]
88. Niu, H.M.; Cai, S.M.; Liu, X.K.; Huang, X.M.; Chen, J.; Wang, S.L.; Zhang, S.H. A novel electrochemical sandwich-like immunosensor based on carboxyl Ti_3C_2Tx MXene and rhodamine b/gold/reduced graphene oxide for Listeria monocytogenes. *Anal. Methods* **2022**, *14*, 843–849. [CrossRef]
89. Jiang, X.; Kuklin, A.V.; Baev, A.; Ge, Y.; Ågren, H.; Zhang, H.; Prasad, P.N. Two-dimensional MXenes: From morphological to optical, electric, and magnetic properties and applications. *Phys. Rep.* **2020**, *848*, 1–58. [CrossRef]
90. Bai, Y.; Zhou, K.; Srikanth, N.; Pang, J.H.L.; He, X.; Wang, R. Dependence of elastic and optical properties on surface terminated groups in two-dimensional MXene monolayers: A first-principles study. *RSC Adv.* **2016**, *6*, 35731–35739. [CrossRef]
91. Zhu, X.; Zhang, Y.; Liu, M.; Liu, Y. 2D titanium carbide MXenes as emerging optical biosensing platforms. *Biosens. Bioelectron.* **2021**, *171*, 112730. [CrossRef]
92. Xu, Q.; Ding, L.; Wen, Y.; Yang, W.; Zhou, H.; Chen, X.; Street, J.; Zhou, A.; Ong, W.-J.; Li, N. High photoluminescence quantum yield of 18.7% by using nitrogen-doped Ti3C2 MXene quantum dots. *J. Mater. Chem. C* **2018**, *6*, 6360–6369. [CrossRef]
93. Xu, G.; Niu, Y.; Yang, X.; Jin, Z.; Wang, Y.; Xu, Y.; Niu, H. Preparation of Ti_3C_2Tx MXene-Derived Quantum Dots with White/Blue-Emitting Photoluminescence and Electrochemiluminescence. *Adv. Opt. Mater.* **2018**, *6*, 1800951. [CrossRef]
94. Zhu, X.; Pang, X.; Zhang, Y.; Yao, S. Titanium carbide MXenes combined with red-emitting carbon dots as a unique turn-on fluorescent nanosensor for label-free determination of glucose. *J. Mater. Chem. B* **2019**, *7*, 7729–7735. [CrossRef]
95. Peng, X.; Zhang, Y.; Lu, D.; Guo, Y.; Guo, S. Ultrathin Ti_3C_2 nanosheets based "off-on" fluorescent nanoprobe for rapid and sensitive detection of HPV infection. *Sens. Actuators B Chem.* **2019**, *286*, 222–229. [CrossRef]
96. Wang, S.; Wei, S.; Wang, S.; Zhu, X.; Lei, C.; Huang, Y.; Nie, Z.; Yao, S. Chimeric DNA-Functionalized Titanium Carbide MXenes for Simultaneous Mapping of Dual Cancer Biomarkers in Living Cells. *Anal. Chem.* **2019**, *91*, 1651–1658. [CrossRef] [PubMed]
97. Kalkal, A.; Kadian, S.; Kumar, S.; Manik, G.; Sen, P.; Kumar, S.; Packirisamy, G. Ti3C2-MXene decorated with nanostructured silver as a dual-energy acceptor for the fluorometric neuron specific enolase detection. *Biosens. Bioelectron.* **2022**, *195*, 113620. [CrossRef] [PubMed]
98. Zhu, X.; Fan, L.; Wang, S.; Lei, C.; Huang, Y.; Nie, Z.; Yao, S. Phospholipid-Tailored Titanium Carbide Nanosheets as a Novel Fluorescent Nanoprobe for Activity Assay and Imaging of Phospholipase D. *Anal. Chem.* **2018**, *90*, 6742–6748. [CrossRef]

99. Shi, Y.E.; Han, F.; Xie, L.; Zhang, C.; Li, T.; Wang, H.; Lai, W.F.; Luo, S.; Wei, W.; Wang, Z.; et al. A MXene of type Ti_3C_2Tx functionalized with copper nanoclusters for the fluorometric determination of glutathione. *Microchim. Acta* **2019**, *187*, 38. [CrossRef]
100. Xue, Q.; Zhang, H.; Zhu, M.; Pei, Z.; Li, H.; Wang, Z.; Huang, Y.; Huang, Y.; Deng, Q.; Zhou, J.; et al. Photoluminescent Ti_3C_2 MXene Quantum Dots for Multicolor Cellular Imaging. *Adv. Mater.* **2017**, *29*. [CrossRef]
101. Guan, Q.; Ma, J.; Yang, W.; Zhang, R.; Zhang, X.; Dong, X.; Fan, Y.; Cai, L.; Cao, Y.; Zhang, Y.; et al. Highly fluorescent Ti3C2 MXene quantum dots for macrophage labeling and Cu(2+) ion sensing. *Nanoscale* **2019**, *11*, 14123–14133. [CrossRef]
102. Guo, Z.; Zhu, X.; Wang, S.; Lei, C.; Huang, Y.; Nie, Z.; Yao, S. Fluorescent Ti3C2 MXene quantum dots for an alkaline phosphatase assay and embryonic stem cell identification based on the inner filter effect. *Nanoscale* **2018**, *10*, 19579–19585. [CrossRef]
103. Liu, M.; Zhou, J.; He, Y.; Cai, Z.; Ge, Y.; Zhou, J.; Song, G. epsilon-Poly-L-lysine-protected Ti3C2 MXene quantum dots with high quantum yield for fluorometric determination of cytochrome c and trypsin. *Microchim Acta* **2019**, *186*, 770. [CrossRef] [PubMed]
104. Chen, X.; Sun, X.; Xu, W.; Pan, G.; Zhou, D.; Zhu, J.; Wang, H.; Bai, X.; Dong, B.; Song, H. Ratiometric photoluminescence sensing based on Ti3C2 MXene quantum dots as an intracellular pH sensor. *Nanoscale* **2018**, *10*, 1111–1118. [CrossRef] [PubMed]
105. Xu, Y.; Wang, X.; Zhang, W.L.; Lv, F.; Guo, S. Recent progress in two-dimensional inorganic quantum dots. *Chem. Soc. Rev.* **2018**, *47*, 586–625. [CrossRef]
106. Shao, B.; Liu, Z.; Zeng, G.; Wang, H.; Liang, Q.; He, Q.; Cheng, M.; Zhou, C.; Jiang, L.; Song, B. Two-dimensional transition metal carbide and nitride (MXene) derived quantum dots (QDs): Synthesis, properties, applications and prospects. *J. Mater. Chem. A* **2020**, *8*, 7508–7535. [CrossRef]
107. Sinha, A.; Dhanjai; Zhao, H.; Huang, Y.; Lu, X.; Chen, J.; Jain, R. MXene: An emerging material for sensing and biosensing. *TrAC Trends Anal. Chem.* **2018**, *105*, 424–435. [CrossRef]
108. Wang, Z.; Xuan, J.; Zhao, Z.; Li, Q.; Geng, F. Versatile Cutting Method for Producing Fluorescent Ultrasmall MXene Sheets. *ACS Nano* **2017**, *11*, 11559–11565. [CrossRef]
109. Desai, M.L.; Basu, H.; Singhal, R.K.; Saha, S.; Kailasa, S.K. Ultra-small two dimensional MXene nanosheets for selective and sensitive fluorescence detection of Ag+ and Mn_2+ ions. *Colloids Surf. A Physicochem. Eng. Asp.* **2019**, *565*, 70–77. [CrossRef]
110. Zhang, J.; Kerr, E.; Usman, K.A.S.; Doeven, E.H.; Francis, P.S.; Henderson, L.C.; Razal, J.M. Cathodic electrogenerated chemiluminescence of tris(2,2′-bipyridine)ruthenium(ii) and peroxydisulfate at pure Ti3C2Tx MXene electrodes. *Chem. Commun.* **2020**, *56*, 10022–10025. [CrossRef] [PubMed]
111. Fang, Y.; Yang, X.; Chen, T.; Xu, G.; Liu, M.; Liu, J.; Xu, Y. Two-dimensional titanium carbide (MXene)-based solid-state electrochemiluminescent sensor for label-free single-nucleotide mismatch discrimination in human urine. *Sens. Actuators B Chem.* **2018**, *263*, 400–407. [CrossRef]
112. Zhuang, T.T.; Zhang, H.X.; Wang, L.; Yu, L.H.; Wang, Z.H. Anchoring luminol based on Ti_3C_2-mediated in situ formation of Au NPs for construction of an efficient probe for miRNA electrogenerated chemiluminescence detection. *Anal. Bioanal. Chem.* **2021**, *413*, 6963–6971. [CrossRef]
113. Yao, B.; Zhang, J.; Fan, Z.; Ding, Y.; Zhou, B.; Yang, R.; Zhao, J.; Zhang, K. Rational Engineering of the DNA Walker Amplification Strategy by Using a Au@Ti_3C_2@PEI-Ru(dcbpy)3(2+) Nanocomposite Biosensor for Detection of the SARS-CoV-2 RdRp Gene. *ACS Appl. Mater. Interfaces* **2021**, *13*, 19816–19824. [CrossRef] [PubMed]
114. Zhang, K.; Fan, Z.; Huang, Y.; Ding, Y.; Xie, M. A strategy combining 3D-DNA Walker and CRISPR-Cas12a trans-cleavage activity applied to MXene based electrochemiluminescent sensor for SARS-CoV-2 RdRp gene detection. *Talanta* **2022**, *236*, 122868. [CrossRef] [PubMed]
115. Sun, Y.; Zhang, Y.; Zhang, H.; Liu, M.; Liu, Y. Integrating Highly Efficient Recognition and Signal Transition of g-C3N4 Embellished Ti_3C_2 MXene Hybrid Nanosheets for Electrogenerated Chemiluminescence Analysis of Protein Kinase Activity. *Anal. Chem.* **2020**, *92*, 10668–10676. [CrossRef]
116. Zhang, H.; Wang, Z.; Zhang, Q.; Wang, F.; Liu, Y. Ti_3C_2 MXenes nanosheets catalyzed highly efficient electrogenerated chemiluminescence biosensor for the detection of exosomes. *Biosens. Bioelectron.* **2019**, *124–125*, 184–190. [CrossRef]
117. Zhang, H.; Wang, Z.; Wang, F.; Zhang, Y.; Wang, H.; Liu, Y. In Situ Formation of Gold Nanoparticles Decorated Ti_3C_2 MXenes Nanoprobe for Highly Sensitive Electrogenerated Chemiluminescence Detection of Exosomes and Their Surface Proteins. *Anal. Chem.* **2020**, *92*, 5546–5553. [CrossRef] [PubMed]
118. Mi, X.; Li, H.; Tan, R.; Feng, B.; Tu, Y. The TDs/aptamer cTnI biosensors based on HCR and Au/Ti_3C_2-MXene amplification for screening serious patient in COVID-19 pandemic. *Biosens. Bioelectron.* **2021**, *192*, 113482. [CrossRef] [PubMed]
119. Luo, W.; Ye, Z.; Ma, P.; Wu, Q.; Song, D. Preparation of a disposable electrochemiluminescence sensor chip based on an MXene-loaded ruthenium luminescent agent and its application in the detection of carcinoembryonic antigens. *Analyst* **2022**, *147*, 1986–1994. [CrossRef] [PubMed]
120. Wei, W.; Lin, H.; Hao, T.; Su, X.; Jiang, X.; Wang, S.; Hu, Y.; Guo, Z. Dual-mode ECL/SERS immunoassay for ultrasensitive determination of Vibrio vulnificus based on multifunctional MXene. *Sens. Actuators B Chem.* **2021**, *332*, 129525. [CrossRef]
121. Yan, Z.; Wang, Z.; Miao, Z.; Liu, Y. Dye-Sensitized and Localized Surface Plasmon Resonance Enhanced Visible-Light Photoelectrochemical Biosensors for Highly Sensitive Analysis of Protein Kinase Activity. *Anal. Chem.* **2016**, *88*, 922–929. [CrossRef]
122. Li, M.; Wang, H.; Wang, X.; Lu, Q.; Li, H.; Zhang, Y.; Yao, S. Ti_3C_2/Cu_2O heterostructure based signal-off photoelectrochemical sensor for high sensitivity detection of glucose. *Biosens. Bioelectron.* **2019**, *142*, 111535. [CrossRef]

123. Chen, G.; Wang, H.; Wei, X.; Wu, Y.; Gu, W.; Hu, L.; Xu, D.; Zhu, C. Efficient Z-Scheme heterostructure based on TiO2/Ti3C2Tx/Cu2O to boost photoelectrochemical response for ultrasensitive biosensing. *Sens. Actuators B Chem.* **2020**, *312*, 127951. [CrossRef]
124. Fu, Y.; Ding, F.; Chen, J.; Liu, M.; Zhang, X.; Du, C.; Si, S. Label-free and near-zero-background-noise photoelectrochemical assay of methyltransferase activity based on a Bi2S3/Ti3C2 Schottky junction. *Chem. Commun.* **2020**, *56*, 5799–5802. [CrossRef] [PubMed]
125. Qiu, Z.L.; Fan, D.C.; Xue, X.H.; Zhang, J.Y.; Xu, J.L.; Lyu, H.X.; Chen, Y.T. Ti_3C_2 MXene-anchored photoelectrochemical detection of exosomes by in situ fabrication of CdS nanoparticles with enzyme-assisted hybridization chain reaction. *RSC Adv.* **2022**, *12*, 14260–14267. [CrossRef]
126. Liu, L.; Yao, Y.; Ma, K.J.; Shangguan, C.J.; Jiao, S.L.; Zhu, S.Y.; Xu, X.X. Ultrasensitive photoelectrochemical detection of cancer-related miRNA-141 by carrier recombination inhibition in hierarchical Ti_3C_2@ReS_2. *Sens. Actuators B Chem.* **2021**, *331*, 129470. [CrossRef]
127. Zheng, Y.L.; Zhou, Y.L.; Cui, X.T.; Yin, H.S.; Ai, S.Y. Enhanced photoactivity of CdS nanorods by MXene and $ZnSnO_3$: Application in photoelectrochemical biosensor for the effect of environmental pollutants on DNA hydroxymethylation in wheat tissues. *Mater. Today Chem.* **2022**, *24*, 100878. [CrossRef]
128. Chen, X.; Li, J.; Pan, G.; Xu, W.; Zhu, J.; Zhou, D.; Li, D.; Chen, C.; Lu, G.; Song, H. Ti_3C_2 MXene quantum dots/TiO_2 inverse opal heterojunction electrode platform for superior photoelectrochemical biosensing. *Sens. Actuators B Chem.* **2019**, *289*, 131–137. [CrossRef]
129. Kurtoglu, M.; Naguib, M.; Gogotsi, Y.; Barsoum, M.W. First principles study of two-dimensional early transition metal carbides. *MRS Commun.* **2012**, *2*, 133–137. [CrossRef]
130. Kim, J.; Campbell, A.S.; de Ávila, B.E.-F.; Wang, J. Wearable biosensors for healthcare monitoring. *Nat. Biotechnol.* **2019**, *37*, 389–406. [CrossRef]
131. An, H.; Habib, T.; Shah, S.; Gao, H.; Radovic, M.; Green, M.J.; Lutkenhaus, J.L. Surface-agnostic highly stretchable and bendable conductive MXene multilayers. *Sci. Adv.* **2018**, *4*, eaaq0118. [CrossRef]
132. Nah, J.S.; Barman, S.C.; Zahed, M.A.; Sharifuzzaman, M.; Yoon, H.; Park, C.; Yoon, S.; Zhang, S.; Park, J.Y. A wearable microfluidics-integrated impedimetric immunosensor based on Ti_3C_2Tx MXene incorporated laser-burned graphene for noninvasive sweat cortisol detection. *Sens. Actuators B Chem.* **2021**, *329*, 129206. [CrossRef]
133. Lei, Y.; Zhao, W.; Zhang, Y.; Jiang, Q.; He, J.-H.; Baeumner, A.J.; Wolfbeis, O.S.; Wang, Z.L.; Salama, K.N.; Alshareef, H.N. A MXene-Based Wearable Biosensor System for High-Performance In Vitro Perspiration Analysis. *Small* **2019**, *15*, 1901190. [CrossRef]
134. Li, M.; Wang, L.; Liu, R.; Li, J.; Zhang, Q.; Shi, G.; Li, Y.; Hou, C.; Wang, H. A highly integrated sensing paper for wearable electrochemical sweat analysis. *Biosens. Bioelectron.* **2021**, *174*, 112828. [CrossRef]
135. Zhang, S.; Zahed, M.A.; Sharifuzzaman, M.; Yoon, S.; Hui, X.; Chandra Barman, S.; Sharma, S.; Yoon, H.S.; Park, C.; Park, J.Y. A wearable battery-free wireless and skin-interfaced microfluidics integrated electrochemical sensing patch for on-site biomarkers monitoring in human perspiration. *Biosens. Bioelectron.* **2021**, *175*, 112844. [CrossRef]
136. Saleh, A.; Wustoni, S.; Bihar, E.; El-Demellawi, J.K.; Zhang, Y.; Hama, A.; Druet, V.; Yudhanto, A.; Lubineau, G.; Alshareef, H.N.; et al. Inkjet-printed Ti3C2Tx MXene electrodes for multimodal cutaneous biosensing. *J. Phys. Mater.* **2020**, *3*, 044004. [CrossRef]
137. Chen, L.; Chen, F.; Liu, G.; Lin, H.; Bao, Y.; Han, D.; Wang, W.; Ma, Y.; Zhang, B.; Niu, L. Superhydrophobic Functionalized Ti_3C_2Tx MXene-Based Skin-Attachable and Wearable Electrochemical pH Sensor for Real-Time Sweat Detection. *Anal. Chem.* **2022**, *94*, 7319–7328. [CrossRef]
138. Ma, Y.; Liu, N.; Li, L.; Hu, X.; Zou, Z.; Wang, J.; Luo, S.; Gao, Y. A highly flexible and sensitive piezoresistive sensor based on MXene with greatly changed interlayer distances. *Nat. Commun.* **2017**, *8*, 1207. [CrossRef] [PubMed]
139. Cai, Y.; Shen, J.; Ge, G.; Zhang, Y.; Jin, W.; Huang, W.; Shao, J.; Yang, J.; Dong, X. Stretchable Ti_3C_2Tx MXene/Carbon Nanotube Composite Based Strain Sensor with Ultrahigh Sensitivity and Tunable Sensing Range. *ACS Nano* **2018**, *12*, 56–62. [CrossRef]
140. Gao, Y.; Yan, C.; Huang, H.; Yang, T.; Tian, G.; Xiong, D.; Chen, N.; Chu, X.; Zhong, S.; Deng, W.; et al. Microchannel-Confined MXene Based Flexible Piezoresistive Multifunctional Micro-Force Sensor. *Adv. Funct. Mater.* **2020**, *30*, 1909603. [CrossRef]
141. Mathew, M.; Rout, C.S. Electrochemical biosensors based on Ti_3C_2Tx MXene: Future perspectives for on-site analysis. *Curr. Opin. Electrochem.* **2021**, *30*, 100782. [CrossRef]
142. Heikenfeld, J.; Jajack, A.; Rogers, J.; Gutruf, P.; Tian, L.; Pan, T.; Li, R.; Khine, M.; Kim, J.; Wang, J.; et al. Wearable sensors: Modalities, challenges, and prospects. *Lab. Chip* **2018**, *18*, 217–248. [CrossRef]
143. Lei, Y.; Cui, Y.; Huang, Q.; Dou, J.; Gan, D.; Deng, F.; Liu, M.; Li, X.; Zhang, X.; Wei, Y. Facile preparation of sulfonic groups functionalized Mxenes for efficient removal of methylene blue. *Ceram. Int.* **2019**, *45*, 17653–17661. [CrossRef]
144. Sarycheva, A.; Makaryan, T.; Maleski, K.; Satheeshkumar, E.; Melikyan, A.; Minassian, H.; Yoshimura, M.; Gogotsi, Y. Two-Dimensional Titanium Carbide (MXene) as Surface-Enhanced Raman Scattering Substrate. *J. Phys. Chem. C* **2017**, *121*, 19983–19988. [CrossRef]
145. Xie, H.; Li, P.; Shao, J.; Huang, H.; Chen, Y.; Jiang, Z.; Chu, P.K.; Yu, X.F. Electrostatic Self-Assembly of Ti_3C_2Tx MXene and Gold Nanorods as an Efficient Surface-Enhanced Raman Scattering Platform for Reliable and High-Sensitivity Determination of Organic Pollutants. *ACS Sens.* **2019**, *4*, 2303–2310. [CrossRef] [PubMed]

146. Zheng, F.; Ke, W.; Shi, L.; Liu, H.; Zhao, Y. Plasmonic Au-Ag Janus Nanoparticle Engineered Ratiometric Surface-Enhanced Raman Scattering Aptasensor for Ochratoxin A Detection. *Anal. Chem.* **2019**, *91*, 11812–11820. [CrossRef]
147. Liu, R.Y.; Jiang, L.; Yu, Z.Z.; Jing, X.F.; Liang, X.; Wang, D.; Yang, B.; Lu, C.X.; Zhou, W.; Jin, S.Z. MXene (Ti_3C_2Tx)-Ag nanocomplex as efficient and quantitative SERS biosensor platform by in-situ PDDA electrostatic self-assembly synthesis strategy. *Sens. Actuators B Chem.* **2021**, *333*, 129581. [CrossRef]
148. Su, S.; Sun, Q.; Gu, X.; Xu, Y.; Shen, J.; Zhu, D.; Chao, J.; Fan, C.; Wang, L. Two-dimensional nanomaterials for biosensing applications. *TrAC Trends Anal. Chem.* **2019**, *119*, 115610. [CrossRef]
149. Wu, Q.; Li, N.; Wang, Y.; Liu, Y.; Xu, Y.; Wei, S.; Wu, J.; Jia, G.; Fang, X.; Chen, F.; et al. A 2D transition metal carbide MXene-based SPR biosensor for ultrasensitive carcinoembryonic antigen detection. *Biosens. Bioelectron.* **2019**, *144*, 111697. [CrossRef]
150. Wu, Q.; Li, N.; Wang, Y.; Xu, Y.; Wu, J.; Jia, G.; Ji, F.; Fang, X.; Chen, F.; Cui, X. Ultrasensitive and Selective Determination of Carcinoembryonic Antigen Using Multifunctional Ultrathin Amino-Functionalized Ti3C2-MXene Nanosheets. *Anal. Chem.* **2020**, *92*, 3354–3360. [CrossRef] [PubMed]
151. Chen, R.Y.; Kan, L.; Duan, F.H.; He, L.H.; Wang, M.H.; Cui, J.; Zhang, Z.H.; Zhang, Z.H. Surface plasmon resonance aptasensor based on niobium carbide MXene quantum dots for nucleocapsid of SARS-CoV-2 detection. *Microchim. Acta* **2021**, *188*, 316. [CrossRef] [PubMed]

MDPI
St. Alban-Anlage 66
4052 Basel
Switzerland
Tel. +41 61 683 77 34
Fax +41 61 302 89 18
www.mdpi.com

Biosensors Editorial Office
E-mail: biosensors@mdpi.com
www.mdpi.com/journal/biosensors

www.ingramcontent.com/pod-product-compliance
Lightning Source LLC
LaVergne TN
LVHW070148170726
843515LV00007B/1875

9783036578743